International Series in Operations Research & Management Science

Volume 163

AF786659

This book was recommended by Dr. Price

For further volumes:
http://www.springer.com/series/6161

Marida Bertocchi · Giorgio Consigli ·
Michael A.H. Dempster
Editors

Stochastic Optimization Methods in Finance and Energy

New Financial Products and Energy Market Strategies

Springer

Editors
Marida Bertocchi
University of Bergamo
Department of Mathematics, Statistics,
Computer Science and Applications
Via dei Caniana, 2
24127 Bergamo
Italy
marida.bertocchi@unibg.it

Giorgio Consigli
University of Bergamo
Department of Mathematics, Statistics,
Computer Science and Applications
Via dei Caniana, 2
24127 Bergamo
Italy
giorgio.consigli@unibg.it

Michael A.H. Dempster
University of Cambridge
Dept. Pure Mathematics and
Mathematical Statistics
Mill Lane 16
CB2 1SB Cambridge
United Kingdom
mahd2@cam.ac.uk

ISSN 0884-8289
ISBN 978-1-4614-3027-8 ISBN 978-1-4419-9586-5 (eBook)
DOI 10.1007/978-1-4419-9586-5
Springer New York Dordrecht Heidelberg London

Printed on acid-free paper

Springer is part of Springer Science+Business Media (www.springer.com)

To Clara, Gaia, Marco and Silvia – from Marida

To Adriana and Gabriele – from Giorgio

To Anna and Antony – from Michael

Preface

This volume is a collection of research contributions to applied problems in finance and energy which are formulated and solved in the framework of stochastic optimization. The invited authors represent a group of academics and practitioners who have facilitated the growing penetration of stochastic programming techniques into real-world applications in recent years. Their present contributions represent significant advances over a large spectrum of complex decision problems.

As a result of the recent widespread liberalization of the energy sector in Europe and the unprecedented growth of energy prices in international commodity markets, we have witnessed a significant convergence of strategic decision problems in the energy and financial sectors. This has resulted in many common open issues, calling forth a remarkable effort by the industrial and scholarly communities to forward the practical use of advanced analytical and decision tools. The main concerns of the financial community over the past decade have suddenly been taken up by the energy sector in addressing previously intractable management problems.

This volume includes for the first time in a unified framework an extensive set of contributions addressing real-world problems in finance and energy in a common methodological framework. In many cases the chapters have similar underlying economic and financial assumptions.

During the spring and the summer of 2007, the School of Stochastic Programming held in Bergamo (www.unibg.it/sps2007) and the eleventh International Symposium on Stochastic Programming held in Vienna (http://www.univie.ac.at/spxi) offered two venues for the presentation of the chapters included in the volume.

The volume is structured in three parts: devoted respectively to *financial applications* – Part I with six chapters, *energy applications* – Part II with seven chapters, and *theoretical and computational issues* – Part III with five chapters. The contributions in Part III have all been developed recently and are explicitly related to the applied problems tackled in Parts I and II.

The 13 chapters in Parts I and II address modeling and implementation issues arising in multistage stochastic programming formulations. They focus on real-world problems in finance and energy, clearly identifying common solutions and open problems. In each part the topics of individual chapters move from the general to the specific, evolving in a logical and connected order and indicating the way forward in applied stochastic optimization.

Part I: Financial Applications

Chapter 1, by Ziemba and MacLean, revisits the classical problem of capital accumulation under uncertainty which has occupied a central role in economic and financial theory over many years. The authors analyze the *actual* performance of portfolio management schemes relying on the so-called *Kelly-*, or *capital growth-*, investment strategies whose long-run asymptotic properties lead to optimal strategies for "sufficiently patient" investors. The *optimality* of the Kelly strategy is related to expected values of either long-run wealth or first passage times. In the presence of probabilistic security constraints the problem of maximizing the rate of growth of wealth leads to nonconvex optimization problems which are difficult to solve. After reviewing the key theoretical results regarding the adoption of complete or partial Kelly strategies, the authors present an alternative solution approach based on the introduction of a convex penalty on violations of portfolio drawdown conditions and they analyze the comparative performance of different modeling approaches for a long-term equity investor.

Chapter 2, by Dempster et al., addresses from a large institutional investor's perspective the provision of *guaranteed investment products* for *defined contribution pension savings plans*. These latter have recently been forced on individual households by governments and corporations, respectively, withdrawing pay-as-you-go public and funded defined benefit pension schemes as financially unsustainable. The popularity of such products has increased as investors rightly worry about the downside potential of financial markets. This chapter introduces a simplified dynamic stochastic optimization model for the design and risk management of closed end funds backing guaranteed investment products which underlay that actually used to manage the open-ended funds backing more complex products. The authors describe in detail the pricing of minimum guarantees as well as the valuation of a portfolio of bonds using a three-factor term structure model. Upside for the investor is provided by the equity component of the portfolio. Back tests through the Internet bubble and crash show that effective upside provision and downside risk management can be improved by evaluating risk in the model at a higher frequency than portfolio rebalancing and by modeling equity indices with a (downside) asymmetric jump process.

Chapter 3, by Mulvey et al., considers the problem of institutional investors optimally managing their global portfolios funding *defined benefit (DB) pension plans* through the use of dynamic long–short *duration enhancing overlay* strategies involving equal value positions in long treasury bonds and short treasury bills in order to protect asset wealth and surplus through periods of poor market performance. The proposed overlay strategies do not require capital but must be tightly risk managed. In light of the serious underfunding of DB schemes at present and after the previous Internet bubble and crash, wealth protection has become a critical problem on both sides of the Atlantic. A multistage stochastic program provides the ideal setup for constructing an adaptive dynamic optimal DB investment portfolio and the authors also consider the approximation of its optimal recommendations by derived decision rules. Several illustrative examples are presented in the chapter and the performance of the recommended strategies is evaluated by both back tests and a forward-looking ALM system.

Chapter 4, by Beraldi et al., discusses the problem of dynamically *hedging* a portfolio exposed to *interest rate and default risk* in an integrated manner. No overlays are considered in this case, but rather the potential of a stochastic programming formulation for indirect hedging is tested on a large-scale corporate portfolio. The effectiveness of the multistage strategy is evaluated through the recent 2008–2009 credit crisis, which has provided an ideal stressed period for bond portfolio managers worldwide. It is shown that the dynamic stochastic optimization framework provides an effective alternative to hedging using credit derivatives throughout a period when credit derivative markets became highly illiquid.

Chapter 5, by Consigli et al., describes a case study developed in conjunction with a large global *property and casualty insurer* preparing for the implementation of the EU Solvency II regulations on insurance companies. The chapter develops a *multi-objective* (over short-, medium-, and long-term horizons) dynamic stochastic programming ALM model for the management of an investment portfolio subject to premium inflows and catastrophic claim outflows. As opposed to current practice, the chapter stresses the benefits of a dynamic stochastic ALM formulation of the problem and investigates its performance in a *stressed claim liability* scenario. The reported results serve as a basis for further work in this important area.

Chapter 6, by Consiglio and De Giovanni, concerns *reinsurance contracts*. Optimal portfolio management by pension funds with minimum guarantees combines the optimal tracking and the option valuation problems in a unique, often complex, mathematical framework. Similarly, reinsurance products are usually equipped with *minimum and maximum guarantees* and *surrender options* and their pricing is vital for the insurance industry. In the discrete time, discrete state framework used by stochastic programming formulations the optimal payoff replication problem translates into an option valuation problem in *incomplete markets*. Risk management, strategic asset allocation, and product design depend on the correct evaluation of these contingent claims. To overcome Black and Scholes complete market limitations, the authors have developed a stochastic programming *super-hedging* model to determine the fair price of reinsurance contract provisions. An extensive empirical analysis is presented to highlight the effect of incompleteness on the fair values of the optionalities and to show how the proposed framework can be used as a valuable normative tool for insurance companies and regulators.

In summary, the six chapters of Part I move from a classical financial management problem to a set of state-of-the-art applications in finance and insurance within the dynamic stochastic programming paradigm. The key issues of a stochastic programming mathematical formulation, scenario generation for a wide set of financial management problems, risk characterization, and optimal portfolio control, are all considered. From individual (Chapter 1) to institutional (Chapters 2, 3, 4, and 5) investors, such as pension funds and insurance companies, financial management problems are presented in various case studies which provide a rich set of evidence and practical suggestions for both academia and industry.

Part II: Energy Applications

Many of the methodological issues addressed in Part I of the volume are reprised in the chapters dedicated to energy applications. The advent of liberalized energy markets has brought about an increasing need for new modeling approaches and efficient risk management tools. This has strengthened the cooperation between the scientific community and the energy industry as is reflected in the problems and case studies analyzed in this second part of the volume. The growing environmental concern related to energy sector development worldwide has also forced the adoption of new and more efficient decision support systems which take into account the implications and economic sustainability of alternative energy policies. In an optimization framework this results in additional constraints in the models of the problems treated. Unlike financial applications, where securities are held over time and investors' inter-temporal strategies rely explicitly on inventory balance constraints linking period to period, optimal policies by energy producers confronting significant market risks are often determined by taking into account the physical nature of non-storable energy. In competitive markets with a limited number of energy suppliers and state-controlled network connections, energy producers must further employ optimal supply policies which take into account competitors' behavior. These considerations add complexity to the definition and solution of stochastic optimization problems in the energy sector and are fully reflected in the chapters of Part II.

Chapter 7, by Ramos et al., formulates and solves an optimal *resource allocation* problem for *thermal and hydropower plants* with multiple basins and multiply connected reservoirs. The stochastic factors in the problem are model natural hydro inflows. A multivariate scenario tree is developed taking into account these stochastic inputs and their spatial and temporal dependencies. Hydropower plant efficiency depends in general on *water head* and *reservoir volume*. Here the dependence on water head is modeled as nonlinear, leading to a large-scale *stochastic nonlinear optimization* problem whose formulation and solution are described by the authors.

Chapter 8, by Giacometti et al., analyzes an optimal *short-term hydro-thermal coordination* problem for the case of a small *price-taking producer* seeking to maximize expected profits and reduce market risk. Day-ahead market prices and reservoir inflows are characterized by the uncertainty caused by market fluctuations and unpredictable weather conditions. A multistage stochastic programming model framework is appropriate, as in the real-world information evolves over time and uncertainties are disclosed in stages. The overall objective of the stochastic optimization problem is to establish a *one-day production plan* in order to optimally determine the trade-off between current and future expected profits subject to operational constraints.

Chapter 9, by Triki et al., considers the problem of *optimal bidding* in the *spot electricity markets* by each power producer with the aim of maximizing his profits. The multiplicity of bidding opportunities (auctions) and the limitations imposed by technical constraints and market rules make the problem difficult to model and solve. Further difficulties are represented by the dynamic and stochastic nature of the decision process. The authors focus on the use of stochastic programming to deal with the bidding problem and survey the different modeling paradigms for its formulation and the solution methods currently available for its solution.

Chapter 10, by Alonso-Ayuso et al., proposes a *multistage mixed integer full recourse model* for the similar problem for price-taking agents of *structuring energy contract portfolios* in *competitive markets*. The main uncertain parameters represent spot energy prices, water exogenous inflows to the hydrosystem, and fuel and gas costs and availabilities. The objective is given by the maximization of expected (over scenarios) bilateral and spot market trading profits along the time horizon. The problem is formulated in terms of a *mixed 0–1 deterministic equivalent* model of the stochastic problem which imposes Kyoto protocol-based *regulations for pollutant emissions*. Only 0–1 variables appear in the first stage of the constraint system and all continuous variables appear in the formulation of later stages.

Chapter 11, by Fodstad et al., introduces a comprehensive *network model* to solve a *tactical* (up to 3 years) *planning* problem in the *natural gas supply chain* of a Norwegian company. The authors present a decision support tool for a *large producer* with a portfolio of production fields and access to the transportation grid. The system takes a global view of the supply chain, including such elements as production fields, transportation pipelines, storage, bilateral contracts, and spot markets. The need for a stochastic optimization framework in this case is due to the inclusion of *bilateral contracts* in which the buyer's obligations and prices are treated as uncertain parameters. Spot energy prices are also uncertain. The goal for the producer is to allocate gas equitably and use market flexibility to ensure that delivery obligations and expected profit maximization are both attained. The authors address in an integrated way short-term expected profit maximization under long-term transportation bottleneck and other infrastructural constraints.

Chapter 12, by Drapkin et al., is devoted to a dynamic formulation of the *risk management* problem in *liberalized energy markets*. Under conditions of increasing commodity price volatility the need to address in a coherent financial framework an efficient hedging strategy is considered. The authors propose a new approach to risk management in energy optimization by employing the concept of *stochastic dominance*. This leads to a new class of large-scale block-structured mixed-integer linear programs for which the two authors present a decomposition algorithm involving *Lagrangean relaxation* and *cutting plane techniques*. This methodology is applied to stochastic optimization problems related to *operation and investment planning* in energy systems with dispersed generation.

Chapter 13, the final chapter of Part II by Ehrenmann and Smeers, analyzes probably one of the oldest applications of optimization with a long-term perspective, the *capacity expansion* problem. The solution of this problem needs to be reconsidered for the energy industry after its profound restructuring in recent years and the resulting energy price liberalization. The authors analyze the implications of optimal investment strategies within an *equilibrium model* with *risk-averse agents*. A new model paradigm is presented taking into account the endogenous nature of the cost of *investment capital*. The behavior of agents is described by a multistage risk function. They use a multistage version of the *Cochrane good deal* to obtain a risk function that has good time consistency properties and allows one to tie up their problem formulation with standard models of corporate finance such as the CAPM and APT. The problem is formulated as a *stochastic complementarity problem*. Although it has so far been impossible to establish any monotonicity property of the model, more general fixed-point theorems can be invoked to prove existence of an equilibrium. The authors' contribution appropriately links the second part of this volume with the remaining chapters which are dedicated to an important set of theoretical and computational issues.

The seven chapters of Part II provide a unique and well-integrated set of contributions which address valuation and decision problems in the energy sector after the recent policy changes in the sector. The parallels between short-term optimal dynamic portfolio policies in financial markets and optimal management of electricity contracts under energy price volatility are clear. In both cases, the need from a methodological viewpoint to combine an accurate model of uncertainty with a decision tool leading to a practical efficient decision process emphasizes the similar requirements of the corresponding decision models. The contributions in this second part of the volume move from the tactical to the strategic horizon, with increasingly relevant policy implications.

Part III: Theory and Computation

The case studies in Parts I and II of the book call for efficient approximation methods for the underlying generally continuous probability distributions associated with the stochastic nature of the decision problems, together with effective solution schemes for typically computationally intensive problems. Part III consists of five theoretical developments explicitly addressing the issues of unbiased scenario tree generation and the numerical solution of linear and nonlinear stochastic programs. The first three chapters of this part, taken together, provide an accurate assessment of the scenario generation procedures adopted in the case studies presented previously from both the theoretical and computational viewpoints. The last two chapters consider stochastic programming universal problems in, respectively, scenario generation and risk measure specification, namely, statistical dependence between risk factors and the implications of risk measures for optimal decisions. All five chapters contain a mixture of theoretical results and their computational evaluation.

Chapter 14, by Römisch and Heitsch, presents an overview of recent developments in *scenario generation* techniques for financial and energy problems. The authors analyze a set of theoretical results related to the approximation of random processes in appropriate functional spaces using *stability theory* with a view stabilizing the *small sample behavior of objective function values* of dynamic stochastic programming problems. The concepts are illustrated in the context constructing scenario trees of demand and prices for the *electricity portfolio management of a municipal power utility*.

Chapter 15, by Pflug and Pichler, continues the theme of the previous chapter. The authors first consider the basic problem of measuring the *distance between two multivariate probability distributions* and then apply the results to find a set of good single period scenarios before treating the corresponding multi-period problem. They conclude by addressing the important problem of generated *scenario tree reduction* which is minimally objective biased in the smaller sample tree and leads to reduced stochastic optimization problem solution times. Numerical examples are given throughout the chapter to illustrate the concepts discussed.

Chapter 16, by Dempster et al., goes a step further than its two predecessor chapters by focussing on the small sample stability of optimal *implementable* (first stage or root node) *decisions* in dynamic stochastic programming problems. This is *the* important problem in financial applications, in which the implementable decisions represent an optimal initial portfolio robust to alternative future market and liability scenarios, but one upon which little work has previously appeared. The authors propose a fast heuristic for minimizing the *Kantorovich–Wasserstein distance* between probability distributions and go on to compare its effectiveness in application to the objective and implementable decision stability of the guaranteed return fund management dynamic optimization problem of Chapter 2 with *Monte Carlo* and *moment-matched Monte Carlo* scenario generation techniques. This is a computationally intensive study which shows that *algebraic two-moment matching* is the preferred technique. An equally important finding is that much larger scenario trees than have previously been used in the practical literature are needed to remove small sample solution bias, particularly when risk is tightly constrained.

Chapter 17, by Henrion and Strugarek, establishes the convexity properties of stochastic optimization problems with *chance constraints*. The authors extend the classical theory of chance-constrained problems with *independent* random processes to the case of *codependence* modeled through *copulas*. The related theoretical results are presented and illustrated by a number of examples.

Chapter 18, the final chapter of the volume by Fabian et al., presents a theoretical and computational study of *second-order stochastic dominance* risk measures for single period portfolio choice models (cf. Chapter 12 which applies dominance concepts to control risk in the design and operation of power systems). The authors survey recent results regarding the problem addressed and then undertake an extensive computational study to investigate the effects of various risk measures on *portfolio return distributions*. They conclude that finding an acceptable portfolio strategy and then computing one which maximally dominates it is an alternative to the more usual trade-off of risk, however defined, and expected return.

This concludes a summary of the contents of this volume which we hope will be useful to both the reader and prospective reader. All in all we are proud of the collection of timely and practically relevant contributions the book represents and we hope that the reader will gain as much enjoyment and enlightenment from it as we did in compiling and editing it.

Bergamo, Italy — Marida Bertocchi
Bergamo, Italy — Giorgio Consigli
Cambridge, UK — Michael A.H. Dempster

Contents

Part II Energy Applications

Part III Theory and Computation

Contributors

Giovanni Barone Adesi Swiss Finance Department, University of Lugano, Lugano, Switzerland, giovanni.baroneadesi@usi.ch

Antonio Alonso-Ayuso Departamento de Estadística e Investigación Operativa, Universidad Rey Juan Carlos, Móstoles (Madrid), Spain, antonio.alonso@urjc.es

Thomas Bauerfeind PROTINUS Beratungsgesellschaft mbH & Co. KG, Munich, Germany, thomas.bauerfeind@protinus.de

Rafael Bellido Iberdrola, Madrid, Spain, rafael.bellido@iberdrola.es

Patrizia Beraldi Department of Electronics, Informatics and Systems, University of Calabria, Rende, Italy, beraldi@deis.unical.it

Marida Bertocchi University of Bergamo, Department of Mathematics, Statistics, Computer Science and Applications, Via dei Caniana, 2, Bergamo, Italy, marida.bertocchi@unibg.it

Santiago Cerisola Universidad Pontificia Comillas, Madrid, Spain, santiago.cerisola@upcomillas.es

Antonio J. Conejo University of Castilla-La Mancha, Ciudad Real, Spain, antonio.conejo@uclm.es

Giorgio Consigli University of Bergamo, Department of Mathematics, Statistics, Computer Science and Applications, Via dei Caniana, 2, Bergamo, Italy, giorgio.consigli@unibg.it

Andrea Consiglio Department of Statistics and Mathematics "Silvio Vianelli", University of Palermo, Palermo, Italy, consiglio@unipa.it

Edit Csizmás Institute of Informatics, Kecskemét College, Kecskemét, Hungary, csizmas.edit@gamf.kefo.hu

Domenico De Giovanni Department of Management Science, University of Calabria, Rende, Italy, ddegiovanni@unical.it

Francesco De Simone Department of Electronics, Informatics and Systems, University of Calabria, Rende, Italy, fdesimone@deis.unical.it

Michael A.H. Dempster Centre for Financial Research, Statistical Laboratory, Department of Pure Mathematics and Mathematical Statistics, University of Cambridge, Cambridge, UK; Cambridge Systems Associates Ltd., Cambridge, UK, mahd2@cam.ac.uk

Nico di Domenica Value Lab., Milano, Italy, nico.didomenica@valuelab.it

Massimo di Tria Allianz Investment Management, Allianz Group, AIM, Corso Italia 23, Milan, Italy, massimo.di.tria@allianz.it

Dimitri Drapkin Faculty of Mathematics, University of Duisburg-Essen, Duisburg, Germany, dimitri.drapkin@uni-due.de

Andreas Ehrenmann Center of Expertise in Economic Modeling and Studies, GDF SUEZ, Brussels, Belgium, andreas.ehrenmann@gdfsuez.com

Laureano F. Escudero Departamento de Estadística e Investigación Operativa, Universidad Rey Juan Carlos, Móstoles (Madrid), Spain, laureano.escudero@urjc.es

Csaba I. Fábián Institute of Informatics, Kecskemét College, Kecskemét, Hungary; Department of OR, Eötvös Loránd University, Budapest, Hungary, fabian.csaba@gamf.kefo.hu

Marte Fodstad Department of Industrial Economics and Technology Management, Norwegian University of Science and Technology, Trondheim, Norway; SINTEF Technology and Society, Trondheim, Norway, marte.fodstad@sintef.no

Michele Gaffo Allianz Investment Management, Allianz Group, Königinstr. 28, München, 089 38000 Germany, michele.gaffo@allianz.com

Lina P. Garcés Paulista State University, Ilha Solteira, Brazil, linaneg@aluno.feis.unesp.br

Matteo Germano Pioneer Investment Management Ltd., Dublin 2, Eire, matteo.germano@pioneerinvest.ie

Rosella Giacometti Department of Mathematics, Statistics, Computing and Applications, University of Bergamo, Bergamo, Italy, rosella.giacometti@unibg.it

Ralf Gollmer Faculty of Mathematics, University of Duisburg-Essen, Duisburg, Germany, ralf.gollmer@uni-due.de

Uwe Gotzes E.ON Gastransport GmbH, Essen, Germany, uwe.gotzes@eon-gastransport.com

Holger Heitsch Institute of Mathematics, Humboldt University Berlin, Berlin, Germany, heitsch@math.hu-berlin.de

René Henrion Weierstrass Institute for Applied Analysis and Stochastics, Berlin, Germany, henrion@wias-berlin.de

Gaetano Iaquinta Department of Mathematics, Statistics and Computer Science, University of Bergamo, Bergamo, Italy, gaetano.iaquinta@unibg.it

Jesus M. Latorre Universidad Pontificia Comillas, Madrid, Spain, jesus.latorre@upcomillas.es

Elena Lopez Iberdrola, Madrid, Spain, eleloper@iberdrola.es

Leonard C. MacLean Herbert Lamb Chair School of Business, Dalhousie University, Halifax, B3H 3J5, Canada, l.c.maclean@dal.ca

Elena A. Medova Centre for Financial Research, University of Cambridge, Cambridge, UK; Cambridge Systems Associates Ltd., Cambridge, UK, e.medova@jbs.cam.ac.uk; eam28@cam.ac.uk

Kjetil T. Midthun SINTEF Technology and Society, Trondheim, Norway, kjetil.midthun@sintef.no

Gautam Mitra School of Information Systems, Computing and Mathematics, The Centre for the Analysis of Risk and Optimisation Modelling Applications, Brunel University, Uxbridge, Middlesex, UK; OptiRisk Systems, Uxbridge, Middlesex, UK, Gautam.Mitra@brunel.ac.uk

Vittorio Moriggia Department of Mathematics, Statistics and Computer Science, University of Bergamo, Bergamo, Italy, vittorio.moriggia@unibg.it

John M. Mulvey Princeton University, Princeton, NJ, USA, mulvey@princeton.edu

Frederike Neise E.ON Ruhrgas AG, Essen, Germany, frederike.neise@eon-ruhrgas.com

Olga Papp Institute of Informatics, Kecskemét College, Kecskemét, Hungary; Doctoral School in Applied Mathematics, Eötvös Loránd University, Budapest, Hungary, papp.olga@gamf.kefo.hu

Alejandro Perea Iberdrola, Madrid, Spain, alejandro.perea@iberdrola.es

Georg Ch. Pflug Department of Statistics and Operations Research, University of Vienna, Wien – Vienna, Austria, georg.pflug@univie.ac.at

Alois Pichler Department of Statistics and Operations Research, University of Vienna, Wien – Vienna, Austria, Alois.Pichler@univie.ac.at

Celeste Pizarro Departamento de Estadística e Investigación Operativa, Universidad Rey Juan Carlos, Móstoles (Madrid), Spain, celeste.pizarro@urjc.es

Andres Ramos Universidad Pontificia Comillas, Madrid, Spain, andres.ramos@upcomillas.es

Muriel I. Rietbergen Securitisation & Asset Monetisation Group, Morgan Stanley, London, UK, muriel.rietbergen@morganstanley.com

Diana Roman School of Information Systems, Computing and Mathematics, The Centre for the Analysis of Risk and Optimisation Modelling Applications, Brunel University, Uxbridge, Middlesex, UK, Diana.Roman@brunel.ac.uk

Werner Römisch Institute of Mathematics, Humboldt University Berlin, Berlin, Germany, romisch@math.hu-berlin.de

Frode Rømo SINTEF Technology and Society, Trondheim, Norway, frode.romo@sintef.no

Francesco Sandrini Pioneer Investment Management Ltd., Dublin 2, Eire, francesco.sandrini@pioneerinvest.ie

Rüdiger Schultz Faculty of Mathematics, University of Duisburg-Essen, Duisburg, Germany, ruediger.schultz@uni-due.de; schultz@mail.math.uni-duisburg.de

Mike Scrowston Pioneer Investment Management Ltd., Dublin 2, Eire, mark.scrowston@pioneerinvest.ie

Koray D. Simsek Sabanci School of Management, Sabanci University, Tuzla, Istanbul, Turkey, ksimsek@sabanciuniv.edu

Yves Smeers School of Engineering and CORE, Université catholique de Louvain, Louvain-la-Neuve, Belgium, smeers@core.ucl.ac.be

Cyrille Strugarek Credit Portfolio Management, Calyon Credit Agricole CIB, Paris, France, cyrille.strugarek@calyon.com

Asgeir Tomasgard Department of Industrial Economics and Technology Management, Norwegian University of Science and Technology, Trondheim, Norway; SINTEF Technology and Society, Trondheim, Norway, asgeir.tomasgard@sintef.no; asgeir.tomasgard@iot.ntnu.no

Chefi Triki University of Salento, Lecce, Italy, chefi.triki@unisalento.it

Angelo Uristani Department of Mathematics, Statistics and Computer Science, University of Bergamo, Bergamo, Italy, Angelo.Uristani@unibg.it

Tibor Vajnai Institute of Informatics, Kecskemét College, Kecskemét, Hungary, vajnai.tibor@gamf.kefo.hu

Maria Teresa Vespucci Department of Mathematics, Faculty of Engineering, University of Bergamo, Bergamo, Italy, maria-teresa.vespucci@unibg.it

Antonio Violi Department of Electronics, Informatics and Systems, University of Calabria, Rende, Italy, violi@si.deis.unical.it

Mehmet T. Vural ORFE Department, Princeton University, Princeton, NJ, USA, mvural@Princeton.edu

Yee Sook Yong Credit Suisse AG, Singapore, Singapore, yeesook@gmail.com; yeesook@cantab.net

William T. Ziemba University of British Columbia, Vancouver, BC, Canada; Mathematical Institute, Oxford University, Oxford, UK; ICMA Centre, University of Reading, Reading, UK; University of Bergamo, Bergamo, Italy, wtzimi@mac.com

Victor Zverovich School of Information Systems, Computing and Mathematics, The Centre for the Analysis of Risk and Optimisation Modelling Applications, Brunel University, Uxbridge, Middlesex, UK; OptiRisk Systems, Uxbridge, Middlesex, UK, viktar.zviarovich@brunel.ac.uk

Part I
Financial Applications

Chapter 1
Using the Kelly Criterion for Investing

William T. Ziemba and Leonard C. MacLean

Abstract This chapter describes the use of the Kelly capital growth model. This model, dubbed *Fortune's Formula* by Thorp and used in the title by Poundstone (Fortune's Formula: The Untold Story of the Scientific System That Beat the Casinos and Wall Street, 2005), has many attractive features such as the maximization of asymptotic long-run wealth; see Kelly (Bell System Technical Journal 35:917–926, 1956), Breiman (Proceedings of the 4th Berkely Symposium on Mathematical Statistics and Probability 1:63–68, 1961), Algoet and Cover (Annals of Probability 16(2):876–898, 1988) and Thorp (Handbook of Asset and Liability Management, 2006). Moreover, it minimizes the expected time to reach asymptotically large goals (Breiman, Proceedings of the 4th Berkeley Symposium on Mathematical Statistics and Probability 1:63–68, 1961) and the strategy is myopic (Hakansson, Journal of Business 44:324–334, 1971). While the strategy to maximize the expected logarithm of expected final wealth computed via a nonlinear program has a number of good short- and medium-term qualities (see MacLean, Thorp, and Ziemba, The Kelly Capital Growth Investment Critria, 2010b), it is actually very risky short term since its Arrow–Pratt risk aversion index is the reciprocal of wealth and that is essentially zero for non-bankrupt investors. The chapter traces the development and use of this strategy from the log utility formulation in 1738 by Bernoulli (Econometrica 22:23–36, 1954) to current use in financial markets, sports betting, and other applications. Fractional Kelly wagers that blend the E log maximizing strategy with cash tempers the risk and yield smoother wealth paths but with generally less final wealth. Great sensitivity to parameter estimates, especially the means, makes the strategy dangerous to those whose estimates are in error and leads them to poor betting and possible bankruptcy. Still, many investors with repeated investment periods and considerable wealth, such as Warren Buffett and George Soros, use strategies that approximate full Kelly which tends to place most of one's wealth in a few assets and lead to many monthly losses but large final wealth most of the time. A simulation study is presented that shows the possibility of huge gains most of the time, possible losses no matter how good the investments appear to be, and possible extreme losses from overbetting when bad scenarios occur. The study and discussion shows that

W.T. Ziemba (✉)
University of British Columbia, Vancouver, BC, Canada; Mathematical Institute, Oxford University, Oxford, UK; ICMA Centre, University of Reading, Reading, UK; University of Bergamo, Bergamo, Italy
e-mail: wtzimi@mac.com

M. Bertocchi et al. (eds.), *Stochastic Optimization Methods in Finance and Energy*, International Series in Operations Research & Management Science 163,
DOI 10.1007/978-1-4419-9586-5_1,

Samuelson's objections to E log strategies are well understood. In practice, careful risk control or financial engineering is important to deal with short-term volatility and the design of good wealth paths with limited drawdowns. Properly implemented, the strategy used by many billionaires has much to commend it, especially with many repeated investments.

Keywords Kelly investment criterion · Long-range investing · Logarithmic utility functions · Fractional Kelly strategies

1.1 Introduction

The Kelly capital growth strategy is defined as *allocate your current wealth to risky assets so that the expected logarithm of wealth is maximized period by period.* So it is a one-period static calculation that can have transaction costs and other market imperfections considered. Log utility dates to Daniel Bernoulli in 1738 who postulated that marginal utility was monotone increasing but declined with wealth and, specifically, is equal to the reciprocal of wealth, w, which yields the utility of wealth $u(w) = \log w$. Prior to this it was assumed that decisions were made on an expected value or linear utility basis. This idea ushered in declining marginal utility or *risk aversion* or *concavity* which is crucial in investment decision making. In his chapter, in Latin, he also discussed the St. Petersburg paradox and how it might be analyzed using $\log w$.

The St. Petersburg paradox actually originates from Daniel's cousin, Nicolas Bernoulli, a professor at the University of Basel where Daniel was also a professor of mathematics. In 1708, Nicolas submitted five important problems to Professor Pierre Montmort. This problem was how much to pay for the following gamble:

> A fair coin with $\frac{1}{2}$ probability of heads is repeatedly tossed until heads occurs, ending the game. The investor pays c dollars and receives in return 2^{k-1} with probability 2^{-k} for $k = 1, 2, \ldots$ should a head occur. Thus, after each succeeding loss, assuming a head does not appear, the bet is doubled to 2, 4, 8, ... etc. Clearly the expected value is $\frac{1}{2} + \frac{1}{2} + \frac{1}{2} + \ldots$ or infinity with linear utility.

Bell and Cover (1980) argue that the St. Petersburg gamble is attractive at any price c, but the investor wants less of it as $c \to \infty$. The proportion of the investor's wealth invested in the St. Petersburg gamble is always positive but decreases with the cost c as c increases. The rest of the wealth is in cash.

Bernoulli offers two solutions since he felt that this gamble is worth a lot less than infinity. In the first solution, he arbitrarily sets a limit to the utility of very large payoffs. Specifically, any amount over 10 million is assumed to be equal to 2^{24}. Under that bounded utility assumption, the expected value is

$$\frac{1}{2}(1) + \frac{1}{4}(2) + \frac{1}{8}(4) + \cdots + \left(\frac{1}{2}\right)^{24}(2^{24}) + \left(\frac{1}{2}\right)^{25}(2^{24}) + \left(\frac{1}{2}\right)^{26}(2^{24}) + \ldots = 12 + \text{ the original } 1 = 13.$$

When utility is log the expected value is

$$\frac{1}{2}\log 1 + \frac{1}{4}\log 2 + \cdots + \frac{1}{8}\log 4 + \cdots = \log 2 = 0.69315.$$

Use of a concave utility function does not eliminate the paradox.

For example, the utility function $U(x) = x/\log(x + A)$, where $A > 2$ is a constant, is strictly concave, strictly increasing, and infinitely differentiable yet the expected value for the St. Petersburg gamble is $+\infty$.

As Menger (1967) pointed out in 1934, the log, the square root, and many other, but not all, concave utility functions eliminate the original St. Petersburg paradox but it does not solve one where the payoffs grow faster than 2^n. So if log is the utility function, one creates a new paradox by having the payoffs increase at least as fast as log reduces them so one still has an infinite sum for the expected utility. With exponentially growing payoffs one has

$$\frac{1}{2}\log(e^1) + \frac{1}{4}\log(e^2) + \cdots = \infty.$$

The super St. Petersburg paradox, in which even $E \log X = \infty$ is examined in Cover and Thomas (2006: p. 181, 182) where a satisfactory resolution is reached by looking at relative growth rates of wealth. Another solution to such paradoxes is to have bounded utility. To solve the St. Petersburg paradox with exponentially growing payoffs, or any other growth rate, a second solution, in addition to that of bounding the utility function above, is simply to choose a utility function which, though unbounded, grows "sufficiently more" slowly than the inverse of the payoff function, e.g., like the log of the inverse function to the payoff function. The key is whether the valuation using a utility function is finite or not; if finite, the specific value does not matter since utilities are equivalent to within a positive linear transformation ($V = aU + b, a > 0$). So for any utility giving a finite result there is an equivalent one that will give you any specified finite value as a result. Only the behavior of $U(x)$ as $x \to \infty$ matters and strict monotonicity is necessary for a paradox. For example, $U(x) = x,\ x \leqslant A$, will not produce a paradox. But the continuous concave utility function

$$U(x) = \frac{x}{2} + \frac{A}{2}, \qquad x > A$$

will have a paradox. Samuelson (1977) provides an extensive survey of the paradox; see also Menger (1967) and Aase (2001).

Kelly (1956) is given credit for the idea of using log utility in gambling and repeated investment problems and it is known as the Kelly criterion. Kelly's analyses use Bernoulli trials. Not only does he show that log is the utility function which maximizes the long-run growth rate, but that this utility function is myopic in the sense that period by period maximization based only on current capital is optimal. Working at Bell Labs, Kelly was strongly influenced by information theorist Claude Shannon.

Kelly defined the long-run growth rate of the investor's fortune using

$$G = \lim_{N\to\infty} \log \frac{W_N}{W_0},$$

where W_0 is the initial wealth and W_N is the wealth after N trials in sequence. With Bernoulli trials, one wins = $+1$ with probability p and losses -1 with probability $q = 1 - p$. The wealth with M wins and $L = N - M$ losses is

$$W_N = (1+f)^M (1-f)^{N-M} W_0,$$

where f is the fraction of wealth wagered on each of the N trials. Substituting this into G yields

$$G = \lim_{N\to\infty} \left(\frac{M}{N} \log(1+f) + \left(\frac{N-M}{N} \right) \log(1-f) \right) = p \log(1+f) + q \log(1-f) = E \log W$$

by the strong law of large numbers.

Maximizing G is equivalent to maximizing the expected value of the log of each period's wealth. The optimal wager for this is

$$f^* = p - q, \qquad p \geqslant q > 0,$$

which is the expected gain per trial or the edge. If there is no edge, the bet is zero.

If the payoff is $+B$ for a win and -1 for a loss, then the edge is $Bp - q$, the odds are B, and

$$f^* = \frac{Bp - q}{B} = \frac{\text{edge}}{\text{odds}}.$$

Latané (1978) introduced log utility as an investment criterion to the finance world independent of Kelly's work. Focussing, like Kelly, on simple intuitive versions of the expected log criteria he suggested that it had superior long-run properties. Hakansson and Ziemba (1995) survey economic analyses and applications.

Kelly bets can be very large and quite risky short term. For example, if $p = 0.99$ and $q = 0.01$ then $f^* = 0.98$ or 98% of one's current wealth. A real example of this is by Mohnish and Pabrai (2007) who won the bidding for the 2008 lunch with Warren Buffett paying more than \$600,000. He had the following investment in Stewart Enterprises as discussed by Thorp (2008). Over a 24-month period, with probability 0.80 the investment at least doubles, with 0.19 probability the investment breaks even, and with 0.01 probability all the investment is lost. The optimal Kelly bet is 97.5% of wealth and half Kelly is 38.75%. Pabrai invested 10%. While this seems rather low, other investment opportunities, miscalculation of probabilities, risk tolerance, possible short-run losses, bad scenario Black Swan events, price pressures, buying in and exiting suggest that a bet a lot lower than 97.5% is appropriate.

Risk aversion is generally measured by the Arrow–Pratt risk aversion index, namely

$$R_{A(w)} = \frac{-u''(w)}{u'(w)}$$

for absolute wagers and $R_A = wR_{A(w)}$ for proportional wagers.

For log, $R_A = 1/w$ which is close to zero for non-bankrupt investors, so we will argue that log is the most risky utility function one should ever consider. Positive power utility functions like $w^{1/2}$ lead to overbetting and are growth-security dominated. That means that growth and security both decrease.

Breiman (1961), following his earlier intuitive paper Breiman (1960), established the basic mathematical properties of the expected log criterion in a rigorous fashion. He proves three basic asymptotic results in a general discrete time setting with intertemporally independent assets.

Suppose in each period, N, there are K investment opportunities with returns per unit invested $X_{N_1}, \ldots, X_{N_K}$. Let $\Lambda = (\Lambda_1, \ldots, \Lambda_K)$ be the fraction of wealth invested in each asset. The wealth at the end of period N is

$$W_N = \left(\sum_{i=1}^{K} \Lambda_i X_{N_i} \right) W_{N-1}.$$

Property 1 In each time period, two portfolio managers have the same family of investment opportunities, X, and one uses a Λ^* which maximizes $E \log W_N$ whereas the other uses an *essentially different* strategy, Λ, so they differ infinitely often, that is,

$$E \log W_N(\Lambda^*) - E \log W_N(\Lambda) \to \infty.$$

Then

$$\lim_{N\to\infty} \frac{W_N(\Lambda^*)}{W_N(\Lambda)} \to \infty.$$

So the wealth exceeds that with any other strategy by more and more as the horizon becomes more distant.

This generalizes the Kelly Bernoulli trial setting to intertemporally independent and stationary returns.

Property 2 The expected time to reach a preassigned goal A is asymptotically least as A increases with a strategy maximizing $E \log W_N$.

Property 3 Assuming a fixed opportunity set, there is a fixed fraction strategy that maximizes $E \log W_N$, which is independent of N.

1.2 Risk Aversion

We can break risk aversion, both absolute and relative, into categories of investors as Ziemba (2010) has done in his response to letters he received from Professor Paul A Samuelson (2006, 2007, 2008) (Table 1.1).

Ziemba named Ida after Ida May Fuller who paid $24.75 into US social security and received her first social security check numbered 00-000-001 on January 31, 1940, the actual first such check. She lived in Ludlow, Vermont, to the age of 100 and collected $22,889. Such are the benefits and risks of this system; see Bertoccchi, Schwartz, and Ziemba (2010) for more on this. Victor is named for the hedge fund trader Victor Niederhoffer who seems to alternate between very high returns and blowing up; see Ziemba and Ziemba (2007) for some but not all of his episodes. The other three investors are the overbetting Tom who is growth-security dominated in the sense of MacLean, Ziemba, and Blazenko (1992), our E log investor Dick and Harriet, approximately half Kelly, who Samuelson says fits the data well. We agree that in practice, half Kelly is a toned down version of full Kelly that provides a lot more security to compensate for its loss in long-term growth. Figure 1.1 shows this behavior in the context of Blackjack where Thorp first used Kelly strategies.

The edge for a successful card counter varies from about -5 to $+10\%$ depending upon the favorability of the deck. By wagering more in favorable situations and less or nothing when the deck is unfavorable, an average weighted edge is about 2%. An approximation to provide insight into the long-run behavior of a player's fortune is to assume that the game is a Bernoulli trial with a probability of success $= 0.51$ and probability of loss $1 = 0.49$.

Figure 1.1 shows the relative growth rate $f \ln(1 + p) + (1 - f)\ln(1 - p)$ versus the fraction of the investor's wealth wagered, f. This is maximized by the Kelly log bet $f^* = p - q = 0.02$. The growth rate is lower for smaller and for larger bets than the Kelly bet. Superimposed on this graph is also the probability that the investor doubles or quadruples the initial wealth before losing half of this initial wealth. Since the growth rate and the security are both decreasing for $f > f^*$, it follows that it is never advisable to wager more than f^*.

Observe that the E log investor maximizes long-run growth and that the investor who wagers exactly twice this amount has a growth rate of zero plus the risk-free rate of interest. The fractional Kelly strategies are on the left and correspond to

Table 1.1 Samuelson's three investors plus Ziemba's two tail investors

	Victor	Tom	Dick	Harriet	Ida
	w linear	$w^{1/2}$ positive power	$\log w$ geometric mean optimizer	$-\frac{1}{w}$ half Kelly	$-\frac{N}{w}, N \to \infty$ finite risk averse
Absolute $R_A - \frac{u'}{u'w}$	0	$\frac{1}{2w}$	$\frac{1}{2}$	$\frac{2}{w}$	∞
Relative $R_A - \frac{wv''(w)}{u'(w)}$	0	$\frac{1}{2}$	1	2	∞

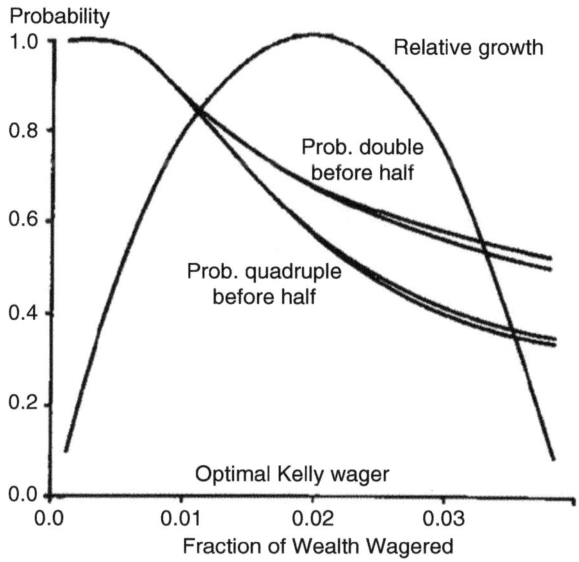

Fig. 1.1 Probability of doubling and quadrupling before halving and relative growth rates versus fraction of wealth wagered for Blackjack (2% advantage, $p = 0.51$ and $q = 0.49$)
Source: MacLean, Ziemba, and Blazenko (1992)

various negative power utility functions αw^{α} for $\alpha < 0$ such as 1/2 Kelly, $\alpha = -1$, and 1/4 Kelly, $\alpha = -3$. These values come from the handy formula for the fractional Kelly

$$f = \frac{1}{1-\alpha} = \frac{1}{R_R},$$

which is exactly correct for lognormal assets and approximately correct otherwise; see MacLean, Ziemba, and Li (2005) for proof. Thorp (2008) shows that this approximation can be very poor.

1.3 Understanding the Behavior of *E* log Strategies

There are many possible investment situations and *E* log Kelly wagering is useful for some of them. Good uses of the strategy are in situations with many repeated bets that approximate an infinite sequence as in the Breiman, etc., theory. See the papers in MacLean, Thorp, and Ziemba (2010b) for such extensions; MacLean, Thorp, and Ziemba (2010a) for good and bad Kelly and fractional Kelly properties; and MacLean, Thorp, Zhao, and Ziemba (2011) for simulations of typical behavior. Luenberger (1993) looks at long-run asymptotic behavior. Futures and options trading, sports betting, including horseracing, are good examples. The policies tend to non-diversify, plunge on a small number of the best assets, have a lot of volatility,

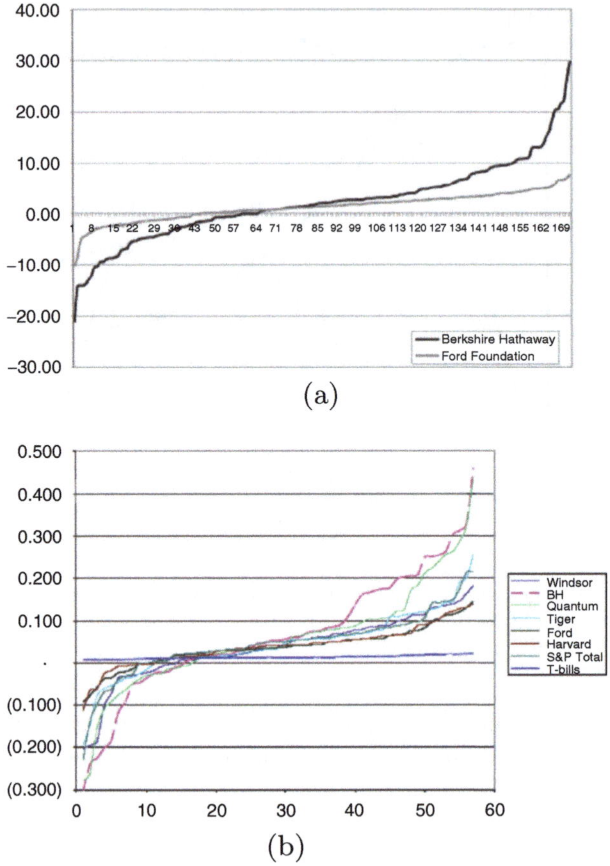

Fig. 1.2 Monthly returns for some funds ranked worst to best. **(a)** Berkshire Hathaway versus Ford Foundation, monthly returns distribution, January 1977 to April 2000. **(b)** Return distributions of all the funds, quarterly returns distribution, December 1985 to March 2000
Source: Ziemba (2005)

and produce more total wealth in the end than other strategies. Notable investors who use such strategies are Warren Buffett of Berkshire Hathaway, George Soros of the Quantum funds, and John Maynard Keynes who ran the King's College Cambridge endowment from 1927 to 1945. Figure 1.2a, b shows the best and worst months for the Buffett and Soros funds. Observe that Buffett and Soros are asymptotically equivalent in both the left and right tails. Figure 1.3 shows their wealth graphs. These correspond to typical Kelly behavior. Some Kelly advocates with a gambling background have produced nice smooth graphs such as those of Hong Kong racing guru Bill Benter, famed hedge fund traders Ed Thorp and Jim Simons; see Fig. 1.4a–d for the various wealth graphs.

According to Ziemba (2005), Keynes was approximately an 80% Kelly bettor with a utility function of $-w^{-0.25}$. In Ziemba (2005) it is argued that Buffett and

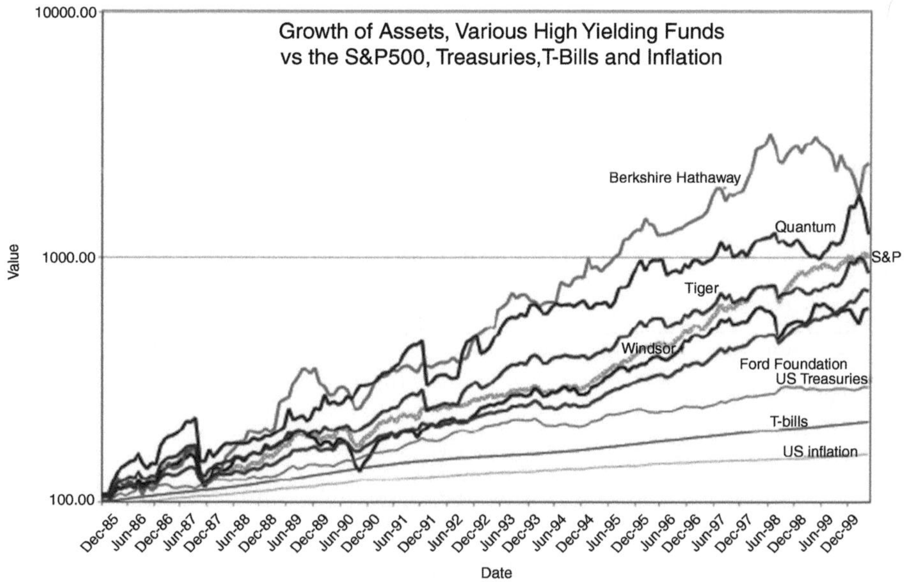

Fig. 1.3 The wealth levels from December 1985 to April 2000 for the Windsor Fund of George Neff, the Ford Foundation, the Tiger Fund of Julian Robertson, the Quantum Fund of George Soros, and Berkshire Hathaway, the fund run by Warren Buffett, as well as the S&P500 total return index *Source*: Ziemba (2005)

Soros are full Kelly bettors. They focus on long run wealth gains, not worrying about short term monthly losses. They tend to have few positions and try not to lose on any of them and not focusing on diversification. Table 1.2 supports this showing their top 10 equity holdings on September 30, 2008. Soros is even more of a plunger with more than half his equity portfolio in just one position.

The basic optimization of an E log strategy is to maximize the expected utility of a logarithmic utility function of final wealth subject to its constraints. Figure 1.5 shows a model formulation where transaction costs are present. Here in this horseracing example q_i is the probability that horse i wins a given race. The probability of an ijk finish is approximated using the Harville (1973) formulas as shown under the assumption that the probability the j wins a race that does not contain i equals $\frac{q_i}{1-q_i}$, etc. In practice these q_i are modified because in reality favorites who do not win do not come second or third as often as these formulas indicate. See Hausch, Lo, and Ziemba (1994, 2008) for these discounted Harville formulas and other approaches to this problem.

The final wealth W inside $E\log(W)$ is the amount not bet plus the winnings from place and show bets to come first or second, or first, second, or third, respectively, namely, the p_i and s_i where the P_i and S_i are the bets of other people. So the expression computes the payoffs after our bets are made assuming we bet last.

There are a number of practical details in current racetrack betting. First, there is rebate so when you bet $B = \sum p_i$ and $\sum s_i$ you receive back a percent, say ΔB

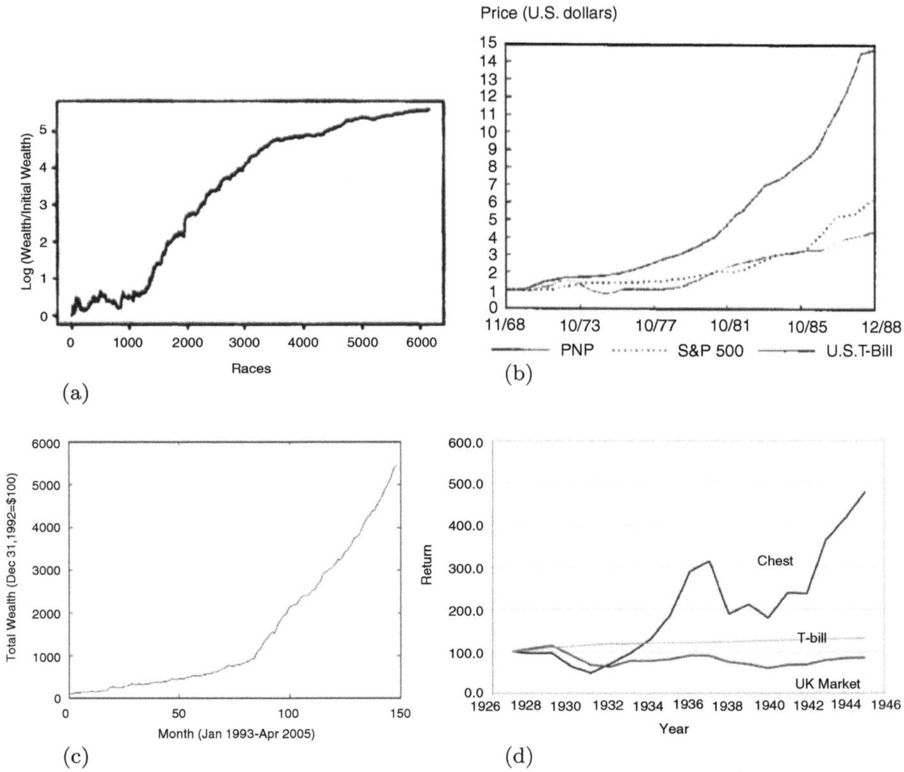

Fig. 1.4 The records of Bill Benter, Edward O. Thorp, Jim Simons, and John Maynard Keynes. **(a)** Benter's record in the Hong Kong Racing Syndicate; **(b)** Thorp's record in Princeton-Newport; **(c)** Jim Simon's record in Renaissance Medallion; and **(d)** John Maynard Keynes' record at King's College Cambridge Endowment
Source: Ziemba (2005) and Ziemba and Ziemba (2007)

where Δ varies depending on the bet, track, etc. The net effect is that the track take instead of being $1 - Q$ =0.13 to 0.30 is actually about 0.1–0.12. So professional bettors have lower transaction costs. Second, this model is just an approximation since about half the money bet does not get recorded in the pools until the race is running because of delays in reporting off track betting. Hence the probabilities must be estimated. Nevertheless, the old 1981 system modified in Hausch and Ziemba (1985) and discussed in the trade books Ziemba and Hausch (1986, 1987) does still seem to work and produce profits when rebate is included. Figure 1.6 shows a 2004 application performed by John Swetye and William Ziemba. A 5000 dollar initial wealth was churned into $1.5 million of bets. The system lost 7% but gained 2% after an average 9% rebate so 2%(1.5 million) = $30,000 profit for full Kelly. Observe that half and one-third Kelly have slightly smoother wealth paths but less final wealth.

Table 1.2 Top 10 equity holdings of Soros Fund Management and Berkshire Hathaway, September 30, 2008

Company	Current value × 1000	Shares	% portfolio
Soros fund management			
Petroleo Brasileiro SA	$ 1,673,048	43,854,474	50.53
Potash Corp Sask Inc.	378,020	3,341,027	11.58
Wal Mart Stores Inc.	195,320	3,791,890	5.95
Hess Corp	115,001	2,085,988	4.49
ConocoPhillips	96,855	1,707,900	3.28
Research in Motion Ltd.	85,840	1,610,810	2.88
Arch Coal Inc.	75,851	2,877,486	2.48
iShares TR	67,236	1,300,000	2.11
Powershares QQQ Trust	93,100	2,000,000	2.04
Schlumberger Ltd.	33,801	545,000	1.12
Berkshire Hathaway			
ConocoPhillips	$ 4,413,390	7,795,580	8.17
Procter & Gamble Co.	4,789,440	80,252,000	8.00
Kraft Foods Inc.	3,633,985	120,012,700	5.62
Wells Fargo & Co.	1,819,970	66,132,620	3.55
Wesco Finl Corp.	1,927,643	5,703,087	2.91
US Bancorp	1,1366,385	49,461,826	2.55
Johnson & Johnson	1,468,689	24,588,800	2.44
Moody's	1,121,760	48,000,000	2.34
Wal Mart Stores, Inc.	1,026,334	19,944,300	1.71
Anheuser Busch Cos, Inc.	725,201	13,845,000	1.29

Source: SEC Filings.

$$\underset{\{p_l\}\{s_l\}}{\text{Maximize}} \sum_{i=1}^{n} \sum_{\substack{j=1 \\ j\neq i}}^{n} \sum_{\substack{k=1 \\ k\neq i,j}}^{n} \frac{q_i q_j q_k}{(1-q_i)(1-q_i-q_j)} \log \left[\begin{array}{l} \dfrac{Q\left(P+\sum_{l=1}^{n} p_l\right)-(p_i+p_j+P_{ij})}{2} \\ \quad \times \left[\dfrac{p_i}{p_i+P_i}+\dfrac{p_j}{p_j+P_j}\right] \\ + \dfrac{Q\left(S+\sum_{l=1}^{n} s_l\right)-(s_i+s_j+s_k+S_{ijk})}{3} \\ \quad \times \left[\dfrac{s_i}{s_i+S_i}+\dfrac{s_j}{s_j+S_j}+\dfrac{s_k}{s_k+S_k}\right] \\ + \omega_0 - \sum\limits_{\substack{l=1 \\ l\neq i,j,k}}^{n} s_l - \sum\limits_{\substack{l=1 \\ l\neq i,j}}^{n} p_l \end{array} \right]$$

$$\text{s.t.} \quad \sum_{l=1}^{n} (p_l + s_l) \leq \omega_0, \qquad p_l \geq 0, \qquad s_l \geq 0, \qquad l = 1, \ldots, n,$$

Fig. 1.5 *E* log transaction cost model for place and show wagering
Source: Hausch, Ziemba, and Rubinstein (1981)

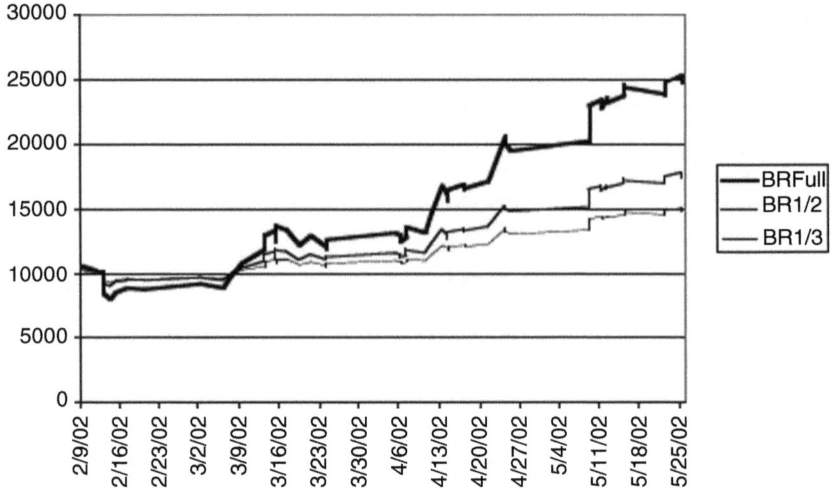

Fig. 1.6 Results of Dr. Z place and show betting with real money in 2004

1.4 A Simulated Example – Equity Versus Cash

In our experiment based on a similar example in Bicksler and Thorp (1973), there are two assets: US equities and US T-bills.[1] According to Siegel (2002), during 1926–2001 US equities returned 10.2% with a yearly standard deviation of 20.3%, and the mean return was 3.9% for short-term government T-bills with zero standard deviation. We assume the choice is between these two assets in each period. The Kelly strategy is to invest a proportion of wealth $x = 1.5288$ in equities and sell short the T-bill at $1 - x = -0.5228$ of current wealth. With the short selling and levered strategies, there is a chance of substantial losses. For the simulations, the proportion λ of wealth invested in equities[2] and the corresponding Kelly fraction f are

λ	0.4	0.8	1.2	1.6	2.0	2.4
f	0.26	0.52	0.78	1.05	1.31	1.57

[1] This example was modified from one in MacLean, Thorp, Zhao, and Ziemba (2011).
[2] The formula relating λ and f for this example is as follows. For the problem

$$\text{Max}_x \left\{E(\ln(1 + r + x(R - r)\right\},$$

where R is assumed to be Gaussian with mean μ_R and standard deviation σ_R, and $r =$ the risk-free rate. The solution is given by Merton (1990) as

$$x = \frac{\mu_R - r}{\sigma_R}.$$

Since $\mu_R = 0.102, \sigma_R = 0.203, r = 0.039$, the Kelly strategy is $x = 1.5288$.

Table 1.3 Final wealth statistics by Kelly fraction for the equity versus cash example

Fraction						
Statistic	0.26 k	0.52 k	0.78 k	1.05 k	1.31 k	1.57 k
Max	65,842.09	673,058.45	5,283,234.28	33,314,627.67	174,061,071.4	769,753,090
Mean	12,110.34	30,937.03	76,573.69	182,645.07	416,382.80	895,952.14
Min	2367.92	701.28	– 4969.78	– 133,456.35	– 6,862,762.81	– 102,513,723.8
St. Dev.	6147.30	35,980.17	174,683.09	815,091.13	3,634,459.82	15,004,915.61
Skewness	1.54	4.88	13.01	25.92	38.22	45.45
Kurtosis	4.90	51.85	305.66	950.96	1755.18	2303.38
$>5 \times 10$	3000	3000	2998	2970	2713	2184
10^2	3000	3000	2998	2955	2671	2129
$>5 \times 10^2$	3000	3000	2986	2866	2520	1960
$>10^3$	3000	2996	2954	2779	2409	1875
$>10^4$	1698	2276	2273	2112	1794	1375
$>10^5$	0	132	575	838	877	751
$>10^6$	0	0	9	116	216	270

Bicksler and Thorp used 10 and 20 yearly decision periods, and 50 simulated scenarios. MacLean et al. used 40 yearly decision periods, with 3000 scenarios.

The results from the simulations appear in Table 1.3 and Figs. 1.7, 1.8 and 1.9. The striking aspects of the statistics in Table 1.3 are the sizable gains and losses. In his lectures, Ziemba always says when in doubt bet less – that is certainly borne out in these simulations. For the most aggressive strategy (1.57 k), it is possible to lose 10,000 times the initial wealth. This assumes that the shortselling is permissible through the decision period at the horizon $T = 40$.

The highest and lowest final wealth trajectories are presented in Fig. 1.7. In the worst case, the trajectory is terminated to indicate the timing of vanishing wealth. There is quick bankruptcy for the aggressive overbet strategies.

The substantial downside is further illustrated in the distribution of final wealth plot in Fig. 1.8. The normal probability plots are almost linear on the upside

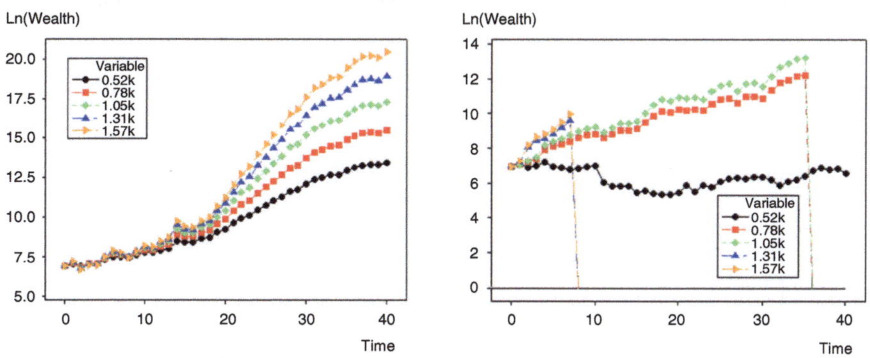

(a) (b)

Fig. 1.7 Trajectories with final wealth extremes for the equity versus cash example II: **(a)** maximum trajectories and **(b)** minimum trajectories

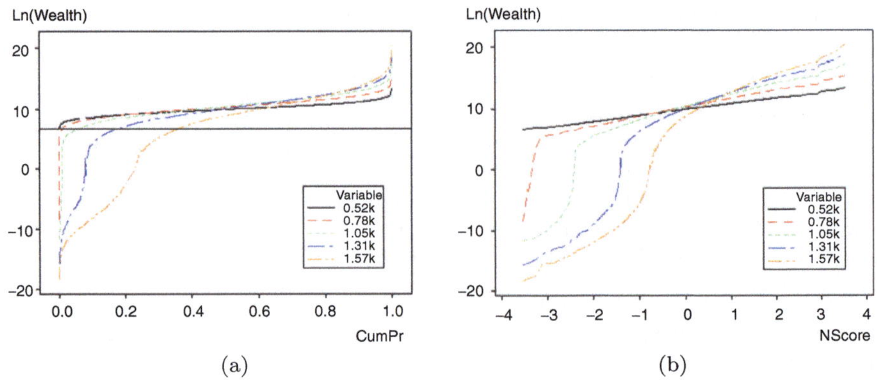

Fig. 1.8 Final ln(wealth) distributions by fraction for the equity versus cash example: **(a)** inverse cumulative and **(b)** normal plot

(log normality), but the downside is much more extreme than log-normal for all strategies except for 0.52 k. Even the full Kelly is very risky in this example largely because the basic position is levered. The inverse cumulative distribution shows a high probability of large losses with the most aggressive strategies. In constructing these plots the negative growth was incorporated with the formula, growth $= \left[\text{sign } W_T\right] \ln(|W_T|)$.

The mean–standard deviation trade-off in Fig. 1.9 provides more evidence concerning the riskiness of the high proportion strategies. When the fraction exceeds the full Kelly, the drop-off in growth rate is sharp, and that is matched by a sharp increase in the standard deviation.

The results of this experiment lead to the following conclusions:

1. The statistics describing the end of the horizon ($T = 40$) wealth are monotone in the fraction of wealth invested in the Kelly portfolio. Specifically (i) the

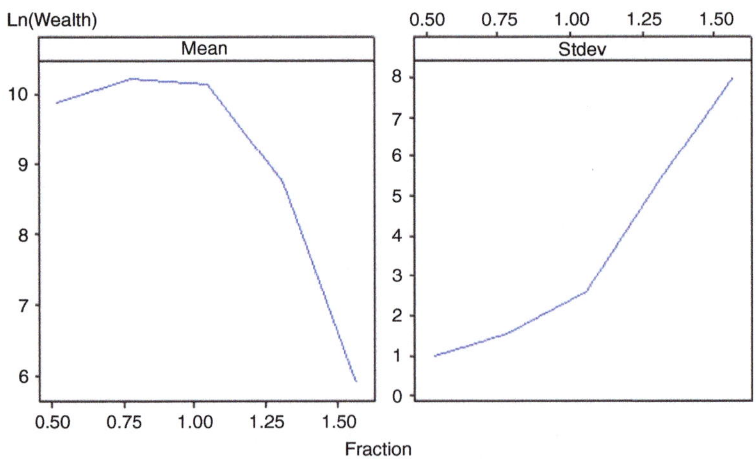

Fig. 1.9 Mean–standard deviation trade-off in the equity versus cash example

maximum terminal wealth and the mean terminal wealth increase in the Kelly fraction and (ii) the minimum wealth decreases as the fraction increases and the standard deviation grows as the fraction increases. The growth and decay are pronounced and it is possible to have extremely large losses. The fraction of the Kelly optimal growth strategy exceeds 1 in the most levered strategies and this is very risky. There is a trade-off between return and risk, but the mean for the levered strategies is growing far less than the standard deviation. The disadvantage of leveraged investment is illustrated with the cumulative distributions in Fig. 1.8. The log normality of final wealth does not hold for the levered strategies.

2. The maximum and minimum final wealth trajectories show the return – risk of levered strategies. The worst and best scenarios are not the same for all Kelly fractions. The worst scenario for the most levered strategy shows a rapid decline in wealth. The mean–standard deviation trade-off confirms the extreme riskiness of the aggressive strategies.

1.5 Final Comments

The Kelly optimal capital growth investment strategy is an attractive approach to wealth creation. In addition to maximizing the asymptotic rate of long-term growth of capital, it avoids bankruptcy and overwhelms any essentially different investment strategy in the long run. See MacLean, Thorp, and Ziemba (2010a) for a discussion of the good and bad properties of these strategies. However, automatic use of the Kelly strategy in any investment situation is risky and can be very dangerous. It requires some adaptation to the investment environment: rates of return, volatilities, correlation of alternative assets, estimation error, risk aversion preferences, and planning horizon are all important aspects of the investment process. Chopra and Ziemba (1993) show that in typical investment modeling, errors in the means average about 20 times in importance in objective value than errors in co-variances with errors in variances about double the co-variance errors. This is dangerous enough but they also show that the relative importance of the errors is risk aversion dependent with the errors compounding more and more for lower risk aversion investors and for the extreme log investors with essentially zero risk aversion the errors are worth about 100:3:1. So log investors must estimate means well if they are to survive. This is compounded even more by the observation that when times move suddenly from normal to bad the correlations/co-variances approach 1 and it is hard to predict the transition from good times to bad. Poundstone's (2005) book, while a very good read with lots of useful discussions, does not explain these important investment aspects and the use of Kelly strategies by advisory firms such as *Morningstar* and *Motley Fools* is flawed; see, for example, Fuller (2006) and Lee (2006). The experiments in Bicksler and Thorp (1973), Ziemba and Hausch (1986), and MacLean, Thorp, Zhao, and Ziemba (2011) and that described here represent some of the diversity in the investment environment. By considering the Kelly and its variants we get a concrete look at the plusses and minuses of the capital growth model. We can conclude that

- The wealth accumulated from the full Kelly strategy does not stochastically dominate fractional Kelly wealth. The downside is often much more favorable with a fraction less than 1.
- There is a trade-off of risk and return with the fraction invested in the Kelly portfolio. In cases of large uncertainty, from either intrinsic volatility or estimation error, security is gained by reducing the Kelly investment fraction.
- The full Kelly strategy can be highly levered. While the use of borrowing can be effective in generating large returns on investment, increased leveraging beyond the full Kelly is not warranted as it is growth-security dominated. The returns from over-levered investment are offset by a growing probability of bankruptcy.
- The Kelly strategy is not merely a long-term approach. Proper use in the short and medium run can achieve wealth goals while protecting against draw-downs. MacLean, Sanegre, Zhao, and Ziemba (2004) and MacLean, Zhao, and Ziemba (2009) discuss a strategy to reduce the Kelly fraction to stay above a pre-specified wealth path with high probability and to be penalized for being below the path.

The great economist Paul Samuelson was a long-time critic of the Kelly strategy which maximizes the expected logarithm of final wealth; see, for example, Samuelson (1969, 1971, 1979) and Merton and Samuelson (1974). His criticisms are well dealt with in the simulation example in this chapter and we see no disagreement with his various analytic points:

1. The Kelly strategy maximizes the asymptotic long-run growth of the investor's wealth, and we agree;
2. The Kelly strategy maximizes expected utility of only logarithmic utility and not necessarily any other utility function, and we agree;
3. The Kelly strategy always leads to more wealth than any essentially different strategy; this we know from the simulation in this chapter is not true since it is possible to have a large number of very good investments and still lose most of one's fortune.

Samuelson seemed to imply that Kelly proponents thought that the Kelly strategy maximizes for other utility functions but this was neither argued nor implied.

It is true that the expected value of wealth is higher with the Kelly strategy but bad outcomes are very possible.

We close this chapter with the main conclusions of the simulation studies

1. that the great superiority of full Kelly and close to full Kelly strategies over longer horizons with very large gains a large fraction of the time;
2. that the short-term performance of Kelly and high fractional Kelly strategies is very risky;
3. that there is a consistent trade-off of growth versus security as a function of the bet size determined by the various strategies; and

4. that no matter how favorable the investment opportunities are or how long the finite horizon is, a sequence of bad scenarios can lead to very poor final wealth outcomes, with a loss of most of the investor's initial capital.

Hence, in practice, financial engineering is important to deal with the short-term volatility and long-run situations with a sequence of bad scenarios. But properly used, the strategy has much to commend it, especially in trading with many repeated investments.

References

Aase, K. K. (2001). On the St. Petersburg Paradox. *Scandinavian Actuarial Journal 3* (1), 69–78.

Bell, R. M. and T. M. Cover (1980). Competitive optimality of logarithmic investment. *Math of Operations Research 5*, 161–166.

Bertocchi, M., S. L. Schwartz, and W. T. Ziemba (2010). *Optimizing the Aging, Retirement, Pensions Dilemma.* Wiley, Hoboken, NJ.

Bicksler, J. L. and E. O. Thorp (1973). The capital growth model: an empirical investigation. *Journal of Financial and Quantitative Analysis 8* (2), 273–287.

Breiman, L. (1960). Investment policies for expanding businesses optimal in a long run sense. *Naval Research Logistics Quarterly 4* (4), 647–651.

Breiman, L. (1961). Optimal gambling system for favorable games. *Proceedings of the 4th Berkeley Symposium on Mathematical Statistics and Probability 1*, 63–68.

Chopra, V. K. and W. T. Ziemba (1993). The effect of errors in mean, variance and co-variance estimates on optimal portfolio choice. *Journal of Portfolio Management 19*, 6–11.

Cover, T. M. and J. Thomas (2006). *Elements of Information Theory* (2nd ed.). Wiley, New York, NY.

Fuller, J. (2006). Optimize your portfolio with the Kelly formula. morningstar.com, October 6.

Hakansson, N. H. and W. T. Ziemba (1995). Capital growth theory. In R. A. Jarrow, V. Maksimovic, and W. T. Ziemba (Eds.), *Finance, Handbooks in OR & MS*, pp. 65–86. North Holland, Amsterdam.

Harville, D. A. (1973). Assigning probabilities to the outcome of multi-entry competitions. *Journal of the American Statistical Association 68*, 312–316.

Hausch, D. B., V. Lo, and W. T. Ziemba (Eds.) (1994). *Efficiency of Racetrack Betting Markets.* Academic, San Diego.

Hausch, D. B., V. Lo, and W. T. Ziemba (Eds.) (2008). *Efficiency of Racetrack Betting Markets* (2 ed.). World Scientific, Singapore.

Hausch, D. B. and W. T. Ziemba (1985). Transactions costs, extent of inefficiencies, entries and multiple wagers in a racetrack betting model. *Management Science 31*, 381–394.

Hausch, D. B., W. T. Ziemba, and M. E. Rubinstein (1981). Efficiency of the market for racetrack betting. *Management Science XXVII*, 1435–1452.

Kelly, Jr., J. R. (1956). A new interpretation of the information rate. *Bell System Technical Journal 35*, 917–926.

Latané, H. (1978). The geometric-mean principle revisited – a reply. *Journal of Banking and Finance 2* (4), 395–398.

Lee, E. (2006). How to calculate the Kelly formula. fool.com, October 31.

Luenberger, D. G. (1993). A preference foundation for log mean-variance criteria in portfolio choice problems. *Journal of Economic Dynamics and Control 17*, 887–906.

MacLean, L. C., R. Sanegre, Y. Zhao, and W. T. Ziemba (2004). Capital growth with security. *Journal of Economic Dynamics and Control 28* (4), 937–954.

MacLean, L. C., E. O. Thorp, Y. Zhao, and W. T. Ziemba (2011). How does the Fortunes Formula-Kelly capital growth model perform? *Journal of Portfolio Management 37* (4).

MacLean, L. C., E. O. Thorp, and W. T. Ziemba (2010a). The good and properties of the Kelly and fractional Kelly capital growth criterion. *Quantitative Finance* (August–September), 681–687.

MacLean, L. C., E. O. Thorp, and W. T. Ziemba (2010b). *The Kelly Capital Growth Investment Criteria*. World Scientific, Singapore.

MacLean, L. C., Y. Zhao, and W. T. Ziemba (2009). Optimal capital growth with convex loss penalties. Working paper, Oxford University.

MacLean, L. C., W. T. Ziemba, and G. Blazenko (1992). Growth versus security in dynamic investment analysis. *Management Science 38*, 1562–1585.

MacLean, L. C., W. T. Ziemba, and Li (2005). Time to wealth goals in capital accumulation and the optimal trade-off of growth versus security. *Quantitative Finance 5* (4), 343–357.

Menger, K. (1967). The role of uncertainty in economics. In M. Shubik (Ed.), *Essays in Mathematical Economics in Honor of Oskar Morgenstern*. Princeton University Press, Princeton, NJ.

Merton, R. C. (1990). *Continuous Time Finance*. Basil Blackwell, Oxford, UK.

Merton, R. C. and P. A. Samuelson (1974). Fallacy of the log-normal approximation to optimal portfolio decision-making over many periods. *Journal of Financial Economics 1*, 67–94.

Pabrai, M. (2007). *The Dhandho Investor*. Wiley, Hoboken, NJ.

Poundstone, W. (2005). *Fortune's Formula: The Untold Story of the Scientific System That Beat the Casinos and Wall Street*. Hill and Wang, New York, NY.

Samuelson, P. A. (1969). Lifetime portfolio selection by dynamic stochastic programming. *Review of Economics and Statistics 51*, 239–246.

Samuelson, P. A. (1971). The fallacy of maximizing the geometric mean in long sequences of investing or gambling. *Proceedings National Academy of Science 68*, 2493–2496.

Samuelson, P. A. (1977). St. Petersburg paradoxes: Defanged, dissected and historically described. *Journal of Economic Literature 15* (1), 24–55.

Samuelson, P. A. (1979). Why we should not make mean log of wealth big though years to act are long. *Journal of Banking and Finance 3*, 305–307.

Samuelson, P. A. (2006–2009). Letters to William T. Ziemba, correspondence December 13, 2006, May 7, 2008, and May 12, 2008.

Siegel, J. (2002). *Stocks for the Long Run*. Wiley, McGraw-Hill, New York.

Thorp, E. O. (2008). Understanding the Kelly criterion. *Wilmott*, May and September.

Ziemba, W. T. (2005). The symmetric downside risk Sharpe ratio and the evaluation of great investors and speculators. *Journal of Portfolio Management Fall, 32* (1), 108–120.

Ziemba, W. T. (2010). A tale of five investors: response to Paul A. Samuelson letters. Working Paper, University of Oxford.

Ziemba, W. T. and D. B. Hausch (1986). *Betting at the Racetrack*. Dr Z Investments, San Luis Obispo, CA.

Ziemba, W. T. and D. B. Hausch (1987). *Dr Z's Beat the Racetrack*. William Morrow, New York, NY.

Ziemba, R. E. S. and W. T. Ziemba (2007). *Scenarios for Risk Management and Global Investment Strategies*. Wiley, Chichester.

Designing Minimum Guaranteed Return Funds

Michael A.H. Dempster, Matteo Germano, Elena A. Medova, Muriel I. Rietbergen, Francesco Sandrini, and Mike Scrowston

Abstract In recent years there has been a significant growth of investment products aimed at attracting investors who are worried about the downside potential of the financial markets. This paper introduces a dynamic stochastic optimization model for the design of such products. The pricing of minimum guarantees as well as the valuation of a portfolio of bonds based on a three-factor term structure model are described in detail. This allows us to accurately price individual bonds, including the zero-coupon bonds used to provide risk management, rather than having to rely on a generalized bond index model.

Keywords Dynamic Stochastic Programming · Asset and Liability Management · Guaranteed Returns · Yield Curve · Economic Factor Model

1 Introduction

In recent years there has been a significant growth of investment products aimed at attracting investors who are worried about the downside potential of the financial markets for pension investments. The main feature of these products is a minimum guaranteed return together with exposure to the upside movements of the market.

There are several different guarantees available in the market. The one most commonly used is the nominal guarantee which guarantees a fixed percentage of the initial investment. However there also exist funds with a guarantee in real terms which is linked to an inflation index. Another distinction can be made between fixed and flexible guarantees, with the fixed guarantee linked to a particular rate and the flexible to for instance a capital market index. Real guarantees are a special case of flexible guarantees. Sometimes the guarantee of a minimum rate of return is even set relative to the performance of other pension funds.

M.A.H. Dempster (✉)
Centre for Financial Research, Statistical Laboratory, Department of Pure Mathematics and Mathematical Statistics, University of Cambridge, Cambridge, UK; Cambridge Systems Associates Ltd., Cambridge, UK
e-mail: mahd2@cam.ac.uk

M. Bertocchi et al. (eds.), *Stochastic Optimization Methods in Finance and Energy*, International Series in Operations Research & Management Science 163,
DOI 10.1007/978-1-4419-9586-5_2, © Springer Science+Business Media, LLC 2011

Return guarantees typically involve hedging or insuring. Hedging involves eliminating the risk by sacrificing some or all of the potential for gain, whereas insuring involves paying an insurance premium to eliminate the risk of losing a large amount.

Many government and private pension schemes consist of defined benefit plans. The task of the pension fund is to guarantee benefit payments to retiring clients by investing part of their current wealth in the financial markets. The responsibility of the pension fund is to hedge the client's risk, while meeting the solvency requirements in such a way that all benefit payments are met. However at present there are significant gaps between fund values, contributions made by employees, and pension obligations to retirees.

One way in which the guarantee can be achieved is by investing in zero-coupon Treasury bonds with a maturity equal to the time horizon of the investment product in question. However using this option foregoes all upside potential. Even though the aim is protect the investor from the downside, a reasonable expectation of returns higher than guaranteed needs to remain.

In this paper we will consider long-term nominal minimum guaranteed return plans with a fixed time horizon. They will be closed end guarantee funds; after the initial contribution there is no possibility of making any contributions during the lifetime of the product. The main focus will be on how to optimally hedge the risks involved in order to avoid having to buy costly insurance.

However this task is not straightforward, as it requires long-term forecasting for all investment classes and dealing with a stochastic liability. *Dynamic stochastic programming* is the technique of choice to solve this kind of problem as such a model will automatically hedge current portfolio allocations against the future uncertainties in asset returns and liabilities over a long horizon (see e.g. Dempster *et al.*, 2003). This will lead to more robust decisions and previews of possible future benefits and problems contrary to, for instance, static portfolio optimization models such as the Markowitz (1959) mean-variance allocation model.

Consiglio *et al.* (2007) have studied fund guarantees over single investment periods and Hertzog *et al.* (2007) treat dynamic problems with a deterministic risk barrier. However a practical method should have the flexibility to take into account multiple time periods, portfolio constraints such as prohibition of short selling and varying degrees of risk aversion. In addition, it should be based on a realistic representation of the dynamics of the relevant factors such as asset prices or returns and should model the changing market dynamics of risk management. All these factors have been carefully addressed here and are explained further in the sequel.

The rest of the chapter is organized as follows. In Section 2 we describe the stochastic optimization framework, which includes the problem set up, model constraints and possible objective functions. Section 3 presents a three-factor term structure model and its application to pricing the bond portfolio and the liability side of the fund on individual scenarios. As our portfolio will mainly consist of bonds, this area has been extensively researched. Section 4 presents several historical backtests to show how the framework would have performed had it been implemented in practice, paying particular attention to the effects of using different objective functions and varying tree structures. Section 5 repeats the backtest when the stock index is modeled as a jumping diffusion so that the corresponding returns exhibit fat tails and Section 6 concludes. Throughout this chapter boldface is used to denote random entities.

2 Stochastic Optimization Framework

In this section we describe the framework for optimizing minimum guaranteed return funds using stochastic optimization. We will focus on risk management as well as strategic asset allocation concerned with allocation across broad asset classes, though we will allow specific maturity bond allocations.

2.1 Set Up

This chapter looks at several methods to optimally allocate assets for a minimum guaranteed return fund using expected average and expected maximum shortfall risk measures relative to the current value of the guarantee. The models will be applied to eight different assets: coupon bonds with maturity equal to 1, 2, 3, 4, 5, 10 and 30 years and an equity index, and the home currency is the euro. Extensions incorporated into these models are the presence of coupon rates directly dependent on the term structure of bond returns and the annual rolling of the coupon-bearing bonds.

We consider a discrete time and space setting. The time interval considered is given by $\left\{0, \frac{1}{12}, \frac{2}{12}, \dots, T\right\}$, where the times indexed by $t = 0, 1, \dots, T-1$ correspond to decision times at which the fund will trade and T to the planning horizon at which no decision is made, see Figure 1. We will be looking at a five-year horizon.

Uncertainty Ω is represented by a *scenario tree*, in which each path through the tree corresponds to a *scenario* ω in Ω and each node in the tree corresponds to a time along one or more scenarios. An example scenario tree is given in Figure 2. The probability $p(\omega)$ of scenario ω in Ω is the reciprocal of the total number of scenarios as the scenarios are generated by Monte Carlo simulation and are hence equiprobable.

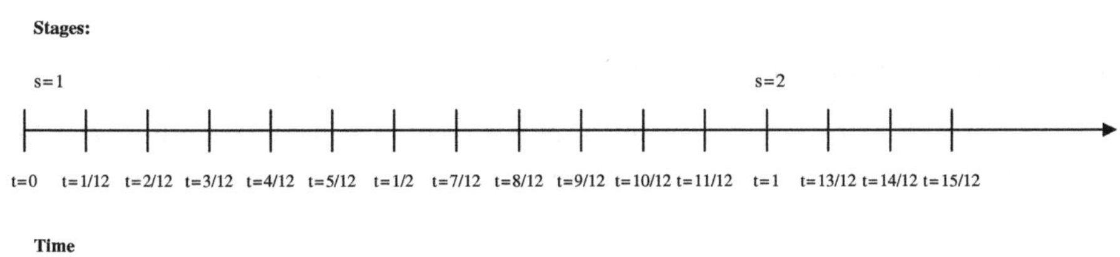

Fig. 1 Time and stage setting

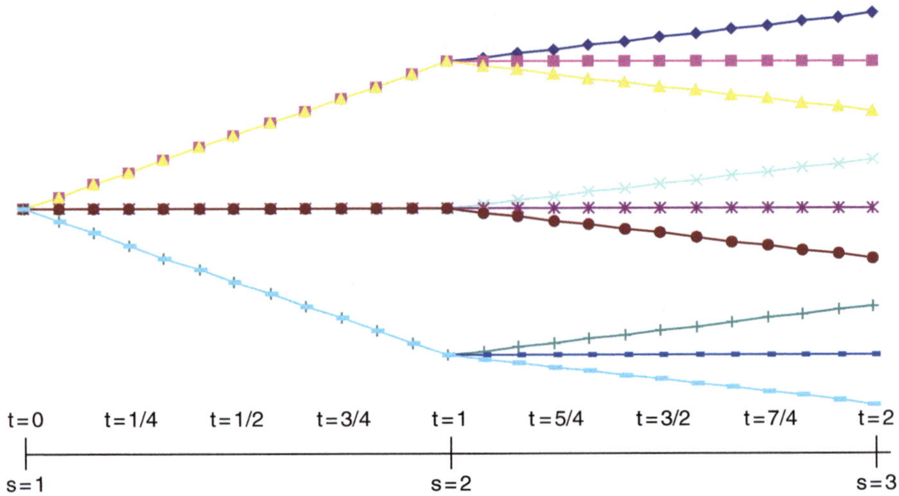

Fig. 2 Graphical representation of scenarios

The stock price process $\mathbf{S}$ is (initially) assumed to follow a geometric Brownian motion, i.e.

$$\frac{d\mathbf{S}_t}{\mathbf{S}_t} = \mu_S dt + \sigma_S d\mathbf{W}_t^S, \tag{1}$$

where $d\mathbf{W}_t^S$ is correlated with the $d\mathbf{W}_t$ terms driving the three term structure factors discussed in Section 3.

2.2 Model Constraints

Let (see Table 1)

- $h_t(\omega)$ denote the *shortfall* at time t and scenario ω, i.e.

$$h_t(\omega) := \max\left(0, L_t(\omega) - W_t(\omega)\right) \quad \forall\omega \in \Omega \quad t \in T^{\text{total}} \tag{2}$$

- $H(\omega) := \max_{t \in T^{\text{total}}} h_t(\omega)$ denote the *maximum shortfall* over time for scenario ω.

The constraints considered for the minimum guaranteed return problem are:

- *cash balance constraints*. These constraints ensure that the net cash flow at each time and at each scenario is equal to zero

$$\sum_{a \in A} f P_{0,a}^{\text{buy}}(\omega) x_{0,a}^{+}(\omega) = W_0 \qquad \omega \in \Omega \tag{3}$$

$$\sum_{a\in A\backslash\{S\}} \frac{1}{2}\delta^a_{t-1}(\omega)F^a x^-_{t,a}(\omega) + \sum_{a\in A} gP^{\text{sell}}_{t,a}(\omega)x^-_{t,a}(\omega) = \sum_{a\in A} fP^{\text{buy}}_{t,a}(\omega)x^+_{t,a}(\omega) \tag{4}$$
$$\omega \in \Omega \quad t \in T^d\backslash\{0\}.$$

In (4) the left-hand side represents the cash freed up to be reinvested at time $t \in T^d\backslash\{0\}$ and consists of two distinct components. The first term represents the semi-annual coupons received on the coupon-bearing Treasury bonds held between time $t-1$ and t, the second term represents the cash obtained from selling part of the portfolio. This must equal the value of the new assets bought given by the right hand side of (4).

Table 1 Variables and parameters of the model

Time Sets	
$T^{\text{total}} = \left\{0, \frac{1}{12}, \ldots, T\right\}$	*set of all times* considered in the stochastic programme
$T^d = \{0, 1, \ldots, T-1\}$	*set of decision times*
$T^i = T^{\text{total}}\backslash T^d$	*set of intermediate times*
$T^c = \left\{\frac{1}{2}, \frac{3}{2}, \ldots, T-\frac{1}{2}\right\}$	*times when a coupon is paid out in-between decision times*
Instruments	
$S_t(\omega)$	*Dow Jones EuroStoxx 50 index level* at time t in scenario ω
$B^T_t(\omega)$	*EU Treasury bond with maturity* T at time t in scenario ω
$\delta^{B^T}_t(\omega)$	*coupon rate* of EU Treasury bond with maturity T at time t in scenario ω
F^{B^T}	*face value* of EU Treasury bond with maturity T
$Z_t(\omega)$	*EU zero-coupon Treasury bond price* at time t in scenario ω
Risk Management Barrier	
$y_{t,T}(\omega)$	*EU zero-coupon Treasury yield* with maturity T at time t in scenario ω
G	*annual guaranteed return*
$L^N_t(\omega)$	*nominal barrier* at time t in scenario ω
Portfolio Evolution	
A	*set of all assets*
$P^{buy}_{t,a}(\omega) / P^{sell}_{t,a}(\omega)$	*buy/sell price* of asset $a \in A$ at time t in scenario ω
f/g	*transaction costs* for buying / selling
$x_{t,a}(\omega)$	*quantity held* of asset $a \in A$ between time t and $t + 1/12$ in scenario ω
$x^+_{t,a}(\omega) / x^-_{t,a}(\omega)$	*quantity bought/sold* of asset $a \in A$ at time t in scenario ω
W_0	*initial portfolio wealth*
$W_t(\omega)$	*portfolio wealth before rebalancing* at time $t \in T$ in scenario ω
$w_t(\omega)$	*portfolio wealth after rebalancing* at time $t \in T^c \cup T^d\backslash\{T\}$ in scenario ω
$h_t(\omega) := \max(0, L_t(\omega) - W_t(\omega))$	*shortfall* at time t in scenario ω

- *short sale constraints*. In our model we assume no short selling of any stocks or bonds

$$x_{t,a}(\omega) \geq 0 \qquad a \in A \quad \omega \in \Omega \quad t \in T^{\text{total}} \tag{5}$$

$$x^{+}_{t,a}(\omega) \geq 0 \qquad \forall a \in A \quad \forall \omega \in \Omega \quad \forall t \in T^{\text{total}} \setminus \{T\} \tag{6}$$

$$x^{-}_{t,a}(\omega) \geq 0 \qquad \forall a \in A \quad \forall \omega \in \Omega \quad \forall t \in T^{\text{total}} \setminus \{0\}. \tag{7}$$

- *information constraints*. These constraints ensure that the portfolio allocation can not be changed during the period from one decision time to the next and hence that no decisions with perfect foresight can be made

$$x^{+}_{t,a}(\omega) = x^{-}_{t,a}(\omega) = 0 \qquad a \in A \quad \omega \in \Omega \quad t \in T^{i} \setminus T^{c}. \tag{8}$$

- *wealth constraint*. This constraint determines the portfolio wealth at each point in time

$$w_t(\omega) = \sum_{a \in A} P^{\text{buy}}_{t,a}(\omega) x_{t,a}(\omega) \qquad \omega \in \Omega \quad t \in T^{\text{total}} \setminus \{T\} \tag{9}$$

$$W_t(\omega) = \sum_{a \in A} P^{\text{sell}}_{t,a}(\omega) x_{t-\frac{1}{12},a}(\omega) \qquad \omega \in \Omega \quad t \in T^{\text{total}} \setminus \{0\} \tag{10}$$

$$w_T(\omega) = \sum_{a \in A} g P^{\text{sell}}_{T,a}(\omega) x_{T-\frac{1}{12},a}(\omega) + \sum_{a \in A \setminus \{S\}} \frac{1}{2} \delta^{a}_{T-1}(\omega) F^{a} x_{T-\frac{1}{12},a}(\omega) \qquad \omega \in \Omega. \tag{11}$$

- *accounting balance constraints*. These constraints give the quantity invested in each asset at each time and for each scenario

$$x_{0,a}(\omega) = x^{+}_{0,a}(\omega) \qquad a \in A \quad \omega \in \Omega \tag{12}$$

$$x_{t,a}(\omega) = x_{t-\frac{1}{12},a}(\omega) + x^{+}_{t,a}(\omega) - x^{-}_{t,a}(\omega) \qquad a \in A \quad \omega \in \Omega \quad t \in T^{\text{total}} \setminus \{0\}. \tag{13}$$

The total quantity invested in asset $a \in A$ between time $t$and $t + \frac{1}{12}$ is equal to the total quantity invested in asset $a \in A$ between time $t - \frac{1}{12}$ and t plus the quantity of asset $a \in A$ bought at time t minus the quantity of asset $a \in A$ sold at time t.

- *annual rolling constraint*. This constraint ensures that at each decision time all the coupon-bearing Treasury bond holdings are sold

$$x^{-}_{t,a}(\omega) = x_{t-\frac{1}{12},a}(\omega) \qquad a \in A \setminus \{S\} \qquad \omega \in \Omega \qquad t \in T^{d} \setminus \{0\}. \tag{14}$$

- *coupon re-investment constraints*. We assume that the coupon paid every six months will be re-invested in the same coupon-bearing Treasury bond

$$x^+_{t,a}(\omega) = \frac{\frac{1}{2}\delta^a_t(\omega) F^a x_{t-\frac{1}{12},a}(\omega)}{f P^{\text{buy}}_{t,a}(\omega)} \qquad x^-_{t,a}(\omega) = 0$$

$$x^+_{t,S}(\omega) = x^-_{t,S}(\omega) = 0 \tag{15}$$

$$a \in A \backslash \{S\} \quad \omega \in \Omega \quad t \in T^c.$$

- *barrier constraints.* These constraints determine the shortfall of the portfolio at each time and scenario as defined in Table 1

$$h_t(\omega) + W_t(\omega) \geq L_t(\omega) \qquad \omega \in \Omega \quad t \in T^{\text{total}} \tag{16}$$

$$h_t(\omega) \geq 0 \qquad \omega \in \Omega \quad t \in T^{\text{total}}. \tag{17}$$

As the objective of the stochastic programme will put a penalty on any shortfall, optimizing will ensure that $h_t(\omega)$ will be zero if possible and as small as possible otherwise, i.e.

$$h_t(\omega) = \max(0, L_t(\omega) - W_t(\omega)) \qquad \forall \omega \in \Omega \quad \forall t \in T^{\text{total}} \tag{18}$$

which is exactly how we defined $h_t(\omega)$ in (2).

To obtain the maximum shortfall for each scenario, we need to add one of the following two sets of constraints:

$$H(\omega) \geq h_t(\omega) \qquad \forall \omega \in \Omega \quad \forall t \in T^{\text{d}} \cup \{T\} \tag{19}$$

$$H(\omega) \geq h_t(\omega) \qquad \forall \omega \in \Omega \quad \forall t \in T^{\text{total}} \tag{20}$$

Constraint (19) needs to be added if the max shortfall is to be taken into account on a yearly basis and constraint (20) if max shortfall is on a monthly basis.

2.3 Objective Functions: Expected Average Shortfall and Expected Maximum Shortfall

Starting with an initial wealth W_0 and an annual *nominal guarantee* of G, the liability at the planning horizon at time T is given by

$$W_0 (1 + G)^T. \tag{21}$$

To price the liability at time $t < T$ consider a zero-coupon Treasury bond, which pays 1 at time T, i.e. $Z_T(\omega) = 1$, for all scenarios $\omega \in \Omega$. The *zero-coupon Treasury bond price* at time t in scenario ω assuming continuous compounding is given by

$$Z_t(\omega) = e^{-y_{t,T}(\omega)(T-t)}, \tag{22}$$

where $y_{t,T}(\omega)$ is the *zero-coupon Treasury yield* with maturity T at time t in scenario ω.

This gives a formula for the value of the *nominal or fixed guarantee barrier* at time t in scenario ω as

$$L_t^N(\omega) := W_0(1+G)^T Z_t(\omega) = W_0(1+G)^T e^{-y_{t,T}(\omega)(T-t)}. \tag{23}$$

In a minimum guaranteed return fund the objective of the fund manager is twofold; firstly to manage the investment strategies of the fund and secondly to take into account the guarantees given to all investors. Investment strategies must ensure that the guarantee for all participants of the fund is met with a high probability.

In practice the guarantor (the parent bank of the fund manager) will ensure the investor guarantee is met by forcing the purchase of the zero coupon bond of (22) when the fund is sufficiently near the barrier defined by (23). Since all upside potential to investors is thus foregone, the aim of the fund manager is to fall below the barrier with acceptably small if not zero probability.

Ideally we would add a constraint limiting the probability of falling below the barrier in a VaR-type minimum guarantee constraint, i.e.

$$P\left(\max_{t\in T^{\text{total}}} h_t(\boldsymbol{\omega}) > 0\right) \leq \alpha \tag{24}$$

for α small. However, such scenario-based probabilistic constraints are extremely difficult to implement, as they may without further assumptions convert the convex large-scale optimization problem into a non-convex one. We therefore use the following two convex approximations in which we trade off the risk of falling below the barrier against the return in the form of the expected sum of wealth.

Firstly, we look at the *expected average shortfall* (EAS) model in which the objective function is given by:

$$\begin{aligned}
&\max_{\left\{\substack{x_{t,a}(\omega),x^+_{t,a}(\omega),x^-_{t,a}(\omega):\\ a\in A,\omega\in\Omega,t\in T^d\cup\{T\}}\right\}} \left\{\sum_{\omega\in\Omega}\sum_{t\in T^d\cup\{T\}} p(\omega)\left((1-\beta)\,W_t(\omega) - \beta\frac{h_t(\omega)}{|T^d\cup\{T\}|}\right)\right\}\\
&= \max_{\left\{\substack{x_{t,a}(\omega),x^+_{t,a}(\omega),x^-_{t,a}(\omega):\\ a\in A,\omega\in\Omega,t\in T^d\cup\{T\}}\right\}} \left\{(1-\beta)\left(\sum_{\omega\in\Omega} p(\omega)\sum_{t\in T^d\cup\{T\}} W_t(\omega)\right)\right.\\
&\qquad\left. -\beta\left(\sum_{\omega\in\Omega} p(\omega)\sum_{t\in T^d\cup\{T\}}\frac{h_t(\omega)}{|T^d\cup\{T\}|}\right)\right\}.
\end{aligned} \tag{25}$$

In this case we maximize the expected sum of wealth over time while penalizing each time the wealth falls below the barrier. For each scenario $\omega \in \Omega$ we can calculate the average shortfall over time and then take expectations over all scenarios.

In this case only shortfalls at decision times are taken into account and any serious loss in portfolio wealth in-between decision times is ignored. However from the fund manager's and guarantor's perspective the position of the portfolio wealth relative to the fund's barrier is significant on a continuous basis and serious or repeated drops below this barrier might force the purchase of expensive insurance. To capture this feature specific to minimum guaranteed return funds, we also consider an objective function in which the shortfall of the portfolio is considered on a monthly basis.

For the *expected average shortfall with monthly checking* (EAS MC) model the objective function is given by

$$\max_{\left\{\substack{x_{t,a}(\omega),x_{t,a}^{+}(\omega),x_{t,a}^{-}(\omega):\\ a\in A,\omega\in\Omega,t\in T^{d}\cup\{T\}}\right\}} \left\{(1-\beta)\left(\sum_{\omega\in\Omega} p(\omega) \sum_{t\in T^{d}\cup\{T\}} W_t(\omega)\right) - \beta\left(\sum_{\omega\in\Omega} p(\omega) \sum_{t\in T^{\text{total}}} \frac{h_t(\omega)}{\left|T^{\text{total}}\right|}\right)\right\}. \tag{26}$$

Note that although we still only rebalance once a year shortfall is now being measured on a monthly basis in the objective and hence the annual decisions must also take into account the possible effects they will have on the monthly shortfall.

The value of $0 \leq \beta \leq 1$ can be chosen freely and sets the level of risk aversion. The higher the value of β, the higher the importance given to shortfall and the less to the expected sum of wealth, and hence the more risk-averse the optimal portfolio allocation will be. The two extreme cases are represented by $\beta = 0$, corresponding to the 'unconstrained' situation, which is indifferent to the probability of falling below the barrier, and $\beta = 1$, corresponding to the situation in which the shortfall is penalized and the expected sum of wealth ignored.

In general short horizon funds are likely to attract more risk-averse participants than long horizon funds, whose participants can afford to tolerate more risk in the short run. This natural division between short and long-horizon funds is automatically incorporated in the problem set up, as the barrier will initially be lower for long-term funds than for short-term funds as exhibited in Figure 3. However the importance of closeness to the barrier can be adjusted by the choice of β in the objective.

The second model we consider is the *expected maximum shortfall* (EMS) model given by:

$$\max_{\left\{\substack{x_{t,a}(\omega),x_{t,a}^{+}(\omega),x_{t,a}^{-}(\omega):\\ a\in A,\omega\in\Omega,t\in T^{d}\cup\{T\}}\right\}} \left\{(1-\beta)\left(\sum_{\omega\in\Omega} p(\omega) \sum_{t\in T^{d}\cup\{T\}} W_t(\omega)\right) - \beta\left(\sum_{\omega\in\Omega} p(\omega) H(\omega)\right)\right\} \tag{27}$$

using the constraints (19) to define $H(\omega)$.

For the *expected maximum shortfall with monthly checking* (EMS MC) model the objective function remains the same but $H(\omega)$ is now defined by (20).

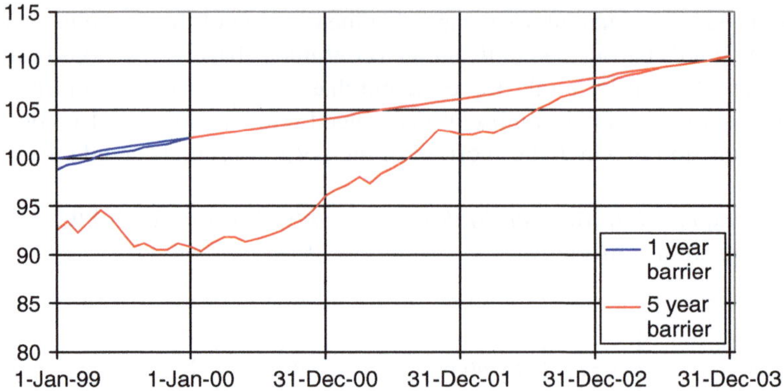

Fig. 3 Barrier for one-year and five-year 2% guaranteed return fund

In both variants of this model we penalize the expected maximum shortfall, which ensures that $H(\omega)$ is as small as possible for each scenario $\omega \in \Omega$. Combining this with constraints (19)/(20) it follows that $H(\omega)$ is exactly equal to the maximum shortfall.

The constraints given in Section 2.2 apply to both the expected average shortfall and expected maximum shortfall models.

The EAS model incurs a penalty every time portfolio wealth falls below the barrier, but it does not differentiate between a substantial shortfall at one point in time and a series of small shortfalls over time. The EMS model on the other hand, focusses on limiting the maximum shortfall and therefore does not penalize portfolio wealth falling just slightly below the barrier several times. So one model limits the *number of times* portfolio wealth falls below the barrier while the other limits *any shortfall substantially*.

3 Bond Pricing

In this section we present a three-factor term structure model which we will use to price both our bond portfolio and the fund's liability. Many interest-rate models, like the classic one-factor Vasicek (1977) and Cox, Ingersoll, and Ross (1985) class of models and even more recent multi-factor models like Anderson and Lund (1997), concentrate on modeling just the short-term rate.

However for the minimum guaranteed return funds we have to deal with a long-term liability and bonds of varying maturities. We therefore must capture the dynamics of the whole term structure. This has been achieved by using the economic factor model described below in Section 3.1. In Section 3.2 we describe the pricing of coupon-bearing bonds and Section 3.3 investigates the consequences of rolling the bonds on an annual basis.

3.1 Yield Curve Model

To capture the dynamics of the whole term structure, we will use a Gaussian *economic factor model* (EFM) (see Campbell (2000) and also Nelson and Siegel (1987)) whose evolution under the risk-neutral measure Q is determined by the stochastic differential equations

$$\mathbf{dX}_t = (\mu_X - \lambda_X X_t)\, dt + \sigma_X \mathbf{dW}_t^X \tag{28}$$

$$\mathbf{dY}_t = (\mu_Y - \lambda_Y Y_t)\, dt + \sigma_Y \mathbf{dW}_t^Y \tag{29}$$

$$\mathbf{dR}_t = k\,(X_t + Y_t - R_t)\, dt + \sigma_R \mathbf{dW}_t^R \tag{30}$$

where the $\boldsymbol{dW}$ terms are correlated. The three unobservable Gaussian factors $\mathbf{R}$, $\mathbf{X}$ and $\mathbf{Y}$ represent respectively a *short rate*, a *long rate* and (minus) the *slope* between an *instantaneous short rate* and the long rate. Solving these equations the following formula for the yield at time t with time to maturity equal to $T - t$ is obtained (for a derivation, see Medova *et al.*, 2005)

$$y_{t,T} = \frac{A\,(t,T)\,R_t + B\,(t,T)\,X_t + C\,(t,T)\,Y_t + D\,(t,T)}{T}, \tag{31}$$

where

$$A\,(t,T) := \frac{1}{k}\left(1 - e^{-k(T-t)}\right) \tag{32}$$

$$B\,(t,T) := \frac{k}{k-\lambda_X}\left\{\frac{1}{\lambda_X}\left(1 - e^{-\lambda_X(T-t)}\right) - \frac{1}{k}\left(1 - e^{-k(T-t)}\right)\right\} \tag{33}$$

$$C\,(t,T) := \frac{k}{k-\lambda_Y}\left\{\frac{1}{\lambda_Y}\left(1 - e^{-\lambda_Y(T-t)}\right) - \frac{1}{k}\left(1 - e^{-k(T-t)}\right)\right\} \tag{34}$$

$$\begin{aligned}
D\,(t,T) := &\left(T - t - \frac{1}{k}\left(1 - e^{-k(T-t)}\right)\right)\left(\frac{\mu_X}{\lambda_X} + \frac{\mu_Y}{\lambda_Y}\right) - \frac{\mu_X}{\lambda_X} B\,(t,T) - \frac{\mu_Y}{\lambda_Y} C\,(t,T) \\
&- \frac{1}{2}\sum_{i=1}^{3}\left\{\frac{m_{X_i}^2}{2\lambda_X}\left(1 - e^{-2\lambda_X(T-t)}\right) + \frac{m_{Y_i}^2}{2\lambda_Y}\left(1 - e^{-2\lambda_Y(T-t)}\right)\right. \\
&+ \frac{n_i^2}{2k}\left(1 - e^{-2k(T-t)}\right) + p_i^2\,(T-t) + \frac{2m_{X_i} m_{Y_i}}{\lambda_X + \lambda_Y}\left(1 - e^{-(\lambda_X+\lambda_Y)(T-t)}\right) \\
&+ \frac{2m_{X_i} n_i}{\lambda_X + k}\left(1 - e^{-(\lambda_X+k)(T-t)}\right) + \frac{2m_{X_i} p_i}{\lambda_X}\left(1 - e^{-\lambda_X(T-t)}\right) \\
&+ \frac{2m_{Y_i} n_i}{\lambda_Y + k}\left(1 - e^{-(\lambda_Y+k)(T-t)}\right) + \frac{2m_{Y_i} p_i}{\lambda_Y}\left(1 - e^{-\lambda_Y(T-t)}\right) \\
&\left. + \frac{2n_i p_i}{k}\left(1 - e^{-k(T-t)}\right)\right\}
\end{aligned} \tag{35}$$

and

$$
\begin{aligned}
m_{X_i} &:= -\frac{k\sigma_{X_i}}{\lambda_X\,(k-\lambda_X)} \\
m_{Y_i} &:= -\frac{k\sigma_{Y_i}}{\lambda_Y\,(k-\lambda_Y)} \\
n_i &:= \frac{\sigma_{X_i}}{k-\lambda_X} + \frac{\sigma_{Y_i}}{k-\lambda_Y} - \frac{\sigma_{R_i}}{k} \\
p_i &:= -\left(m_{X_i} + m_{Y_i} + n_i\right).
\end{aligned} \tag{36}
$$

Bond pricing must be effected under the *risk-neutral* measure Q. However, for the model to be used for forward simulation the set of stochastic differential equations must be adjusted to capture the model dynamics under the real-world or *market* measure P. We therefore have to model the market prices of risk which take us from the risk-neutral measure Q to the real-world measure P.

Under the market measure P we adjust the drift term by adding the *risk premium* given by the *market price of risk* γ in terms of the quantity of risk. The effect of this is a change in the long-term mean, e.g. for the factor $\mathbf{X}$ the long-term mean now equals $\frac{\mu_X+\gamma_X\sigma_X}{\lambda_X}$. It is generally assumed in a Gaussian world that the quantity of risk is given by the *volatility* of each factor.

This gives us the following set of processes under the market measure

$$
d\mathbf{X}_t = (\mu_X - \lambda_X X_t + \gamma_X\sigma_X)dt + \sigma_X d\mathbf{W}_t^X \tag{37}
$$

$$
d\mathbf{Y}_t = (\mu_Y - \lambda_Y Y_t + \gamma_Y\sigma_Y)dt + \sigma_Y d\mathbf{W}_t^Y \tag{38}
$$

$$
d\mathbf{R}_t = \{k(X_t + Y_t - R_t) + \gamma_R\sigma_R\}dt + \sigma_R d\mathbf{W}_t^R, \tag{39}
$$

where all three factors contain a market price of risk γ in volatility units.

The yields derived in the economic factor model are continuously compounded while most yield data are annually compounded. So for appropriate comparison when estimating the parameters of the model we will have to convert the annually compounded yields into continuously compounded yields using the transformation

$$
y^{\text{(continuous)}} = \ln(1 + y^{\text{(annual)}}). \tag{40}
$$

In the limit as T tends to infinity it can be shown that expression (31) derived for the yield does not tend to the 'long rate' factor X, but to a constant. This suggests that the three factors introduced in this term structure model may really be unobservable. To handle the unobservable state variables we formulate the model in state space form, a detailed description of which can be found in Harvey (1993) and use the Kalman filter to estimate the parameters (see e.g. Dempster *et al.*, 1999).

3.2 Pricing Coupon-Bearing Bonds

As sufficient historical data on Euro coupon-bearing Treasury bonds is difficult to obtain we use the zero-coupon yield curve to construct the relevant bonds. Coupons on newly-issued bonds are generally closely related to the corresponding spot rate at the time, so the current zero-coupon yield with maturity T is used as a proxy for the coupon rate of a coupon-bearing Treasury bond with maturity T. For example, on scenario ω the coupon rate $\delta_2^{B^{10}}(\omega)$ on a newly issued 10-year Treasury bond at time $t = 2$ will be set equal to the projected 10-year spot rate $y_{2,10}(\omega)$ at time $t = 2$.

Generally

$$\delta_t^{B^{(T)}}(\omega) = y_{t,T}(\omega) \qquad \forall t \in T^d \qquad \forall \omega \in \Omega \tag{41}$$

$$\delta_t^{B^{(T)}}(\omega) = \delta_{\lfloor t \rfloor}^{(T)}(\omega) \qquad \forall t \in T^i \qquad \forall \omega \in \Omega, \tag{42}$$

where $\lfloor . \rfloor$ denotes integral part. This ensures that as the yield curve falls, coupons on newly-issued bonds will go down correspondingly and each coupon cash flow will be discounted at the appropriate zero-coupon yield.

The bonds are assumed to pay coupons semi-annually. Since we roll the bonds on an annual basis, a coupon will be received after six months and again after a year just before the bond is sold. This forces us to distinguish between the price at which the bond is sold at rebalancing times and the price at which the new bond is purchased.

Let $P_{t,B^{(T)}}^{(\text{sell})}$ denote the selling price of the bond $B^{(T)}$ at time t, assuming two coupons have now been paid out and the time to maturity is equal to $T - 1$, and let $P_{t,B^{(T)}}^{(\text{buy})}$ denote the price of a newly issued coupon-bearing Treasury bond with a maturity equal to T.

The 'buy' bond price at time t is given by

$$\begin{aligned} B_t^T(\omega) = {} & F^{B^T} e^{-(T+\lfloor t \rfloor - t) y_{t,T+\lfloor t \rfloor - t}(\omega)} \\ & + \sum_{s=\frac{\lfloor 2t \rfloor}{2}+\frac{1}{2}, \frac{\lfloor 2t \rfloor}{2}+1, \ldots, \lfloor t \rfloor + T} \frac{\delta_t^{B^T}(\omega)}{2} F^{B^T} e^{-(s-t) y_{t,(s-t)}(\omega)} \\ & \omega \in \Omega \quad t \in T^{\text{total}}, \end{aligned} \tag{43}$$

where the principal of the bond is discounted in the first term and the stream of coupon payments in the second.

At rebalancing times t the sell price of the bond is given by

$$\begin{aligned} B_t^T(\omega) = {} & F^{B^T} e^{-(T-1) y_{t,T-1}(\omega)} + \sum_{s=\frac{1}{2},1,\ldots,T-1} \frac{\delta_{t-1}^{B^T}(\omega)}{2} F^{B^T} e^{-(s-t) y_{t,(s-t)}(\omega)} \\ & \omega \in \Omega \quad t \in \left\{ T^d \setminus \{0\} \right\} \cup \{T\} \end{aligned} \tag{44}$$

with coupon rate $\delta_{t-1}^{B^T}(\omega)$. The coupon rate is then reset for the newly-issued Treasury bond of the same maturity. We assume that the coupons paid at six months are re-invested in the off-the-run bonds. This gives the following adjustment to the amount held in bond B^T at time t

$$x_{t,B^T}(\omega) = x_{t-\frac{1}{12},B^T}(\omega) + \frac{\frac{1}{2}\delta_t^{B^T}(\omega) F^{B^T} x_{t-\frac{1}{12},B^T}(\omega)}{f P_{t,B^T}^{\text{buy}}(\omega)} \qquad t \in T^c \quad \omega \in \Omega. \tag{45}$$

4 Historical Backtests

We will look at an *historical backtest* in which statistical models are fitted to data up to a trading time t and scenario trees are generated to some chosen horizon $t+T$. The optimal root node decisions are then implemented at time t and compared to the historical returns at time t + 1. Afterwards the whole procedure is rolled forward for T trading times.

Our backtest will involve a *telescoping horizon* as depicted in Figure 4.

At each decision time t the parameters of the stochastic processes driving the stock return and the three factors of the term structure model are re-calibrated using historical data up to and including time t and the initial values of the simulated scenarios are given by the actual historical values of the variables at these times. Re-calibrating the simulator parameters at each successive initial decision time t captures information in the history of the variables up to that point.

Although the optimal second and later-stage decisions of a given problem may be of "what-if" interest, manager and decision maker focus is on the implementable first-stage decisions which are hedged against the simulated future uncertainties. The reasons for implementing stochastic optimization programmes in this way are twofold. Firstly, after one year has passed the actual values of the variables realized may not coincide with any of the values of the variables in the simulated scenarios. In this case the optimal investment policy would be undefined, as the model only has

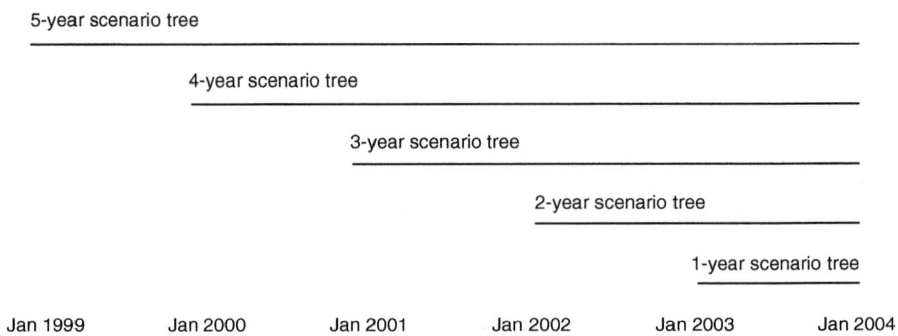

Fig. 4 Telescoping horizon backtest schema

optimal decisions defined for the nodes on the simulated scenarios. Secondly, as one more year has passed new information has become available to re-calibrate the simulator's parameters. Relying on the original optimal investment strategies will ignore this information. For more on backtesting procedures for stochastic optimization models see Dempster *et al.* (2003).

For our backtests we will use three different tree structures with approximately the same number of scenarios, but with an increasing initial branching factor. We first solve the five-year problem using a 6.6.6.6.6 tree, which gives 7776 scenarios. Then we use 32.4.4.4.4 = 8192 scenarios and finally the extreme case of 512.2.2.2.2 = 8192 scenarios.

For the subsequent stages of the telescoping horizon backtest we adjust the branching factors in such a way that the total number of scenarios stays as close as possible to the original number of scenarios and the same ratio is maintained. This gives us the tree structures set out in Table 2.

Historical backtests show how specific models would have performed had they been implemented in practice. The reader is referred to Rietbergen (2005) for the calibrated parameter values employed in these tests. Figures 5 to 10 show how the various optimal portfolios' wealth would have evolved historically relative to the barrier. It is also important to determine how the models performed in-sample on the generated scenario trees and whether or not they had realistic forecasts with regard to future historical returns. To this end one-year-ahead in-sample expectations of

Table 2 Tree structures for different horizon backtests

Jan 1999	6.6.6.6.6 = 7776	32.4.4.4.4 = 8192	512.2.2.2.2 = 8192
Jan 2000	9.9.9.9 = 6561	48.6.6.6 = 10368	512.2.2.2 = 4096
Jan 2001	20.20.20 = 8000	80.10.10 = 8000	768.3.3 = 6912
Jan 2002	88.88 = 7744	256.32 = 8192	1024.8 = 8192
Jan 2003	7776	8192	8192

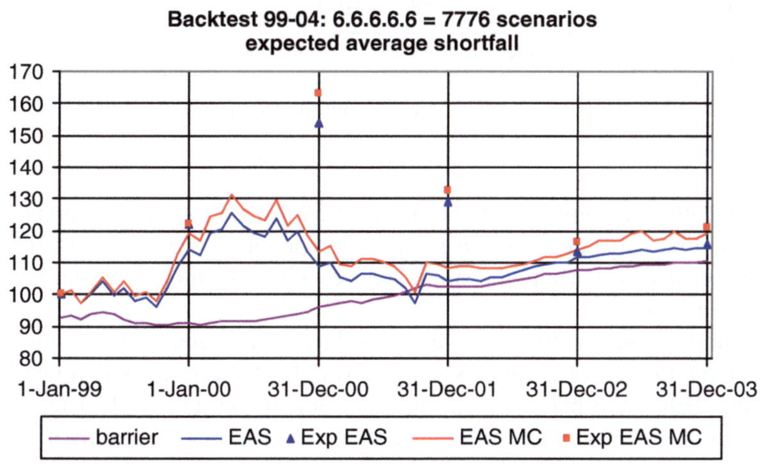

Fig. 5 Backtest 1999-2004 using expected average shortfall for the 6.6.6.6.6 tree

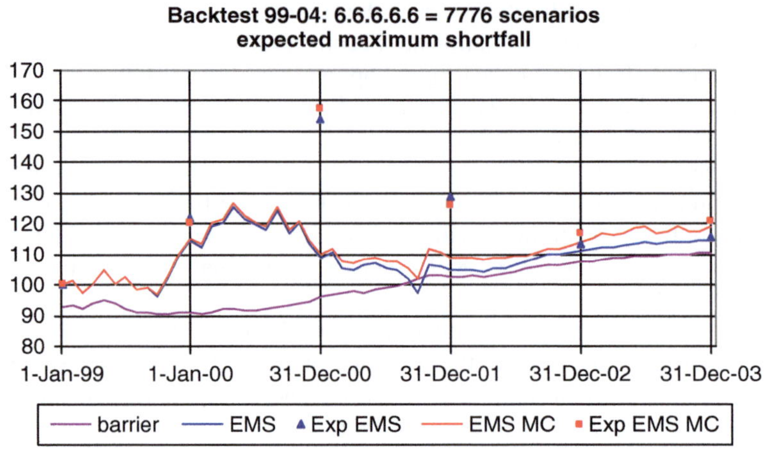

Fig. 6 Backtest 1999-2004 using expected maximum shortfall for the 6.6.6.6.6 tree

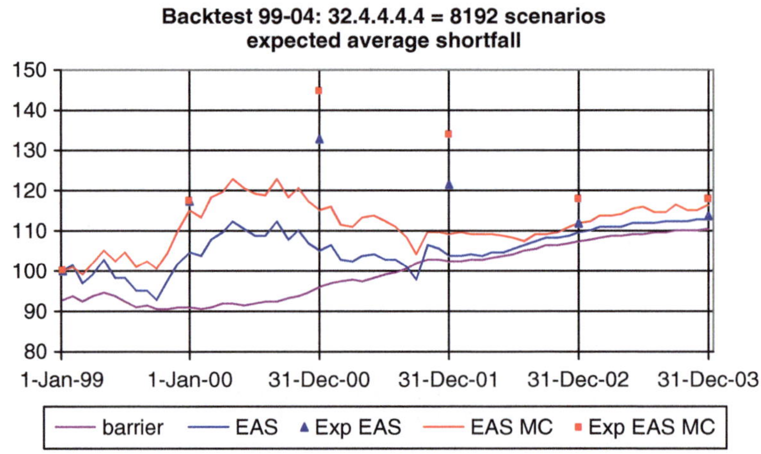

Fig. 7 Backtest 1999-2004 using expected average shortfall for the 32.4.4.4.4 tree

portfolio wealth are shown as points in the backtest performance graphs. Implementing the first-stage decisions *in-sample* the portfolio's wealth is calculated one year later for each scenario in the simulated tree after which an expectation is taken over the scenarios.

From these graphs a first observation is that the risk management monitoring incorporated into the model appears to work well. In all cases the only time portfolio wealth dips below the barrier, if at all, is on September 11, 2001. The initial in-sample wealth overestimation of the models is likely to be due mainly to the short time series available for initial parameter estimation which led to hugely inflated stock return expectations during the equity bubble. However as time progresses and

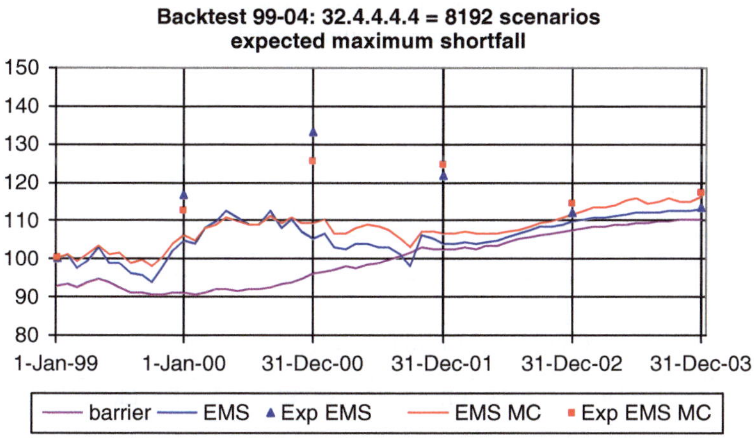

Fig. 8 Backtest 1999-2004 using expected maximum shortfall for the 32.4.4.4.4 tree

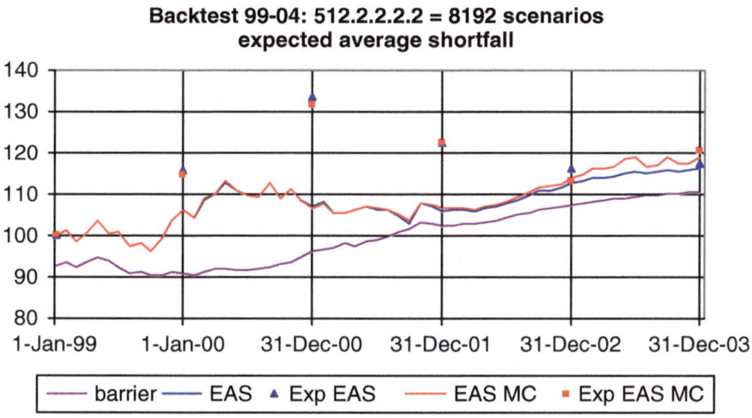

Fig. 9 Backtest 1999-2004 using expected average shortfall for the 512.2.2.2.2 tree

more data points to re-calibrate the model are obtained, the models' expectations and real-life realizations very closely approximate each other.

For reference we have included the performance of the EuroStoxx 50 in Figure 11 to indicate the performance of the stock market over the backtesting period. Even though this was a difficult period for the optimal portfolios to generate high historical returns, it provides an excellent demonstration that the risk management incorporated into the models operates effectively. It is in periods of economic downturn that one wants the portfolio returns to survive.

Tables 3 and 4 give the optimal portfolio allocations for the 32.4.4.4.4 tree using the two maximum shortfall objective functions. In both cases we can observe a tendency for the portfolio to move to the safer, shorter-term assets as time progresses. This is naturally built into the model as depicted in Figure 3.

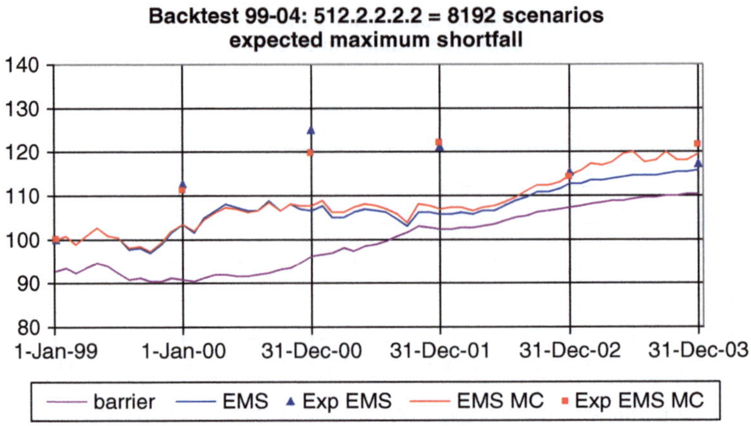

Fig. 10 Backtest 1999-2004 using expected maximum shortfall for the 512.2.2.2.2 tree

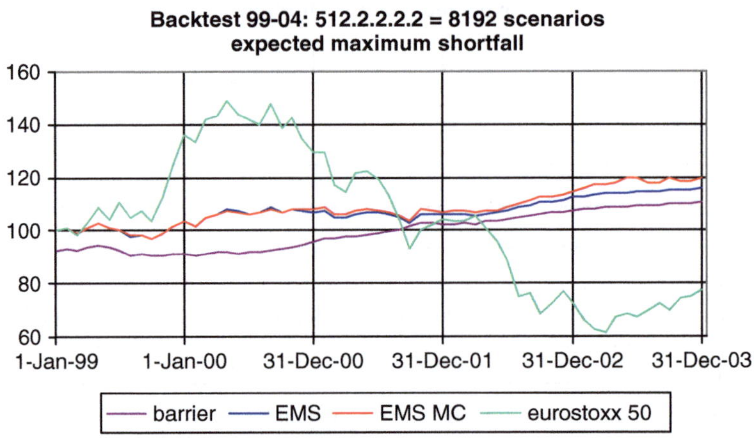

Fig. 11 Comparison of the fund's portfolio performance to the eurostoxx 50

Table 3 Portfolio allocation expected maximum shortfall using the 32.4.4.4.4 tree

	1y	2y	3y	4y	5y	10y	30y	Stock
Jan 99	0	0	0	0	0	**0.23**	**0.45**	**0.32**
Jan 00	0	0	0	0	0	0	**0.37**	**0.63**
Jan 01	**0.04**	0	0	0	0	**0.39**	**0.53**	**0.40**
Jan 02	**0.08**	**0.16**	**0.74**	0	0	0	0	**0.01**
Jan 03	**0.92**	0	0	0	0	**0.07**	0	**0.01**

Table 4 Portfolio allocation expected Maximum shortfall with monthly checking using the 32.4.4.4.4 tree

	1y	2y	3y	4y	5y	10y	30y	Stock
Jan 99	0	0	0	0	**0.49**	**0.27**	0	**0.24**
Jan 00	0	0	0	0	**0.25**	**0.38**	0	**0.36**
Jan 01	0	0	0	0	**0.49**	**0.15**	0	**0.36**
Jan 02	0	0	0	**0.47**	**0.44**	0	0	**0.10**
Jan 03	0	0	**0.78**	**0.22**	0	0	0	**0.01**

For the decisions made in January 2002/2003, the portfolio wealth is significantly closer to the barrier for the EMS model than it is for the EMS MC model. This increased risk for the fund is taken into account by the EMS model and results in an investment in safer short-term bonds and a negligible equity component. Whereas the EMS model stays in the one to three year range the EMS MC model invests mainly in bonds with a maturity in the range of three to five years and for both models the portfolio wealth manages to stay above the barrier.

From Figures 5 to 10 it can be observed that in all cases the method with monthly checking outperforms the equivalent method with just annual shortfall checks. Similarly as the initial branching factor is increased, the models' out-of-sample performance is generally improved. For the $512.2.2.2.2 = 8192$ scenario tree, all four objective functions give optimal portfolio allocations which keep the portfolio wealth above the barrier at all times, but the models with the monthly checking still outperform the others. The more important difference however seems to lie in the deviation of the expected in-sample portfolio's wealth from the actual historical realization of the portfolio value. Table 5 displays this annual deviation averaged over the five rebalances and shows a clear reduction in this deviation for three of the four models as the initial branching factor is increased. Again the model that uses the expected maximum shortfall with monthly checking as its objective function outperforms the rest.

Overall the historical backtests have shown that the described stochastic optimization framework carefully considers the risks created by the guarantee. The EMS MC model produces well-diversified portfolios that do not change drastically from one year to the next and results in optimal portfolios which even through a period of economic downturn and uncertainty remained above the barrier.

Table 5 Average annual deviation

	EAS	EAS MC	EMS	EMS MC
6.6.6.6.6	9.87%	13.21%	9.86%	10.77%
32.4.4.4.4	10.06%	9.41%	9.84%	7.78%
512.2.2.2.2	10.22%	8.78%	7.78%	6.86%

5 Robustness of Backtest Results

Empirical equity returns are now well known not to be normally distributed but rather to exhibit complex behaviour including fat tails. To investigate how the EMS MC model performs with more realistic asset return distributions we report in this section experiments using a geometric Brownian motion with Poisson jumps to model equity returns. The stock price process $\mathbf{S}$ is now assumed to follow

$$\frac{d\mathbf{S}_t}{\mathbf{S}_t} = \tilde{\mu}_S dt + \tilde{\sigma}_S d\tilde{\mathbf{W}}_t^S + \mathbf{J}_t d\mathbf{N}_t, \tag{46}$$

where $\mathbf{N}$ is an independent Poisson process with intensity λ and the jump saltus $\mathbf{J}$ at Poisson epochs is a normal random variable.

As the EMS MC model and the 512.2.2.2.2 tree provided the best results with Gaussian returns the backtest is repeated for this model and treesize. Figure 12 gives the historical backtest results and Tables 5 and 6 represent the allocations for the 512.2.2.2.2 tests with the EMS MC model for the original GBM process and the GBM with Poisson jumps process respectively. The main difference in the two tables is that the investment in equity is substantially lower initially when the equity index volatility is high (going down to 0.1% when the volatility is 28% in 2001), but then increases as the volatility comes down to 23% in 2003. This is born out by Figure 12 which shows much more realistic in-sample one-year-ahead portfolio wealth predictions (*cf.* Figure 10) and a 140 basis point increase in terminal historical fund return over the Gaussian model. These phenomena are the result of the calibration of the normal jump saltus distributions to have negative means and hence more downward than upwards jumps resulting in downwardly skewed equity index return distributions, but with the same compensated drift as in the GBM case.

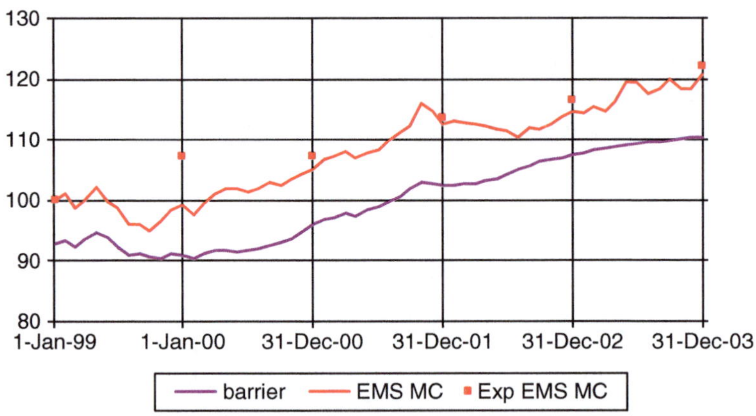

Fig. 12 Expected maximum shortfall with monthly checking using 512.2.2.2.2 tree for the GBM with jumps equity index process

Table 6 Portfolio allocation expected maximum shortfall with monthly checking using the 512.2.2.2.2 tree

	1y	2y	3y	4y	5y	10y	30y	Stock
Jan 99	0	0	0	0	**0.69**	**0.13**	0	**0.18**
Jan 00	0	0	0	0	**0.63**	0	0	**0.37**
Jan 01	0	0	0	0	**0.37**	**0.44**	0	**0.19**
Jan 02	0	0	0	0	**0.90**	0	0	**0.10**
Jan 03	0	0	**0.05**	0	**0.94**	0	0	**0.01**

Table 7 Portfolio allocation expected maximum shortfall with monthly checking using the 512.2.2.2.2 tree for the GBM with poisson jumps equity index process

	1y	2y	3y	4y	5y	10y	30y	Stock
Jan 99	0	0	0	0	**0.12**	**0.77**	0	**0.11**
Jan 00	0	0	0	0	0	**0.86**	0	**0.14**
Jan 01	0	0	0	0	**0.43**	**0.56**	0	**0.01**
Jan 02	0	0	0	0	**0.70**	**0.11**	0	**0.19**
Jan 03	0	0	0	0	**0.04**	**0.81**	0	**0.15**

As a consequence the optimal portfolios are more sensitive to equity variation and take benefit from its lower predicted value in the last year.

Although much more complex equity return processes are possible, these results show that the historical backtest performance of the EMS MC model is only improved in the presence of downwardly skewed asset equity return distributions possessing fat tails due to jumps.

6 Conclusions

This chapter has focussed on the design of funds to support investment products which give a minimum guaranteed return. We have concentrated here on the design of the liability side of the fund, paying particular attention to the pricing of bonds using a three-factor term structure model with reliable results for long-term as well as the short-term yields. Several objective functions for the stochastic optimization of portfolios have been constructed using expected average shortfall and expected maximum shortfall risk measures to combine risk management with strategic asset allocation. We also introduced the concept of monthly shortfall checking which improved the historical backtesting results considerably. In addition to the standard GBM model for equity returns we reported experiments using a GBM model with Poisson jumps to create downwardly skewed fat tailed equity index return distributions. The EMS MC model responded well with more realistic expected portfolio wealth predictions and the historical fund portfolio wealth staying significantly above the barrier at all times.

The models of this paper have been extended in practice to open ended funds which allow for contributions throughout the lifetime of the corresponding investment products. In total funds of the order of 10 billion euros have been managed with these extended models. In future research we hope to examine open multi-link pension funds constructed using several unit linked funds of varying risk aversion in order to allow the application of individual risk management to each client's portfolio.

Acknowledgments The authors wish to thank Yee Sook Yong and Ning Zhang for their able assistance with computational matters. We also acknowledge the helpful comments of two anonymous referees which led to material improvements in the paper.

References

Andersen, T.G. and Lund, J. (1997). Estimating continuous-time stochastic volatility models of the short-term interest rate. *Journal of Econometrics*, **77** (2), 343–377.

Campbell, R. (2000). The economic factor model: Theory and applications. Lehman Brothers Presentation to Europlus Research and Management, Dublin, 31 March 2000.

Consigli, G. and Dempster, M.A.H. (1998). The CALM stochastic programming model for dynamic asset-liability management. In *Worldwide Asset and Liability Modeling*, eds. J.M. Mulvey and W.T. Ziemba, Cambridge University Press, Cambridge, 464–500.

Consiglio, A., Cocco, F. & Zenios, S.A. (2007). The PROMETEIA model for managing insurance policies with guarantees. In *Handbook of Asset and Liability Management*, Vol. 2, eds. S. Zenios and W.T. Ziemba, North-Holland, Oxford, UK, 663–705.

Cox, J.C., Ingersoll, J.E. and Ross, S.A. (1985). A theory of the term structure of interest rates. *Econometrica*, **53**, 385–407.

Dempster, M.A.H., Germano, M., Medova, E.A. and Villaverde, M. (2003). Global asset liability management. *British Actuarial Journal*, **9** (1), 137–216.

Dempster, M.A.H., Hicks-Pedron, N., Medova, E.A., Scott, J.E. and Sembos, A. (2000). Planning logistics operations in the oil industry. *Journal of the Operational Research Society*, **51**, 1271–1288.

Dempster, M.A.H., Jones, C.M., Khokhar, S.Q. and Hong, G.S-S. (1999). Implementation of a model of the stochastic behaviour of commodity prices. Report to Rio Tinto, Centre for Financial Research, Judge Institute of Management, University of Cambridge.

Harvey, A.C. (1993). *Time Series Models*. Harvester Wheatsheaf, Hemel Hempstead, UK.

Hertzog, F., Dondi, G., Keel, S., Schumann, L.M. and Geering, H.P. (2007). Solving ALM problems via sequential stochastic programming. *Quantitative Finance*, **7** (2), 231–244.

Hull, J.C. (1997). *Options, Futures and Other Derivatives*. Prentice Hall, Upper Saddle River, NJ.

Markowitz, H.M. (1959). *Portfolio Selection*. John Wiley & Sons, NY.

Medova, E.A., Rietbergen, M.I., Villaverde, M. and Yong, Y.S. (2005). Modelling the long-term dynamics of the yield curve. Working Paper, Centre for Financial Research, Judge Business School, University of Cambridge.

Nelson, C.R. and Siegel, A.F. (1987). Parsimonious modeling of yield curves. *Journal of Business*, **60**, 473–489.

Rietbergen, M.I. (2005). Long term asset liability management for minimum guaranteed return funds. PhD Dissertation, Centre for Financial Research, Judge Business School, University of Cambridge.

Vasicek, O. (1977). An equilibrium characterization of the term structure. *Journal of Financial Economics*, **5** (2), 177–188.

Chapter 3
Performance Enhancements for Defined Benefit Pension Plans

John M. Mulvey, Thomas Bauerfeind, Koray D. Simsek, and Mehmet T. Vural

Abstract Over the next several decades, traditional corporate and government pension plans will encounter increasingly severe problems in many countries. Contributing factors include underfunding status, demographic trends, low savings rates, and inefficient investment/saving strategies. This chapter takes up the last point, showing that a systematic forward-looking asset–liability management model can improve performance across many reward and risk measures. The model takes the form of a multi-stage stochastic program. We approximate the stochastic program via a set of state-dependent policy rules. A duration-enhancing overlay rule improves performance during economic contractions. The methodology is evaluated via historical backtests and a highly flexible, forward-looking financial planning tool.

Keywords Asset and liability management · Financial optimization · Pension plans · Risk management · Asset allocation · Surplus optimization

3.1 Introduction

Traditional pension trusts, called defined benefit (DB) plans herein, are threatened by a number of forces. The factors include (1) the loss of funding surpluses occurring over the past 10 years and current underfunded ratios, (2) a demographic nightmare – long-lived retirees and a shrinking workforce, (3) changing regulations, (4) greater emphasis on individual responsibilities for managing personal affairs such as retirement, and (5) inefficient financial planning. The days of someone working for a single organization–IBM for example–for their entire career and then retiring with comfort supported by the company's contributions are largely gone (except for public sector employees in certain cases).

This chapter takes up the lack of effective risk management and financial planning by DB pension plans. The 2001/2002 economic contraction showed that the ample pension plan surpluses that existed in 1999 could be lost during an equity market downturn and a commensurate drop in interest rates which raises

---

J.M. Mulvey (✉)
Princeton University, Princeton, NJ, USA
e-mail: mulvey@princeton.edu

M. Bertocchi et al. (eds.), *Stochastic Optimization Methods in Finance and Energy*, International Series in Operations Research & Management Science 163,
DOI 10.1007/978-1-4419-9586-5_3,

the market value of liabilities (Mulvey et al. 2005b; Ryan and Fabozzi 2003). The loss of surplus could have been largely avoided by applying modern asset and liability management models to the problem of DB pension plans. Boender et al. (1998), Bogentoft et al. (2001), Cariño et al. (1994), Dempster et al. (2003, 2006), Dert (1995), Hilli et al. (2003), Kouwenberg and Zenios (2001), Mulvey et al. (2000, 2008), Zenios and Ziemba (2006), and Ziemba and Mulvey (1998) describe the methodology and show examples of successful applications. The Kodak pension plan (Olson 2005), for example, implemented an established ALM system for pensions in 1999, protecting its surplus over the subsequent recession. The situation repeated itself during the 2008 crash when most pension plan funding ratios dropped further. Again, systematic risk management via ALM models would have largely protected the pension plans.

Over the past decade, there has been considerable debate regarding the appropriate level of risk for a DB pension plan. On one side, advocates of conservative investments, called liability-driven investing or LDI in this chapter, have proposed a portfolio tilted to fixed income securities, similar to the portfolio of an insurance company. These proponents argue that a pension plan must fulfill its obligations to the retirees over long-time horizons and accordingly should reduce risks to the maximum degree possible.

To minimize risks, pension liabilities are "immunized" by the purchase of assets with known (or predictable) cash flows which are "adequate" to pay future liabilities. The goal is to maintain a surplus for the pension plan: Surplus/deficit = value(assets) − PV(liabilities), where the liability discount rate is prescribed by regulations such as promulgated by the Department of Labor in the United States. To protect the pension surplus[1] requires an analysis of the future liabilities for the pension plan, i.e., expected payments to the plan retirees throughout a long time period – 40 or 50 or even 60 years in the future. Clearly with an ongoing organization, these liabilities are uncertain due to longevity risks, to future inflation, to possible modifications of payments for changing policy, and to other contingencies. Importantly, interest rate uncertainty plays a major role since the value of the liabilities (and many assets) will depend directly on interest rate movements. For these reasons, the asset mix will need to be modified at frequent intervals under LDI, for example, as part of the annual valuation exercises conducted by qualified actuaries. Similar to an insurance company, the duration of assets and the duration of liabilities should be matched (approximately) at least if the pension plan is to ensure that the surplus does not disappear due to interest rate movements.

As an alternative perspective, other experts suggest that a pension plan should be willing and able to accept risk-bearing investments since it has a long horizon with only a small percentage of cash flow payments occurring during the near term. An apt analogy is a university endowment. Here, in fact, there has been a decided shift

[1] We define the term "pension surplus" to indicate the difference between the market value of assets and the present value of liabilities (positive or negative). A related term is the funding ratio: market value of assets/present value of liabilities.

over the past 10–15 years by leading universities, promoted by the efforts of David Swensen (2000) at Yale University and others, from employing primarily market-traded assets to private instruments such as hedge funds, private equity, venture capital, and hard assets such as timber lands. Supporting this shift is the theory that private markets will provide greater returns than publically traded markets, e.g., due to an illiquidity premium. Until the 2008 crash, the alternative investment domain did indeed provide superior returns. However, it became apparent in the latter part of 2008 that private markets are not immune to dramatic downturns. During this period, the value of the overall portfolio for many university endowments dropped 25–35%. Despite these large draw downs, university administrators state that the previously high returns would not have been possible if they had followed conservative investment strategies. Clearly, time-dependent risks and rewards play a role in the investment decision process.

From the standpoint of a pension plan, there is another issue to consider. All defined benefit pension plans in the USA and many other countries are protected against default by government (or quasi-government) organizations. The Pension Benefit Guarantee Corporation (PBGC) provides a safety net when a US company is no longer able to make adequate contributions typically during a bankruptcy. A fee is assessed of all DB pensions in order to fund the PBGC. Therefore, the risks to most pensioners are largely mitigated. However, there are systemic risks to taxpayers if and when the entire DB pension system is threatened (Parks 2003).

We focus on issues involving the protection of a pension surplus. The investment world has experienced two severe downturns over the past 10 years – the 2001/2002 tech-bubble and the 2008–2009 great recession. Over the same period, interest rates have gradually declined causing the present value of liabilities to increase. As mentioned, the relatively health surpluses for pension plans occurring in 1999 have disappeared today due to these two causes. Any attempt to manage pension plan risk must consider both asset valuation and the risk of interest rate decreases (Mulvey et al. 2006).

Importantly, to address risk mitigation approaches, we must recognize that the complicated nature of pension planning – numerous stakeholders, long time periods, conflicting measures of risk – requires approaches that strive to achieve a number of simultaneous objectives (Arnott and Bernstein 1990; Bader 2003; Black 1995; Mulvey et al. 2008; Muralidhar and van der Wouden 1999). For example, we can claim that a sponsoring organization should increase its contribution to render the pension plan utterly safe and riskless. However, the sponsors will have other productive uses for its capital and may very well be in distress when asked or required to make substantial contributions. A large contribution may cause the pension plan to go bankrupt. Thus, while there is an immediate trade-off between contribution today and risk reduction for the pension plan, the long-term health of the organization must also be considered.

There is unanimity about the long-term joint survivability of the sponsoring organization and the pension plan. A healthy organization will be able to fulfill its promises in the future–in terms of the DB pension plan. Thus, strategies which can reduce future contributions *and* protect the surplus should be implementable. More will be said about this issue in the empirical tests (Section 3.5). For our purposes, we aim to please all (or a majority) of stakeholders with our proposed investment strategy.

To highlight the benefits of a method to protect the asset wealth and pension surplus, we must understand the characteristics of a market crash, for instance, as took place in 2000/2001 and in 2008/2009. The first observation is that the equity markets dropped dramatically in short- or mid-term speed (certainly within a year). Thus, any protection must be able to adjust dynamically in weeks or months or the protection must be in place well before the crash occurs. In addition to stock market drops, there are several other common characteristics to a crash, including higher volatility, contagion across asset categories (very high correlation coefficients), and lower government interest rates (in situations when the government is considered safe and sound). Thus, any model with a fixed correlation and single-period structure, such as the traditional Markowitz model, will be unlikely to provide much diversification benefits since the market's behavior during a crash is very different from behavior during normal times.

An effective ALM model must take into account short time steps – months, weeks, or even days. And the investment strategy must aim to protect both the asset wealth and the pension surplus. We suggest that an efficient approach for dealing with pension-surplus protection is to implement dynamic strategies involving long-term government bonds (or strips during crises). We call our strategy – duration enhancing overlay (DEO) – for defined benefit pension plans. DEO provides the best of both worlds since, as we will demonstrate in Section 3.5, it protects asset wealth and the pension plan's surplus (which has an impact on future expected contribution).

The basic concept is straightforward: Increase the duration of the assets in a dynamic fashion as conditions in equities and other risk-bearing assets deteriorate (Mulvey et al. 2010). To accomplish this objective, we take positions in long-duration government bonds (or strips) and take an equal amount of short positions in short-term government securities. The strategy is implemented by means of a quasi-protection process. The motivating principle involves the "flight to quality" during stressful economic environments. The strategy is efficiently implemented via the futures/swap markets (Mulvey et al. 2007). Thus, the futures strategy does not impact the investor's core asset portfolio; optimal asset allocation remains mostly intact. At the strategic level, futures instruments do not take capital, but possess risks and thereby require risk allocation procedures.

The remainder of the chapter is organized as follows. The next section defines the asset–liability management model for a DB pension plan. The structure involves integrating the sponsoring organization with the pension plan as a single entity (enterprise risk management) across a large number of stochastic scenarios. The generic model can be readily specialized for a standalone pension plan. Section 3.3 defines alternative objective functions, depicting the goals of the various stakeholders – retirees, sponsoring company, governments, and so on. A short discussion of solution strategies is included. The DEO overlay strategy appears in Section 3.4, along with several variants. Section 3.5 describes the empirical results of applying the DEO strategy. Both backtests and a forward-looking ALM system are evaluated showing the benefits of a global DEO strategy. Conclusions follow in Section 3.6.

3.2 An Asset–Liability Management Model for DB Pension Plans

This section defines our asset and liability management (ALM) model for a defined benefit pension plan. We follow the framework established in Mulvey et al. (2005a, 2008) via a multi-stage stochastic program. This framework allows for realistic conditions to be modeled such as the requirement for funding contributions when the pension deficit exceeds a specified limit and addressing transaction costs. However, as with any multi-stage stochastic optimization model, the number of decision variables grows exponentially with the number of stages and state variables. To compensate and to reduce the execution time to a manageable amount, we will apply a set of policy rules within a Monte Carlo simulation. The quality of the policy rules can be evaluated by means of an "equivalent" stochastic program. See Mulvey et al. (2008) and Section 3.5 for further details.

To start, we establish a sequence of time stages for the model: $t = [1, 2, \ldots, T]$. Typically, since a pension plan must maintain solvency and be able to pay its liabilities over long time periods, we generate a long-horizon model – over 10–40 years with annual or quarterly time steps. To defend the pension plan over short time periods, we employ the DEO overlay strategies – which are dynamically adjusted over days or weeks. However, the target level of DEO is set by the strategic ALM model. In effect, the DEO provides a tactical rule for protecting the pension plan during turbulent conditions.

We define a set of generic asset categories $\{A\}$ for the pension plan. The categories must be well posed so that either a passive index can be invested in, or so that a benchmark can be established for an active manager. In the ALM model, the investment allocation is revised at the end of each time period with possible transaction costs. For convenience, dividends and interest payments are reinvested in the originating asset classes. Also, we assume that the variables depicting asset categories are non-negative. Accordingly, we include "investment strategies" in $\{A\}$, such as long–short equity or buy–write strategies in the definition of "asset categories." The need for investment strategies in $\{A\}$ has become evident as standard long-only securities in 2008 became almost completely correlated (massive contagion). The investment strategies themselves may take action (revise their own investment allocations) more frequently and dynamically than as indicated by the strategy ALM model.

Next, a set of scenarios $\{S\}$ is generated as the basis of the forward-looking financial planning system. The scenarios should be built along several guiding principles. First, importantly, the time paths of economic variables should be plausible and should to the degree possible depict a comprehensive range of plausible outcomes. Second, the historical data should provide evidence of reasonable statistical properties, for example, the historical volatilities of the stock returns over the scenarios should be consistent with historical volatilities. Third, current market conditions should be considered when calibrating the model's parameters. As an example, interest rate models ought to begin (time = 0) with the current spot rate or forward rate curves. Third, as appropriate, expert judgment should be taken into account. The expected returns for each asset category should be evaluated by the institution's economists. There should be consistency among the various parties in a financial organization or at least the differences should be explainable. A number of scenario generators have been successfully applied over the past 25 years for asset and liability management models (Dempster et al. 2003; Høyland and Wallace 2001; Mulvey 1996).

For each $i \in \{A\}, t = [1, 2, \ldots, T], s \in \{S\}$, we define the following parameters and decision variables in the basic ALM model:

Parameters

$r_{i,t,s} = 1 + \rho_{i,t,s}$, where $\rho_{i,t,s}$ is the rate of return for asset i, in period t, under scenario s

$g_{t,s} = 1 + \gamma_{t,s}$, where $\gamma_{t,s}$ is the percent growth rate of the organization in period t, under scenario s

$b_{t,s}$ Payments to beneficiaries in period t, under scenario s

π_s Probability that scenario s occurs - $\sum_{s \in S} \pi_s = 1$

$x^{\mapsto}_{i,0,s}$ Amount allocated to asset class i, at the end of period 0, under scenario s, before first rebalancing

$y^{\mapsto}_0$ Value of the organization at the end of time period 0

$e_{t,s}$ Borrowing costs for period t, under scenario s

$\sigma_{i,t}$ Transaction costs for rebalancing asset i, period t (symmetric transaction costs are assumed)

Decision variables: Except as indicated, these variables are non-negative

$x_{i,t,s}$ Amount allocated to asset class i, at the beginning of period t, under scenario s, after rebalancing

$x^{\mapsto}_{i,t,s}$ Amount allocated to asset class i, at the end of period t, under scenario s, before rebalancing

$x^{\text{BUY}}_{i,t,s}$ Amount of asset class i purchased for rebalancing in period t, under scenario s

$x^{\text{SELL}}_{i,t,s}$ Amount of asset class i sold for rebalancing in period t, under scenario s

$x^{\text{TA}}_{t,s}$ Total amount of assets in pension plan at the beginning of time period t, under scenario s

$y_{t,s}$ Value of the organization after a contribution is made in period $t - 1$, under scenario s

$y^{\mapsto}_{t,s}$ Value of the organization at the end of period t, before contribution is made in period t, under scenario s

$y^{\text{CONT}}_{t,s}$ Amount of cash contributions made at the end of period t, under scenario s

$x^{\text{BORR}}_{t,s}$ Amount of borrowing by the organization at the end of period t for use in pension plan, under scenario s

Given these decision variables and parameters, we can define the general multi-objective optimization model as follows:

$$\text{Maximize} \qquad U\{\, Z_1,\; Z_2, \ldots,\; Z_k \,\}, \qquad \text{[ALM]}$$

where the goals are defined as functions of the decision variables (below): $Z_k = f_k(\boldsymbol{x}, \boldsymbol{y})$
Subject to

$$\sum_{i \in \mathrm{A}} x_{i,t,s} = x_{t,s}^{\mathrm{TA}} \qquad \forall s \in S,\; t = 1, \ldots, T+1, \tag{3.1}$$

$$x_{i,t,s}^{\mapsto} = r_{i,t,s} x_{i,t,s} \qquad \forall s \in S,\; t = 1, \ldots, T,\; i \in \mathrm{A}, \tag{3.2}$$

$$y_{t,s}^{\mapsto} = g_{t,s} y_{t,s} \qquad \forall s \in S,\; t = 1, \ldots, T+1, \tag{3.3}$$

$$y_{t,s} = y_{t-1,s}^{\mapsto} - y_{t-1,s}^{\mathrm{CONT}} - e_{t-1,s}(x_{t-2,s}^{\mathrm{BORR}}) \qquad \forall s \in S,\; t = 1, \ldots, T+1, \tag{3.4}$$

$$x_{i,t,s} = x_{i,t-1,s}^{\mapsto} + x_{i,t-1,s}^{\mathrm{BUY}}(1-\sigma_{i,t-1}) - x_{i,t-1,s}^{\mathrm{SELL}} \qquad \forall s \in S,\; i \neq 1,\; t = 1, \ldots, T+1, \tag{3.5}$$

$$x_{1,t,s} = x_{1,t-1,s}^{\mapsto} + \sum_{i \neq 1} x_{i,t-1,s}^{\mathrm{SELL}}(1-\sigma_{i,t-1}) - \sum_{i \neq 1} x_{i,t-1,s}^{\mathrm{BUY}} - b_{t-1,s} + y_{t-1,s}^{\mathrm{CONT}} + x_{t-1,s}^{\mathrm{BORR}}$$
$$\forall s \in \mathrm{S},\; t = 1, \ldots, T+1, \tag{3.6}$$

$$x_{i,t,s} = x_{i,t,s'} \quad \text{and } y_{t,s}^{\mathrm{CONT}} = y_{t,s'}^{\mathrm{CONT}},\, x_{t,s}^{\mathrm{BORR}} = x_{t,s'}^{\mathrm{BORR}},\, x_{t,s}^{\mathrm{BUY}} = x_{t,s'}^{\mathrm{BUY}},\, x_{t,s}^{\mathrm{SELL}} = x_{t,s'}^{\mathrm{SELL}}$$

$$\forall s \text{ and } s' \text{ with identical historical path to time } t, \text{ and } \forall t, \forall s. \tag{3.7}$$

Numerous additional constraints are required to manage the pension plan environment. For example, we will limit the amount of risks allowed in the portfolio by constraints:

$$\mathrm{Risk}\{\, Z_1,\; Z_2, \ldots,\; Z_k \,\} \quad \leq \quad \mathrm{Risk}_{\max}. \tag{3.8}$$

A network graph (Fig. 3.1) depicts, for a given scenario, the pertinent cash flows for each asset and the company's contribution. The graphical form allows managers to readily comprehend the model's structure for the main cash flow constraints. For simplicity, we have excluded the non-network constraints in our ALM formulation (e.g., Rauh 2006). The non-anticipativity constraints (3.7) depend upon the format of the scenario tree; these conditions are addressed in our tests by means of "policy

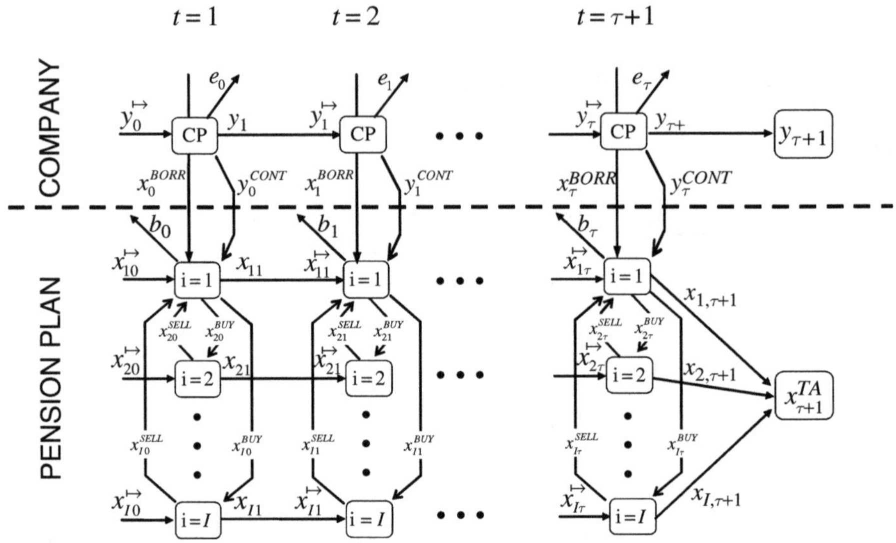

Fig. 3.1 Network representation of program ALM: This graph depicts cash flows for a given scenario for the pension plan and the sponsoring organization. At each time step, the model allows for rebalancing across the asset categories and between the sponsor and the pension plan. The overlay variables and additional constraints are not shown

rules" over computable state variables. We do not assume any Markov property in our formulation.

3.3 Multi-objective Functions and Solution Strategies

Setting an investment strategy for a DB pension plan is complicated by conflicting requirements and the diverse goals of the stakeholders. Each of the interested groups is served by several of the defined Z-objective functions. Especially relevant is the relationship between the pension plan and the sponsoring organization. In the USA, DB pension plans fall under the auspices of the Departments of Labor and Tax, and the requirement of the 1974 Employee Retirement and Security Act ERISA (with ongoing modifications by changing regulations and Congressional action). Thus, a US-based DB pension plan must undergo annual valuations by certified actuaries, who compute the various ratios including the accumulated benefit obligations (ABO), the projected benefit obligations (PBO), and funding ratios. These valuation exercises help determine the requirements for contributions by the sponsoring organization and the fees to be paid to the quasi-governmental organization PBGC (whose job is to take over pensions from bankrupt companies).

We will employ the following five objective functions (Mulvey et al. 2005a, 2008).

3.3.1 Economic Value

The first function, called economic value, is a combination of the expected risk-adjusted discounted value of future contributions (Black 1995) and the discounted value of the surplus/deficit of the pension plan at the horizon, time $= T$. The first part of this objective provides a measure for the long-run cost of the pension trust:

$$Z_{1\_A} = \sum_{s \in S} \pi_s \sum_{t \in T} y_{t,s}^{\mathrm{CONT}} / (1 + r_{t,s}),$$

where the risk-adjusted discount rate equals $r_{t,s}$ and is based on actuarial and economic judgment. The second part involves the discounted value of the pension's surplus wealth at the end of the planning horizon:

$$Z_{1\_B} = \sum_{s \in S} \pi_s \, Sw_{\tau+1,s}.$$

This part focuses on the investment strategy and contribution policy of the pension trust so that the highest average surplus value is achieved. Thus, the first objective function is to maximize economic value:

$$Z_1 = Z_{1\_B} - Z_{1\_A}$$

3.3.2 Volatility of Z_1

The second objective function indicates the volatility of the cost of running the pension plan. It is a risk measure for the most important objective function Z_1: $Z_2 = \mathrm{Std}(Z_1)$.

3.3.3 Worst Case of Z_1

Next, we compute the probability that the sponsoring organization will make an excess contribution, defined as a second risk-based objective. This function is calculated as follows: $Z_3 = \sum_{s \in S} \pi_s I_s$, where the indicator is $I_s = \{1$ if there exists an excess contribution under scenario s, and 0 otherwise$\}$. An alternative is to employ the more tractable conditional value at risk measure (Bogentoft et al. 2001).

3.3.4 Probability of a Significant Contribution

As a fourth measure, we focus on the likelihood of a large payment, called excess contribution, to the pension trust at any point during the planning period:

$Z_4 = E\left[\left(\left(y_{t,s}^{\mathrm{CONT}} - \alpha \cdot y_{t,s}^{\mapsto}\right)^+\right)^2\right]$, where α equals a substantial fraction of the company's total capital at the time of the contribution (Bogentoft et al. 2001). This function approximates the company's ability to service its pension trust under severe circumstances. This risk measure highlights situations in which the organization may have difficulty paying its contributions (Ippolito 2002).

3.3.5 Volatility of Contribution

For many companies, the volatility of the contribution can be an important factor since it has a direct impact on the volatility of earnings for the sponsoring firm (Peskin 1997): $Z_5 = \mathrm{Std}(Z_{1\_A})$.

Other objective functions have been applied in practice (see Mulvey et al. 2008, for example). However, to keep the discussion manageable we assume the aforementioned five objective functions. Since these objectives are clearly conflicting, we develop a multi-objective financial planning model in which the trade-offs among the goals are illustrated. The goal of the overall model is to maximize the health of the pension plan and the sponsoring company. We will show that the integrated entity can benefit by applying the developed global DEO strategy (discussed in the next section).

Due to the massive size of the ALM model, we approximate its solution by reference to a set of state-dependent policy rules. These rules have been developed and evaluated by executing a large set of stochastic programming models. It can be shown that for many pension plans (Mulvey et al. 2008), the results of the stochastic programs compare favorably with the performance of the developed policy rules, thereby we apply the policy rules in our empirical tests in Section 3.5.

3.4 Advantages of Futures Market Strategies

At the strategic level, the futures markets does not require any *direct* capital investment and is thereby distinguished from traditional asset variables $\{A\}$. A prominent example involves commitments made in the futures/forward/swap markets (Mulvey et al. 2007). Here, for example, the investor may engage in a contract to purchase or sell a designated amount of a commodity such as corn at a designated date in the future. The investors (buyer and seller) must, of course, conform to exchange requirements for initial and margin capital and must participate in the mark-to-the-market mechanisms at the end of each trading day. We do not model these tactical issues on our strategic ALM model. Rather, we treat securities in the futures market as adjuncts to the core assets and assume that all margin calls are managed in the standard way to avoid margin calls. These investments are defined via "overlay variables." Implicitly we assume that the size of the overlay variables is relatively modest in scale and diverse enough to treat them at the strategic level without the need for tactical issues.

There are several advantages to futures market investments. First, it is straightforward to "go" short or long on a particular contract without the burden and costs of borrowing the security (traditional shorting). Second, long/short investments in commodities, currencies, and fixed income combinations can assist the investor in achieving the goal of achieving wide diversification. For a DB pension plan, there are additional advantages. In particular, a pension plan must maintain a health funding ratio in order to minimize contributions from the sponsoring organization to the pension plan. As mentioned, the funding ratio and pension surplus depend upon not only upon the market value of assets but also on the discounted value of estimated future cash flows (liabilities to pay retirees). The discount rate has a large impact on the funding ratio. During major economic downturns, for instance, the risk-free rate can drop by substantial amounts – with a possible commensurate decrease in the funding ratio. Accordingly, the duration of assets and the duration of liabilities will contribute to the management of a DP pension plan. A duration mismatch can be addressed by engaging in swaps or other futures/forward market operations. Mulvey et al. (2007, 2010) provide further discussions of overlay strategies.

We add futures strategies via a new set $\{A\text{–}O\}$ and associated decision variables – $x_{j,t,s}$ for $j\varepsilon\{A\text{–}O\}$ – to the ALM model:

$$\sum_{j\in A-O} x_{j,t,s}{}^{*}(r_{j,t,s}) = x_{t,s}^{\text{Overlay}} \qquad \forall s \in S, t = 1,\ldots,T+1. \tag{3.9}$$

The total return of the futures variables in time t and under scenario s is defined by $r_{j,t,s}$. We include the return from these variables in the cash flow constraint (3.6) as follows:

$$x_{1,t,s} = x_{1,t-1,s}^{\mapsto} + \sum_{i\neq 1} x_{i,t-1,s}^{\text{SELL}}(1-\sigma_{i,t-1}) - \sum_{i\neq 1} x_{i,t-1,s}^{\text{BUY}} - b_{t-1,s} + y_{t-1,s}^{\text{CONT}} + x_{t-1,s}^{\text{BORR}} + x_{t-1,s}^{\text{overlay}}$$

$$\forall s \in \text{S},\ t = 1,\ldots,T+1. \tag{3.10}$$

The futures variables provide much flexibility to establish a widely diversified and dynamic portfolio for a DB pension plan. A prime example is the global duration-enhancing strategy – called global DEO described in detail in this chapter. This strategy extends a single country (the USA) and fixed-mix version of DEO (Mulvey and Kim 2010) by modeling the process in a dynamic fashion across several sovereign debt markets.

The DEO concept increases the commitment to long-duration government bonds, e.g., strips, during stressful market conditions. Figure 3.2 further motivates the strategy. Here, the correlation between equity returns and government bond returns becomes negative during economic downturns. Why? This relationship is caused by investors seeking the safest investment during high volatility, crash periods. As a consequence, the optimal level of government long-term bonds in a portfolio

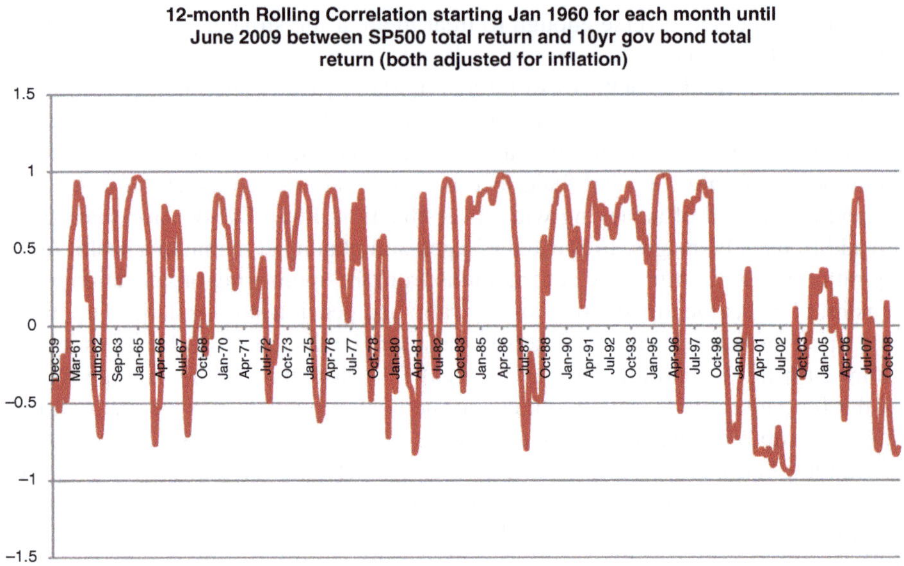

Fig. 3.2 Rolling correlation of US equity and government bond returns over extended time periods

becomes much higher than during time periods in which the correlation is positive (normal periods). We can model this process with multiple regimes.

As mentioned, the futures strategies are included in the ALM model as expanded variables in the multi-stage stochastic program. To simplify the process, we apply a non-anticipative policy rule. In particular, the approach for investing in a long/short strategy based on the flight-to-quality condition. The investment is made under a set of three triggers. Whenever the tests are active (engaged), the strategy takes action in the respective countries by "purchasing" a long-duration government bond and by shorting a short-term government security (*t*-bills in the USA). The long/short strategy can be implemented in several equivalent ways. A swap can be activated. Alternatively, a futures market such as the CME's ultra-bond index can be employed. In the latter case, a long position in the futures market is the sole investment needed since the costs of the short side is taken care of in the price of the futures contract. For investors who do not have access to futures markets, the DEO strategy can be applied with exchange-traded funds, such as long-government bond funds (e.g., TLT and EDV).

The first trigger involves tracking volatility of equity markets in the respective country. Whenever short-term volatility exceeds a long-term moving average, the volatility trigger is engaged. A second trigger is aimed at short-term indicators. This trigger will be disengaged for a certain time period whenever a sufficiently large loss occurs. The third trigger takes into account the spread in the yield curve between long and short instruments. When the spread exceeds a fixed parameter, this trigger is engaged. The three triggers must all be positive before a position is established. The target amount of underlying securities is a fixed percentage of the investor's capital at any time t. For the implementation in the next section, we evaluate DEO in three regions: (a) the USA, (b) Japan, and (c) Europe.

The time step for the DEO strategy is much shorter than the annual time periods in the strategic ALM model. We analyze the DEO strategy on a daily and monthly basis, whereas the strategic model renders decisions yearly. In this manner, we combine the strategic ALM model with a tactical policy rule. The combination provides a dynamic and responsive process to protect the pension surplus during crash regimes, while at the same time linking to the strategic ALM model – which is consistent with the decision-making setting for large institutional investors (annual meetings to determine allocation decisions). Other tactical processes can be added in a similar fashion.

3.5 Empirical Tests

The empirical tests have been conducted in two ways. First, we evaluate the DEO strategies over historical backtest periods. Then, we apply a forward-looking ALM system.

3.5.1 Historical Backtests

We selected the decade January 1, 2000, – December 31, 2009. This period has had two major downturns with very large drawdown values (almost 60% in the majority equity markets). Also, global co-integration has taken place during this time period. Markets are connected throughout the world as never before. Table 3.1 depicts summary statistics for the three regions tested – the USA, Europe, and Japan. Note that the DEO strategies all display positive returns with modest drawdown in the US version, and much higher drawdown values in Europe and Japan. A combined equal-weighted strategy for the three regions with monthly rebalancing (global DEO 3X) gives excellent returns, but with commensurately high risks in terms of volatility and drawdown values. A less leveraged global DEO strategy has very good returns and very modest volatility and drawdown values. Note the comparison with three traditional assets: The S&P 500 equity index and two commodity indices (AIG and Goldman Sachs). The global DEO strategy outperforms these three benchmarks over the historical decade.

The DEO strategies are based on overlay concepts. Hence, the performance numbers should provide almost additive performance for a traditional portfolio. This issue is made more important when the positive returns of the DEO strategy occurs mostly during downturns in equity and other traditional assets. To this point, the correlations in monthly returns among the seven asset classes in Table 3.1 (excluding global DEO 3X) are given in Table 3.2. We present the correlation coefficients for the full data period as well as for two subsets: S&P500 losing months and winning months. Overall, the DEO strategies have very low correlation with the traditional

Table 3.1 Performance of DEO strategies in three countries and comparison with other assets January 1, 2000–December 31, 2009

	US DEO	Euro DEO	Japan DEO	Global DEO (1/3 each)	Global DEO (100% each)	SP500	DJ AIG commodity index	GS commodity index
Annual geometric return	5.39%	6.85%	4.72%	5.89%	16.65%	−2.68%	4.20%	5.05%
Annual volatility	8.71%	11.44%	11.38%	8.05%	24.16%	16.28%	17.34%	25.31%
Maximum drawdown (MDD)	7.58%	21.32%	17.75%	9.25%	25.80%	52.80%	54.50%	67.65%
Return per volatility	0.62	0.60	0.41	0.73	0.69	−0.16	0.24	0.20
Return per MDD	0.71	0.32	0.27	0.64	0.65	−0.05	0.08	0.07

Table 3.2 Monthly return correlations: First panel is for the full period of January 2000 to December 2009 and the second and third panels are for S&P500 losing and winning months, respectively

All data	*US DEO*	*Euro DEO*	*Japan DEO*	*Global DEO*	*SP500*	*DJ AIG CI*
Euro DEO	0.36					
Japan DEO	0.31	0.44				
Global DEO	0.68	0.81	0.79			
SP500	−0.08	0.15	0.04	**0.06**		
DJ AIG CI	−0.22	0.30	0.01	**0.07**	0.29	
GSCI	−0.23	0.17	−0.09	**−0.05**	0.20	0.90
S&P Neg	*US DEO*	*Euro DEO*	*Japan DEO*	*Global DEO*	*SP500*	*DJ AIG CI*
Euro DEO	0.15					
Japan DEO	0.20	0.42				
Global DEO	0.54	0.76	0.81			
SP500	−0.15	0.13	−0.14	**−0.06**		
DJ AIG CI	−0.24	0.44	−0.02	0.11	0.30	
GSCI	−0.25	0.35	−0.13	0.01	0.40	0.92
S&P Pos	*US DEO*	*Euro DEO*	*Japan DEO*	*Global DEO*	*SP500*	*DJ AIG CI*
Euro DEO	0.48					
Japan DEO	0.39	0.45				
Global DEO	0.75	0.83	0.78			
SP500	−0.16	0.05	0.21	**0.06**		
DJ AIG CI	−0.21	0.16	0.03	0.01	0.24	
GSCI	−0.23	0.02	−0.07	−0.10	0.10	0.89

asset classes. Furthermore, global DEO has a slightly negative correlation with the S&P500 when the latter is losing and a slightly positive one when the markets are up.

As we take up the issue of optimal asset allocation with respect to various objectives in the next section, our backtesting analyses demonstrate the added benefits

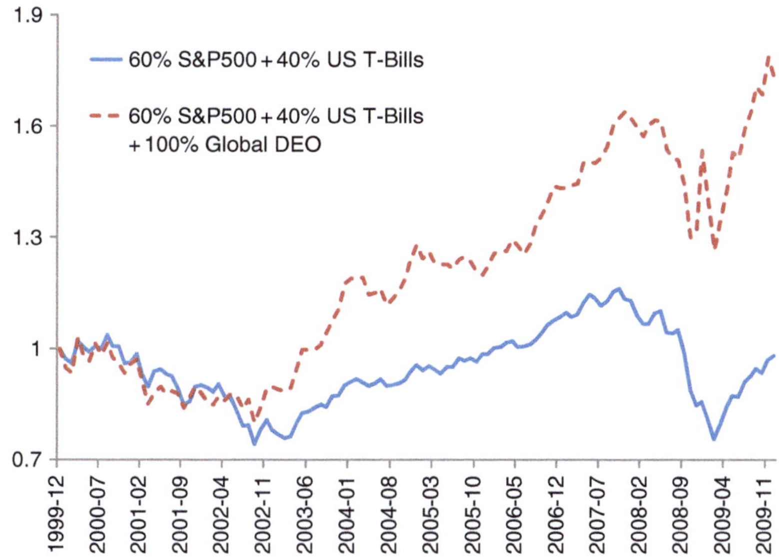

Fig. 3.3 Time series plot of $1 invested with a traditional strategy compared with the added benefits of DEO

of overlay strategies for investors with traditional asset allocations which are not necessarily optimal. Such an investor may have allocated 60% in S&P500 and 40% in US Treasury Bills with monthly rebalancing during the period January 2000 to December 2009. In Fig. 3.3, we plot how this investor may have fared during this decade without and with global DEO. It is apparent that with the DEO, the investor does not underperform in bull markets and can move in the opposite direction in bear markets (e.g., recent downturn at the end of 2008).

Another hypothetical investor might have picked the best-performing fixed-mix strategy during that decade by allocating about 70% in long-term (25+ years) US government bonds and 30% in GSCI. This strategy would have provided more than 7% annual geometric return with a maximum drawdown of about 20%. Even this performance-concerned fixed-mix investor with perfect foresight would have benefited from DEO strategies. As depicted in Fig. 3.4, given the perfect foresight, the base strategy performs quite well. Nonetheless, a modest amount of global DEO brings a remarkable improvement without increasing the drawdown.

To summarize the performance of some traditional strategies in risk-reward framework and how they are improved by DEOs, we present in Fig. 3.5 annual geometric return and maximum drawdown values in a scatter plot analysis. The unfilled markers correspond to traditional strategies whereas the filled ones with the same geometric shape are for the same strategies with a modest amount of global DEO.

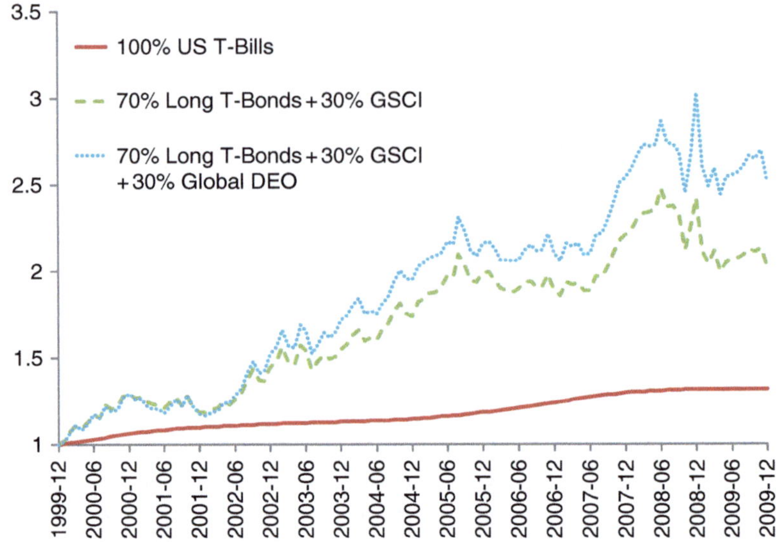

Fig. 3.4 Time series plot of $1 invested with a perfect foresight compared with the added benefits of DEO

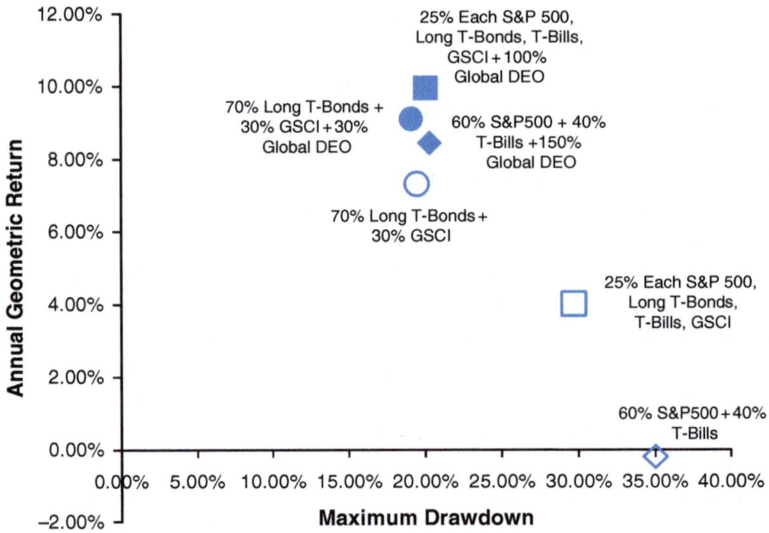

Fig. 3.5 Risk-reward plots of various fixed-mix strategies with and without global DEO

3.5.2 Forward-Looking ALM Tests

From a strategic vantage point for an investor or pension sponsor, rule-based simulators are the instruments of choice for ongoing strategic risk control and management. These approaches allow for covering with very high degree of accuracy all aspects of the complex, individual situation in which strategic financial decisions are made. Realistic simulations of an investments vehicle's stochastic behavior is a rich and reliable source for the information needed in ongoing strategic risk control activities as well for regularly revising decisions on risk optimal strategies allocation.

Most strategic decisions occur across a multi-periodic context, often with complex, path-dependent rules for rebalancing, contribution, and withdrawals. In most cases such conditional rebalancing rules can be found for each single asset; it is necessary to include them in order to represent each asset's marginal contribution. Such rules are not only an essential aspect of asset allocation, but also the rules offer the opportunity to design and optimize them as an integral part of the strategy. In that sense one has to define strategic asset allocation not only as the composition of a portfolio, which would be sufficient if the world were a static one, but rather as an asset allocation strategy, which includes portfolio compositions and management rules. On the level of the strategy, these rules should not depend on exogenous conditions of single markets, but rather on the overall goal achievement compared to the individual preference structure, e.g., at a certain high funding level, a de-risking of the strategic portfolio happens, since no more risk taking is necessary to achieve the overall goals.

Such management rules always depend on additional sets of evaluation rules, internally or externally given, to evaluate the development of the results of the strategies, for example, under commercial law balance sheet and profit/loss calculations, taxation, or various other aspects. These rules produce incentives to favor one allocation strategy over another. Thus it is a basic requirement in order to find individually optimal strategies, to work with rule simulators, which represent relevant management and evaluation rules adequately and with the least level of simplification. This requirement is matched by modern rule simulators such as the PROTINUS Strategy Cockpit™ – described below and employed for the forward-looking tests.

When the multi-period environment is represented accurately, it becomes possible to design individually specific sets of objectives, constraints, and bounds. This is a major advantage of rule simulators, for which it is a condition qua sine non to have optimization and simulation in a single model setting. The necessity of optimizing strategies based on individual objective functions comes from the fact that any strategic investment is done to serve not a general common goal, but always to fulfill goals depending on individual conditions, rule sets, and preference structure. Systems like the PROTINUS Strategy Cockpit™ allow for setting up any objective functions derived from the variable of the rule set and perform desired calculation on them.

In this context, a policy-rule simulator can be seen as financial laboratory tool, allowing designing, testing, and optimizing strategies in a real-life setting tailored to the investor's situation and needs and assumptions. Additionally these instruments are perfect to measure regularly and prospectively the development of the strategic risk positions based on updated figures with the same model, with which these positions were developed. This additional application of such simulators has been sparse in the past. Typically they were only employed in allocation studies on a yearly or even longer basis. When used for regular controlling, they give valid information which helps minimize the effect of the possible crises. Ideally, such implementations will increase over time.

Unfortunately, most rule-based optimization problems lead to non-convex models – thus requiring computationally intensive, heuristic algorithms. The length of the vector which is to be optimized is limited since it increases the number of combinations and thus computational time. Therefore, it was standard until recently that one short vector for all s, t was optimized, which can be interpreted as the best compromise solution over all states of the environment. With advanced systems such as the one briefly described herein, this limitation can be overcome to some degree by optimizing on a small number of short vectors which are each assigned to certain conditions, called regimes. Such regimes can be defined by the time step, e.g., every 3 years a new fixed-mix rebalancing vector is defined, or by some state conditions, e.g., there is a vector for a normal funding range and two additional ones for positive and negative funding situations. These approaches have already been successfully implemented; they give a priori recommendations on what to do when arriving in some specific and maybe critical environmental states. Investor can benefit by this type of information.

3.5.2.1 The PROTINUS Strategy Cockpit^{TM}

Protinus Strategy Cockpit is an integrated development environment for modeling multi-period stochastic planning problems in finance. It has been employed successfully as a strategic risk controlling and risk management tool over the past decade. It is a policy-rule simulator and optimizer, working in all MS Windows^{TM} and MS Server^{TM} environments developed by the consulting firm PROTINUS in Munich, Germany. Ziemba and Mulvey (1998) described such systems in general, which they called decision rule approaches in comparison to alternative approaches, at a time when only very small number of implementations of such approaches existed. Predecessors of the PROTINUS Strategy Cockpit^{TM} developed by the former Princeton-based organization Lattice Financial LLC were implemented, for example, as the first and, for several years, primary ALM system employed at the large German corporate Siemens.

The PROTINUS Strategy Cockpit^{TM} generates user-specific rules by assembling a library of building block rules, ranging from simple arithmetic operations to complex accounting standards. The building blocks are put into the desired order via an execution schedule and where necessary a quantitative condition for their execution.

PROTINUS Strategy Cockpit^{TM} includes a powerful meta-heuristic, non-convex optimization algorithm based on early concepts of Glover's Tabu search (Glover and Laguna 1997). The analyses are updated from past runs by saving solution information as templates and by grouping into reports. PROTINUS Strategy Cockpit^{TM} includes project management tools typically needed when working with large scenario spaces and complex rule sets and performing risk measurement and allocation studies. Figures 3.6 and 3.7 are screenshots from the financial engineers' version. There is also a configuration available, which provides a fully automatic version, called PROTINUS Strategy Factory^{TM}. Over the past several years, this system has been implemented to individually optimize on a quarterly basis, in a fully automatic fashion, several tens of thousands of portfolios for unit-linked insurance contracts provided by the Insurance Company Skandia in several countries in central Europe.

3.5.2.2 Evaluating Several Versions of the Global DEO Strategy

This section has two purposes. First, it provides brief examples of how a rule simulator can help determine the effects of addition-specific investment products to a given asset universe, a question investors as well as providers of such products have

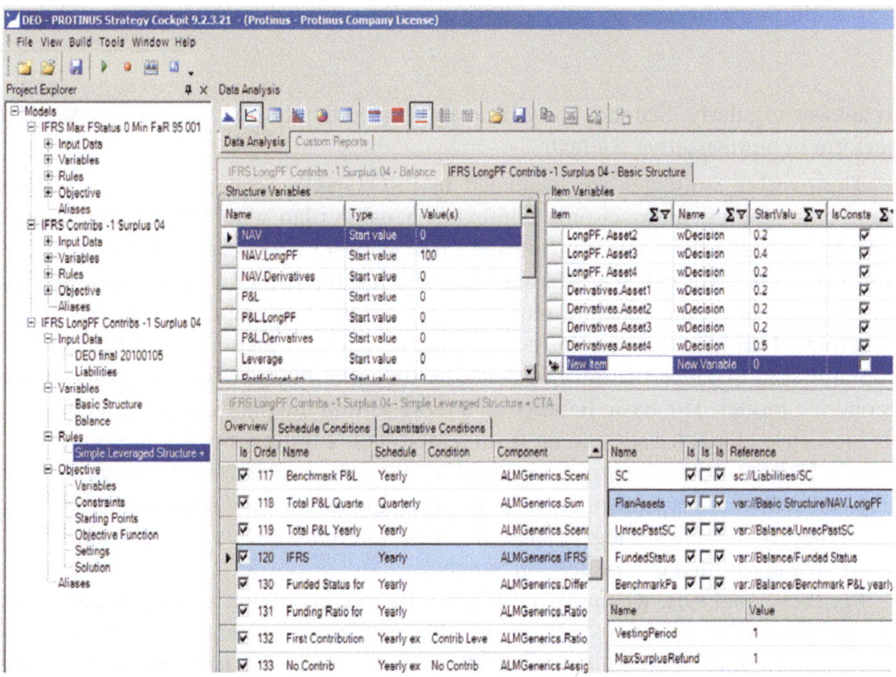

Fig. 3.6 Screenshot 1 showing several models with their basic structure. For the highlighted model its variables, with starting values, and a part of the building blocks, with the links to variables are shown

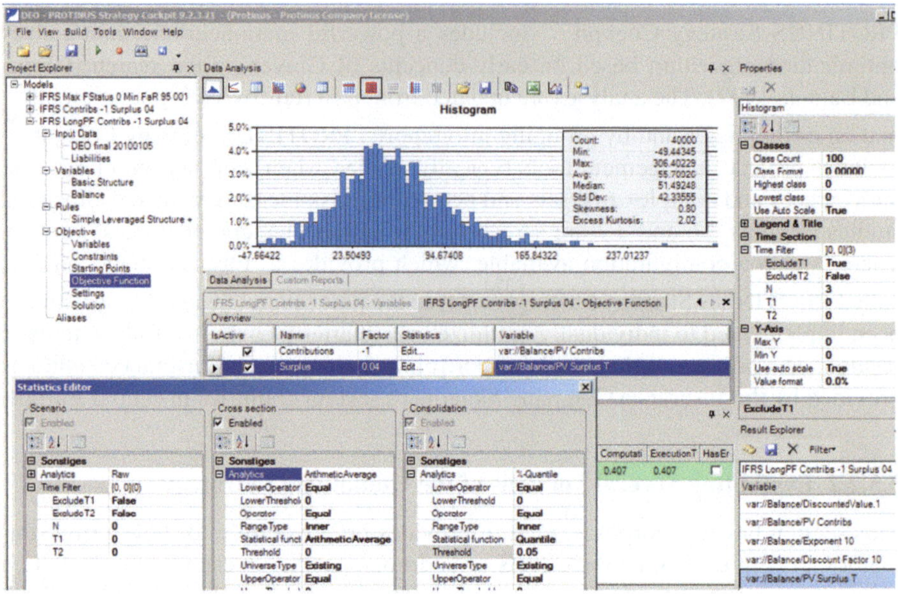

Fig. 3.7 Screenshot 2: One example for integrated statistical analysis. In the middle an objective function is designed, picking a variable from list on the right and defining the calculation performed thereon with the statistics editor

to answer regularly. Second, we will analyze specifically how the DEO strategies perform in a prospective context.

The overlay strategies are evaluated in PROTINUS Strategy Cockpit™ with three case studies: an asset-only context, a pension plan with IFRS balance sheet calculations, and the same pension plan rule with an added conditional contribution rule. All three cases are based on scenarios for the DEO overlay strategies and a group of broad, standard markets, i.e., US and Euro-equity as well as Euro-government and Euro-corporate bonds. For the pension plan cases, IFRS liability scenarios are produced from the scenarios for inflation and discount rates of the economic model. The basic liability data representing the plan's population with all its biometric characteristics come from a modified real data set. We choose to generate a 10-year, 1000-scenario space with a quarterly scaling, since all shorter time steps are not of the typical interest for the strategic investor. In addition, we employ a commodity overlay strategy, called the Princeton Index (Mulvey and Vural 2010), to provide additional diversification benefits.

For these analyses, we apply a cascade structure economic model which allows for the generation of consistent scenario spaces for market returns and fundamentals and liabilities. The economic model is again made out of a group of simple building blocks. Early versions of such models are described by Mulvey (1996). The model includes basic equations with mean reverting assumptions and derives most returns processes implicitly from the cascade of fundamental processes. These return processes are either produced by functional dependencies or via advanced random number drawing techniques. For all processes the first four moments, including skewness and kurtosis, and the interactions among the variables can be defined explicitly. The cascade represents one core macroeconomic environment, typically representing the investor's home region in order to include the local inflation and the rate, especially discount rate environment. These variables, along with scenarios for the liability variables such as IFRS defined benefit obligation (DBO), service cost, and pension payments are produced, so that a complete and fully consistent set of scenario spaces results, which represents accurately relevant exogenous risk factors.

To calibrate the model means for the fundamentals and the standard markets, we construct estimates based on long-term forecasts available from standard sources in combination with sound, standard forecasting approaches (Mulvey 1996; Høyland and Wallace 2001). Higher moments including skewness and excess kurtosis are solely dependent on historic data. For the overlays all values are taken from the historic simulations as described before.

The rules simulate quarterly fixed-mix rebalancing, which is applied to the long portfolio for the standard markets as well as for the four overlay strategies. More complex conditional rebalancing rules are not chosen in order to have a clearer view on the effects of adding the overlay strategies to a standard long portfolio. The long portfolio represents the total asset wealth of the investor. The derivatives (overlay) portfolio contributes to this wealth only through its quarterly change in its value. The IFRS pension model has additionally built in an IFRS building block, which is executed every fourth quarter and performs a full IFRS balance and P&L calculation with the so-called SORIE approach. For the third case, a contribution rule analogous to the Pension Protection Act rule in the USA is implemented.

In all three cases, we optimize first on a portfolio vector including only the four standard markets and then on a vector including the long portfolio of the four standard markets and the four overlay strategies. This factor is the sum of the fractions of the single overlay positions, given by the user or determined by the optimization. Thus the optimization determines the optimal leverage as well as the composition of the overlay portfolio. For the asset-only case we restrict this leverage factor to 2, for the pensions model it is set to 1, to represent some preferences of various types of investors roughly. Each single position is constrained to 100% of the NAV of the long portfolio.

We compare in our evaluations long-only optimizations and those with the overlay strategies to assess the effects on the total result. Due to the short computing time we ran several dozens of optimizations with various non-convex objective functions. The following represents a selection focussing on typical types of objectives, optimized regularly in the consulting practice during last decade.

3.5.2.3 Asset-Only Case

For this case, we work with an objective function and weight therein, which represent a conservative investor, who is not required to obey any institutional legal requirements. The first part of the function calculates in each point in time the

Table 3.3 Allocations for portfolios with and without overlays in the asset-only case

	Portfolio long only (%)	Portfolio with overlays (%)
Euro stocks	8.0	6.0
US stocks	0.0	0.0
Euro gov.	92.0	58.6
Euro corp.	0.0	35.4
DEO Euro	na	0.0
DEO USD	na	3.5
DEO JPN	na	1.9
Princeton index	na	5.1

5-percentile of the total portfolio's profit and loss and arithmetic average of these values along time. The second part calculates the discrete return along each scenario and the arithmetic average thereof at the end of the planning horizon. Examples for a better fit to specific individual investors would be, besides the weight between the two parts, some weighting along time or the introduction of some target return or wealth path of which deviations are minimized. We choose a weight which leads to conservative portfolios in the long positions. The weight remains unchanged for both the long only as well as for the optimization including the derivatives position.

As one finding, including the overlay strategies does not change the distribution among the core portfolio of bonds and equities significantly. When the overlay strategies are added, the long portfolio shifts into corporate bonds to some degree allowing for higher returns and additional diversification (Table 3.3).

Second, the optimal solution in this more risk-averse context does not require high leverage. As expected, an optimum solution when combing the overlay strategies with an existing strategy is not necessarily reached when using the maximum leverage is allowed. In this case, the maximum could have been as high as 200% of the volume of the long portfolio. But already by adding less than 10% overlays clear improvements on development of the portfolios can be observed especially on the level of 5th percentile (Fig. 3.8).

Clearly, a different objective function focussing on higher moments of the result variables would increase the leverage to profit more from the high skewness and kurtosis of the returns of the strategies. But for this quite representative conservative case, low leverage is enough.

3.5.2.4 Pension Plan ALM Example 1

For the pension plan example we optimize on an objective function maximizing the median of funded status across scenarios in each point in time and the arithmetic average thereof and minimizing arithmetic average of the 5-percentile of the funded status along each scenario. With more aggressive weights and a constraint that the Euro component of the portfolio is higher than the US component, which represents the typical local bias, the following fractions result for the two portfolios (Table 3.4).

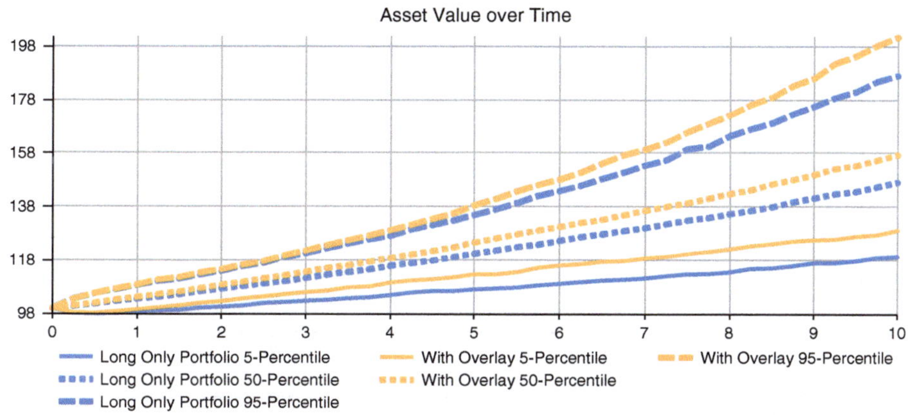

Fig. 3.8 Asset values over time for portfolios with and without overlays in the asset-only case

Table 3.4 Allocations for portfolios with and without overlay variables in ALM case 1

	Portfolio long only (%)	Portfolio with overlays (%)
Euro stocks	24.3	10.1
US stocks	21.9	9.1
Euro gov.	0.0	0.0
Euro corp.	53.8	80.8
DEO Euro	na	0.0
DEO USD	na	22.8
DEO JPN	na	0.0
Princeton index	na	73.9

While minimizing what could also be called "funding at risk" (FaR) we try to reduce the deviations of a value that is calculated as the difference of market value of plan assets minus the DBO. In reality both are driven to some degree by local inflation and local discount factor for the DBO. These dependencies are very accurately covered by the economic model which generates the scenarios for both the asset returns and the liabilities. The resulting portfolios are thus not surprising, with a high fraction in local currency corporate bonds. When the overlay strategies are allowed, an alternative skewed source of return is available, resulting in de-risking the long portfolio and increasing the liability hedge.

The long portfolio alone has a proposed equity fraction which is likely the maximum most sponsors would accept nowadays. Nevertheless, this value is inadequate to achieve a solid funding given the liability requirements on average over time. The lower end of the distribution does show quite such dramatic developments. When adding the overlay strategies to the portfolio, the situation changes significantly. This change is caused by the volume of the overlays occurring at the maximum level allowed (i.e., at 100% of the value of the long portfolios (Fig. 3.9)).

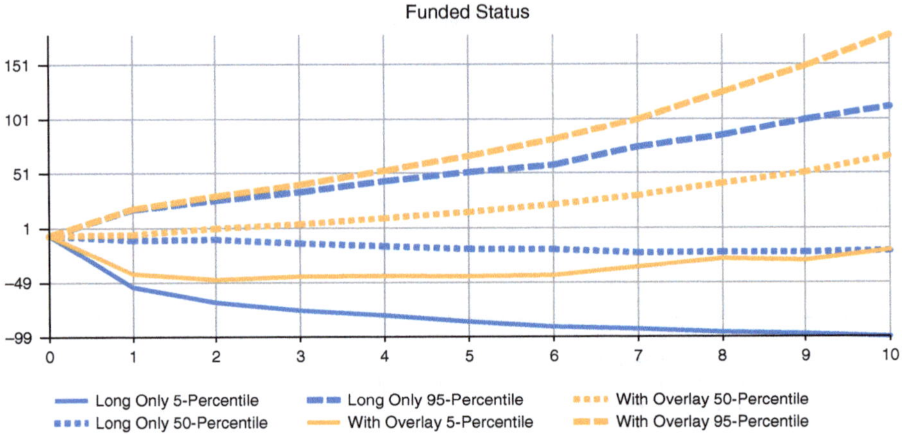

Fig. 3.9 Funded status over time for portfolios with and without overlays in ALM case 1

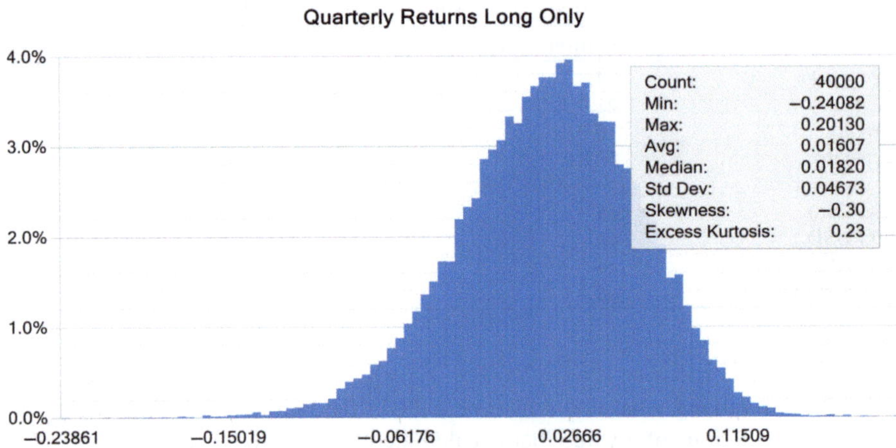

Fig. 3.10 Return distribution of the portfolio without overlays in ALM case 1

When looking at the portfolio returns in a histogram the positive effect of adding the overlays to the portfolio can be seen quite clearly: The shape of the distribution is more skewed with a higher kurtosis (Figs. 3.10 and 3.11).

3.5.2.5 Pension Plan ALM Example 2

For the last analyses we engage an additional feature of the software, which encourages the modeler to run series optimizations automatically, for example, in order to change the weights of components of an objective function to generate a set of results – analogously to an efficient frontier. We expand the IFRS model with an additional contribution building block. This step executes whenever the funding

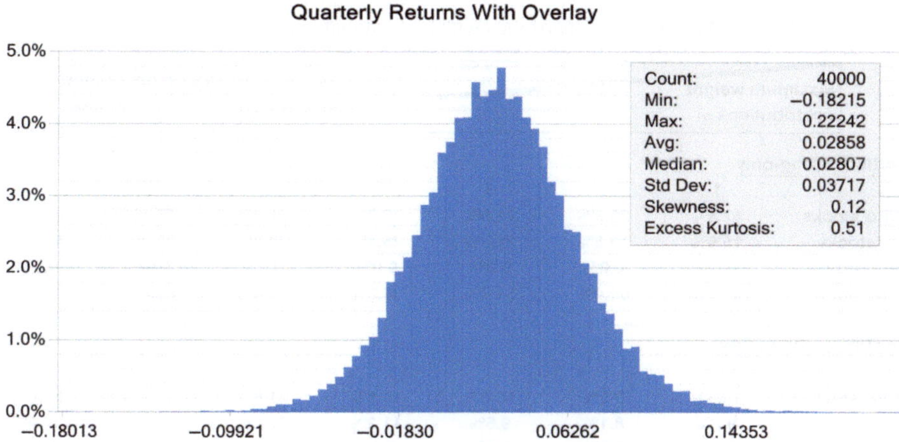

Fig. 3.11 Return distribution of the portfolio with overlay variables in ALM case 1

ratio of plan assets vs. DBO in a scenario falls below 80%. In this case one-seventh of the deficit in the funding status is contributed in cash to the plan. Thus, future contributions in this situation are distributed evenly over the next 7 years; the resulting cash flow vector is discounted with the same rate for discounting the DBOs in this scenario.

We optimize on the objective one Z_1 (Section 3.3) with weight leading to more conservative results. The function consists of average of the present values of the contributions in each knot along a scenario and the average thereof across scenarios and time. The second part is the average of the surplus at the end of each scenario discounted again with the according rate.

The results are again evaluated with and without the overlay strategies. In this analysis, we find again the positive effects from the two prior cases. Adding the overlay strategies to the universe leaves the optimal long portfolio only marginally changed as compared to the long-only case (Table 3.5). As in the other cases, strategies with the overlays are clearly superior to one without, as it can be seen in the two graphs (Figs. 3.12 and 3.13).

Had we seen the advantages of adding the overlay strategies in the first two cases either on the risk or on the reward side, now this last example shows a combined effect on both goals. Again the allowed leverage of 100% is fully engaged, which explains the strong improvements achieved by adding the overlays. As with the previous case, adding the overlays allows for de-risking the long portfolio.

Incorporating the overlays increases the average terminal surplus significantly and at the same time reduces the need for average contributions. This case is representative of many pension plans – the input is based on real data, only slightly modified for modeling purposes. Importantly, we see that the pension plan can only fulfill its long-term goals with fairly high fractions of risk-bearing assets, which probably many trustees would not accept. In the median, the strategy with overlays

Table 3.5 Allocations for portfolios with and without overlays in ALM case 2

Maximum weight contributions → Maximum weight surplus							
Portfolios long-only							
	1	2	3	4	5	6	7
Euro stocks	17.4%	20.8%	24.0%	28.3%	31.2%	35.8%	52.6%
US stocks	15.6%	18.8%	21.6%	25.4%	28.0%	32.2%	47.4%
Euro gov.	0.0%	0.0%	0.0%	0.0%	0.0%	0.0%	0.0%
Euro corp.	67.0%	60.4%	54.3%	46.3%	40.8%	32.0%	0.0%
Portfolios with overlay							
Euro stocks	10.3%	9.0%	10.6%	12.5%	31.0%	38.2%	52.6%
US stocks	9.2%	8.1%	9.5%	11.2%	27.9%	34.4%	47.4%
Euro gov.	2.2%	0.0%	0.0%	0.0%	0.0%	0.0%	0.0%
Euro corp.	78.3%	83.0%	79.9%	76.3%	41.0%	27.4%	0.0%
DEO Euro	12.5%	0.1%	0.0%	0.0%	0.0%	0.0%	0.0%
DEO USD	27.2%	25.4%	24.6%	29.1%	41.0%	34.5%	31.6%
DEO JPN	0.2%	0.0%	0.0%	0.0%	0.0%	0.0%	4.6%
Princeton index	60.1%	74.5%	75.4%	70.9%	78.8%	65.5%	63.9%

is already zero contributions in year 2, whereas the long portfolio alone reaches this point only after year 8. When looking at the 5% worst cases, the overlay strategies give impressive results – in 95% of all cases no contribution is required after year 10 (Figs. 3.12 and 3.13).

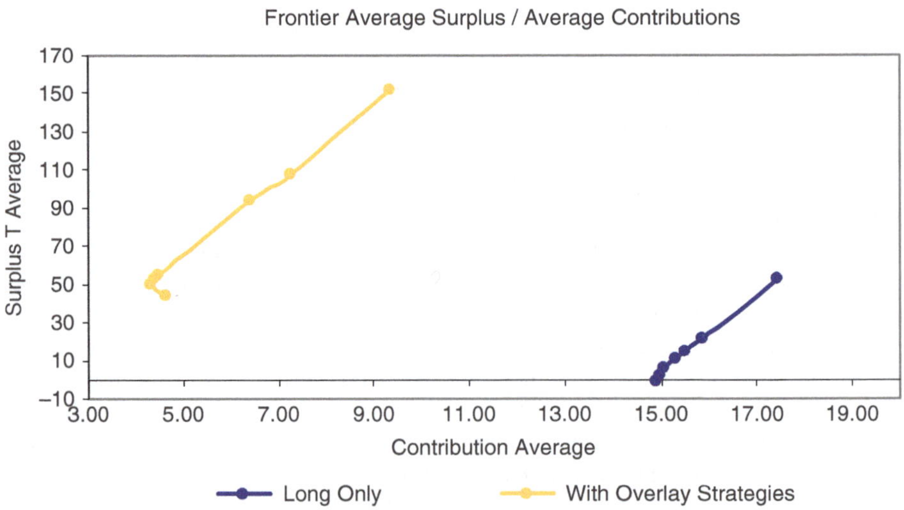

Fig. 3.12 Efficient frontiers with and without overlay variables in ALM case 2

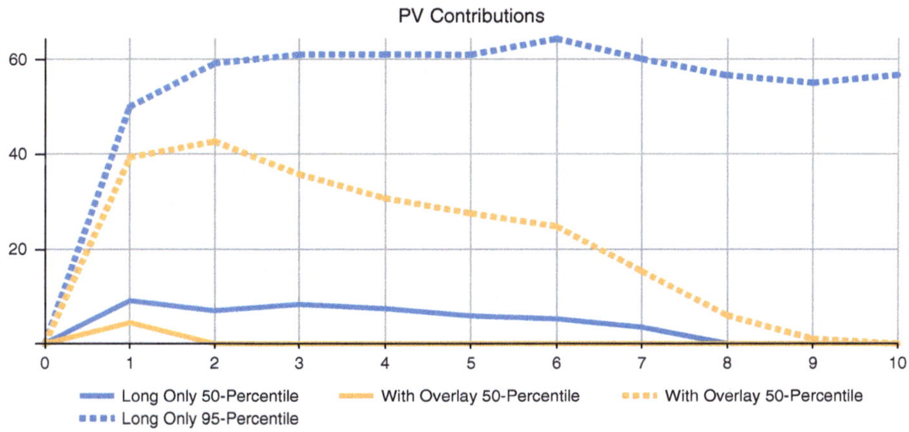

Fig. 3.13 Present value of contributions over time for portfolios with and without overlay variables, ALM case 2

3.6 Conclusions

Defined benefit pension plans will continue to experience severe difficulties in developed countries due to long-term demographic trends, the lack of contribution from sponsoring organizations, and ineffective ALM planning. To address the third issue, we suggest that a systematic forward-looking planning process can assist pension plan administrators. This chapter has shown that a DB pension plan can benefit by applying a duration-enhancing strategy in conjunction with traditional asset allocation and ALM policies. Not only does the strategy improve asset performance but also it enhances economic value in the ALM context. We take advantage of the plan's ability to increase duration in the assets – thereby improving performance across multiple objective functions. The approach allows the pension plan to continue to invest in higher returning assets such as equity and other risk-bearing instruments so that the long-term costs of managing a plan are reduced. At the same time, the risks to the pension surplus are constrained so that the plan's future contributions are relatively low. This is relevant during an economic contraction when rendering a substantial contribution can be onerous for the sponsoring organization.

The global duration enhancement strategy provides an effective mechanism for carrying out the basic concepts and extends previous research (Mulvey et al. 2010). A global perspective can be helpful in providing diversification benefits – as compared with a single country system such as developed in the USA. The strategy is pertinent for a worldwide organization possessing pension plans across countries. The issue of balancing the needs of pension plans of subsidiaries in individual countries with the global enterprise remains difficult to resolve – and will continue to be a topic for future research.

The 2008 economic crash showed that traditional approaches to investment planning, such as embodied by the static Markowitz model, possess severe limitations. The assumption of a constant covariance matrix was widely in error during the October-December 2008 period as correlation coefficients approached 1. In conjunction, the need for dynamic remedies became evident as markets plunged throughout the world in a short order. Single-period models are generally unable to respond in an appropriate manner to protect the investor's wealth or surplus for pension plans. Another contributing factor has been the lack of attention to liquidity concerns. Single-period portfolio models do not properly address transaction costs or liquidity constraints. In contrast, a multi-stage ALM model can include realistic issues such as dynamic policy rules (e.g., DEO), transaction costs, conditional liquidity constraints, and related real-world matters directly in the model. Likewise, multi-period ALM systems can evaluate innovative and dynamic investment strategies in a realistic forward-looking context.

References

Arnott, R., P. Bernstein. 1990. Defining and managing pension fund risk. F. J. Fabozzi, N. Mencher, eds. *Pension Fund Investment Management: A Handbook for Sponsors and their Advisors*. Probus, Chicago, 38–51.

Bader, L. N. 2003. Treatment of pension plans in a corporate valuation. *Financ. Anal. J.* **59**(3), 19–24.

Black, F. 1995. The plan sponsor's goal. *Financ. Anal. J.* **51**(4), 6–7.

Boender, G. E. C., P. C. van Aalst, F. Heemskerk. 1998. Modelling and management of assets and liabilities of pension plans in the Netherlands. W. T. Ziemba, J. M. Mulvey, eds. *Worldwide Asset and Liability Modeling*. Cambridge University Press, Cambridge, 561–580.

Bogentoft, E., H. E. Romeijn, S. Uryasev. 2001. Asset/liability management for pension funds using CVaR constraints. *J. Risk Finance* **3**(1), 57–71.

Cariño, D. R., T. Kent, D. H. Myers, C. Stacy, M. Sylvanus, A. Turner, K. Watanabe, W. Ziemba. 1994. The Russel-Yasuda Kasai model: An asset/liability model for a Japanese insurance organizationusing multistage stochastic programming. *Interfaces*. **24**(1), 29–49.

Dempster, M. A. H., M. Germano, E. A. Medova, M. Villaverde, 2003. Global asset liability management. *Br. Actuarial J.* **9**, 137–216.

Dempster, M. A. H., M. Germano, E. A. Medova, M. I. Rietbergen, F. Sandrini, M. Scrowston. 2006. Managing guarantees. *J. Portfolio Manage*. **32**, 51–61.

Dert, C. L. 1995. Asset liability management for pension funds. Ph.D. thesis, Erasmus University.

Glover, F., M. Laguna. 1997. *Tabu Search*. Kluwer, Boston.

Hilli, P., M. Koivu, T. Pennanen, A. Ranne. 2003. A stochastic programming model for asset liability management of a Finnish pension company. *Stochastic Program*. (E-Print Series)

Høyland, K., S. Wallace. 2001. Generating scenario trees for multistage problems. *Manage. Sci.* **47**(2), 295–307.

Ippolito, R. A. 2002. Replicating default risk in a defined-benefit plan. *Financ. Anal. J.* **58**(6), 31–40.

Kouwenberg, R., S. A. Zenios. 2001. Stochastic programming models for asset liability management. Working Paper 01-01, HERMES Center, University of Cyprus.

Mulvey, J. M. 1996. Generating scenarios for the Towers Perrin investment system. *Interfaces*. **26**(2), 1–15.

Mulvey, J. M., W. C. Kim. 2010. The role of alternative assets for optimal portfolio construction. E. L. Melnick, B. S. Everitt, eds. *Encyclopedia of Quantitative Risk Analysis and Assessment*, Volume 4. J Wiley, Chichester, 1574–1583.

Mulvey, J. M., M. T. Vural. 2010. A robust strategy for overlay strategies in commodities. Princeton University Report.

Mulvey, J. M., G. Gould, C. Morgan. 2000. An asset and liability management system for Towers Perrin-Tillinghast. *Interfaces.* **30**(1), 96–114.

Mulvey, J. M., K. D. Simsek, Z. Zhang, M. Holmer. 2005a. Analysis of defined benefit pension plans via an integrated corporate/pension planning system. Report to U.S. Department of Labor.

Mulvey, J. M., F. J. Fabozzi, B. Pauling, K. D. Simsek, Z. Zhang. 2005b. Modernizing the defined-benefit pension system. *J. Portfolio Manage.* **31**, 73–82.

Mulvey, J. M., K. D. Simsek, Z. Zhang. 2006. Improving investment performance for pension plans. *J. Asset Manage.* **7**, 93–108.

Mulvey J. M., C. Ural, Z. Zhang. 2007. Improving Performance for long-term investors: wide diversification, leverage and overlay strategies. *Quant. Finance* **7**(2), 1–13.

Mulvey, J. M., K. D. Simsek, Z. Zhang, F. Fabozzi. 2008. Assisting defined-benefit pension plans. *Oper. Res.* **56**(5), 1066–1078.

Mulvey, J. M., W. C. Kim, Y. Ma. 2010. Duration-enhancing overlay strategies for defined-benefit pension plans. *J. Asset Manage.* **11**, 136–162.

Muralidhar, A. S., R. J. P. van der Wouden. 1999. Optimal ALM strategies for defined benefit pension plans. *J. Risk* **2**, 2.

Olson, R. L. 2005. *The School of Hard Knocks: The Evolution of Pension Investing at Eastman Kodak.* RIT Cary Graphic Arts Press, Rochester, NY.

Parks, J. P. 2003. Safeguarding America's retirement security: An examination of defined benefit pension plans and the Pension Benefit Guaranty Corporation. U. S. Senate Committee on Governmental Affairs Subcommittee on Financial Management, the Budget, and International Security Hearing.

Peskin, M. W. 1997. Asset allocation and funding policy for corporate-sponsored defined-benefit pension plans. *J. Portfolio Manage.* **23**(2), 66–73.

Rauh, J. D. 2006. Investment and financing constraints: Evidence from the funding of corporate pension plans. *J. Finance* **61**(1), 33–71.

Ryan, R., F. J. Fabozzi. 2003. Pension fund crisis revealed. *J. Investing.* **Fall 2003**, 43–48.

Swensen, D. F. 2000. *Pioneering Portfolio Management.* The Free Press, New York, NY.

Zenios, S. A., W. T. Ziemba. 2006. *Handbook of Asset and Liability Management: Theory and Methodology.* North-Holland, Amsterdam.

Ziemba, W. T., J. M. Mulvey. 1998. *Worldwide Asset and Liability Modeling.* Cambridge University Press, Cambridge.

Chapter 4
Hedging Market and Credit Risk in Corporate Bond Portfolios

Patrizia Beraldi, Giorgio Consigli, Francesco De Simone, Gaetano Iaquinta, and Antonio Violi

Abstract The European market for corporate bonds has grown significantly over the last two decades to become a preferable financing channel for large corporations in the local and Eurobond markets. The 2008 credit crisis has, however, dramatically changed corporations funding opportunities with similar effects on borrowing policies of sovereigns as well. Accordingly institutional and individual investors have progressively reduced the share of credit risky instruments in their portfolios. This chapter investigates the potential of multistage stochastic programming to provide the desired market and credit risk control for such portfolios over the recent, unprecedented financial turmoil. We consider a Eurobond portfolio, traded in the secondary market, subject to interest and credit risk and analyse whether a jump-to-default risk model and a dynamic control policy would have reduced the impact of severe market shocks on the portfolios during the crisis, to limit the systemic impact of investment strategies. The methodology is shown to provide an effective alternative to popular hedging techniques based on credit derivatives at a time in which such markets became extremely illiquid during the Fall of 2008.

Keywords Multistage stochastic programming · Bond portfolio optimization · Credit risk

4.1 Introduction

The 2007–2009 financial turmoil witnessed an unprecedented market downturn for financial instruments carrying credit risk (Abaffy et al. 2007; Berndt 2004; Crouhy et al. 2000; Duffie and Singleton 1999) spanning both the secondary markets for corporate and sovereign securities and the markets for derivatives based on such instruments. The crisis propagated in US markets from mortgage-backed securities (MBS) and collateralized debt obligations (CDO) to credit instruments traded over-the-counter and thus into international portfolios. Widespread lack of liquidity in the secondary market induced first the Federal Reserve, then the European Central Bank, to adopt an expansive monetary policy through a sequence of base rate

P. Beraldi (✉)
Department of Electronics, Informatics and Systems, University of Calabria, Rende, Italy
e-mail: beraldi@deis.unical.it

M. Bertocchi et al. (eds.), *Stochastic Optimization Methods in Finance and Energy*, International Series in Operations Research & Management Science 163,
DOI 10.1007/978-1-4419-9586-5_4,

reductions. Such policy partially limited the fall of bond prices but could do very little against instability in the corporate equity and fixed income markets.

The debate over the causes and possible remedies of such a prolonged financial crisis involved, from different perspectives, policy makers, financial intermediaries, and economists (European Central Bank 2010), all interested in analyzing the equilibrium recovery conditions, possibly within a quite different market architecture.

Financial investors, on the other hand, suffered dramatic portfolio losses within increasingly illiquid money, secondary stock and bond and credit derivative markets. In this chapter we present a stochastic model for interest rate and credit risk applied to a portfolio of corporate bonds traded in the Eurobond market with portfolio strategies tested over the 2008–2009 crisis. The portfolio management problem is formulated as a dynamic stochastic program with recourse (Consigli and Dempster 1998; Pflug and Römisch 2007; Zenios and Ziemba 2007). This chapter provides evidence of the potential offered by dynamic policies during a dramatic market crisis. Key to the results presented are the definitions of

- a statistical model capturing common and bond-specific credit risk factors that will determine jointly with the yield curve the defaultable bonds price behavior (Dai and Singleton 2003; Das and Tufano 1996; Duffie and Singleton 2000; Jarrow and Turnbull 2000; Kijima and Muromachi 2000; Longstaff et al. 2005) and
- a multistage strategy determined by a risk-reward objective function explicitly considering an extreme risk measure (Bertocchi et al. 2007; Consigli et al. 2010; Dempster et al. 2003; Jobst et al. 2006; Jobst and Zenios 2005; Rockafellar and Uryasev 2002). The problem considers a constant investment universe excluding credit derivatives and exogenous hedging strategies.

The stochastic model for bond returns and the optimization model are strictly related and their accuracy will determine the quality of the resulting portfolio strategy. The return model leads to the definition of bond price scenarios used as inputs to a stochastic programming model over a 1-year time horizon with monthly rebalancing decisions. The stochastic program includes buying, holding, and selling decisions regarding an extended universe of corporate bonds, spanning several economic sectors and all rating classes under Standard and Poor's classification (Abaffy et al. 2007; Moody's Investors Service 2009). Dynamic bond portfolio management approaches, since the seminal work by (Bradley and Crane 1972), have been widely adopted in the past for Treasury portfolios, while limited experience has been reported for corporate bond portfolios (Consigli et al. 2010; Jobst and Zenios 2005). This motivates our study. We show that, under the given market conditions, a dynamic policy would have provided an effective contingency plan without the need to hedge against credit losses with derivative contracts.

The chapter evolves as follows. Section 4.2 describes the market and credit risk model which represents the bond dynamics. In Section 4.3 a mathematical model for the optimal bond portfolio management is introduced, while Section 4.4 reports on the scenario generation procedure for the simulation of input data for the decision model. Computational experiments for the validation of the proposed approach are reported in Section 4.5. Some concluding remarks end the chapter.

4.2 Corporate Bonds Risk Exposure

The recent credit crisis witnessed an unprecedented credit spread increase across all maturities in the Eurobond market while, first in the USD monetary area and then in the UK and Europe, interest rates were rapidly, though ineffectively, driven down to try to facilitate a liquidity recovery in the markets. In this work we assume a risk model incorporating both common and specific risk factors and allow the investor to determine her optimal policy by exploiting a scenario representation of the credit spreads evolution which may be affected at random times by severe market shocks. We consider a universe of Euro-denominated bonds with ratings from AAA to $CCC - C$ (see (Abaffy et al. 2007)) plus one default-free government bond. Ratings are alphanumeric indicators specified by international rating agencies such as standard and poor, moody and fitch, defining the credit merit of the bond issuer and thus the likelihood of possible defaults over a given risk horizon. Rating revisions can be induced by market pressures or expert opinions and only occur infrequently. Bonds trading in the secondary market on the other hand generate a continuous information flow on expected interest rates and yield movements.

Bonds belonging to the same rating class may be distinguished according to their maturity and more importantly the activity sector to which the issuer belongs. Bond issuers within the same rating class and industry sector may finally be distinguished according to their market position and financial strength: the riskier the issuer, according to the above classification, the higher the credit spread that will be requested by investors to include a specific fixed income security in the portfolio.

Corporate prices are thus assumed to be driven by

- a common factor affecting every interest-sensitive security in the market (Cox et al. 1985), related to movements of the yield curve,
- a credit risk factor, related to movements of the credit curves (Abaffy et al. 2007; Duffie and Singleton 1999; Jobst and Zenios 2005), one for each rating class, and
- a specific bond factor, related to the issuer's economic sector and its general financial health.

A degree of complexity may be added to this framework when trying to take into account the possible transition over the portfolio lifetime of bond issuers across different rating classes (Jarrow et al. 1997) and bond-specific liquidity features resulting in the bid-ask spread widening in the secondary market. In this work we will focus on the solution of a 1-year corporate bond portfolio management problem with monthly portfolio revision, not considering either transition risk or liquidity risk explicitly.

4.2.1 Market and Credit Risk Model

We consider a credit-risky bond market with a set I of *securities*. For each bond $i \in I$, we denote by T_i its *maturity*. Let K be the set of *rating classes*. Assuming a canonical S&P risk partition from AAA to CCC-C and D (the default state), we will denote by $k = 1, 2, \ldots, 8$ the credit risk indicator associated, respectively, with default-free sovereign and corporate AAA, AA, A, BBB, BB, B, CCC-C, and D ratings. A difference is thus postulated between a sovereign and a corporate AAA rating: the first one will be associated with the prevailing market yield for default-free securities with null credit spread over its entire market life, while the second may be associated with a positive, though negligible, credit spread. The security i credit risk class is denoted by k_i, while I_k is the set of bonds belonging to rating class k. The *price* of security i at time t will be denoted by v^i_{t,T_i}. For given security price and payment structure the *yield* y^i_{t,T_i} can be inferred from the classical price–yield relationship:

$$v^i_{t,T_i} = \sum_{t<m\leq T_i} c^i_m e^{-y^i_{t,T_i}(m-t)}, \tag{4.1}$$

where the *cash payments* over the security residual life are denoted by c^i_m up to and including the maturity date. In (4.1) the yield y^i_{t,T_i} will in general reflect the current term structure of risky interest rates for securities belonging to class k_i. We assume y^i_{t,T_i} generated by a bond-relevant *credit spread* π^i_{t,T_i} and a *default-free interest rate* r_{t,T_i}, i.e. $y^i_{t,T_i} = r_{t,T_i} + \pi^i_{t,T_i}$.

A one-factor model is considered for both the default-free interest rate curve and the security-specific credit spread. The latter is assumed to be generated by a rating-specific factor π^k_t and an idyosincratic factor η^i_t.

The three state variables of the model are assumed to follow the s.d.e.'s:

$$dr_t(\omega) = \mu_r dt + \sigma_r(t,r)dW^r_t(\omega), \tag{4.2}$$

$$d\pi^k_t(\omega) = \mu_k dt + \sigma_k(t,\pi^k)\sum_{l\in K} q^{kl} dW^l_t(\omega) \quad \forall k, \tag{4.3}$$

$$d\eta^i_t(\omega) = \mu_\eta dt + \sigma_i dW^i_t(\omega) + \beta^i(\omega)d\Psi^i_t\left(\lambda^i\right) \quad \forall i, \tag{4.4}$$

where ω is used to identify a generic random variable. The first equation describes the short interest rate evolution as a diffusion process with constant drift and possibly state- and time-dependent volatility. $dW^r_t \sim N(0, dt)$ is a normal Wiener increment. The $k = 1, 2, \ldots, 7$ credit spread processes in (4.3) are also modeled as diffusion processes with time- and state-varying volatilities. Credit spreads and the risk-free rate are correlated: the coefficients q^{kl} denote the elements of the correlation matrix lower triangular Choleski factor linking the eight risk factors together, given the independent Wiener increments dW^k_t, $k = 1, 2, \ldots, 7$.

Equation (4.4) assumes an independent security spread process with a diffusion component generated by a Wiener process W^i_t and a jump-to-default component generated by a default process Ψ^i_t with intensity λ^i and jump size β^i. The bond price will either follow a continuous path up to maturity if no default occurs or will default before maturity generating a price shock. All security prices in the fixed-income market are assumed to be driven by the short rate r_t. The credit spreads $\pi_t^{k_i}$ in (4.3) will determine the price behavior of bonds i belonging to class k_i and thus the relative behavior of bonds in different rating classes. The bond-specific marginal spread processes η_t^i will finally determine the relative price dynamics of bonds within the same rating class.

From (4.1), given the current time 0 price v_{0,T_i}^i, price dynamics for $t \in [0, T]$ are derived from the yield scenarios by introducing the canonical duration-convexity update in the real-world, market, probability space:

$$dv_{t,T_i}^i = v_{t,T_i}^i \left[y_t^i dt - \delta_t^i dy_t^i + 0.5\gamma_t^i (dy_t^i)^2 \right], \tag{4.5}$$

where $y_t^i = r_t + \pi_t^{k_i} + \eta_t^i$. Given the current price and duration-convexity pair (δ_t^i, γ_t^i), each yield increment will determine an associated price transition. Then an update of the duration and convexity parameters leads to a new price transition for the given yield increment, and so on up to maturity, unless default occurs.

To specify the spread process in (4.4) and employ the pricing update in (4.5), the marginal spread η_t^i must be identified. Below in the case study we present a model assuming a pure jump process for the marginal spread. In the general case, for a given price history v_{l,T_i}^i, $l \le t$, at time t and cash flow structure, we imply the marginal spread as

$$\begin{aligned} \eta_l^i &= \pi_l^i - \pi_l^{k_i}, \\ \pi_l^i &= -\ln\left(\frac{v_{l,T_i}^i}{v_{l,T_i}^f}\right) \cdot (\delta_l^i)^{-1}. \end{aligned} \tag{4.6}$$

v_{l,T_i}^f in (4.6) denotes the price of a default-free security with the same payment structure as bond i and δ_l^i is the duration of bond i at time l. Such a price is computed from the interest rate term structure history $r_{l,T}$. The *spread* $\pi_l^{k_i}$ denotes instead the CDS spread for the rating class introduced in (4.3). We use the above approximation to differentiate risky price dynamics within a rating class.

Scenario generation for the corporate bond portfolio problem relies on the set of equations (4.5) for all $i \in I$ and the underlying risk factor processes (4.2), (4.3), and (4.4).

4.2.2 Default Model

Rating agencies regularly estimate default frequencies observed in the economy for borrowers belonging to a certain rating class by year cohorts leading to the definition of real-world default probabilities at the rating class level. Moody's 2009 report (Moody's Investors Service 2009) updates previous studies presenting default statistics for the 1982–2008 period that we consider in our case study. We refer to actual default frequencies and expected losses per unit credit published by the rating agency as the natural or real-world, or empirical, default probabilities and recovery rates considered in the model. These probabilities need to be distinguished from market default probabilities implied from bond prices (Berndt et al. 2005; Consigli 2004; Duffie and Singleton 1999; Elton et al. 2001).

Market implied default probabilities are recognized to over estimate default frequencies actually observed in the economy (Berndt et al. 2005). Defaultable spreads do price additional risk factors such as liquidity risk, company evolving financial conditions, and the stability of the default intensities in the economy. Rating assessments moreover occur at discrete times with occasional rating revisions: to overcome this drawback we propose below a simple and practical heuristic for the default intensity coefficient λ^i in the model.

In particular, let Ψ_t^i be the default process affecting security i. Prior to default the process is at 0 and moves to 1 upon default. In a discrete setting for every time stage $t = 1, 2, \ldots, T$ the process transition from 0 to 1 is determined by an intensity λ^i with two components: the 1-year rating-specific default frequency λ^{k_i} and a user-defined multiplier ξ^i to capture the security-specific perspectives as estimated through available market and company information by credit analysts. The security-specific risk coefficient ξ^i is assumed to depend on the company geographic exposure, its economic sector, and the security default premium (Berndt et al. 2005). We assume that credit analysts map the above analysis into eight risk partitions $\xi^i = \kappa * 0.125, \kappa = 4, 5, \ldots, 12$, where

$$\lambda^i = \lambda^{k_i} \times \xi^i. \tag{4.7}$$

This heuristic provides a user-based update of the historical default frequencies. The loss-given default for a bond investor is determined by the market price loss suffered by the investor upon default announcement. Moody's report in (Moody's Investors Service 2009) the historic recovery rates on a large sample of Eurobonds in which the canonical inverse relationship between default frequencies and recovery rates is consistently confirmed across time and economic sectors.

4.2.3 Price–Yield Relationship

Following the intensity specification in (4.7) an individual security can at any time be affected by a *jump-to-default* event attributed by the arrival of new information that will drive the security into a default state. The relationship between the incremental spread behavior and the security price is as follows. Let the value of the default process $d\Psi_\tau^i = 1$ at a random time τ. Then, given a recovery value β^i, we have

$$v^i_{\tau,T_i} = \sum_{\tau<m\leq T_i} c^i_m e^{-\beta^i} e^{-(r_\tau+\pi^{k_i}_\tau)(m-\tau)}. \tag{4.8}$$

At the default announcement, the bond price will reflect the values of all payments over the bond residual life discounted at a flat risky discount rate including the risk-free interest rate and the class k_i credit spread *after default*. Widening credit spreads across all the rating classes will push default rates upward without discriminating between corporate bonds within each class. On the other hand, once default is triggered at τ then the marginal spread upward movement will be absorbed by a bond price drop which is consistent with the postulated recovery rate.

Following (4.5) the price movement will depend on the yield jump induced by η^i_t. This is assumed to be consistent in the mean with rating agencies recovery estimates (Moody's Investors Service 2009) and carries a user-specified uncertainty reflected in a lognormal distribution with stochastic jump size $\beta^i \sim \ln N\left(\mu_{\beta^i}, \sigma_{\beta^i}\right)$. Then $\ln\left(\beta^i\right) \sim N\left(\mu_{\beta^i}, \sigma_{\beta^i}\right)$ so that $\beta^i_t \in (0, +\infty)$ and both $e^{-\beta^i}$ (the recovery rate) and $1 - e^{-\beta^i}$ (the loss rate) belong to the interval (0, 1).

In summary the modeling framework has the following characteristic features:

- default arrivals and loss upon default are defined taking into account rating-specific estimates by Moody's (2009) adjusted to account for a borrower's sector of activity and market-default premium;
- over a 1-year investment horizon default intensities are assumed constant and will determine the random arrival of default information to the market; upon default the marginal spread will suffer a positive shock immediately absorbed by a price-negative movement reflecting the constant partial payments of the security over its residual life;
- the Poisson process and the Wiener processes drive the behavior of an independent ideosyncratic spread process which is assumed to determine the security price dynamics jointly with a short rate and a credit spread process;
- the credit spreads for each rating class and the short interest rate are correlated.

The model accommodates the assumption of correlated risk factors whose behavior will drive all securities across the different rating categories, but a specific default risk factor is included to differentiate securities behavior within each class. No contagion phenomena are considered and the exogenous spread dynamics impact specific bond returns depending on their duration and convexity structure.

4.2.4 The Portfolio Value Process

We now extend the statistical model to describe the key elements affecting the portfolio value dynamics over a finite time horizon T. Assuming an investment universe of fixed income securities with annual coupon frequency we will have only one possible coupon payment and capital reimbursement over a 1-year horizon. Securities may default and generate a credit loss only at those times. We assume a wealth equation defined at time 0 by an investment portfolio value and a cash account $\mathcal{W}_0 = \sum_{k \in K} \sum_{i \in I_k} X_{i0} + C_0$, where X_{i0} is the time-0 market value of the position in security i, generated by a given nominal investment x_{i0} and by the current security price v^i_{0,T_i}. For $t = 0, \Delta t, \ldots, T - \Delta t$, we have

$$\mathcal{W}_{t+\Delta t,\omega} = \sum_{k \in K} \sum_{i \in I_k} x_{it} v^i_{t,T_i} (1 + \rho^i_{t+\Delta t}(\omega)) + C(t + \Delta t, \omega). \tag{4.9}$$

In (4.9) the price return $\rho^i_{t+\Delta t}$ is generated for each security according to (4.5) as $\frac{v^i(t+\Delta t,\omega)}{v^i(t)} - 1 = [-\delta^i_t dy^i_t + y^i_t dt + 0.5\gamma^i_t (dy^i_t)^2]$, where y^i_t represents the yield return at time t for security i. For $k = 0$ we have the default-free class and the yield will coincide with the risk-free yield. As time evolves an evaluation of credit losses is made. Over the annual time span as before, we have

$$\begin{aligned} C(t + \Delta t, \omega) = C(t) e^{r_t(\omega)\Delta t} + \sum_{k \in K} \sum_{i \in I_k} c_{i,t+\Delta t} + \\ - \sum_{k \in K} \sum_{i \in I_k} c_{i,t+\Delta t} \left(1 - e^{\left[-\beta^i_{t+\Delta t}(\omega) d\Psi^i_{t+\Delta t}(\omega)\right]}\right). \end{aligned} \tag{4.10}$$

The cash balance $C(t)$ evolution will depend on the expected cash inflows $c_{i,t}$ and the associated credit losses for each position i, the third factor of (4.10). Coupon and capital payments in (4.10) are generated for given nominal investment by the security-specific coupon rates and expiry date. If a default occurs then, for a given recovery rate, over the planning horizon the investor will face both a cash loss in (4.10) and the market value loss of (4.9) reflecting an assumption of constant partial payments over the security residual time span within the default state.

Equation (4.10) considers payment defaults explicitly. However, the mathematical model introduced below considers an *implicit* default loss definition. This is obtained by introducing the coefficient specification

$$f_{i,t+\Delta t}(\omega) = c_i \left(e^{-\left[\beta^i_{t+\Delta t}(\omega) d\Psi^i_{t+\Delta t}(\omega)\right]}\right) \tag{4.11}$$

and, thus, $c_{i,t+\Delta t} = x_{it} f_{i,t+\Delta t}$, where c_i denotes the coupon rate on security i.

In this framework the corporate bond manager will seek an optimal dynamic strategy looking for a maximum price and cash return on his portfolio and at the same time trying to avoid possible default losses. The model is quite general and alternative specifications of the stochastic differential equations (4.2), (4.3), and (4.4) will lead to different market and loss scenarios accommodating a wide range of possible wealth distributions at the horizon.

4.3 Dynamic Portfolio Model

In this section we present a multistage portfolio model adapted to formulate and solve the market and credit risk control problem. It is widely accepted that stochastic programming provides a very powerful paradigm in modeling all those applications characterized by the joint presence of uncertainty and dynamics. In the area of portfolio management, stochastic programming has been successfully applied as witnessed by the large number of scientific contributions appearing in recent decades (Bertocchi et al. 2007; Dempster et al. 2003; Hochreiter and Pflug 2007; Zenios and Ziemba 2007). However, few contributions deal with the credit and market risk portfolio optimization problem within a stochastic programming framework. Here we mention (Jobst et al. 2006) (see also the references therein) where the authors present a stochastic programming index tracking model for fixed income securities.

Our model is along the same lines, however, introducing a risk-reward trade-off in the objective function to explicitly account for the downside generated by corporate defaults.

We consider a 1-year planning horizon divided in the periods $t = 0, \ldots, T$ corresponding to the trading dates. For each time t, we denote by $\mathcal{N}_t$ the set of nodes at stage t. The *root* node is labeled with 0 and corresponds to the initial state. For $t \geq 1$ every $n \in \mathcal{N}_t$ has a unique *ancestor* $a_n \in \mathcal{N}_{t-1}$, and for $t \leq T - 1$ a non-empty set of *child* nodes $H_n \in \mathcal{N}_{t+1}$. We denote by $\mathcal{N}$ the whole set of nodes in the scenario tree. In addition, we refer by t_n the time stage of node n. A scenario is a path from the root to a leaf node and represents a joint realization of the random variables along the path to the planning horizon. We shall denote by S the set of scenarios. Figure 4.1 depicts an example scenario tree.

The scenario probability distribution $\mathcal{P}$ is defined on the leaf nodes of the scenario tree so that $\sum_{n \in \mathcal{N}_T} p_n = 1$ and for each non-terminal node $p_n =$

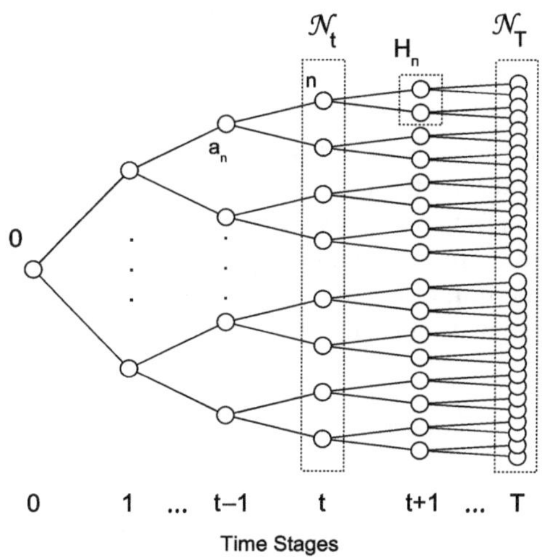

Fig. 4.1 A sample scenario tree

$\sum_{m \in H_n} p_m, \forall n \in \mathcal{N}_t, t = T - 1, .., 0$, that is each node receives a conditional probability mass equal to the combined mass of all its descendant nodes. In our case study portfolio revisions imply a transition from the previous time $t - 1$ portfolio allocation at the ancestor node to a new allocation through holding, buying, and selling decisions on individual securities, for $t = 1, \ldots, T - 1$. The last possible revision is at stage $T - 1$ with one period to go. Consistent with the fixed-income portfolio problem, decisions are described by nominal, face-value, positions in the individual security $i \in I$ of rating class $k_i \in K$.

4.3.1 Parameters

Let $\bar{x}_i$ be the initial holding in bond $i \in I$ and let C_0 and $\bar{g}$ denote the initial cash and debt, respectively. The random coefficients in the optimization problem include

r_n the one period interest rate for cash deposits available at node n;
π_n^k the credit spread of rating class k at node n;
η_n^i the bond-specific incremental spread of bond i at node n;
b_n the one period interest rate on borrowing decisions taken at node n;
v_{in} the market price of bond i at node n, generated given the initial price v_{i0} as a function of the security-specific yield dynamics (cf. (4.5));
f_{in} the cash flow generated by bond i at node n following (4.11) which takes into account possible cash defaults over the planning horizon.

With r_n, π_n^k, and η_n^i as state variables in the statistical model, for given initial values $v_{i0}, r_0, \pi_0^k, \eta_0^i, i \in I$, for $t = 1, 2, .., T$ and $n \in \mathcal{N}_t$ we have

$$v_{in} = v_{ia_n} \left[1 - \delta_{a_n}^i dy_{a_n}^i + y_{a_n}^i dt + 0.5 \gamma_{a_n}^i (dy_{a_n}^i)^2 \right], \tag{4.12}$$

where $dy_{a_n}^i = y_n^i - y_{a_n}^i = \left[(r_n - r_{a_n}) + \left(\pi_n^{k_i} - \pi_{a_n}^{k_i} \right) + \left(\eta_n^i - \eta_{a_n}^i \right) \right]$ is defined up to the end of the last stage. A default event in node n will simultaneously affect the $d\eta_{a_n}^i$ increment and thus the bond return in the given subperiod, and the cash account through the coefficient f_{in} specified in the previous section.

For a given tree structure, following the corresponding set of conditional sample paths, an update of the duration and convexity coefficients is needed at each node before the next price transition is generated according to

$$\delta_n^i = \frac{\sum_{t_n \le m \le T_i} \left((m - t_n)\, c_{in} e^{-y_n^i (m - t_n)} \right)}{\left(v_n^i \right)}, \tag{4.13}$$

$$\gamma_n^i = \frac{\sum_{t_n \le m \le T_i} \left((m - t_n)^2\, c_{in} e^{-y_n^i (m - t_n)} \right)}{\left(v_n^i \right)}, \tag{4.14}$$

where t_n denotes the time stage of node n and m defines the expected payment dates for bond i up to maturity T_i.

A set of stage-independent upper bounds is considered to reflect exogenous policy constraints imposed by the investment manager to limit the portfolio risk exposure. The maximum investment-grade portfolio fraction is denoted by ϕ, while ζ identifies the maximum speculative-grade portfolio fraction. Parameter ν_k represents the maximum portfolio fraction for bonds belonging to rating class k. An upper bound for the debt level with the bank is also introduced in the model by the parameter γ. Finally, χ^+ and χ^- are proportional buying and selling transaction costs.

Decision Variables

For each node n of the scenario tree and each bond i we define

x_{in} nominal amount (in euros) *held* in bond i at node n;
x_{in}^+ *buying* decision (nominal amount) on bond i at node n;
x_{in}^- *selling* decision on bond i at node n.

In addition, we denote by g_n and z_n the nominal *debt* and the amount invested in *cash* at node n, respectively. All the decision variables are constrained to be non-negative.

4.3.2 Constraints

The portfolio composition is optimized under various restrictions. The model includes the classical inventory balance constraints on the nominal amount invested in each bond, for each node of the scenario tree:

$$x_{i0} = \bar{x}_i + x_{i0}^+ - x_{i0}^- \qquad \forall i \in I, \tag{4.15}$$

$$x_{in} = x_{ia_n} + x_{in}^+ - x_{in}^- \qquad \forall i \in I, \ \forall n \in \mathcal{N}. \tag{4.16}$$

The cash balance constraint is imposed for the first and later stages:

$$\sum_{i \in I} v_{i0} x_{i0}^+ (1 + \chi^+) + z_0 - g_0 = C_0 - \bar{g} + \sum_{i \in I} v_{i0} x_{i0}^- (1 - \chi^-), \tag{4.17}$$

$$\sum_{i \in I} v_{in} x_{in}^+ (1 + \chi^+) + z_n - g_n = \sum_{i \in I} v_{in} x_{in}^- (1 - \chi^-) + z_{a_n} e^{r_{a_n}} - g_{a_n} e^{b_{a_n}} + \sum_{i \in I} f_{in} x_{ia_n} \qquad \forall n \in \mathcal{N}. \tag{4.18}$$

The model also includes constraints bounding the amount invested in each rating class (4.19) as well as in investment grade (4.20) and speculative grade (4.21) classes, respectively, as fractions of the current portfolio value:

$$\sum_{i \in I_k} v_{in} x_{in} \leq \nu_k \sum_{i \in I} v_{in} x_{in} \qquad \forall k \in K, \quad \forall n \in \mathcal{N}, \tag{4.19}$$

$$\sum_{k=0}^{4} \sum_{i \in I_k} v_{in} x_{in} \leq \phi \sum_{i \in I} v_{in} x_{in} \qquad \forall n \in \mathcal{N}, \tag{4.20}$$

$$\sum_{k=5}^{7} \sum_{i \in I_k} v_{in} x_{in} \leq \zeta \sum_{i \in I} v_{in} x_{in} \qquad \forall n \in \mathcal{N}. \tag{4.21}$$

Finally, a limit on the debt level for each node of the scenario tree is imposed:

$$g_n \leq \gamma (C_0 + \textstyle\sum_{i \in I} v_{i0} \bar{x}_i - \bar{g}) \qquad \forall n \in \mathcal{N}. \tag{4.22}$$

4.3.3 Objective Function

The goal of financial planning is twofold: maximize the expected wealth generated by the investment strategy while controlling the market and credit risk exposure of the portfolio. This trade-off can be mathematically represented by adopting a *risk-reward* objective function:

$$\max (1 - \alpha) E[\mathcal{W}_n] - \alpha \theta_\epsilon [\mathcal{W}_n], \tag{4.23}$$

where α is a user-defined parameter accounting for the risk aversion attitude and n are the leaf nodes ($n \in \mathcal{N}_T$) of the scenario tree. The higher the α the more conservative, but also the less profitable, the suggested financial plan. The first term of (4.23) denotes the expected value of terminal wealth, computed as

$$E[\mathcal{W}_n] = \sum_{n \in \mathcal{N}_T} p_n \mathcal{W}_n, \tag{4.24}$$

where the wealth at each node n is

$$\mathcal{W}_n = \sum_{i \in I} v_{in} x_{in} + z_n - g_n. \tag{4.25}$$

The second term in (4.23) accounts for risk. In particular, we have considered the conditional value at risk (CVaR) at a given confidence level ϵ (usually 95%). CVaR measures the expected value of losses exceeding the value at risk (VaR). It is a "coherent" risk measure, suitable for asymmetric distributions and thus able to control the downside risk exposure. In addition, it enjoys nice computational properties (Andersson et al. 2001; Artzner et al. 1999; Rockafellar and Uryasev 2002) and admits a simple linear reformulation. In tree notation the CVaR of the portfolio terminal wealth can be defined as

$$\theta_\epsilon = \xi_\epsilon + \frac{1}{1-\epsilon} \sum_{n \in \mathcal{N}_T} p_n \left[L_n - \xi_\epsilon \right]_+ , \tag{4.26}$$

where ξ_ϵ denotes the VaR at the same confidence level. Here L_n represents the loss at node n, measured as the negative deviation from a given target value of the portfolio terminal wealth:

$$L_n = \max \left[0, \widetilde{\mathcal{W}}_n - \mathcal{W}_n \right], \tag{4.27}$$

where $\widetilde{\mathcal{W}}_n$ represents a reference value, computed on the initial wealth 1-year compounded value for given current (known) risk-free rate:

$$\mathcal{W}_0 = (C_0 + \textstyle\sum_{i \in I} v_{i0} \bar{x}_i - \bar{g}), \tag{4.28}$$

$$\widetilde{\mathcal{W}}_n = \mathcal{W}_0 e^{r_0 t_n} \qquad \forall n \neq 0. \tag{4.29}$$

The overall objective function can be linearized through a set of auxiliary variables (ς_n) and the constraints as follows:

$$\theta_\epsilon = \xi_\epsilon + \frac{1}{1-\epsilon} \sum_{n \in \mathcal{N}_T} p_n \varsigma_n, \tag{4.30}$$

$$\varsigma_n \geq L_n - \xi_\epsilon \qquad \forall n \in \mathcal{N}_T, \tag{4.31}$$

$$\varsigma_n \geq 0 \qquad \forall n \in \mathcal{N}_T. \tag{4.32}$$

This model specification leads to an easily solvable large-scale linear multi-stage stochastic programming problem. Decisions at any node explicitly depend on the corresponding postulated realization for the random variables and have a direct impact on the decisions at descendant nodes. Depending on the number of nodes in the scenario tree, the model can become very large calling for the use of specialized solution methods.

4.4 Scenario Generation

The bond portfolio model implementation requires the definition of an extended set of random coefficients (Dupačová et al. 2001; Heitsch and Römisch 2003; Pflug 2001). We focus in this section on the relationship between the statistical model actually implemented to develop the case study and the set of scenario-dependent coefficients included in the stochastic programming formulation.

The statistical model drives the bond returns over the planning horizon $0, \ldots, T$. At time 0 all coefficients in (4.2), (4.3), (4.4), (4.5), (4.7), and (4.8) have been estimated and, for a given initial portfolio, Monte Carlo simulation can be used to estimate the credit risk exposure of the current portfolio. In this application a simple random sampling algorithm for a fixed, pre-defined, tree structure is adopted to define the event tree structure underlying the stochastic programming formulation.

A method of moments (Campbell et al. 1997) estimation is first performed on (4.2), (4.3), and (4.4) from historical data, then Moody's statistics (Moody's Investors Service 2009) are used to estimate the natural default probability and the recovery rates, which are calibrated following (4.7), to account for recent market evidence and economic activity, and (4.8), allowing a limited dispersion from the average class-specific recovery rates.

In the case study implementation the risk-free rate and the credit spread processes are modeled as correlated square root processes according to the tree specification, for $s \in S, t = 1, \ldots, T, n \in \mathcal{N}_t$, where h_n denotes the child node in the given scenario s. For each $k \in K$ and initial states r_0 and π_0^k, we have

$$\begin{aligned} \Delta r_n &= \mu^r (t_{h_n} - t_n) + \sigma^r \sqrt{r_n} \sqrt{t_{h_n} - t_n} e_n, \\ \Delta \pi_n^k &= \mu^k (t_{h_n} - t_n) + \sigma^k \sqrt{\pi_n^k} \sqrt{t_{h_n} - t_n} \sum_{l \in K} q_n^{kl} e_n^l. \end{aligned} \tag{4.33}$$

In (4.33), $e_n^l \sim N(0, 1)$, for $l = 0, 1, .., 7$, independently and q^{kl} denote the Choleski coefficients in the lower triangular decomposition of the correlation matrix. The nodal realizations of the risk-free rate and the credit spreads are also used to identify the investor's borrowing rate $b_n = r_n + \pi_n^{\tilde{k}}$ in the dynamic model implementation, where $\tilde{k}$ denotes the investors specific rating class.

The incremental spread η_n^i for security i has been implemented in the case study as a pure jump-to-default process with null mean and volatility. The associated idlosyncratic tree processes, all independent from each other, will in this case for all i follow the dynamic

$$d\eta_n^i = \beta_n^i d\Psi^i(\lambda^i, n) \tag{4.34}$$

with constant default intensity and random market impact on the security-specific spread. For given initial security prices and any specified tree processes it is possible to derive the input coefficients to the portfolio problem.

The procedure can be sketched as follows:

Corporate Bonds Scenario Generation

given $v_0^i, y_0^k = r_0 + \pi_0^k, y_0^i = y_0^k + \eta_0^i, \forall i, k$
  for $t = 1, \ldots, T - 1$
    for $n \in \mathcal{N}_t$
      generate r_n, π_n^k correlated
      begin
        for $i = 1, \ldots, I$
          sample $d\Psi_n^i \in Poisson(\lambda^i)$
          if $d\Psi_n^i = 0, d\eta_n^i = 0, y_n^i = y_n^k$
            derive $dy_n^i, v_n^i, \delta_n^i, \gamma_n^i, f_{in}$
          elseif $d\Psi_n^i = 1$
            sample $\beta_n^i \in Ln(\beta^i, \sigma_{\beta^i})$
            generate $d\eta_n^i = \beta_n^i, dy_n^i = dy_n^k + d\eta_n^i$
            derive $v_n^i, \delta_n^i, \gamma_n^i, f_{in}$
          end if
        end for i
    end for n
  end for t

Following the above assumptions we can summarize the key steps supporting the scenario generation algorithm for the credit risk management problem. Inputs to the procedure are

- the investment universe prices at time 0 and the associated duration and convexity pairs;
- the value at time 0 of 1-year risk-free rate and credit spreads, one for every rating class $k \in K$ together with the set of statistical coefficients in (4.33);
- the time 0 security-specific marginal spreads and the shock distributions (jump magnitude distribution and intensity).

4.5 Computational Experiments

In this section, we present the results obtained by considering a 1-year planning horizon (starting from June 2, 2008) with monthly rebalancing. The universe of financial instruments contains a risk-free treasury security and 36 corporate Eurobonds spanning the Standard and Poor's rating classes. Details are reported in Table 4.1. In particular, for each bond $i \in I$, we report the ISIN code, the issuer rating class, the sector of activity and geographic area, the coupon rate, its frequency, and maturity.

The implementation relies on two main phases strictly interconnected: scenario generation, which has been performed by using MATLAB R2009B,[1] and the for-

[1] www.mathworks.com

Table 4.1 Bonds set for test case

ISIN	Guarantor	Sector	Country	Rating	CPN rate (%)	CPN frq	Maturity	Def. intensity
XS0140280644	Koninklijke Ahold NV	Food	The Netherlands	BBB	5.875	Annual	14-Mar-2012	0.0048
FR0010070805	Alcatel Lucent	Telecommunications	France	B	6.375	Annual	07-Apr-2014	0.134
DE000A0EUB86	BASF AG	Chemicals	Germany	AA	3.375	Annual	30-May-2012	0.0002
XS0225369403	Bayer AG	Chemicals	Germany	BBB	5.00	Annual	29-Jul-2015	0.0048
XS0162732951	BMW AG	Auto manufacturers	Germany	A	4.625	Annual	20-Feb-2013	0.0005
FR0000480691	Carrefour SA	Food	France	A	6.125	Annual	26-May-2010	0.0004
FR0010455626	Casino Guichard Perrachon ET CIE	Food	France	BBB	4.875	Annual	10-Apr-2014	0.0048
FR0010333385	Compagnie de Saint Gobain SA	Building materials	France	BBB	4.875	Annual	31-May-2016	0.0048
DE000A0JQA39	Deutsche Lufthansa AG	Airlines	Germany	BBB	4.625	Annual	06-May-2013	0.006
XS0273230572	Dexia SA	Diversified finan serv	Belgium	BB	4.892	Annual	02-Nov-2016	0.036
XS0171325557	Dow Chemical CO	Chemicals	USA	BBB	4.375	Annual	25-Jun-2010	0.006
XS0172308594	EDF Energy PLC	Electric	UK	A	4.375	Annual	15-Dec-2010	0.0003
XS0181582056	Edison SPA	Electric	Italy	BBB	5.125	Annual	10-Dec-2010	0.1
XS0253995368	FIAT SPA	Auto manufacturers	Italy	BB	5.625	Six monthly	15-Nov-2011	0.03
XS0215093534	Finmeccanica SPA	Aerospace/defense	Italy	BBB	4.875	Annual	24-Mar-2025	0.0048
FR0000471948	France Telecom SA	Telecommunications	France	A	7.25	Annual	28-Jan-2013	0.0004
FR0000472326	Gaz de France	Gas	France	A	4.25	Annual	19-Feb-2013	0.0004
XS0171942757	General Motors Corp	Auto manufacturers	USA	CCC−C	7.25	Annual	03-Jul-2013	0.453
DE0001135242	Germany (Govt)	Sovereign	Germany	AAA	4.25	Annual	04-Jan-2014	0

Table 4.1 (continued)

ISIN	Guarantor	Sector	Country	Rating	CPN rate (%)	CPN frq	Maturity	Def. intensity
XS0222377300	GlaxoSmithKline PLC	Pharmaceuticals	UK	A	3.00	Annual	18-Jun-2012	0.0005
XS0303396062	ING Groep NV	Insurance	The Netherlands	A	4.75	Annual	31-May-2017	0.0006
XS0284840542	JPMorgan Chase	Diversified finan serv	USA	A	4.375	Annual	30-Jan-2014	0.0008
XS0196776214	Koninklijke KPN NV	Telecommunications	The Netherlands	BBB	4.5	Annual	21-Jul-2011	0.006
XS0259604329	Linde Finance BV	Engin/construction	The Netherlands	BBB	7.375	Annual	14-Jul-2016	0.006
FR0000474223	LVMH Moet Hennessy Louis Vuitton SA	Holding companies-divers	France	BBB	5.00	Annual	29-Apr-2010	0.0072
XS0282586311	Morgan Stanley	Diversified finan serv	USA	A	4.375	Annual	10-Feb-2012	0.0008
XS0173605311	Petroleos Mexicanos	Oil-gas	Mexico	BBB	6.25	Annual	05-Aug-2013	0.006
FR0010068486	PPR SA	Retail	France	BBB	5.25	Annual	29-Mar-2011	0.0048
XS0202649934	Repsol YPF	Oil-gas	Spain	BBB	4.625	Annual	08-Oct-2014	0.0048
XS0180277393	Reuters Finance PLC	Media	UK	BBB	4.625	Annual	19-Nov-2010	0.006
XS0188009004	Rolls Royce Group PLC	Aerospace/defense	UK	BBB	4.5	Annual	19-Mar-2011	0.006
XS0162513211	RWE AG	Electric	Germany	A	5.75	Annual	14-Feb-2033	0.0003
XS0131224155	Siemens AG	Manufactor	Germany	A	5.75	Annual	04-Jul-2011	0.0004
XS0096330732	Teliasonera Finland OYJ	Telecommunications	Finland	A	4.625	Annual	16-Apr-2009	0.0004
XS0188733207	Thyssenkroup AG	Iron/steel	Germany	BBB	5.00	Annual	29-Mar-2011	0.0048
FR0000474975	Veolia Environment	Water	France	BBB	4.875	Annual	28-May-2013	0.0072

mulation and solution of the stochastic programming problem, which has been performed by GAMS 23.4[2] with CPLEX 12[3] as its solution kernel. All the computational experiments have been carried out on a DELL workstation with an INTEL XEON quadcore 2 GHz processor and 8 GB RAM. The final aim of the extensive testing phase has been to investigate the effectiveness of the stochastic programming formulation as a decision tool for simultaneously controlling market and credit risk, in comparison with other strategies.

In the case study we have considered a maximum debt level $\gamma = 20\%$, a maximum investment grade fraction $\phi = 70\%$, a maximum speculative grade fraction $\zeta = 70\%$, and a maximum fraction of investment in each rating class $\nu_k = 30\%, k \in K$. Transaction costs for both buying and selling are assumed constant and equal to 0.5% of the bond value. The initial holdings consist of a portfolio with an equal allocation in each bond, with no initial cash or debt.

4.5.1 Scenario Analysis

Scenario generation has been performed by implementing the procedures described in Section 4.4. The following tables specify the parameters adopted in the simulation process. In particular, Table 4.2 reports the values of the correlation matrix among the risk-free rate and the spread for each rating class, together with the adopted default intensities.

Further, in Table 4.3, we report the weights used to compute the default intensity of each bond (last column of Table 4.1) as a function of the guarantor economic sector and geographic region, with ξ^i in (4.7) determined by multiplying the specified coefficients for each sector–region pair.

Figure 4.2 shows the scenario trees generated for a low-risk (AA) and high-risk (CCC−C) bond, respectively. Unlike the AA bond, the CCC−C security is affected by a default event on several scenarios.

Scenario generation represents a critical issue in any stochastic programming formulation since the quality of the generated scenarios heavily affects the quality of

Table 4.2 Correlation matrix and Moody's rating homogeneous default intensities

	RF	AAA	AA	A	BBB	BB	B	CCC−C	Default intensity
RF	1	0.0005	0.3044	0.2865	−0.0022	−0.0937	−0.16	−0.153	0
AAA		1	0.903	0.904	0.923	0.835	0.919	0.896	0
AA			1	0.991	0.899	0.833	0.845	0.83	0.0002
A				1	0.915	0.829	0.853	0.844	0.0004
BBB					1	0.863	0.936	0.945	0.0048
BB						1	0.936	0.896	0.0243
B							1	0.977	0.134
CCC−C								1	0.29

[2] www.gams.com
[3] http://www-01.ibm.com/software/integration/optimization/cplex-optimizer/

Table 4.3 Heuristic default intensity weights for economic sector and geographic region

Sector	Weight	Region	Weight
Automotive	1.25	Europe	1
Telecommunications	1	UK	1.25
Utility	0.75	USA	1.25
Manufacture	1	Emerging markets	1.25
Financial	1.5		

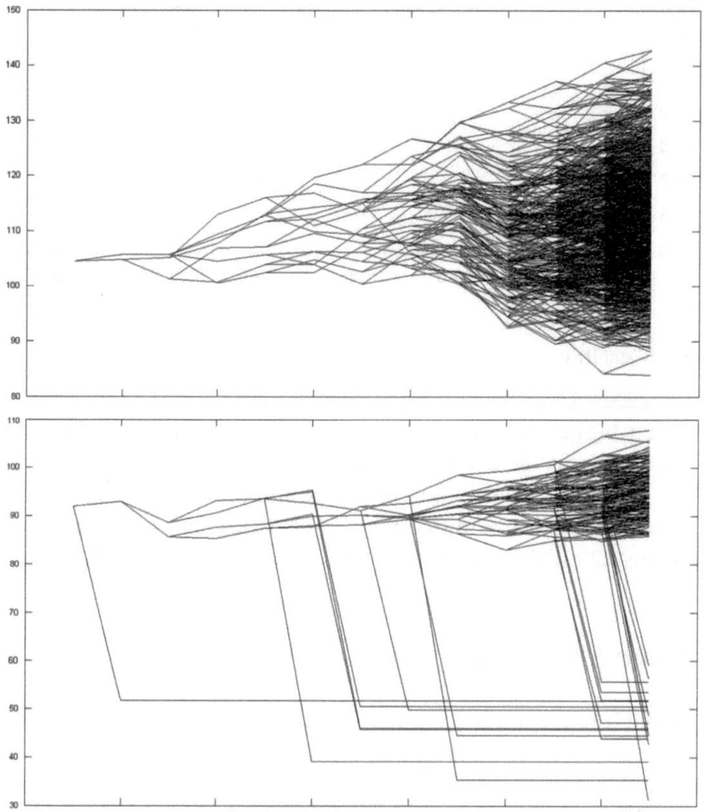

Fig. 4.2 AA and CCC−C bond scenario tree

the recommendations provided by the model. With the aim of assessing the robustness of the stochastic programming approach, additional tests have been performed to measure the so-called in-sample stability. This is evaluated by generating several scenario trees and solving the corresponding stochastic programming problem for each tree.

Figure 4.3 reports, for an increasing set of scenarios, the values of the expected final wealth and CVaR generated by each problem solution. The center of the circle represents the average value, whereas the radius measures the standard deviation.

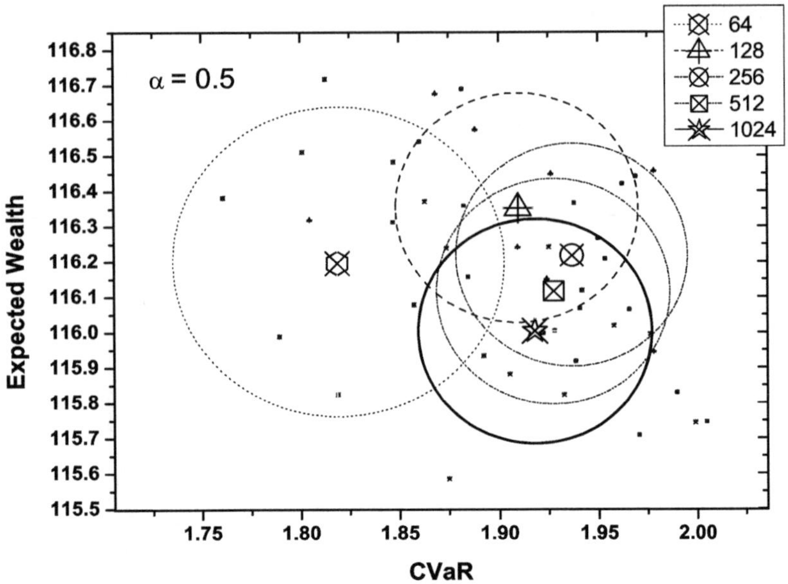

Fig. 4.3 Solutions stability

As expected the bigger the tree, the lower the radius and thus the solution pair dispersion. The analysis clearly shows that robust solutions may be obtained with scenario trees of limited size. Following these preliminary results, we consider in our experiments a 1024 scenario tree with 2071 nodes. The resulting mathematical problem contains 227,810 variables and 97,374 constraints.

4.5.2 Portfolio Strategies

We analyze the optimization output for different levels of the risk aversion parameter α. Figure 4.4 shows the trade-off between the expected terminal wealth and the CVaR risk measure as α increases from 0.01 to 0.99. Each point on the frontier is generated by a stochastic program solution with an underlying 1024 scenario tree over the 12 stages.

In Table 4.4 we show how the portfolio composition determined at the first stage (referred in the following as implementable or here-and-now solution) changes for different values of the risk aversion parameter.

As α increases, the model solution reflects the agent's increasing risk aversion. For $\alpha = 0.99$ in particular the fraction invested in investment grade securities is 60%, whereas for $\alpha = 0.1$, speculative grade securities span 46% of the optimal portfolio value.

Figure 4.5 reports the wealth evolution along specific scenarios prior and after optimization to clarify the theoretical impact of the optimal multistage strategy on the investor wealth evolution. For the given statistical model, the *do nothing* against

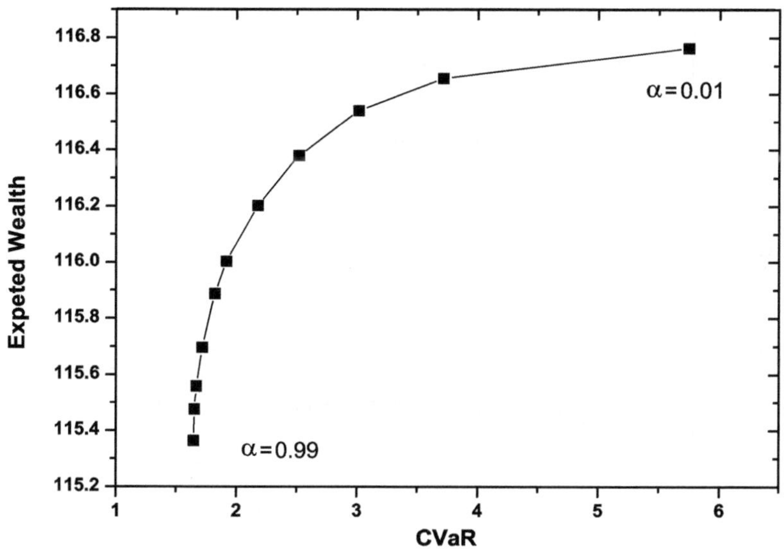

Fig. 4.4 Risk-return trade-off

Table 4.4 Implementable solution for different α values

Rating class	$\alpha = 0.99$	$\alpha = 0.5$	$\alpha = 0.1$
AAA	20%	22%	0%
AA	0%	0%	0%
A	20%	24%	20%
BBB	20%	18%	34%
BB	0%	18%	29%
B	2%	0%	0%
CCC−C	4%	0%	17%
Cash	34%	18%	0%

the *dynamic strategy* outputs are compared in order to clarify the financial improvement expected to be brought about by the dynamic strategy.

Figures 4.3, 4.4, 4.5, and 4.6 and Table 4.4 present a set of results associated with multistage stochastic program defined over a 1-year planning horizon starting on June 2, 2008. Next we consider the results of adopting a 12-month rolling window with monthly forward updates of the scenario tree and the associated stochastic coefficients definition to analyze the risk control performance induced by a sequence of H&N solutions over the most severe period of the crisis. An average risk-averse investor with $\alpha = 0.5$ was considered. Figure 4.7 shows how the optimal H&N portfolio composition changed during the test period. Without imposing specific constraints on the portfolio policy, highly volatile financial markets lead to substantial portfolio revisions from one stage to the other.

Figure 4.8, finally, shows the out-of-sample market performance of the optimal H&N strategies evaluated on realized historical prices. For comparative purposes, the results have been computed for different levels of risk aversion α on a

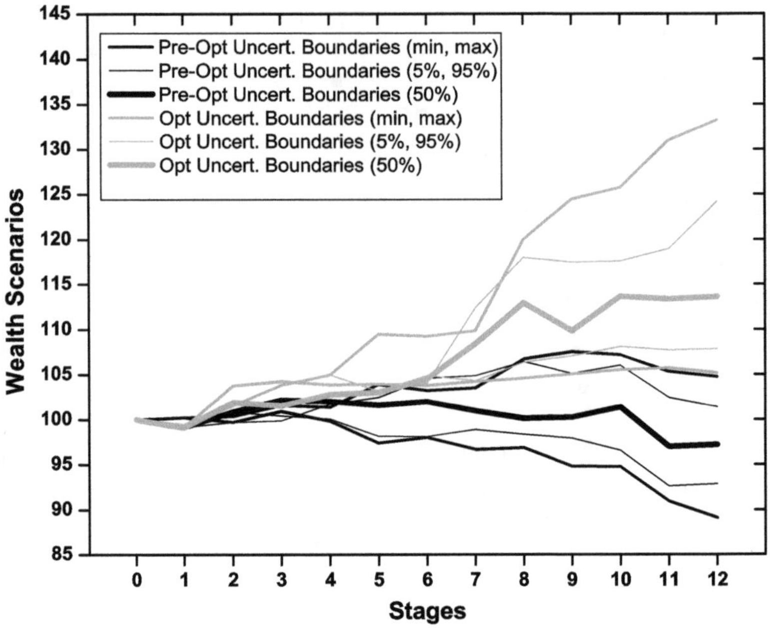

Fig. 4.5 Wealth evolution along the scenario tree

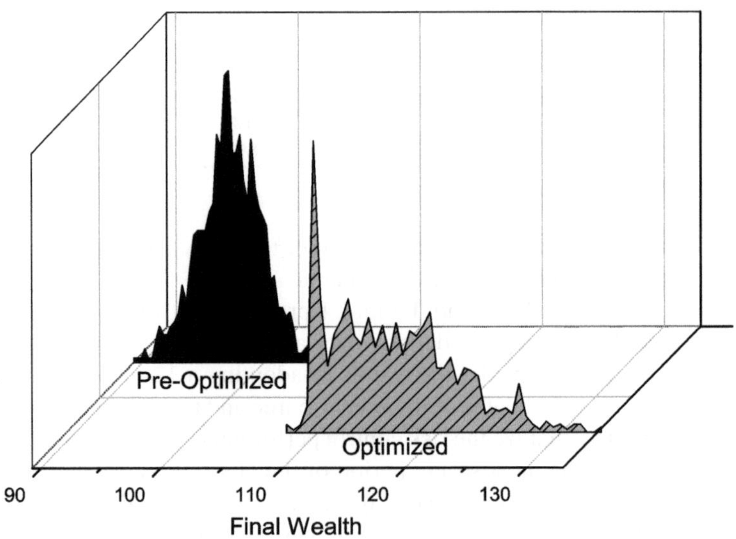

Fig. 4.6 Distribution of the final wealth

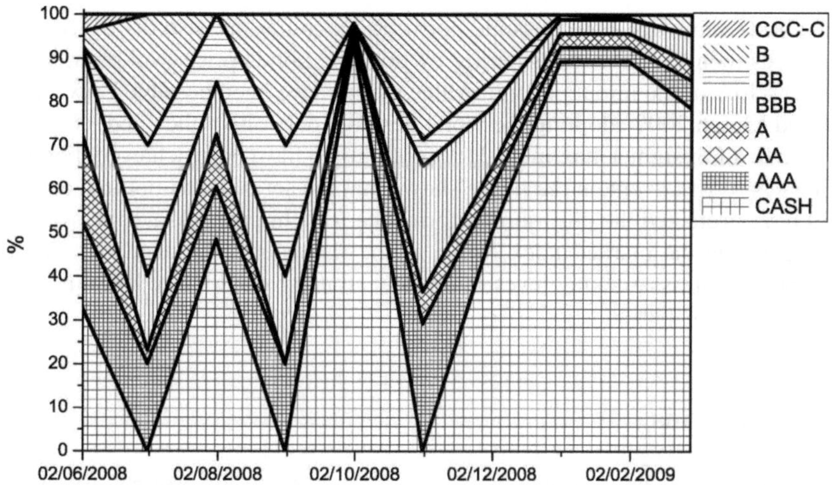

Fig. 4.7 Portfolio evolution

normalized portfolio value of 100 at the beginning of June 2008. In the same Fig. 4.8 we report, for relative valuation purposes, the performance of the initial, perfectly diversified, portfolio and the IBRCPAL corporate bond index, a benchmark for corporate bond traders in the Euro market.[4]

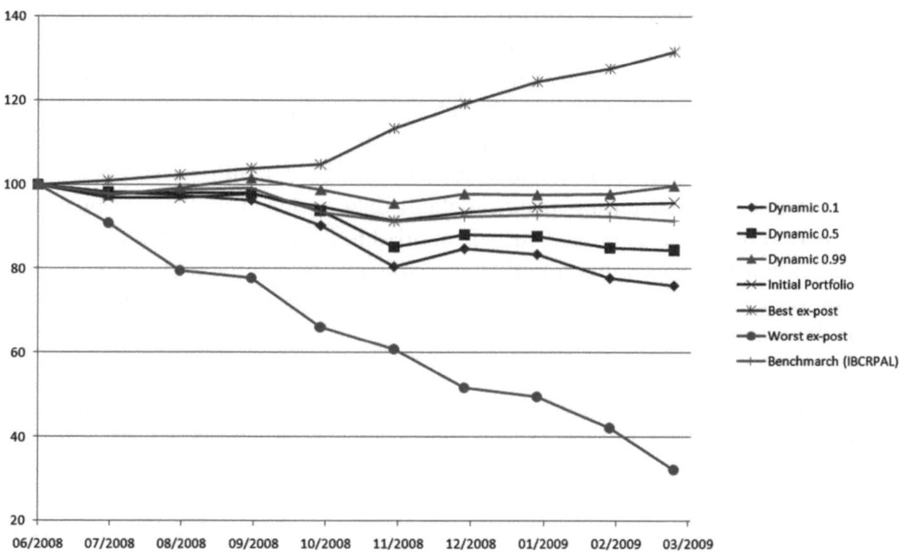

Fig. 4.8 Backtesting analysis

[4] http://kbc-pdf.kbc.be/ng/feed/am/funds/star/fa/BE0168961846.pdf

The best- and worst-case possible portfolio outcomes are also displayed for the given investment universe considering ex post the bonds price evolution: a maximum 70% loss of the initial portfolio value would have been possible between June 2008 and March 2009 adopting the worst possible portfolio policy and, conversely, under perfect foresight a maximum 31% profit might have been achieved. For different degrees of risk aversion the figure shows that the set of H&N optimal decisions would have protected the portfolio performance for a highly risk-averse investor by overperforming the market index, while as the risk coefficient decreases the back test shows that a risk-seeking investor optimal strategy would have led to roughly a 20% loss over the period.

4.6 Conclusions

The chapter focuses on the potential of a multistage stochastic program formulation of a bond portfolio management problem including securities exposed to market and credit risk factors. The recent credit crisis has lead to unprecedented portfolio losses in the market of corporate liabilities and represents an ideal stress period for advanced modeling paradigms at a time in which not only the market suffered a severe crisis but also the credit derivative market became very illiquid and made classical hedging impractical.

The formulation of an optimal dynamic strategy in the corporate bond market has been evaluated over the June 2008 to March 2009 period relying on a risk model in which the specific credit risk carried by individual securities has been separated from the rating-based credit risk factors affecting all securities within a rating class. No transition risk has been considered in the analysis and the bond risk exposure has been modeled through a jump-to-default model calibrated on available Moody's default statistics (Moody's Investors Service 2009). Consistent with industry standards, the final assessment of the individual borrower's default risk has been based on available sector and rating-specific default data enriched with credit analysts' views, here reflected in a simple heuristic applied to the available default estimates.

The evidence suggests that a multistage strategy for a sufficiently large investment universe would have protected the investors from market and credit losses during the worst part of the crisis.

Acknowledgments The authors acknowledge the support given by the research grant PRIN2007 "Optimization of stochastic dynamic systems with applications to finance," no. 20073BZ5A5, sci.resp. G. Consigli.

References

Abaffy J., M. Bertocchi, J. Dupačová, V. Moriggia and G. Consigli (2007): Pricing non diversifiable credit risk in the corporate Eurobond market, *Journal of Banking and Finance* **31**, 2233–2263.

Andersson F., H. Mausser, D. Rosen and S. Uryasev (2001): Credit risk optimization with conditional value-at-risk criterion, *Mathematical Programming Series B* **89**, 273–291.

Artzner P., F. Delbaen, J.-M. Eber and D. Heath (1999): Coherent measures of risk, *Mathematical Finance* **9**(3), 203–228.

Berndt A. (2004): Estimating the term structure of yield spreads from callable corporate bond price data. Working paper, Carnegie Mellon University.

Berndt A., R. Douglas, D. Duffie, M. Ferguson and D.Schranz (2005): Measuring Default Risk Premia from Default Swap Rates and EDFs. Working paper, Stanford University.

Bertocchi M., V. Moriggia and J. Dupačová (2007): Bond Portfolio Management via Stochastic Programming, in S.A. Zenios and W.T. Ziemba Eds, *Handbook of Asset and Liability Management* Vol 1, 305–336. North-Holland, Amsterdam.

Bradley, S.P. and D.B. Crane (1972): A dynamic model for bond portfolio management, *Management Science* **19**, 139–151.

Campbell J.Y., W.A. Lo and A.C. MacKinlay (1997): *The Econometrics of Financial Markets*, Princeton University Press Princeton, NJ.

Cox J. C., J. E. Ingersoll and S.A. Ross (1985): A theory of the term structure of interest rates, *Econometrica* **53**, 385–407.

Consigli G. (2004): Credit default swaps and equity volatility: theoretical modeling and market evidence. Workshop on Portfolio optimisation and option pricing, Dept of Mathematics University Ca'Foscari of Venice, http://www.dma.unive.it/, 45–70.

Consigli G. and M.A.H. Dempster (1998): Dynamic stochastic programming for asset-liability management, *Annals of Operations Research* **81**, 131–161.

Consigli, G., G. Iaquinta and V. Moriggia (2010): Path-dependent scenario trees for multistage stochastic programs in finance, *Quantitative Finance*. DOI:10.1080/14697688.2010.518154.

Crouhy M., D. Galai and R. Mark (2000): A comparative analysis of current credit risk models, *Journal Banking & Finance* **24**, 59–117.

Dai Q. and K. J. Singleton (2003): Term structure dynamics in theory and reality, *Review of Financial Studies* **16**, 631–678.

Das S. and P. Tufano (1996): Pricing credit-sensistive debt when interest rates, credit ratings and credit spreads are stochastic, *Journal of Financial Engineering* **5**, 161–198.

Dempster M.A.H., M. Germano, E.A. Medova and M. Villaverde (2003): Global asset liability management, *British Actuarial Journal* **9**(1), 137–195.

Duffie D. and K. Singleton (1999): Modeling term structures of defaultable bonds, *The Review of Financial Studies* **12**, 687–720.

Dupačová J., G. Consigli and S.W. Wallace (2001): Scenarios for multistage stochastic programmes, *Annals of Operations Research* **100**, 25–53.

Elton, E., M. Gruber, D. Agrawal, and C. Mann (2001): Explaining the rate spread on corporate bonds, *Journal of Finance* **56**, 247–277.

European Central Bank (2010): ECB Research Bulletting No.9, March 2010, http://www.ecb.int/pub/pdf/other/researchbulletin09en.pdf.

Heitsch H. and W. Römisch (2003): Scenario reduction algorithms in stochastic programming, *Computational Optimization and Applications* **24**, 187–206.

Hochreiter R. and G.Ch. Pflug (2007): Financial scenario generation for stochastic multi-stage decision processes as facility location problems, *Annals of Operations Research* **152**(1), 252–272.

Jarrow R. and S. Turnbull (2000): The intersection of market and credit risk, *Journal of Banking and Finance* **24**, 271–299.

Jarrow R, D. Lando and S. Turnbull (1997): A Markov model for the term structure of credit risk spreads, *Review of Financial Studies* **10**, 481–523.

Jobst N.J., G. Mitra and S.A. Zenios (2006): Integrating market and credit risk: A simulation and optimisation perspective, *Journal of Banking & Finance,* **30**(2), 717–742.

Jobst N.J. and S.A. Zenios (2005): On the simulation of portfolios of interest rate and credit risk sensitive securities, *EJOR* **161**(2), 298–324.

Kijima J. and Y. Muromachi (2000): Evaluation of credit risk of a portfolio with stocastic interest rate and default processes, *Journal of Risk* **3**, 5–30.

Longstaff F.A., S. Mithal and E. Neis (2005): Corporate yields spreads: Default risk or liquidity? New evidence from the credit-default swap market, *Journal of Finance* **60**, 2213–2253.

Moody's Investors Service (2009): 1920–2008 Report on Corporate defaults and recovery rates. From www.moodys.com, 1–55.

Pflug G.Ch. (2001): Optimal scenario tree generation for multiperiod financial planning, *Mathematical Programming, Series B* **89**, 251–271.

Pflug G.Ch. and W. Römisch (2007): *Modeling, Measuring and Managing Risk,* World Scientific Publisher.

Rockafellar R.T. and S. Uryasev (2002): Conditional value at risk for general loss distributions, *Journal of Banking and Finance* **26**, 1443–1471.

Zenios S.A. and W.T. Ziemba Eds (2007): *Handbook of Asset and Liability Management: Applications and Case Studies*, volume 2, North-Holland Handbooks Finance Series, Amsterdam.

Chapter 5
Dynamic Portfolio Management for Property and Casualty Insurance

Giorgio Consigli, Massimo di Tria, Michele Gaffo, Gaetano Iaquinta, Vittorio Moriggia, and Angelo Uristani

Abstract Recent trends in the insurance sector have highlighted the expansion of large insurance firms into asset management. In addition to their historical liability risk exposure associated with statutory activity, the growth of investment management divisions has caused increasing exposure to financial market fluctuations. This has led to stricter risk management requirements as reported in the Solvency II 2010 impact studies by the European Commission. The phenomenon has far-reaching implications for the definition of optimal asset–liability management (ALM) strategies at the aggregate level and for capital required by insurance companies. In this chapter we present an ALM model which combines in a dynamic framework an optimal strategic asset allocation problem for a large insurer and property and casualty (P&C) business constraints and tests it in a real-world case study. The problem is formulated as a multistage stochastic program (MSP) and the definition of the underlying uncertainty model, including financial as well as insurance risk factors, anticipates the model's application under stressed liability scenarios. The benefits of a dynamic formulation and the opportunities arising from an integrated approach to investment and P&C insurance management are highlighted in this chapter.

Keywords Property and casualty insurance · Asset–liability management · Multistage stochastic programming · Insurance liabilities

5.1 Introduction

Increasing competition in the insurance sector in developed countries, and more recently record property and casualty (P&C) insurance claims reported by global players (CEA, 2010; Europe Economics, 2010), has generated remarkable

---

Disclaimer The views expressed in this chapter do not necessarily reflect Allianz policy nor do the results presented here represent actual figures and operational approaches currently adopted by the Allianz Group.

G. Consigli (✉)
University of Bergamo, Department of Mathematics, Statistics, Computer Science and Applications, Via dei Caniana, 2, 24127 Bergamo, Italy
e-mail: giorgio.consigli@unibg.it

M. Bertocchi et al. (eds.), *Stochastic Optimization Methods in Finance and Energy*, International Series in Operations Research & Management Science 163,
DOI 10.1007/978-1-4419-9586-5_5,

pressures on the financial stability of P&C divisions within insurance firms, leading to increased technical reserves and requiring a higher capital base (European Parliament, 2009). At the same time we have witnessed a remarkable expansion of investment management divisions, reinforcing the role of insurers as institutional investors competing in fixed income and equity markets with other players such as pension and mutual funds. Increasing market volatility in the last few years has, as a result, affected large insurers' market risk exposure. Regulatory bodies, through the Solvency II regulation (European Parliament, 2009; ISVAP, 2010), have supported risk-based capital allocation measures for insurance firms as a whole. As a response large insurance companies have pursued restructuring aimed from an operational perspective at integrating the historical insurance business with the investment management business. Such an integration is also motivated by the perceived role of the P&C division as a liquidity buffer for the cash deficits generated by fixed income portfolios typically held by risk-averse investment divisions. A trade-off between safe liquidity conditions, key to shareholders' short–medium-term returns, and long-term business sustainability has emerged and lead to strategies based on long-term horizons. This motivates the adoption of a multistage stochastic programming (MSP) problem formulation (Birge and Louveaux, 1997; Cariño et al., 1994; Consigli and Dempster, 1998; Mulvey and Erkan, 2005; Zenios and Ziemba, 2007a) able to capture both short- and long-term goals. Contrary to current standards in insurance-based investment divisions which largely rely on one-period static approaches (de Lange et al., 2003; Mulvey and Erkan, 2003; Zenios and Ziemba, 2007b), the adoption of dynamic approaches allows both the extension of the decision horizon and a more accurate short-term modelling of P&C variables.

In this chapter we present an asset–liability management (ALM) problem integrating the definition of an optimal asset allocation policy over a 10-year planning horizon with the inclusion of liability constraints generated by an ongoing P&C business (de Lange et al., 2003; Dempster et al., 2003; Mulvey and Erkan, 2005; Mulvey et al., 2007). Relying on a simplified P&C income statement we clarify the interaction between the investment and the classical insurance business and introduce an objective function capturing short-, medium-, and long-term goals within a multistage model. The planning horizon is divided in six time periods: four quarters in the first year and then 2 and 7 years to reach the 10-year horizon. Over this period alternative insurance and financial scenarios will affect the insurance management optimal forward policy. We show that integrated management of the insurance liability and asset portfolios is required to protect firm solvency in the presence of unexpectedly increased P&C claims. A relatively simple multivariate Gaussian return model is adopted to generate return scenarios (Dupačová et al., 2001) for a realistic investment universe including fixed income, equity, and real estate investment opportunities.

Multistage stochastic programming techniques have only occasionally been applied to ALM problems for the insurance sector since the seminal work (Cariño et al., 1994; Zenios and Ziemba, 2007b) for the Japanese Yasuda-Kasai Insurance company. Notable examples specifically addressing strategic management problems for P&C insurance through stochastic programming approaches are given by Gaivoronski et al. (2001), de Lange et al. (2003), Mulvey and Erkan (2003, 2005), and Mulvey et al. (2007) for the Norwegian and US markets in particular. Far more extensive is the related literature on institutional ALM problems that have similar features to the analysis conducted in this chapter. The development of a real-world application and the relevant modelling issues induced by the problem motivate this contribution. As for many other global players in the market, the insurance company for which the model has been developed has an historical record of liability-driven strategic asset allocation largely relying on sophisticated statistical models and simulation tools as well as one-period optimization techniques.

The chapter includes in Section 5.2 a description of the key P&C income variables and clarifies the interaction between short-, medium-, and long-term objectives relevant to insurance management. In Section 5.3 we formulate the asset–liability management model and define the data processes underlying the problem. In Section 5.4 we present a real-world application (with modified numbers for confidentiality reasons) and summarize the collected evidence to date. Section 5.5 concludes.

5.2 P&C Income Generation and Asset Allocation

We consider an insurance company holding a property and casualty liability portfolio with annual issuance of insurance contracts and a diversified asset portfolio spanning fixed income, real estate, and equity markets. The insurance management sets a target for the end-of-the-year operating profit on an annual basis and seeks an optimal trade-off between this short-term goal, medium-term industrial plan targets, and a (10-year) long-term sustainability goal. The company's financial soundness will depend on the solidity of both the P&C business division, hereafter also referred to as the *technical* division, and the *investment* division. In a simplified framework, with stable insurance provisions and reserves, we assume that technical profitability primarily depends on collected annual premiums, operating and human resource costs, and, crucially, recorded insurance claims. We assume throughout that the company will fix insurance premiums so as to maintain its market position within a highly competitive market and that the operational efficiency with minimal operating, staff and administrative costs is maintained over time *exogenously* to the optimal financial management problem. In this setting alternative insurance claim scenarios are likely from 1 year to the next to heavily affect the company technical profitability. Increasing technical provisions may, on the other hand, weaken the capital base. This is indeed the risk faced currently by several global players in the insurance sector (Europe Economics, 2010), with record claims in the automotive and increasingly the real estate sector, and regarding catastrophic events (*CAT* risk events). Under such conditions the management will search for an asset portfolio strategy able to preserve the firm's liquidity in the short term and its overall sustainability in the long term. Only the first goal can be accommodated within

Table 5.1 Simplified technical and investment income decomposition for a P&C firm

Income statement
Insurance premiums
– p&c claims
– staff
– administration and other costs
= technical income A
Realised capital gains
+ equity dividends
+ margin of interest
= investment income B
A + B = profit before taxes
– taxes
= net profit

a static 1-year decision problem. The investment profit depends on realized price gains from trading, dividends, and interest cash flows generated by the asset portfolio; see Table 5.1. Risky investment strategies may very well generate sufficient liquidity and financial profit over 1 year but then lead to heavy long-term losses, jeopardizing the company's market position and solvency. Revenues and costs, as shown in Table 5.1, will affect annually the firm's profitability and are considered in the ALM model formulation.

The P&C annual operating profit does not consider the portfolio's gain and losses, which, if unrealized, reflect the asset portfolio's revaluation over time (and can translate into current realized gains or losses if required). If actually realized portfolio profits and losses are nevertheless considered outside the core insurance activity. Net of future liabilities, the maximization of the investment portfolio expected value, can thus be considered a medium-term goal to be pursued by the management. Realized investment and technical profits, together with unrealized gains, will eventually jointly determine the business long-term sustainability reflected in the 10-year target. We assume a 10-year decision horizon with the first year divided into four quarters, a subsequent 2-year stage and a final 7-year stage as shown in Fig. 5.1.

Decisions are allowed at the beginning of each stage respecting the nonanticipativity, i.e. historical dependence only, of the decision process. At the 10-year horizon

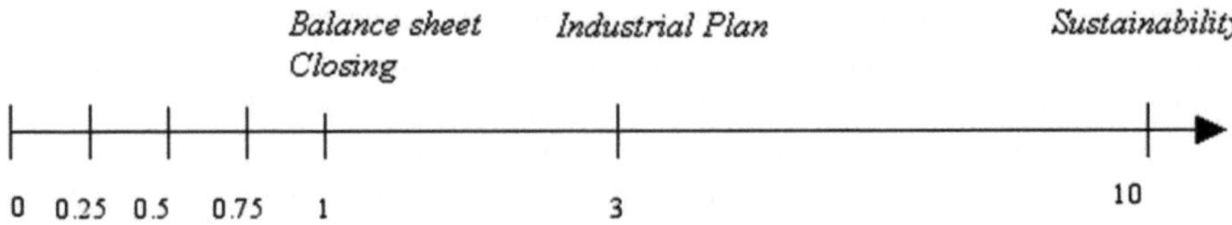

Fig. 5.1 Decision stages over a 10-year horizon

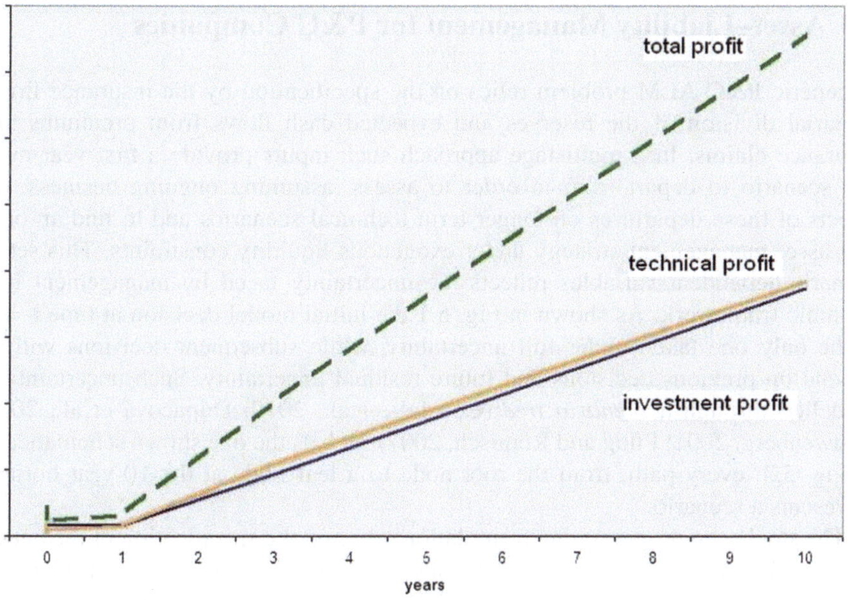

Fig. 5.2 Balanced technical and investment profit generation

the result of the adopted strategy can only be observed, with no further decisions made. The objective function of the MSP problem presented in Section 5.3 reflects the dynamic nature of management's goals over an extended decision horizon. The optimal investment strategy will depend directly on future P&C scenarios and the asset portfolio risk exposure. With balanced technical and investment profit generation, under normal operating conditions, the pattern of Fig. 5.2 over a 10-year horizon results.

We may assume in what follows that Fig. 5.2 displays an average scenario, which ceteris paribus will support the relatively independent management of the technical and the investment divisions. In general, however, years of profitable technical activity may come with negative financial market scenarios and vice versa. Investment strategies will adapt to such scenarios. Consistent with the adopted MSP formulation with recourse, a discrete tree process is adopted for return and liability scenarios affecting the insurance management optimal policy. Uncertainty affects year-by-year premium collection by the firm and associated insurance claims, whose evolution will also depend on current inflation, while on the asset side price and cash returns will depend on financial market behaviour. The results presented in this chapter will clarify the advantages of integrated asset–liability management, in particular under stringent liability scenarios, and the potential of a multistage stochastic programming formulation to capture the fundamental trade-offs affecting the insurance firm's optimal strategy, unlike a one-period static formulation.

5.3 Asset–Liability Management for P&C Companies

A generic P&C ALM problem relies on the specification by the insurance firm's actuarial division of the reserves and expected cash flows from premiums and insurance claims. In a multistage approach such inputs provide a first-year average scenario to depart from in order to assess, assuming ongoing business, the effects of these departures on longer term technical scenarios and to find an optimal asset management strategy under exogenous liquidity constraints. This set of scenario-dependent variables reflects the uncertainty faced by management in a dynamic framework. As shown in Fig. 5.1 the initial model decision at time $t = 0$ is the only one taken under full uncertainty, while subsequent decisions will all depend on previous decisions and future residual uncertainty. Such uncertainty is modelled through a *scenario tree* (Consigli et al., 2010; Dupačová et al., 2001; Kouwenberg, 2001; Pflug and Römisch, 2007) such as the one shown schematically in Fig. 5.3: every path, from the root node to a leaf node at the 10-year horizon represents a scenario.

The stochastic programming formulation relies on the specification of the underlying random factors as *tree processes* endowed with a given branching structure. In the formulation of the ALM problem we denote by $R(t, s)$ the insurance premiums collected in stage t under scenario s. Premiums represent the fundamental cash flows generated every year by the technical business. For given P&C contract renewals, the insurance actuarial division will periodically revise its estimate on forthcoming claims $L(t, s)$ and required statutory and discretional reserves $\Lambda(t, s)$. These quantities for given human resources and administrative costs $C(t, s)$ will determine the behaviour of the company's *combined ratio*: this is the ratio of all insurance

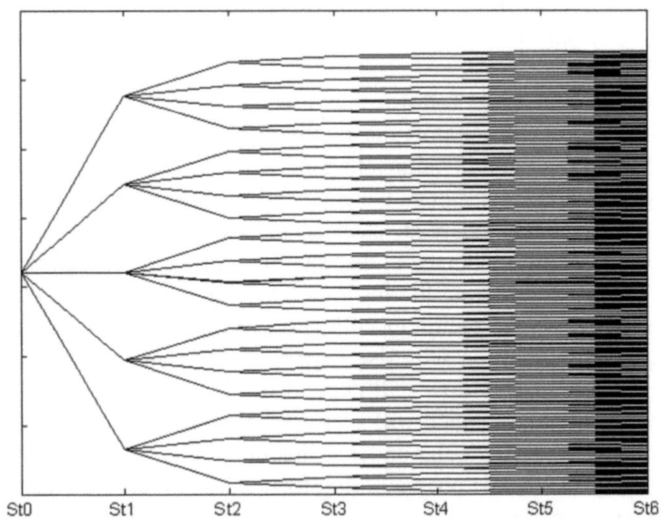

Fig. 5.3 A 5.4.2.2.2.2 scenario tree schema

claims and technical costs to the earned premiums. A bad liability scenario will thus be associated with increasing insurance claims at a time of reducing premium renewals and operational inefficiency. Consequently the ex post combined ratio will increase and may go over the 100% threshold. The firm's investment profit, instead, is generated by realized capital gains $G(t, s)$ in stage t, scenario s; the interest margin, defined as the difference between positive and negative interest cash flows, $M(t, s) = I^+(t, s) - I^-(t, s)$; and by dividends $D(t, s)$ and other financial revenues $Y(t, s)$.

The annual income $\Pi(t, s)$ is determined by the sum of technical and investment profit. The net income, here assumed entirely distributed, is derived by subtracting the corporate tax $\mathcal{T}$ from the profit $D^-(t, s) = \Pi(t, s)(1 - \mathcal{T})$.

5.3.1 The ALM Model

The ALM problem is formulated as a linear MSP recourse problem (Birge and Louveaux, 1997; Consigli 2007; Consigli and Dempster, 1998) over six stages. The optimal root node decision is taken at time $t_0 = 0$ and, from a practical viewpoint, represents the key *implementable* decision. Recourse decisions occur at $t_1 = 0.25$ (after a quarter), $t_2 = 0.5$, $t_3 = 0.75$, $t_4 = 1$ year, and $t_5 = 3$ years. The objective function includes a first-year profit shortfall with respect to a target (at t_4), a 3-year expected excess portfolio value relative to the insurance reserves (at t_5), and a 10-year wealth objective (at $t_6 = H$):

$$\max_{x \in X} \left\{ -\lambda_1 E\left[\widetilde{\Pi}_{t_4} - \Pi_{t_4} | \Pi_{t_4} < \widetilde{\Pi}_{t_4}\right] + \lambda_2 E\left(X_{t_5} - \Lambda_{t_5}\right) + \lambda_3 E\left[W_H\right]\right\}, \quad (5.1)$$

with $\lambda_1 + \lambda_2 + \lambda_3 = 1, 0 \leq \lambda_1, \lambda_2, \lambda_3 \leq 1$.

The first-year horizon reflects the short-term objective to be achieved at the end of the current year: a target net profit is included and for given liabilities and random cash flows over the initial four quarters the model will tend to minimize the expected shortfall with respect to the target (Artzner et al., 1999; Rockafellar and Uryasev, 2002). The intermediate 3-year objective reflects the maximization of the investment portfolio value above the P&C liabilities. The 10-year expected terminal wealth objective, finally, reflects management's need to maximize on average the long-term realized and unrealized profits.

The prospective optimal strategy is determined as a set of scenario-dependent decisions to maximize this objective function subject to several constraints. Decision variables include holding, selling, and buying indices in a representative strategic investment universe. We distinguish the set $i \in I_1$, including all interest-bearing assets with a specified maturity, from $i \in I_2$, which includes real estate and equity assets without an expiry date. The asset universe is $I = I_1 \cup I_2$:

$x^+(i,t,s)$ *investment* in stage t, scenario s, of asset i (with maturity T_i for $i \in I_1$);
$x^-(i,h,t,s)$ *selling* in stage t, scenario s, of asset i (with maturity T_i for $i \in I_1$) that was bought in h;
$x^{\exp}(i,h,t,s)$ *maturity* in stage t, scenario s, of asset $i \in I_1$ that was bought in h;
$x(i,h,t,s)$ *holding* in stage t, scenario s, of asset i (with maturity T_i for $i \in I_1$) that was bought in h;
$z(t,s) = z^+(t,s) - z^-(t,s)$ *cash account* in stage t, scenario s.

The investment epochs $h = t_0, t_1, \ldots, t_{n-1}$ are also introduced in order to estimate the capital gains on specific investments. An extended set of linear constraints will determine the feasibility region of the SP problem (Birge and Louveaux, 1997; Consigli and Dempster, 1998).

5.3.1.1 Inventory Balance Constraints

The inventory balance constraints affect the time evolution of the asset portfolio. Given an initial input portfolio the constraints take care of the stage evolution of the optimal strategy. Starting at time 0 we model the market value of the investment at each node along a scenario, for each asset class i. Unlike real estate and equity investments, fixed income portfolio holdings are assumed to carry a maturity date. At time 0 an initial *input* portfolio $\mathring{x}_i$, prior to any rebalancing decision, is assumed. We distinguish between the initial time 0 constraints and those for the later stages:

$$\begin{aligned} x_{i,0} &= \mathring{x}_i + x^+_{i,0} - x^-_{i,0} \qquad \forall i \in I, \\ X_0 &= \textstyle\sum_{i \in I} x_{i,0}. \end{aligned} \tag{5.2}$$

For $t = t_1, t_2, \ldots, H$, all scenarios $s = 1, 2, \ldots, S$

$$\begin{aligned} x_{i,h,t_j}(s) &= x_{i,h,t_{j-1}}(s)\left(1 + \rho_{i,t_j}(s)\right) - x^-_{i,h,t_j}(s) - x^{\exp}_{i,h,t_j}(s) \qquad \forall i, h < t_j, \\ x_{i,t_j}(s) &= \textstyle\sum_{h<t_j} x_{i,h,t_j}(s) + x^+_{i,t_j}(s) \qquad \forall i, \\ X_t(s) &= \textstyle\sum_i x_{i,t}(s). \end{aligned} \tag{5.3}$$

At each stage the market returns $\rho_{i,t_j}(s)$ for asset i realized in scenario s at time t_j will determine the portfolio revaluation paths: previous stage holdings plus buying decisions minus selling and expiry will define the new portfolio position for the following period. Each rebalancing decision, jointly with all cash flows induced by P&C premium renewals and claims, will determine the cash balance evolution up to the horizon.

5.3.1.2 Cash Balance Constraints

We consider cash outflows due to liability payments, negative interest on cash account deficits, dividend payments to the holding company or to shareholders in the second quarter of each year, corporate taxes, buying decisions, and operating and human resource costs. Cash inflows are due to insurance premiums, equity dividends, fixed income coupon payments, asset expiry (fixed income benchmarks), selling decisions, and interest on cash account surpluses. For $t = 0$, given an initial cash balance $\overset{\circ}{z}$

$$\overset{\circ}{z} + \sum_i x^-_{i,0} - \sum_i x^+_{i,0} + z^+_0 - z^-_0 = 0 \qquad \forall i \in I. \tag{5.4}$$

For $t = t_1, t_2, \ldots, t_n$ and $s = 1, 2, .., S$

$$\begin{aligned} & z^+_{t_{j-1}}(s)\left(1+\zeta^+_{t_j}(s)\right) - z^-_{t_{j-1}}(s)\left(1+\zeta^-_{t_j}(s)\right) - z^+_{t_j}(s) + z^-_{t_j}(s) \\ & \quad + \sum_{i\in I_1}\sum_{h<t_j} x^-_{i,h,t_j}(s)\left(1+\eta_{i,h,t_j}(s)\right) + \sum_{i\in I_2}\sum_{h<t_j} x^-_{i,h,t_j}(s) \\ & \quad + \sum_{i\in I_1}\sum_{h<t_j} x^{\exp}_{i,h,t_j}(s) - \sum_{i\in I} x^+_{i,t_j}(s) + \sum_{i\in I_2} x_{i,t_{j-1}}(s)\delta_{i,t_j}(s) \\ & \quad + R_{t_j}(s) - L_{t_j}(s) - C_{t_j}(s) - D^-_{t_j}(s) - T_{t_j}(s) = 0. \end{aligned} \tag{5.5}$$

Along each scenario, consistent with the assumed tree structure, cash surpluses and deficits will be passed forward to the following stage together with the accrual interest. Very low positive interest rates $\zeta^+_{t_j}(s)$ and penalty negative interest rates $\zeta^-_{t_j}(s)$ will force the investment manager to minimize cash holdings over time. The cash surplus at the end of the horizon is part of company terminal wealth.

5.3.1.3 Bounds on the Asset Portfolio

Rebalancing decisions over the decision horizon are typically constrained to maintain a sufficient portfolio diversification and avoid undesirable portfolio instabilities. For $t = t_1, \ldots, t_n$, $i = 1, \ldots, I$, and $s = 1, \ldots, S$, we have

$$l_i X_{t_j}(s) \le x_{i,t_j}(s) \le u_i X_{t_j}(s), \tag{5.6}$$

where l_i and u_i define percentage lower and upper bounds on holdings in asset i with respect to the current portfolio position. A maximum turnover will limit portfolio rebalancing from one stage to the next. For $t = t_1, t_2, \ldots, t_n$, $i \in I$, and $s = 1, \ldots, S$

$$\sum_i \sum_{h<t} x^-_{i,h,t}(s) + \sum_i x^+_{i,t}(s) \le 0.30 \cdot \left[\sum_i X_{i,t-1}(s) \cdot \left(1+\rho_{i,t}(s)\right)\right]. \tag{5.7}$$

Following Table 5.1 the profit equation and the terminal wealth include a set of random variables capturing both the technical and investment business sides.

5.3.1.4 Profit and Terminal Wealth

For $t = t_1, t_2, t_3$, $s = 1, 2, .., S$, prior to dividend distribution, given $\Pi_0 = 0$ at the start of the year, we compute recursively for every stage along each scenario s an aggregate insurance profit as

$$\Pi_{t_j}(s) = \Pi_{t_{j-1}}(s) + \sum_{i=I_2}\sum_{h<t_j} x^-_{i,h,t_j}(s) \cdot \gamma_{i,h,t_j}(s) + \tag{5.8}$$

$$+ \sum_{i \in I_2} x_{i,t_{j-1}}(s)\delta_{i,t_j}(s) + M_{t_j}(s) + R_{t_j}(s) - L_{t_j}(s) - C_{t_j}(s), \tag{5.9}$$

where $M_{t_j}(s)$, the interest margin at time t_j under scenario s, is clarified below. All other quantities in (5.9) have been introduced already. At business year end, for $t = t_4, t_5, t_6$, we compute the corporate tax payment and the net income results:

$$T_{t_j}(s) = \Pi_{t_j}(s) \cdot \mathcal{T},$$
$$D^-_{t_j}(s) = \Pi_{t_j}(s) - T_{t_j}(s).$$

The terminal wealth at $t_n = H$ in each scenario s is the sum of the holding portfolio, the cash surplus, and the realized profit:

$$W_H(s) = X_H(s) + \Pi_H(s) + z_H(s). \tag{5.10}$$

Following (5.9) and (5.10) which define the quantities in the objective function at $t_6 = 10y$ and $t_4 = 1y$, the optimal strategy will focus in the short term on realized profits, and in the long term jointly on realized and unrealized profits.

5.3.2 Scenario Generation

The optimal ALM strategy in the model depends on an extended set of asset and liability processes. We assume a multivariate normal distribution for monthly price returns of a representative set of market benchmarks including fixed income, equity, and real estate indices. The following $I = 12$ investment opportunities are considered: for $i = 1$: the EURIBOR 3 months; for $i = 2, 3, 4, 5, 6$: the Barclays Treasury indices for maturity buckets 1–3, 3–5, 5–7, 7–10, and 10+ years, respectively; for $i = 7, 8, 9, 10$: the Barclays Corporate indices, again spanning maturities 1–3, 3–5, 5–7, and 7–10 years; finally for $i = 11$: the GPR Real Estate Europe index and $i = 12$: the MSCI EMU equity index. Recall that in our notation I_1 includes all fixed income assets and I_2 the real estate and equity investments.

The benchmarks represent total return indices incorporating over time the securities' cash payments. In the definition of the strategic asset allocation problem, unlike real estate and equity investments, money market and fixed income benchmarks are assumed to have a fixed maturity equal to their average duration (3 months, 2 years, 4, 6, and 8.5 years for the corresponding Treasury and Corporate indices and 12 years for the 10+ Treasury index). Income cashflows due to coupon payments will be estimated ex post through an approximation described below upon selling or expiry of fixed income benchmarks and disentangled from price returns. Equity investments will instead generate annual dividends through a price-adjusted random dividend yield. Finally, for real estate investments a simple model focusing only on price returns is considered. Scenario generation translates the indices return trajectories for the above market benchmarks into a tree process (see, for instance, the tree structure in Fig. 5.3) for the ALM coefficients (Consigli et al., 2010; Chapters 15 and 16): this is referred to in the sequel as the tree *coefficient* process. We rely on the process nodal realizations (Chapter 4) to identify the coefficients to be associated with each decision variable in the ALM model. We distinguish in the decision model between asset and liability coefficients.

5.3.2.1 Asset Return Scenarios

The following random coefficients must be derived from the data process simulations along the specified scenario tree. For each scenario s, at $t = t_0, t_1, \ldots, t_n = H$, we define

$\rho_{i,t}(s)$: the *return* of asset i at time t in scenario s;
$\delta_{i,t}(s)$: the *dividend yield* of asset i at time t in scenario s;
$\eta_{i,h,t}(s)$: a positive *interest* at time t in scenario s per unit investment in asset $i \in I_1$, in epoch h;
$\gamma_{i,h,t}(s)$: the *capital gain* at time t in scenario s per unit investment in asset $i \in I_2$ in epoch h;
$\zeta_t^{+/-}(s)$: the (positive and negative) *interest rates* on the cash account at time t in scenario s.

Price returns $\rho_{i,t}(s)$ under scenario s are directly computed from the associated benchmarks $V_{i,t}(s)$ assuming a multivariate normal distribution, with $\rho := \{\rho_i(\omega)\}$, $\rho \sim N(\mu, \Sigma)$, where ω denotes a generic random element, and μ and Σ denote the return mean vector and variance–covariance matrix, respectively. In the statistical model we consider for $i \in I$ 12 investment opportunities and the inflation rate. Denoting by Δt a monthly time increment and given the initial values $V_{i,0}$ for $i = 1, 2, .., 12$, we assume a stochastic difference equation for $V_{i,t}$:

$$\rho_{i,t}(s) = \frac{V_{i,t}(s) - V_{i,t-\Delta t}(s)}{V_{i,t-\Delta t}(s)}, \tag{5.11}$$

$$\frac{\Delta V_{i,t}(s)}{V_{i,t}(s)} = \mu \Delta t + \Sigma \Delta W_t(s). \tag{5.12}$$

Table 5.2 Asset class statistics

	Asset class	Mean return per annum (%)	Implied income coefficient (%)	Return volatility per annum (%)
x_1	Euribor 3 months	1.21	0.00	1.16
x_2	Barcap Euro Agg Treasuries 1–3 Y	4.12	1.62	1.74
x_3	Barcap Euro Agg Treasuries 3–5 Y	4.89	1.89	3.10
x_4	Barcap Euro Agg Treasuries 5–7 Y	5.40	2.40	3.98
x_5	Barcap Euro Agg Treasuries 7–10 Y	5.62	2.62	4.74
x_6	Barcap Euro Agg Treasuries 10+ Y	6.15	3.15	6.75
x_7	Barcap Euro Agg Corp 1–3 Y	4.45	1.70	1.99
x_8	Barcap Euro Agg Corp 3–5 Y	4.82	1.57	3.41
x_9	Barcap Euro Agg Corp 5–7 Y	5.03	1.78	5.15
x_{10}	Barcap Euro Agg Corp 7–10 Y	5.16	1.91	6.68
x_{11}	GPR Real Estate Europe	5.51	0.00	12.30
x_{12}	MSCI EMU	3.92	2.00	22.19

In (5.12), $\Delta W_t \sim N(0, \Delta t)$. We show in Tables 5.2 and 5.3 the estimated statistical parameters adopted to generate through Monte Carlo simulation (Consigli et al., 2010; Glasserman, 2003) the correlated monthly returns in (5.12) for each benchmark i and scenario s. Monthly returns are then compounded according to the horizon partition (see Fig. 5.1) following a prespecified scenario tree structure. The return scenarios in tree form are then passed on to an algebraic language deterministic model generator to produce the stochastic program deterministic equivalent instance (Consigli and Dempster, 1998).

Equity dividends are determined independently in terms of the equity position at the beginning of a stage as $\sum_{i \in I_2} x_{i,t_{j-1}}(s)\delta_{i,t_j}(s)$, where $\delta_{i,t_j}(s) \sim N(0.02, 0.005)$ is the dividend yield.

Interest margin $M_{t_j}(s) = I^+_{t_j}(s) - I^-_{t_j}(s)$ is computed by subtracting negative from positive interest cash flows. Negative interest $I^-_{t_j}(s) = z^-_{t_{j-1}}(s)\zeta^-_{t_j}(s)$ is generated by short positions on the current account according to a fixed 2% spread over the Euribor 3-month rate for the current period. The positive interest rate for cash surplus is fixed at $\zeta^+(t, s) = 0.5\%$ for all t and s.

Positive interest $I^+_{t_j}(s)$ is calculated from return scenarios of fixed income investments $i \in I_2$ using buying–selling price differences assuming initial unit investments; this may be regarded as a suitable first approximation. In the case of negative price differences, the loss is entirely attributed to price movements and the interest accrual is set to 0.

We recognize that this is a significant simplifying assumption. It is adopted here to avoid the need for an explicit yield curve model, whose introduction would go beyond the scope of a case study. Nevertheless we believe that this simplification allows one to recognize the advantages of a dynamic version of the portfolio management problem.

No coupon payments are otherwise assumed during the life of the fixed income benchmarks:

Table 5.3 Estimated parameter specification for the P&C case study

Correlation matrix													
	1	2	3	4	5	6	7	8	9	10	11	12	13
1	1												
2	0	1											
3	0	0.9613	1										
4	0	0.9132	0.9755	1									
5	0	0.8553	0.9394	0.9844	1								
6	0	0.4620	0.3777	0.2229	0.1253	1							
7	0	0.4931	0.5598	0.5248	0.5247	0.6393	1						
8	0	0.5497	0.5186	0.3887	0.3057	0.9574	0.7906	1					
9	0	0.5675	0.5632	0.4555	0.3880	0.9035	0.8667	0.9861	1				
10	0	0.5598	0.5836	0.5020	0.4567	0.8263	0.9322	0.9424	0.9827	1			
11	0	0.2024	0.2967	0.4097	0.4656	−0.4535	−0.0811	−0.3434	−0.2762	−0.2105	1		
12	0	0.0222	0.1005	0.1890	0.2600	−0.5769	−0.2341	−0.5167	−0.4699	−0.4141	0.6304	1	
13	0	−0.3127	−0.3841	−0.3747	−0.3864	−0.0107	−0.1753	−0.1052	−0.1294	−0.1456	−0.3280	−0.4634	1

$$I^+(t_j, s) = \sum_{i \in I_1} \sum_{h<t_j} x^-_{i,h,t_j}(s) \cdot \eta_{i,h,t_j}(s),$$

$$\eta_i(h, t, s) = \max\left[1 - \frac{1}{\prod_{\tau=h+1,\ldots,t}(1+\rho_i(\tau, s))}; 0\right].$$

Capital gains and losses $\gamma_{i,h,t_j}(s)$ for real estate and equity investments, $i \in I_2$, are generated in the same way, but allowing in this case for negative returns:

$$G(t, s) = \sum_{i=I_2}\left(\sum_{h<t} x^-_{i,h,t_j}(s) \cdot \gamma_{i,h,t_j}(s)\right),$$

$$\gamma_{i,h,t_j}(s) = 1 - \frac{1}{\prod_{\tau=h+1,\ldots,t}(1+\rho_i(\tau, s))}.$$

Scenarios for $\rho_i(t, s)$ are first generated by Monte Carlo simulation, then compounded to identify, for $i \in I_2$, the accrual interest and, for $i \in I_2$, the capital gains or losses according to the above definitions.

5.3.2.2 Liability Scenarios

Annual premium renewals by P&C policyholders represent the key technical income on which the company relies for its ongoing business requirements. Premiums have associated random insurance claims over the years and, depending on the business cycle and the claim class, they might induce different reserve requirements. The latter constitute the insurance company's most relevant liability. In our analysis premiums are assumed to be written annually and will be stationary over the 10-year horizon with limited volatility. Insurance claims are also assumed to remain stable over time in nominal terms, but with an inflation adjustment that in the long term may affect the company's short-term liquidity. For a given estimate of insurance *premiums* R_0 collected in the first year, along scenario s, for $t = t_1, t_2, \ldots, H$, with $e_r \sim N(1, 0.03)$, we assume that

$$R_{t_j}(s) = \left[R_{t_{j-1}}(s) \cdot e_r\right]. \qquad (5.13)$$

Insurance *claims* are assumed to grow annually at the inflation rate and in nominal terms are constant in expectation. For given initial liability L_0, with $e_l \sim N(1, 0.01)$

$$L_{t_j}(s) = \left(L_{t_{j-1}}(s) \cdot e_l\right)(1 + \pi_{t_j}(s)). \qquad (5.14)$$

Every year the liability stream is assumed to vary in the following year according to a normal distribution with the previous year's mean and a 1% volatility per year with a further inflation adjustment. Given $\pi(0)$ at time 0, the inflation process $\pi_t(s)$ is assumed to follow a mean-reverting process

$$\Delta\pi_t(s) = \alpha_\pi(\mu - \pi_t(s))\Delta t + \sigma_\pi \Delta W_t(s), \tag{5.15}$$

with μ set in our case study to the 2% European central bank target and $\Delta W_t \sim N(0, \Delta t)$. As for the other liability variables *technical reserves* are computed as a linear function of the current liability as $\Lambda_t(s) = L_t(s) \cdot \lambda$, where λ in our case study is approximated by $1/0.3$ as estimated by practitioner opinion.

Operational costs $C_{t_j}(s)$ include staff and back-office and are assumed to increase at the inflation rate. For a given initial estimate C_0, along each scenario and stage, we have

$$C_{t_j}(s) = C_{t_{j-1}}(s) \cdot (1 + \pi_{t_j}(s)). \tag{5.16}$$

Bad liability scenarios will be induced by years of premium reductions associated with unexpected insurance claim increases, leading to higher reserve requirements that, in turn, would require a higher capital base. A stressed situation is discussed in Section 5.4 to emphasize the flexibility of dynamic strategies to adapt to bad technical scenarios and compensate for such losses. The application also shows that within a dynamic framework the insurance manager is able to achieve an optimal trade-off between investment and technical profit generation across scenarios and over time, and between risky and less risky portfolio positions.

5.3.2.3 Statistical Estimation

The statistical parameters in Table 5.2 have been estimated on *monthly* observations from January 1999 to December 2009 to provide the inputs to the scenario generator.

The inflation rate process is estimated by the method of moments (Campbell et al., 1997) with long-term mean equal to the European Central Bank target of 2%, with a mean-reversion coefficient of 0.7 and volatility 0.50%.

Price returns are computed taking correlation into account: a canonical Choleski decomposition is applied to the correlation matrix and integrated into a Monte Carlo simulator (Consigli et al., 2010; Glasserman, 2003) to yield price transitions on the given scenario tree structure. Table 5.3 displays the correlation coefficients between the asset returns plus, in row 13, the correlation with the inflation process.

A symmetric tree structure with branching 7.4.3.2.2.2 is considered in the case study leading to 672 scenarios at the horizon. This is a relatively small tree over 10 years, but sufficient to highlight the flexibility of the dynamic stochastic programming approach. The statistical assumptions adopted here are relatively simple and are considered for preliminary testing purposes. The scenario tree process for each benchmark in (5.12) is generated from Monte Carlo simulation based on the above coefficients following the introduced conditional structure. See the other contributions to this volume for scenario generation procedures more suitable for long-term planning problems, in particular (Chapters 14, 15, and 16).

5.4 Case Study: A 10-Year P&C ALM Problem

We present results for a test problem modelling a large P&C company assumed to manage a portfolio worth 10 billion € (bln) at t_0. The firm's management has set a first-year operating profit of 400 million € (mln) and wishes to maximize in expectation the company realized and unrealized profits over 10 years. Company figures have been disguised for confidentiality reasons, though preserving the key elements of the original ALM problem. We name the P&C company *Danni Group*.

The following tree structure is assumed in this case study. The current implementable decision, corresponding to the root node, is set at 2 January 2010 (Table 5.4).

Within the model the P&C management will revise its strategy quarterly during the first year to minimize the expected shortfall with respect to the target.

Initial conditions in the problem formulation include average first-year insurance premiums $R(t)$ estimated at 4.2 bln €, liability reserves $\Lambda(0)$ equal to 6.5 bln €, and expected insurance claims $L(t)$ in the first year of 2.2 bln €. The optimal investment policy, furthermore, is constrained by the following upper and lower bounds relative to the current portfolio value:

- Bond portfolio upper bound: 85%
- Equity portfolio upper bound: 20%
- Corporate portfolio upper bound: 30%
- Real estate portfolio upper bound: 25%
- Desired turnover at rebalanced dates: $\leq$30%
- Cash lower bound: 5%

The results that follow are generated through a set of software modules combining MATLAB 7.4b as the main development tool, GAMS 21.5 as the model generator and solution method provider, and Excel 2007 as the input and output data collector running under a Windows XP operating system with 1 GB of RAM and a dual processor.

The objective function (5.1) of the P&C ALM problem considers a trade-off between short-, medium-, and long-term goals through the coefficients $\lambda_1 = 0.5, \lambda_2 = 0.2$, and $\lambda_3 = 0.3$ at the 10-year horizon. The first coefficient determines a penalty on profit target shortfalls. The second and the third coefficients are associated, respectively, with a medium-term portfolio value decision criterion and long-term terminal wealth. Rebalancing decisions can be taken at decision epochs from time 0 up to the beginning of the last stage; no decisions are allowed at the horizon.

Table 5.4 P&C case study time and space discretization

	10 years	Start date:	2.1.2010				
Planning horizon	Stages	1	2	3	4	5	6
Stage time distribution	h&n	3 m	6 m	9 m	1 y	3 y	10 y
Tree structure	7	4	3	2	2	2	
Scenarios		7	28	84	168	336	672

Over the planning horizon the investment manager may exploit cash surpluses generated by the technical activity or suffer from liquidity deficits due to technical losses. We analyse the first-stage implementable decision and the average portfolio evolution over the 10-year horizon. Along a specific mean scenario the aggregated technical and investment profit dynamics are considered in order to highlight the implications of realized and unrealized profit generation. The analysis is first applied to the case of an average mean liability scenario and then to a stressed scenario induced by severe liability losses, as experienced in recent years by several large insurance firms. No risk capital constraints are explicitly considered in the problem leading, as explained below, to optimal strategies that may model the results *expensive* from the regulatory capital viewpoint (European Parliament, 2009). We show however, even under very negative insurance claim scenarios, that a dynamic investment strategy is able to avoid significant capital impact and preserve the firm's solvency.

The optimal problem solution leads to the terminal wealth distribution of Fig. 5.4 defined as the sum of a cash surplus, the investment portfolio value, and realized profits from both technical and investment activity. Thanks to a structural liquidity surplus from the technical business and an active portfolio strategy, the insurance manager will achieve sufficient terminal wealth even under the less favourable financial scenario.

Best, worst, and average wealth scenarios are defined with respect to the above distribution as those specific scenarios remaining over the 10-year horizon *closest* to the set of highest, lowest, and mean wealth realizations at each stage. We present below a set of optimal strategies along a *mean wealth* scenario $W(\hat{\omega}_t))$, where $\hat{\omega}_t$ is selected by first computing the expectation $E(\omega_t^s))$ at each stage across scenarios and then identifying a specific scenario which, in terms of Euclidean distance, remains over time within a neighbourhood $\mathcal{N}(E(\omega_t^s), \epsilon)$ of radius ϵ of that scenario. In a similar fashion the best and worst scenarios may be identified: first the best and worst wealth realizations in each stage are specified and then those scenarios which (from the root to the leaf nodes) remain closest to those realizations within a given tolerance are identified. The described portfolio strategies are in this way associated with scenarios actually forecast by the scenario generator.

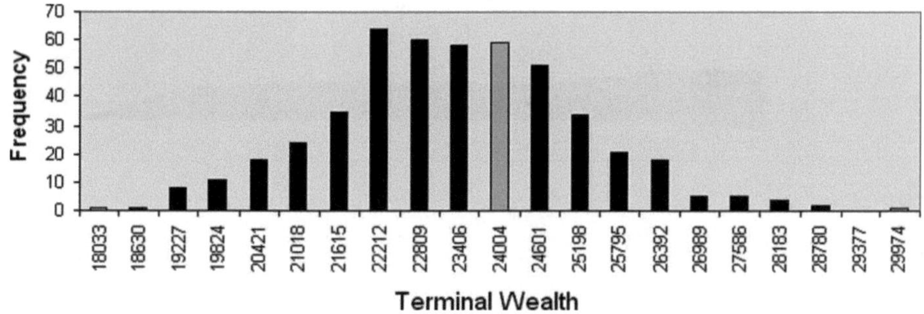

Fig. 5.4 Terminal wealth distribution

5.4.1 Optimal Investment Policy Under P&C Liability Constraints

The optimal portfolio composition at time 0, current time, the only one under full uncertainty regarding future financial scenarios, is displayed in Fig. 5.5.

Relying on a good liquidity buffer at time 0 the insurance manager will allocate wealth evenly across the asset classes with a similar portion of long Treasury and corporate bonds. As shown in Fig. 5.6 the optimal strategy will then progressively

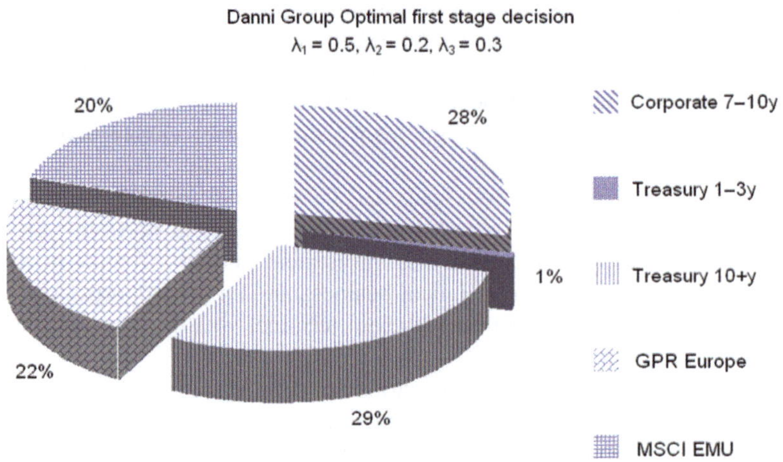

Fig. 5.5 Danni Group – optimal root node decision

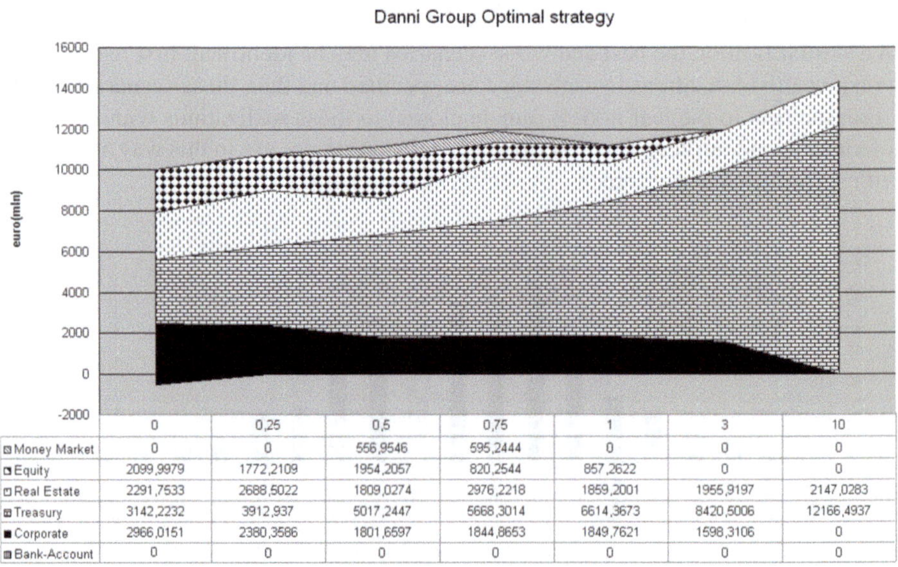

	0	0,25	0,5	0,75	1	3	10
Money Market	0	0	556,9546	595,2444	0	0	0
Equity	2099,9979	1772,2109	1954,2057	820,2544	857,2622	0	0
Real Estate	2291,7533	2688,5022	1809,0274	2976,2218	1859,2001	1955,9197	2147,0283
Treasury	3142,2232	3912,937	5017,2447	5668,3014	6614,3673	8420,5006	12166,4937
Corporate	2966,0151	2380,3586	1801,6597	1844,8653	1849,7621	1598,3106	0
Bank-Account	0	0	0	0	0	0	0

Fig. 5.6 Portfolio composition along the mean wealth scenario over time

reduce holdings in equity and corporate bonds to increase the investment in Treasuries. A non-negligible real estate investment is kept throughout the 10 years. The allocation in the real estate index remains relatively stable as a result of the persistent company liquidity surplus. Real estate investments are overall preferred to equity investments due to higher expected returns per unit volatility estimated from the 1999 to 2009 historical sample. The optimal investment policy is characterized by quarterly rebalancing decisions during the first year with limited profit-taking operations allowed beyond the first business year. The 3-year objective is primarily associated with the medium-term maximization of the portfolio asset value; the strategy will maximize unrealized portfolio gains specifically from long Treasury bonds, real estate, and corporate bonds. Over the 10 years the portfolio value moves along the average scenario from the initial 10 bln € to roughly 14.3 bln €.

The optimal portfolio strategy is not affected by the liquidity constraints from the technical side as shown in Table 5.5. Danni Group has a strong liquidity buffer generated by the core insurance activity, but only sufficient on its own to reach the target profit set at the end of the first business year.

At the year 1 horizon the investment manager seeks to minimize the profit target shortfall while keeping all the upside. Indeed a 1.1 billion € gross profit is achieved prior to the corporate tax payment corresponding to roughly 750 mln € net profit. Thanks to the safe operational environment, no pressure is put on the investment side in terms of income generation and the investment manager is free to focus on the maximization of portfolio value at the 3-year horizon. Over the final 7-year stage the portfolio manager can be expected to concentrate on both realized and unrealized profit maximization, contributing to overall firm business growth, as witnessed in Table 5.5 and Fig. 5.6. Empirical evidence suggests that P&C optimal portfolio strategies, matching liabilities average life time, tend to concentrate on assets with limited duration (e.g. 1–3 years). We show below that such a strategy without an explicit risk capital constraint would penalize the portfolio terminal value.

Table 5.5 Average Danni group income

Expected profit by sources							
Average P&C income scenario							
Stage	1	2	3	4	5	6	7
Time	0	0.25	0.5	0.75	1	3	10
Revenue		1060.88	1062.47	1070.48	1070.25	8491.90	29681.34
Liability stream		–559.30	–559.03	–564.62	–563.87	–4563.54	–16089.6
Operating costs		–200	–200	–200	–200	–1600	–5600
Technical income		301.57	303.43	305.86	306.38	2328.35	7991.74
Dividend inflows		0	31.49	0	16.40	0	0
Margin of interests		0.26	15.14	11.80	18.87	255.84	1837.21
Capital gain		–54.38	–30.56	–30.47	–21.76	286.37	48.08
Investment income		–59.64	16.07	–18.66	13.52	542.22	1885.29
Stage profit		246.93	319.50	287.19	319.89	2870.58	9877.03

5.4.2 Dynamic Asset Allocation with $\lambda_1 = 1$

Remaining within a dynamic framework in the objective function (5.1) of the P&C ALM problem we set $\lambda_1 = 1$, with thus $\lambda_2 = \lambda_3 = 0$ over the 10-year horizon. The optimal policy will in this case be driven by the year 1 operating profit target, carrying on until the end of the decision horizon.

Diving a different view from Fig. 5.6, we display in Figs. 5.7 and 5.8 the optimal strategies in terms of the portfolio's time-to-maturity structure in each stage. Relative to the portfolio composition in Fig. 5.8 which corresponds to giving more weight to medium- and long-term objectives, the portfolio in Fig. 5.7 concentrates from stage 1 on fixed income assets with lower duration. During the first year it shows an active rebalancing strategy and a more diversified portfolio. At the 9-month horizon part of the portfolio is liquidated and the resulting profit will minimize at the year 1 horizon the expected shortfall with respect to the target. Thereafter the strategy, suggesting a buy and hold management approach, will tend to concentrate on those assets that would not expire within the 10-year horizon.

Consider now Fig. 5.8, recalling that no risk capital constraints are included in the model. The strategy remains relatively concentrated on long bonds and assets without a contractual maturity. Nevertheless the portfolio strategy is able to achieve the first target and heavily overperform over the 10 years. At $T = 10$ years the first portfolio in this representative scenario is worth roughly 12.1 billions €, while the second achieves a value of 14.2 billions €.

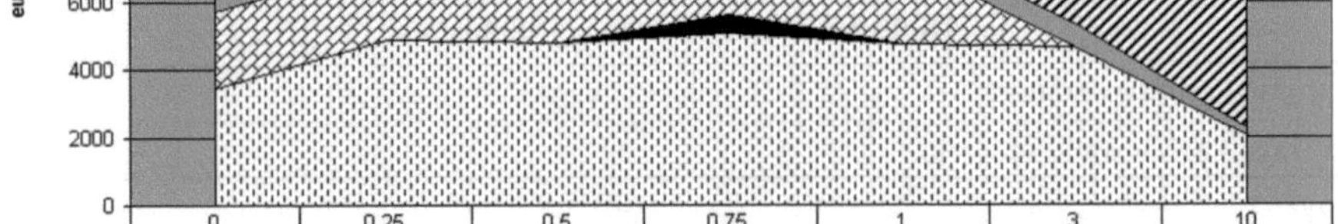

	0	0,25	0,5	0,75	1	3	10
10+ years	767,7763	1719,9246	1788,0766	794,0827	3141,4868	6467,9292	9986,5151
8.5 years	2931,2455	1406,7909	0	828,6303	0	0	0
6 years	0	0	1493,2723	0	0	0	0
4 years	568,0484	576,0094	585,8683	587,5292	603,2236	661,7471	193,1217
2 years	2247,4473	2270,1667	2294,3366	4018,6485	2441,9905	0	0
3 months	0	0	0	594,3044	0	0	0
no expire	3485,4726	4886,9113	4784,3054	5062,893	4781,5405	4676,5106	2054,3628

years

Fig. 5.7 Danni Group – Mean scenario optimal portfolio composition with respect to assets' time to maturity, $\lambda_1 = 1$, $\lambda_2 = \lambda_3 = 0$

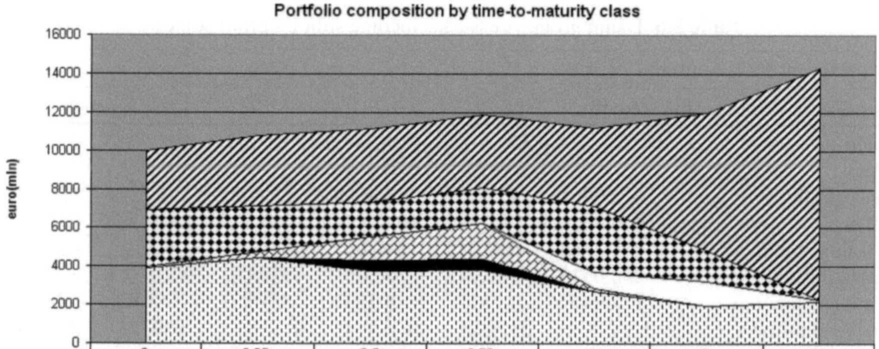

	0	0,25	0,5	0,75	1	3	10
10+ years	3057,2444	3620,1231	3804,9033	3851,305	4037,7563	7135,3666	12034,873
8.5 years	2966,0151	2380,3586	1801,6597	1844,8653	3423,6011	1598,3106	0
6 years	0	0	0	0	866,0937	1285,134	131,6207
4 years	0	0	0	0	0	0	0
2 years	84,9788	292,8139	1212,3414	1816,9964	136,6783	0	0
3 months	0	0	556,9546	595,2444	0	0	0
no expire	3891,7512	4460,7131	3763,2331	3796,4762	2716,4623	1955,9197	2147,0283

years

Fig. 5.8 Danni Group – mean scenario optimal portfolio composition with respect to assets' time to maturity $\lambda_1 = 0.5$, $\lambda_2 = 0.2$, $\lambda_3 = 0.3$

The year 1 horizon is the current standard for insurance companies seeking an optimal risk-reward trade-off typically within a static, one-period, framework. The above evidence suggests that over the same short-term horizon, a dynamic setting would in any case induce a more active strategy and, furthermore, a 10-year extension of the decision horizon would not jeopardize the short–medium-term profitability of the P&C shareholder while achieving superior returns in the long term.

5.4.3 Investment Strategy Under Stressed P&C Scenarios

A different picture arises if a dramatic 50% increase of liability claims is assumed while keeping premium renewals constant. Under these extreme conditions we analyse the impact on the income statement and the portfolio strategy. We will show that the investment policy would in such a case compensate the losses generated by the technical business, whose combined ratio will go well above 100% under such a situation.

Consider the figures in Table 5.6 which report a dramatic increase of insurance claims on the technical side, to severely affect the firm's liquidity over to 10-year horizon. The lower part of the table shows that under such a scenario the portfolio manager will modify her strategy to maintain a sufficient liquidity level, increase the investment realized profit, and reduce the portfolio upside over the long horizon.

The optimal portfolio strategy will keep the overall P&C firm's cumulative profit positive until the end of the third year with a relatively sustainable loss over the

Table 5.6 Danni group stress-case income split by profit source

Expected profit by sources							
Stress-case P&C income scenario							
Stage	1	2	3	4	5	6	7
Time	0	0.25	0.5	0.75	1	3	10
Cum. profit		54.2667	143.1944	238.8708	271.8932	404.1524	–213.813
Revenue		1046.833	1048.887	1042.324	1037.277	8319.786	28263.04
Liability stream		–814.2763	–808.212	–810.66	–821.966	–6458	–23511.2
Operating costs		–250.0513	–250.429	–250.451	–250.808	–2025.61	–7204.99
Technical income		–17.4946	–9.7537	–18.7877	–35.4971	–163.82	–2453.11
Dividend inflows		0	31.4221	0	0	67.9724	439.4421
Margin of interests		71.7613	67.2592	62.5	68.5195	500	1799.854
Capital gain		0	0	51.9641	0	0	0
Investment income		71.7613	98.6813	114.4641	68.5195	567.9724	2239.296
Stage profit		54.2667	88.9276	95.6764	33.0224	404.1524	–213.814

residual horizon (last row in Table 5.6 which show the stage extensions leading to roughly 30 million € annual losses). As shown in Fig. 5.10 the portfolio strategy would yield a terminal portfolio worth 10.8 bln €. At time 0 the suggested optimal investment policy is reported in Fig. 5.9.

The implementable first-stage (root node) decision is primarily concentrated in long Treasury bonds and real estate.

Figure 5.10 shows that under such a stressed liability scenario the portfolio is kept mainly in low-risk fixed income assets and that equity and real estate rebalancing decisions help to preserve sufficient liquidity at the end of the first year. Thereafter the investment manager seeks a portfolio value maximizing expected growth while limiting any downside. Interestingly, compared with the one in Fig. 5.6, the portfolio strategy represents a higher concentration in low-risk instruments on average, but with appropriate relevant portfolio revisions over the horizon.

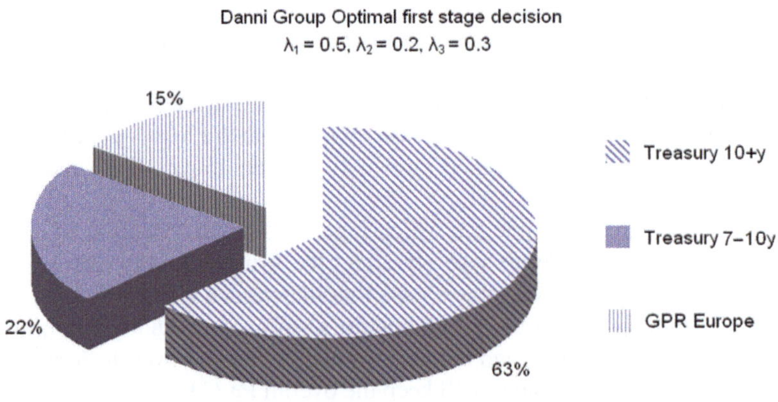

Fig. 5.9 Danni Group – optimal root node decision under the stressed scenario

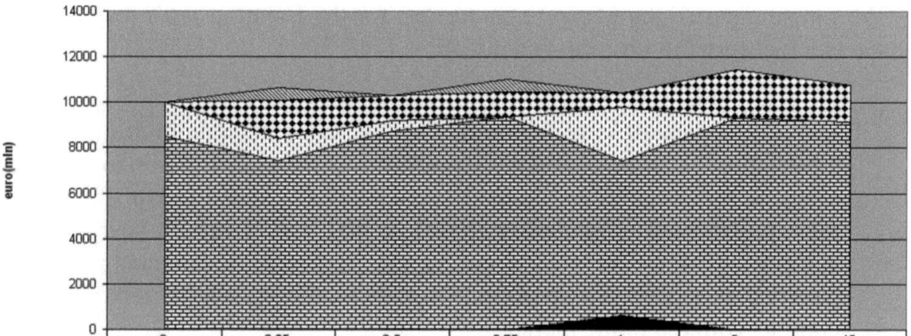

	0	0,25	0,5	0,75	1	3	10
Money Market	0	531,8191	0	551,6354	0	0	0
Equity	0	1697,9927	1082,2855	1103,2708	665,2672	2081,927	1591,5145
Real Estate	1499,9985	1001,6349	462,7342	0	2364,6015	22,5822	23,9907
Treasury	8499,9915	7404,9347	8755,1116	9377,8021	6788,7112	9331,8143	9154,5296
Corporate	0	0	0	0	631,8821	0	0
Bank-Account	0	0	0	0	0	0	0

years

Fig. 5.10 Expected portfolio composition evolution in the stressed scenario

The impact of such a strategy can be clarified by comparing Fig. 5.11 with Fig. 5.2 at the beginning of this chapter. The first-year target is not achieved under the stressed scenario. A positive income is generated, however, and the first year closes in profit. Therafter, the remarkable impact of the negative liability scenario is compensated by the investment strategy employed.

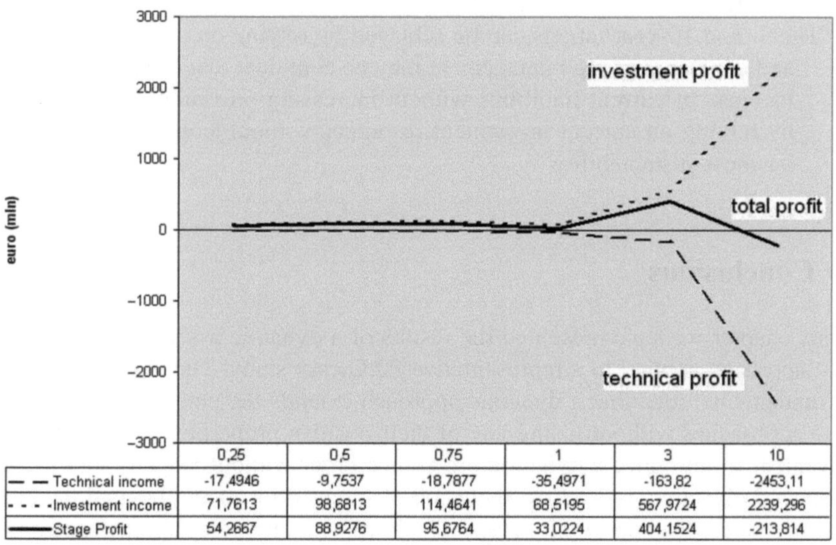

	0,25	0,5	0,75	1	3	10
Technical income	-17,4946	-9,7537	-18,7877	-35,4971	-163,82	-2453,11
Investment income	71,7613	98,6813	114,4641	68,5195	567,9724	2239,296
Stage Profit	54,2667	88,9276	95,6764	33,0224	404,1524	-213,814

years

Fig. 5.11 Expected profit in the stress scenario

We can summarize the general conclusions for the Danni group's management which emerge from this case study solution:

The P&C company has strong liquidity and the estimated premium streams are expected to generate a structural surplus over the planning horizon, greatly reducing the pressure on the investment portfolio. Danni group's combined ratio is expected to remain well below 100% over the 10-year horizon. A more aggressive premium policy can be conducted by reducing premiums per contract type to expand the market and/or enlarge the spectrum of liability contracts offered to the market

The first-year 400 mln € net profit target appears under the forecast market and financial conditions to be underestimating the Danni group's potential. A more ambitious target could be proposed without jeopardizing the company's long-term profitability. The wealth scenarios (combining the investment portfolio market value over time and the cumulative profit generated by the activity) under worst-case market conditions define a final wealth at the 10-year horizon of around 11 bln €

During the first quarters of 2010 the portfolio manager is expected to keep a diversified portfolio with 60% invested in long corporate and treasury bonds and roughly 40% of its wealth in the real estate and the equity market. After the first year, without a risk capital constraint on the asset–liability duration mismatch, an increasing portion of the portfolio is expected to move towards long fixed income instruments under both average and worst-case market conditions. The expected optimal policy obviously depends on the objective function and the generated scenario set: alternative statistical models may lead to different optimal policies

The 3- and 10-year targets can be achieved by relying on current expectations and the Danni group management may be confident that even an unexpected increase of current liabilities without increasing provisions can be handled by relying on current investment resources without jeopardizing long-term business sustainability.

5.5 Conclusions

In this chapter we have presented the results of a dynamic asset–liability management approach applied to a representative P&C case study. The evidence supports the fundamental idea that a dynamic approach extends the potential of one-period static approaches without losing any of their positive properties. It is possible and convenient to introduce a dynamic preference order without jeopardizing the year 1 financial objective. Indeed the inclusion of stages prior to and beyond the year 1 horizon reduces the opportunity costs associated with the static approach and greatly enhances the decision strategy flexibility with respect to contingent insurance factors. The joint consideration of liabilities and revenues across stages and scenarios allows the derivation of alternative optimal policies contingent on the current and forward prevailing conditions. The key trade-off between short- and long-term profitability can be handled across time and also across profit centres (technical versus investment income), easing the assessment of the potential offered by each profit line.

The case study with stressed liability scenarios presented shows that the dynamic approach leads to risk control – hedging strategies – from the very beginning of the planning horizon with relevant effects on the first-stage portfolio decision.

The analysis has been conducted relying on a set of simplifying assumptions, specifically related to the absence of explicit risk capital allocation constraints (with optimal strategies that, contrary to the industry practice, are not restricted from allowing asset–liability duration gaps) and a relatively standard statistical model to capture the core uncertainty of the problem. These limitations, which do not affect the validity of the evidence collected, are currently been relaxed or removed to yield a more accurate representation of the strategic P&C problem and link the dynamic ALM model to the risk management model. The investment universe, furthermore, will be extended to include a set of alternative investments (private equity, direct real estate, renewable energy, and infrastructure) of primary importance for global insurance companies worldwide. We will report the related results in a separate study.

Acknowledgements The authors acknowledge the support given by the research project PRIN2007 "Optimization of stochastic dynamic systems with applications to finance", no. 20073BZ5A5, sci. resp. G. Consigli. The research has been partially supported by MIUR-ex60% 2008–2009, sci. resp. V. Moriggia.

References

Artzner P., Delbaen F., Eber J. and Heath D. (1999): Coherent measures of risk. *Mathematical Finance* **9**(3), 203–228.

Birge J.R. and Louveaux F. (1997): *Introduction to Stochastic Programming*. Springer, New York, NY.

Campbell J.Y., Lo W.A. and MacKinlay A.C. (1997): *The Econometrics of Financial Markets*. Princeton University Press, Princeton, NJ.

Cariño D.R., Kent T., Myers D.H., Stacy C., Sylvanus M., Turner A.L., Watanabe K. and Ziemba W.T. (1994): The Russel-Yasuda-Kasai model: An asset-liability model for a Japanese insurance company using multistage stochastic programming. *Interfaces* **24**(1), 29–49.

CEA (2010): CEA statistics no. 42, *European Insurance in Figures*, www.cea.eu, 1–54.

Consigli G. (2007): Asset and Liability management for Individual Investors, in Zenios S.A. and Ziemba W.T. Eds, *Handbook of Asset and Liability Management: Applications and Case Studies*, volume **2**, 751–828. North-Holland Handbooks Finance Series, Amsterdam.

Consigli G. and Dempster M.A.H. (1998): Dynamic stochastic programming for asset-liability management. *Annals of Operations Research* **81**, 131–161.

Consigli G., Iaquinta G. and Moriggia V. (2010): Path-dependent scenario trees for multistage stochastic programs in finance. *Quantitative Finance*, iFirst, Dec 2010, 1–17.

de Lange P.E., Fleten S-E. and Gaivoronski A. (2003): Modeling financial reinsurance in the casualty insurance business via stochastic programming. PhD thesis, Department of Industrial Economics and Technology Management, N.T.N.U., Trondheim (NO).

Dempster M.A.H., Germano M., Medova E.A. and Villaverde M. (2003): Global asset liability management. *British Actuarial Journal* **9**(1), 137–195.
Dupačová J., Consigli G. and Wallace S.W. (2001): Scenarios for multistage stochastic programmes. *Annals of Operations Research* **100**, 25–53.
Europe Economics (2010): Retail Insurance Market Study, http://ec.europa.eu/internalmarket/insurance/docs/motor/20100302rimen.pdf
European Parliament (2009): European Parliament, Solvency II Directive, http://eur-lex.europa.eu/LexUriServ/LexUriServ.do?uri=OJ:L:2009:335:0001:0155:EN:PDF
Gaivoronski A.A., Hoyland K. and de Lange P.E. (2001): Statutory regulation of casualty insurance companies: an example from Norway with stochastic programming analysis, in *Stochastic Optimization: Algorithms and Applications* (Gainesville, FL, 2000), volume **54** of Appl. Optim., Kluwer, Dordrecht, 55–85.
Glasserman P. (2003): *Monte Carlo Methods in Financial Engineering*. Springer, New York, NY.
ISVAP (2010): ISVAP, Solvency II - Lettera al Mercato, http://www.isvap.it/isvapcms/docs/F27176/SolvencyIIilnuovosistema-di-vigilanza-prudenziale.pdf
Kouwenberg R. (2001): Scenario generation and stochastic programming models for asset liability management. *European Journal of Operational Research* **134** Issue 2, 279–292.
Mulvey J.M. and Erkan H.G. (2003): Simulation for risk management: risk management of a P/C insurance company scenario generation, simulation and optimization. In Proceedings of the 35th conference on Winter simulation: driving innovation, 364–371.
Mulvey J.M. and Erkan H.G. (2005): Decentralized risk management for global property and casualty insurance companies, in Wallace S.W. and W.T. Ziemba Eds, *Applications of Stochastic Programming*, 503–530. MPS-SIAM Series on Optimization, United States.
Mulvey J.M., Pauling B., Britt S. and Morin F. (2007): Dynamic financial analysis for Multinational Insurance Companies, in Zenios S.A. and W.T. Ziemba Eds, *Handbook of Asset and Liability Management: Applications and Case Studies*, volume **2**, 543–589. North-Holland Handbooks Finance Series, Amsterdam.
Pflug G.Ch. and Römisch W. (2007): *Modeling, Measuring and Managing Risk*. World Scientific Publisher, Singapore.
Rockafellar R.T. and S. Uryasev (2002): Conditional value at risk for general loss distributions. *Journal of Banking and Finance* **26**, 1443–1471.
Zenios S.A. and Ziemba W.T. Eds (2007a): *Handbook of Asset and Liability Management: Theory and Methodology*, Volume **1**, North-Holland Handbooks Finance Series, Amsterdam.
Zenios S.A. and Ziemba W.T. Eds (2007b): *Handbook of Asset and Liability Management: Applications and Case Studies*, Volume **2**, North-Holland Handbooks Finance Series, Amsterdam.

Chapter 6
Pricing Reinsurance Contracts

Andrea Consiglio and Domenico De Giovanni

Abstract Pricing and hedging insurance contracts is hard to perform if we subscribe to the hypotheses of the celebrated Black and Scholes model. Incomplete market models allow for the relaxation of hypotheses that are unrealistic for insurance and reinsurance contracts. One such assumption is the tradeability of the underlying asset. To overcome this drawback, we propose in this chapter a stochastic programming model leading to a superhedging portfolio whose final value is at least equal to the insurance final liability. A simple model extension, furthermore, is shown to be sufficient to determine an optimal reinsurance protection for the insurer: we propose a conditional value at risk (VaR) model particularly suitable for large-scale problem instances and rationale from a risk theoretic point of view.

Keywords Reinsurance · Option pricing · Incomplete markets

6.1 Introduction

Hedging a liability is the best practice to mitigate the potential negative impact due to market swings. The issue of pricing and hedging embedded options in insurance contracts is also ruled by the International Financial Reporting Standards (IFRS 4) and by the new regulatory capital framework for insurers Solvency II. The new reporting standards prescribe that the cost of options and guarantees embedded in insurance contracts are measured consistently with the market and, for this purpose, it is suggested to split the risk into hedgeable and non-hedgeable component.

The hedging process, as described in finance textbooks, however is hard to apply in the insurance context. As a primary obstacle, the underlying of an insurance contract is usually a non-tradeable asset, thus making impossible the trading activity needed to hedge the protection seller risk exposure. Moreover, the liabilities generated by an insurance contract are long-term ones, and unexpected shortfalls in the asset returns are not envisaged by the geometric Brownian motion underneath the celebrated Black and Scholes model. These unpredictable events can

A. Consiglio (✉)
Department of Statistics and Mathematics "Silvio Vianelli", University of Palermo, Palermo, Italy
e-mail: consiglio@unipa.it

M. Bertocchi et al. (eds.), *Stochastic Optimization Methods in Finance and Energy*,
International Series in Operations Research & Management Science 163,
DOI 10.1007/978-1-4419-9586-5_6,

generate serious losses and rolling the hedge forward, as witnessed by the Metallgesellschaft default (Mello and Parsons 1995), may lead to unexpected severe losses.

In options theory such limitations are known as sources of *market incompleteness*, and recently, scholars have started coping with pricing and hedging options in incomplete markets. For example, Consiglio and De Giovanni (2008) adopt a super-replication model to determine the hedging portfolio of insurance policies whose final liabilities depend on a minimum guarantee option and a bonus distribution scheme. An extension of such a model (Consiglio and De Giovanni forthcoming) allows the pricing of insurance contract with a surrender option, that is, the option to leave the contract before maturity. An alternative approach is proposed by Coleman et al. (2007). They address and solve a similar incomplete-market hedging problem extending the traditional Black and Scholes price process to include Merton's jump diffusion process. The insurance claim is then hedged, by using the underlying asset and a set of standard options expiring before the claim's maturity. The hedging strategy is here determined by applying the minimum local hedging risk principle by Föllmer and Schweizer (1989).

The aim of this chapter is to extend the model in Consiglio and De Giovanni (2008, forthcoming) to handle not only the primary risk exposure associated with a short insurance position but also the exposure induced by a reinsurance agreement. To this aim we propose a novel approach to the asset–liability management of such contracts. A general definition of *reinsurance contract* is that of a coverage purchased by an insurance company (insurer) from typically another insurance company (reinsurer) to transfer an original risk exposure. The reinsurance agreement specifies the basis for which the reinsurer will pay the insurer's losses (excess of loss or proportional) and the reinsurance premium paid by the reinsured.

The chapter is organized as follows. Section 6.2 introduces the reinsurance problem. Section 6.3 deals with the mathematical formulation of the stochastic programming models, including: (i) the notation, (ii) the pricing of European contingent claims, and (iii) an efficient method to build scenario trees. The asset–liability management problem to manage reinsurance contracts is described in Section 6.4, while implementation notes and discussion of the results are reported in Section 6.5. Section 6.6 concludes the chapter and highlights the major findings.

6.2 Stop-Loss Reinsurance Contracts in the Property and Casualty Market

Broadly speaking, reinsurance is the insurance of insurance liabilities. Insurance firms with a specified risk exposure might decide—or are forced to by regulators—to transfer part of their own risk exposure to third-party companies by buying reinsurance protection.

Popular agreements in the reinsurance market are proportional reinsurance and aggregate excess reinsurance. With proportional contracts, the transferring party cedes to the reinsurance company a fixed proportion of the risk associated with the portfolio, while with aggregate excess treaty, the reinsurance company promises to pay the claims exceeding a given retention (see Straub (1997) for more details on reinsurance contracts).

We denote by L_T the stochastic cash flow representing the future liabilities of the company, and by L_0 the premium income collected. L_T is the sum of all the payments occurring during the time interval $[0, T]$. A typical aggregate excess reinsurance contract is represented by the stop-loss agreements, where the reinsurer pays the part of L_T which exceeds a certain amount, R. The reinsurer obligation is usually limited to a given amount, M. More precisely, the payoff of a stop-loss contract can be summarized as follows:

$$Y_T = \begin{cases} 0, & L_T \leq R \\ (L_T - R), & R < L_T < R + M. \\ M, & L_T \geq R + M \end{cases} \tag{6.1}$$

It is worth mentioning that no standardized contracts can be found on the market. Reinsurance agreements are usually tailored to meet specific requirements of the transferring parts.

Reinsurance agreements are usually evaluated either by standard actuarial methods (see Straub 1997) or by equilibrium techniques, using the so-called *financial reinsurance* method. In the former case, the price is computed as the sum of the expected value of the future payments and a risk premium which is usually proportional to the variance of the distribution of L_T. In the latter case, the insurance liability is assumed to be highly correlated to a traded asset, and the price follows by application of the capital asset pricing model of Sharpe (1964). Application of financial reinsurance and the design of reinsurance contract have been studied in de Lange et al. (2004).

In the sequel of the chapter we describe a stochastic programming model to evaluate reinsurance contract by super-replication. This is done by recognizing that the contract Y_T is a European contingent claim (ECC) written on the non-traded underlying liability L_T.

We also consider an asset and liability management (ALM) problem, where an insurance firm with a future liability L_T needs to determine the optimal investment strategy in order to maximize its expected profit. The asset universe in which the company might invest consists of a set of risky assets (stocks), a risk-free security, and the reinsurance contract. The goal of the company is to maximize its expected profit, given a set of regulatory constraints to be fulfilled. In fact, international accounting standards require that insurance companies meet specific obligations, expressed in terms of risk margins which are usually proportional to the value at risk (VaR) of the loss distribution.

6.3 A Stochastic Programming Model for Super-Replication

The problem of option pricing in incomplete markets is currently an active area of research. The different methodologies proposed over the years differ on the way the option price is defined. In what follows, we describe a superhedging method based on a stochastic programming representation of the hedging problem. An alternative approach is the quadratic hedging of Föllmer and Schweizer (1989) (see also Dahl and Møller 2006, Dahl et al. 2008, for recent applications in the life insurance context). Global risk minimization (see Cerny and Kallsen 2009, and reference therein) based on quadratic hedging is a promisingly new area of research. Utility-based pricing algorithms have been proved to be effective in different area of applications. We refer to Carmona (2008) for a book-length treatment of the topic including application to the insurance sector.

6.3.1 Notation and Probabilistic Structure

We introduce a financial market where security prices and other payments are discrete random variables supported by a finite-dimensional probability space $(\Omega, \mathcal{F}, P)$. The atoms $\omega \in \Omega$ are sequences of real-valued vectors (asset values and payments) over the discrete time periods $t = 0, 1, \ldots, T$. The path histories of the security prices up to time t correspond one-to-one with nodes $n \in \mathcal{N}_t$. The set $\mathcal{N}_0$ consists of the root node $n = 0$ and the leaf nodes $n \in \mathcal{N}_T$ correspond one-to-one with the probability atoms $\omega \in \Omega$. This probabilistic structure can be modeled by a discrete, non-recombining scenario tree. In the scenario tree, every node $n \in \mathcal{N}_t$, $t = 1, \ldots, T$, has a unique ancestor $a(n) \in \mathcal{N}_{t-1}$, and every node $n \in \mathcal{N}_t$, $t = 0, \ldots, T-1$, has a non-empty set of child nodes $\mathcal{C}(n) \subset \mathcal{N}_t$. The collection of all the nodes is denoted by $\mathcal{N} \equiv \bigcup_{t=0}^{T} \mathcal{N}_t$. The information arrival can be modeled by associating a set of σ-algebras $\{\mathcal{F}_t\}_{t=0,\ldots,T}$ with $\mathcal{F}_0 = \{\phi, \Omega\}$, $\mathcal{F}_{t-1} \subseteq \mathcal{F}_t$, and $\mathcal{F}_T = \mathcal{F}$.

We model the probability measure P by attaching weights $p_n > 0$ to each leaf node $n \in \mathcal{N}_T$ so that $\sum_{n \in \mathcal{N}_T} p_n = 1$. The probability of each non-final node $n \in \mathcal{N}_t$, $t \neq T$ is recursively determined by

$$p_n = \sum_{m \in \mathcal{C}(n)} p_m. \tag{6.2}$$

The market consists of $J + 1$ tradeable securities indexed by $j = 0, 1, \ldots, J$ one of which, say security 0, is the risk-free security. We model the risk-free asset by assuming the existence of a market for zero coupon bonds for each trading date $t = 0, \ldots, T-1$. At time $t+1$ the risk-free value in node n, S_n^0, corresponds to the previous value $S_{a(n)}^0$ capitalized with the return given by the bond $B_{t,t+1}^n$ in node n. That is, S_n^0 is a one-dimensional stochastic process defined on the finite probability space $(\Omega, \mathcal{F}, P)$ and $\mathcal{F}_{t+1}$-measurable.[1] This allows us to introduce in the market model the stochastic dynamics of the yield curve. Other securities prices (stocks, future, etc.) are described by a nonnegative-valued vector $S_0 = \left(S_0^1, \ldots, S_0^J\right)$ of initial (known) market prices and nonnegative-valued random vectors $S_n : \Omega \rightarrow \mathbb{R}^J$, $\mathcal{F}_t$-measurable, $\forall n \in \mathcal{N}_t$ and $t = 0, \ldots, T$.

The yield curve plays an important role in the insurance industry. It is used to determine the *best estimate* of future liabilities as provided by the European accounting standards Solvency II. In the context of this chapter, the term structure is used to include in the universe of investment opportunities an asset in which investors can trade with no risk.

In presence of risk factors other than the traded securities, the process S_t is augmented by $J + 1$ real-valued variables $\xi_t = \left(\xi_0^1, \ldots, \xi_t^J\right)$ whose path histories match the nodes $n \in \mathcal{N}_t$, for each $t = 0, 1, 2, \ldots, T$. This is certainly the case of the property and casualty market, where liabilities are generated by factors exogenous to the financial markets (car accidents, earthquakes, etc.). In the framework described above, an ECC is a security whose owner is entitled to receive the $\mathcal{F}_t$-adapted stochastic cash flow $F = \{F_t\}_{t=0,\ldots,T}$. The above definition of ECC is general enough to encompass a large variety of derivatives, including futures and exotic options. Options written on non-traded underlying variables—which is one of the major source of market incompleteness—are also embraced. The stop-loss reinsurance contract described in (6.1) clearly falls in the class of ECCs, with $F_n = Y_n$ for $n \in \mathcal{N}_T$ and $F_n = 0$ for $n \in \mathcal{N}_t$, $t = 0, \ldots, T - 1$. Here Y_n is the value of the reinsurance contract given as a function of the liability L_n occurred at node n and corresponds to the underlying asset of the ECC. In the property and casualty markets, contrary to the financial markets, one cannot trade with the underlying liabilities. This makes the reinsurance contract difficult to hedge by trading in the underlying, and thus we must switch to the theory of option pricing in incomplete markets.

In the general framework of incomplete markets, contingent claims cannot be uniquely priced by no-arbitrage arguments. Rather, there exists the so-called *arbitrage interval* which describes the range of all arbitrage-free evaluations. Lower and upper bounds of the arbitrage interval are determined by the buyer price and seller price, respectively. Informally, the buyer price is the maximum amount of money an investor is willing to pay in order to buy the contingent claim. Likewise, the seller price is the minimum amount of money that is needed to the writer in order to build a portfolio whose payout is at least as equal as to that of the contingent claim. In this chapter we assume that the policyholder are price-taker, thus concentrating our attention on the seller problem only. For details about the construction of the buyer price and the relation between the arbitrage interval and the set of equivalent martingale measure we refer to King (2002).

[1] This means that, at node n, we already know the ending-period value S_m^0, $\forall m \in \mathcal{C}(n)$ but, $\forall n, m \in \mathcal{N}_t$ with $n \neq m$, S_n^1 and S_m^0 could be different.

6.3.2 Super-Replication of ECCs

The reinsurer (protection seller) receives an amount, V, corresponding to the premium of the reinsurance contract and she agrees to pay the protection buyer the payoff of the reinsurance contract. The seller's objective is to select a portfolio of tradeable assets that enables her to meet her obligations without risk (i.e., on all nodes $n \in \mathcal{N}$). The portfolio process must be *self-financing*, i.e., the amount of assets bought has to offset the amount of assets sold. Put it differently, there are no inflows or outflows of money, since the amount of money available at node n is funded only by price movements at the ancestor node, $a(n)$, and at the node itself.

Using the tree notation, we denote by Z_n^j the number of shares held at each node $n \in \mathcal{N}$ and for each security $j = 0, 1, \ldots, J$.

At the root node, the value (price) of the hedging portfolio plus any payout to be covered is the price of the option, that is

$$\sum_{j=0}^{J} S_0^j Z_0^j = V. \tag{6.3}$$

At each node $n \in \mathcal{N}_t$, $t = 1, 2, \ldots, T$, the self-financing constraints are defined as follows:

$$\sum_{j=0}^{J} S_n^j Z_n^j = \sum_{j=0}^{J} S_n^j Z_{a(n)}^j. \tag{6.4}$$

To sum up, the stochastic programming model for the insurance evaluation problem can be written as follows:

Problem 3.1 (Writer problem for ECCs)

$$\underset{Z_n^j \in}{\text{Minimize}}\ V \tag{6.5}$$

$$\sum_{j=0}^{J} S_0^j Z_0^j = V, \tag{6.6}$$

$$\sum_{j=0}^{J} S_n^j Z_n^j = \sum_{j=0}^{J} S_n^j Z_{a(n)}^j \qquad \text{for all } n \in \mathcal{N}_t,\ t = 1, 2, \ldots, T-1, \tag{6.7}$$

$$\sum_{j=0}^{J} S_n^j Z_n^j + L_n = \sum_{j=0}^{J} S_n^j Z_{a(n)}^j \qquad \text{for all } n \in \mathcal{N}_T, \tag{6.8}$$

$$\sum_{j=0}^{J} S_n^j Z_n^j \geq 0 \qquad \text{for all } n \in \mathcal{N}_T, \tag{6.9}$$

$$Z_n^j \in \mathbb{R} \qquad \text{for all } n \in \mathcal{N},\ j = 0, 1, \ldots, J. \tag{6.10}$$

We highlight here some important points:

1. Problem 3.1 is a linear programming model where the objective function minimizes the value of the hedging portfolio or rather the price of the option.
2. The payout process $\{F_n\}_{n \in \mathcal{N}}$ is a parameter of Problem 3.1. This implies that any complicated structure for F_n does not change the complexity of the model.
3. Constraints (6.9) ensure that at each final node the total position of the hedging portfolio is not short. In other words, if short positions are allowed, the portfolio process must end up with enough long positions so that a positive portfolio value is delivered.

6.3.3 Tree Generation

We generate the tree for the underlying price process S by matching the first M moments of its unknown distribution. Our approach is based on the moment matching method of Høyland and Wallace (2001), where the user provides a set of moments, $\mathcal{M}$, of the underlying distribution (mean, variance, skewness, covariance, or quantiles), and then, prices and probabilities are jointly determined by solving a non-linear system of equations or a non-linear optimization problem. The method also allows for intertemporal dependencies, such as mean reverting or volatility clumping effect. For a review on alternative scenario generation methods, see Dupačová et al. (2000) and references therein.

The following system of non-linear equations formalizes the moment matching problem:

Problem 3.2 (Moment matching model)

$$f_k(S, p) = \tau_k \quad \forall k \in \mathcal{M}, \tag{6.11}$$

$$\sum_{m \in \mathcal{C}(n)} p_m = 1, \tag{6.12}$$

$$p_m \geq 0, \quad m \in \mathcal{C}(n), \tag{6.13}$$

where $f_k(S, p)$ is the algebraic expression for the statistical property $k \in \mathcal{M}$ and τ_k is its target value. For example, let us consider the problem of matching the expected values $\mu_{S^0}, \ldots, \mu_{S^J}$, the variances, $\sigma^2_{S^0}, \ldots, \sigma^2_{S^J}$, and the correlations, $\rho_{j,k}$, $j = 0, 1, \ldots, J$, and $k \neq j$, of the log-returns of each asset. The moment matching problem becomes

Problem 3.3 (Moment matching model: an example)

$$\sum_{m \in \mathcal{C}(n)} r^j_m p_m = \mu_{S^j}, \quad j = 0, \ldots, J, \tag{6.14}$$

$$\sum_{m \in \mathcal{C}(n)} \left(r_m^j - \mu_{S^j}\right)^2 p_m = \sigma_{S^j}^2, \quad j = 0, 1, \ldots, J, \tag{6.15}$$

$$\sum_{m \in \mathcal{C}(n)} \frac{\left(r_m^j - \mu_{S^j}\right)\left(r_m^k - \mu_{S^k}\right) p_m}{\sigma_{S^j}\sigma_{S^k}} = \rho_{j,k}, \quad j = 0, 1, \ldots, J \quad \text{and} \quad k \neq j, \tag{6.16}$$

$$\ln\left(\frac{S_m^j}{S_n^j}\right) = r_m^j, \quad j = 1, 2, \ldots, n, \tag{6.17}$$

$$\sum_{m \in \mathcal{C}(n)} p_m = 1, \tag{6.18}$$

$$p_m \geq 0, \quad m \in \mathcal{C}(n). \tag{6.19}$$

The non-linearity of Problem 3.2 could lead to infeasibility. An alternative strategy is to formulate Problem 3.2 as a goal programming model. We can minimize the weighted distance between the statistical properties of the tree, and their target values, that is

Problem 3.4 (Moment matching model: goal programming)

$$\underset{S,p}{\text{Minimize}} \sum_{k \in \mathcal{M}} \omega_k \left(f_k(S, p) - \tau_k\right)^2, \tag{6.20}$$

$$\sum_{m \in \mathcal{C}(n)} p_m = 1, \tag{6.21}$$

$$p_m \geq 0, \qquad m \in \mathcal{C}(n), \tag{6.22}$$

where ω_k is the weight which measures the importance associated with the statistical property $k \in \mathcal{M}$. Problem 3.4 is easier to solve with respect to the previous non-linear system, but a perfect match is generally difficult to meet. A hybrid model, in which only some statistical properties are required to be perfectly matched, may be adopted to avoid bad match of the specifications. In detail, we split the set $\mathcal{M}$ of parameters to fit in two subsets, $\mathcal{M}_1$ and $\mathcal{M}_2$, and implement the following non-linear programming problem:

Problem 3.5 (Moment matching model: mixed goal programming)

$$\underset{S,p}{\text{Minimize}} \sum_{k \in \mathcal{M}_1} \omega_k \left(f_k(S, p) - \tau_k\right)^2, \tag{6.23}$$

$$f_i(S, p) = \tau_i, \quad i \in \mathcal{M}_2, \tag{6.24}$$

$$\sum_{m \in \mathcal{C}(n)} p_m = 1, \tag{6.25}$$

$$p_m \geq 0, \quad m \in \mathcal{C}(n). \tag{6.26}$$

Although these three different strategies can be implemented, there is no general rule that suggests the choice of a particular one. Each of these strategies has its own pros and cons, and the choice depends on the specific problem to be faced with.

To be consistent with financial asset pricing theory, arbitrage opportunities should be avoided. In the setting described in Section 6.3, the no-arbitrage condition is fundamental. In fact, if the market allows for arbitrages, the stochastic programming problem described in this chapter will end up with an unbounded solution.

Following Gulpinar et al. (2004), we can preclude arbitrages by imposing the existence of a strictly positive martingale measure such that

$$S_n^j = e^{-r_n} \sum_{m \in \mathcal{C}(n)} S_m^j \pi_m, \tag{6.27}$$

where r_n is the risk-free interest rate observed in state n. We refer to Klaassen (2002) for alternative, more stringent, no-arbitrage conditions.

6.4 Asset and Liability Management with Reinsurance Contracts

In Section 6.3 we have considered the position of the reinsurance company whose objective is to determine the optimal hedging strategy for the contract and to price it accordingly. We now turn to the position of the buyer of the protection. The goal is to manage the liabilities she has to face by making use of both financial investments and the reinsurance contract. We assume that the price of the reinsurance contract is determined by the protection seller and that the company has to satisfy regulatory constraints expressed in terms of value at risk of the loss distribution. To be more specific, we define the loss function

$$\Xi(n) = -\sum_{j=0}^{J} S_n^j Z_n^j - xY_n + L_n, \qquad n \in \mathcal{N}_T, \tag{6.28}$$

which corresponds to the value of the portfolio after the payments for the liability L_n have been made. According to (6.28): $\Xi(n) = -\Psi(n)$, where $\Psi(n)$ is the profit realized at node n after the liability has been paid. Following Rockafellar and Uryasev (2000), the conditional value at risk (CVar) of the loss distribution at the probability level α is defined by the following set of equations:

$$\zeta + \frac{1}{(1-\alpha)} \frac{1}{|\mathcal{N}_T|} \sum_{n \in \mathcal{N}_T} \phi(n), \tag{6.29}$$

$$\phi(n) = \Xi(n) - \zeta, \qquad n \in \mathcal{N}_T, \tag{6.30}$$

$$\phi(n) \geq 0, \qquad n \in \mathcal{N}_T. \tag{6.31}$$

The endogenous variable ζ can be shown to represent the value at risk of the loss distribution. Notice that the shortfalls in (6.30) are defined to be non-negative. As a consequence, in order to bound ζ, an upper limit on (6.29) is sufficient. The following linear program determines the optimal investment strategy for an insurance company seeking to maximize its expected profits, given that the α-CVar of the loss distribution cannot exceed the limit ω:

Problem 4.1 (Conditional VaR model)

$$\underset{Z_n^j, x, \zeta}{\text{Maximize}} \frac{1}{|\mathcal{N}_T|} \sum_{n \in \mathcal{N}_T} \Psi(n), \tag{6.32}$$

$$\sum_{j=0}^{J} S_0^j Z_0^j + x Y_0 = L_0, \tag{6.33}$$

$$\sum_{j=0}^{J} S_n^j Z_n^j x Y_n - L_n = \Psi(n), \qquad n \in \mathcal{N}_T, \tag{6.34}$$

$$\sum_{j=0}^{J} S_n^j \left(Z_n^j - Z_{a(n)}^j \right) = 0 \quad \text{for all } n \in \mathcal{N}_t, \; t = 1, 2, \ldots, T, \tag{6.35}$$

$$- \sum_{n \in \mathcal{N}_T} \Psi(n) - \zeta \leq \phi(n), \quad n \in \mathcal{N}_T, \tag{6.36}$$

$$\zeta + \frac{1}{(1-\alpha)} \frac{1}{|\mathcal{N}_T|} \sum_{n \in \mathcal{N}_T} \phi(n) \leq \omega, \tag{6.37}$$

$$\phi(n) \geq 0, \tag{6.38}$$

$$x, Z_n^j, \geq 0. \tag{6.39}$$

6.5 Implementation Notes and Results

We consider a tree structure with six stages and perform all the experiments by assuming a time horizon $T = 10$ years. The time step between two consecutive stages is thus fixed to 1.67 years. In the tree, each non-final node branches into five successor nodes. The resulting tree has exactly 19,531 total nodes and $5^6 = 15{,}625$ final nodes.

The financial market consists of three risky securities (stocks) plus a risk-free asset. The risk-free asset has initial value 100 and is assumed to grow at the continuously compounded annual rate of 3%. We generate the risky assets by using the tree generation model described in Problem 3.4 and by restricting the moment matching problem to fit the expected values, standard deviations, and correlations of the continuously compounded asset returns as displayed in Table 6.1. For all the assets, the skewness parameter and kurtosis are set, respectively, to $\beta_1 = 0$ and

Table 6.1 Statistical properties for the returns of securities used in the experiments

			Correlations		
	Mean	Std. Dev	Asset 1	Asset 2	Asset 3
Asset 1	0.04	0.13	1	–	–
Asset 2	0.045	0.15	0.6	1	–
Asset 3	0.054	0.2	0.45	0.24	1

We built scenario trees by matching means, standard deviations, and correlations shown in the table. Skewness and kurtosis parameters are set, respectively, to $\beta_1 = 0$ and $\beta_2 = 3$

$\beta_2 = 3$. This is equivalent to assume a market where asset returns are Gaussian with parameters specified as in Table 6.1. This is a simplifying assumption that does not affect the relevance of our framework. In a more general setup, we can determine probability distributions that match higher moments (see Consiglio and De Giovanni, forthcoming). Finally, the initial value of all the assets is set equal to 100.

6.5.1 Pricing Reinsurance Contracts

Our experiments are based on four case studies (CS). More specifically, we consider four different distributions for the random variable L_T that represents the actuarial claim the insurance company will face with. Accordingly, we generate four different scenario trees for L_T, one for each CS, using Problem 3.4. Table 6.2 displays the expected values and the standard deviation of each CS. The skewness and kurtosis parameter are set, respectively, to $\beta_1 = 0$ and $\beta_2 = 3$, while the financial assets and the actuarial claims are assumed to be uncorrelated.

Table 6.3 displays the prices of the reinsurance contract with fixed values of R and M. The pattern shown in Table 6.3 is not surprising. The higher the expected value and the variance of the liability distribution, the higher the price of the reinsurance contract. It is, however, interesting to look at the initial composition of the strategies produced to super-replicate the reinsurance contract. Such results are displayed in Fig. 6.1. We observe that to super-replicate the contract the reinsurer is long in the risk-free security and short in the risky assets. The cause of this strategy

Table 6.2 Statistical properties for the actuarial claims used in the experiments

	Mean	Std. Dev
CS-1	180	22
CS-2	190	40
CS-3	195	45
CS-4	200	50

We built scenario trees by matching means and standard deviations as displayed. The skewness and kurtosis parameter are set, respectively, to $\beta_1 = 0$ and $\beta_2 = 3$. No correlation between the financial market and the actuarial risk is assumed

Table 6.3 Prices of the reinsurance contract for two different levels of the parameters R and M

	$R = 160, M = 200$	$R = 140, M = 220$
CS_1	21.264	36.473
CS_2	49.494	64.921
CS_3	59.051	74.542
CS_4	64.356	79.835

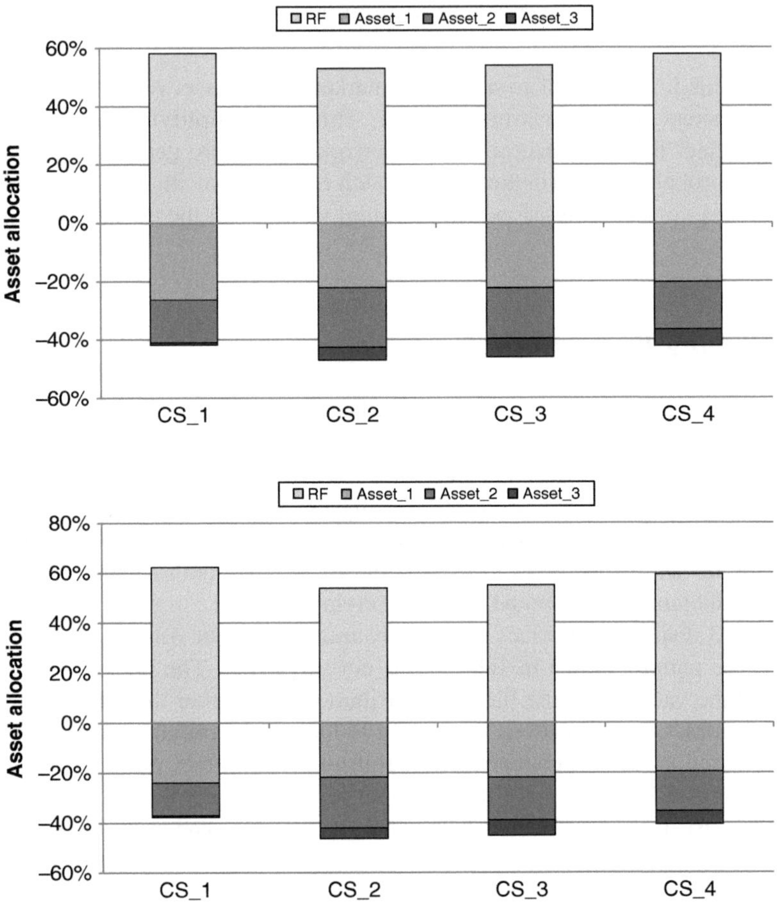

Fig. 6.1 Initial values, in percentage, of the super-replicating portfolio for $R = 160$, $M = 200$ (*top panel*) and $R = 140$, $M = 220$ (*bottom panel*)

is that the underlying asset (the insurance liability L_T) and the risky assets are uncorrelated. This prevents the reinsurer to exploit the co-movements between the traded assets and the underlying (see the discussion in Consiglio and De Giovanni, forthcoming, section 5). We also observe that the amount of asset 1 and asset 2 to short-sell in the super-replicating strategy is much higher than that of asset 3 which corresponds to the most risky asset in the market. The latter comes into play when the riskiness of the liability exceeds a certain level that in this experiment is given by CS_3 and CS_4.

6.5.2 Risk Management with Reinsurance Contracts

We use the liability structure based on the simulation CS_1 in Table 6.1, and set the initial premium collected by the insurance company to $L_0 = 145$. Two different parametrizations of the reinsurance contract are considered with (1) $R = 160$ and $M = 200$ and (2) $R = 140$ and $M = 220$. The prices of the two contracts are set to their super-replication values: $Y_0 = 21.264$ and $Y_0 = 36.473$, respectively. We then run Problem 4.1 with increasing levels of the risk tolerance, from $\omega = 0$ to $\omega = 3$. The resulting efficient frontiers are displayed in Fig. 6.2, while the optimal portfolios for each level of tolerance are displayed in Fig. 6.3. Note that the problem allows for a feasible optimal solution when $\omega = 0$. This is because the reinsurance contract completely offset any possible loss, thus eliminating any risk.

From the experiment we learn some important facts. First, the optimal portfolios include a high percentage of reinsurance in both cases and for all levels of ω (Fig. 6.3) even if the contracts themselves are evaluated at their super-replication prices. This empirical finding is supported by theoretical arguments in King (2002, section 8), where the author proves that in the presence of a liability structure the investors are willing to trade in the derivative. Second, the amount of reinsurance declines with the increase of the level of riskiness allowed ω. This is a confirmation of the rational choice of the model which reduces the level of the reinsurance if the risk exposure is increased by the decision maker. We strongly believe this could be

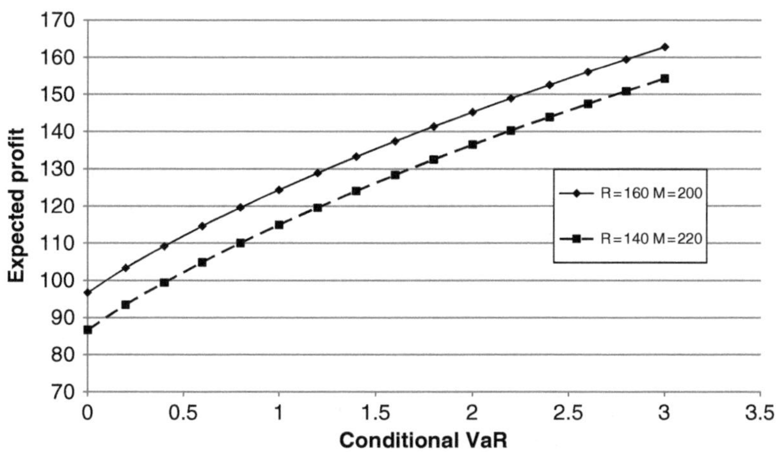

Fig. 6.2 Efficient frontiers. Conditional VaR of the losses vs expected profit of the portfolio for confidence level $\alpha = 99\%$. The efficient frontiers start at $\omega = 0$ as the losses can be totally offset by purchasing a reinsurance contract

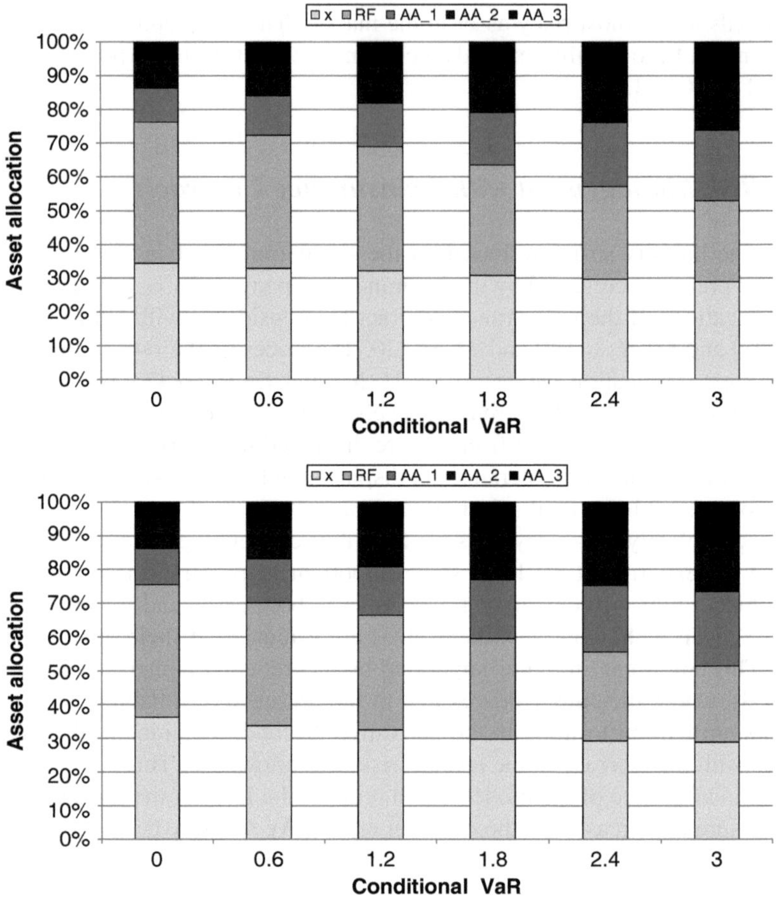

Fig. 6.3 Optimal portfolios for $R = 160$, $M = 200$ (*top panel*) and $R = 140$, $M = 220$ (*bottom panel*)

a useful tool for regulators which can ascertain whether a given level of reinsurance is safe for the liability structure of a company, and in case of negative result forces the company toward a more consistent level of protection.

6.6 Conclusions

We propose in this chapter a stochastic programming model to cope with the price and management of reinsurance contracts. We show that the pricing in incomplete markets is a feasible alternative to the stronger hypotheses of completeness, where, in this case, the main obstacle in hedging the insurance contract is the non-tradeability of the underlying asset. The model is flexible enough to be embedded in a wider asset–liability model to determine the optimal level of trade-off between expected profit and conditional VaR.

References

Renè Carmona, editor. *Indifference Pricing: Theory and Applications*, Princeton Series in Financial Engineering. Princeton University Press, Princeton, NJ, 2008.

A. Cerny and J. Kallsen. Hedging by sequential regressions revisited. *Mathematical Finance*, 19(4):591–617, 2009.

T.F. Coleman, Y. Kim, Y. Li, and M. Patron. Robustly hedging of variable annuities with guarantees under jump and volatility risks. *Journal of Risk and Insurance*, 74(2):347–376, 2007.

A. Consiglio and D. De Giovanni. Evaluation of insurance products with guarantee in incomplete markets. *Insurance: Mathematics & Economics*, 42(1):332–342, 2008.

A. Consiglio and D. De Giovanni. Pricing the option to surrender in incomplete markets. *Journal of Risk and Insurance*, 77(4):935–957, 2010.

M. Dahl and T. Møller. Valuation and hedging of life insurance liabilities with systematic mortality risk. *Insurance: Mathematics & Economics*, 39(203):193–217, March 2006.

M. Dahl, M. Melchior, and T. Møller. On systematic mortality risk and risk-minimization with survivor swaps. *Scandinavian Actuarial Journal*, 2008(2&3):114–146, 2008.

P.E. de Lange, S.E. Fleten, and A.A. Gaivoronski. Modeling financial reinsurance in the casualty insurance business via stochastic programming. *Journal of Economic Dynamics & Control*, 28(5):991–1012, 2004.

J. Dupačová, G. Consigli, and S.W. Wallace. Scenarios for multistage stochastic programs. *Annals of Operations Research*, 100:25–53, 2000.

H. Föllmer and M. Schweizer. Hedging by sequential regression: An introduction to the mathematics of option trading. *Astin Bulletin*, 1:147–160, 1989.

N. Gulpinar, B. Rustem, and R. Settergren. Simulation and optimization approaches to scenario tree generation. *Journal of Economic Dynamics and Control*, 28(7):1291–1315, 2004.

K. Høyland and S.W. Wallace. Generation scenario trees for multistage decision problems. *Management Science*, 47(2):295–307, 2001.

A.J King. Duality and martingales: A stochastic programming perspective on contingent claims. *Mathematical Programming, Series B*, 91:543–562, 2002.

P. Klaassen. Comment on "Generating scenario trees for multistage decision problems". *Management Science*, 48(11):1512–1516, 2002.

S. Mello and E. Parsons. Maturity structure of a hedge matters: lessons from the Metallgesellschaft debacle. *Journal of Applied Corporate Finance*, 8(1):106–121, 1995.

R.T. Rockafellar and S. Uryasev. Optimization of conditional Value–at–Risk. *Journal of Risk*, 2(3):21–41, 2000.

W.F. Sharpe. Capital asset prices: A theory of market equilibrium under conditions of risk. *Journal of Finance*, 19(3):425–442, 1964.

E. Straub. *Non–Life Insurance Mathematics*. Springer, New York, NY, 1997.

Part II
Energy Applications

Chapter 7
A Decision Support Model for Weekly Operation of Hydrothermal Systems by Stochastic Nonlinear Optimization

Andres Ramos, Santiago Cerisola, Jesus M. Latorre, Rafael Bellido, Alejandro Perea, and Elena Lopez

Abstract This chapter formulates and solves an optimal resource allocation problem of thermal and hydropower plants with multiple basins and multiple connected reservoirs. The stochastic factors of the problem are here represented by natural hydro inflows. A multivariate scenario tree is in this case obtained taking into account the stochastic inputs and their spatial and temporal dependencies. The hydropower plant efficiency depends on its water head and the reservoir volume depends nonlinearly on the headwater elevation, leading to a large-scale stochastic nonlinear optimization problem, whose formulation and solution are detailed in the case study. An analysis of exhaustive alternatives of computer implementation is also discussed.

Keywords Medium-term hydrothermal scheduling · Nonlinear optimization · Water head · Stochastic optimization · Scenario tree · Cost minimization · Hydro reservoirs

Notation

For clarity purposes parameters are represented in uppercase letters and variables are represented by lowercase letters.

Indices

p	time period
p'	time subperiod
t	thermal unit
h	storage hydro or pumped storage hydro plant

A. Ramos (✉)
Universidad Pontificia Comillas, 28015 Madrid, Spain
e-mail: andres.ramos@upcomillas.es

M. Bertocchi et al. (eds.), *Stochastic Optimization Methods in Finance and Energy*, International Series in Operations Research & Management Science 163,
DOI 10.1007/978-1-4419-9586-5_7,

r	hydro reservoir
$rr(r)$	reservoir upstream of reservoir r
$rh(r)$	reservoir r upstream of storage hydro plant
$hr(r)$	storage hydro plant upstream of reservoir r
ω	inflow scenario
$a(\omega)$	ancestor scenario of scenario ω in previous period

Parameters

$D_{pp'}$	demand of subperiod p' in period p (MW)
$\mathrm{DU}_{pp'}$	duration of subperiod p' in period p (h)
P_p^{ω}	probability of scenario ω in period p (p.u.)
$\underline{TP}_{pt}, \overline{TP}_{pt}$	minimum and maximum output of thermal unit t in period p
VC_t	Variable cost of thermal unit t (€/MWh)
FOR_t	Forced outage rate of thermal unit t (p.u.)
$\underline{H}_t, \overline{H}_t$	Minimum and maximum yearly operation hours of thermal unit t (h)
$\mathrm{PDH}_t, \mathrm{PEH}_t$	Penalty by deficit or surplus of yearly operation hours of thermal unit t (€/kh)
$\underline{\mathrm{HP}}_{ph}, \overline{\mathrm{HP}}_{ph}$	Minimum and maximum output of storage hydro plant h (MW)
$\underline{\mathrm{PP}}_{ph}, \overline{\mathrm{PP}}_{ph}$	Minimum and maximum consumption of pumped storage hydro plant h (MW)
η_h	Efficiency of pumped storage hydro plant h (p.u)
$\underline{R}_{pr}, \overline{R}_{pr}$	Minimum and maximum operational reserve volume of reservoir r in period p (hm^3) (GWh)
$\underline{C}_{pr}, \overline{C}_{pr}$	Minimum and maximum capacity of reservoir r in period p (hm^3) (GWh)
$\mathrm{IR}_r, \mathrm{FR}_r$	Initial and final reserve volume of reservoir r (hm^3) (GWh)
$\mathrm{PDFR}_r, \mathrm{PEFR}_r$	Penalty by deficit in final reserve (and in minimum and artificial reserve) and surplus in final reserve (and in maximum reserve) of reservoir r (k€/hm^3) (€/MWh)
$\underline{G}_{pr}$	Minimum release of reservoir r in period p (hm^3) (GWh). This accounts for other uses of water for water supply, environmental and ecological concerns like fish and wildlife maintenance and recreational activities
I_{pr}^{ω}	Unregulated inflow of reservoir r in period p of scenario ω (m^3/s) (GWh)
TH_h	Tailrace elevation of hydro plant h(m)
RH_r	Reference elevation of reservoir r(m)
A_h, A'_h	Fixed and linear term of production function of plant (hWh/m^3) (hWh/m^4)
B_r, B'_r, B''_r	Fixed, linear, and quadratic terms of reserve volume of reservoir r (hm^3) (hm^3/m) (hm^3/m^2)

Variables

$tp^{\omega}_{pp't}$	Output of thermal unit t in subperiod p' of period p of scenario ω (MW)
$hp^{\omega}_{pp'h}$	Output of storage hydro plant h in subperiod p' of period p of scenario ω (MW)
$pp^{\omega}_{pp'h}$	Consumption of pumped storage hydro plant h in subperiod p' of period p of scenario ω (MW)
s^{ω}_{pr}	Spillage of reservoir r in period p of scenario ω (hm^3) (GWh)
ar^{ω}_{pr}	Artificial reserve of reservoir r in period p of scenario ω (hm^3) (GWh)
r^{ω}_{pr}	Reserve volume of reservoir r at the end of period p of scenario ω (hm^3) (GWh)
dfr^{ω}_{r}, efr^{ω}_{r}	Deficit and surplus of final reserve of reservoir r in scenario ω (hm^3) (GWh)
dmr^{ω}_{pr}, emr^{ω}_{pr}	Deficit of minimum reserve and surplus of maximum reserve of reservoir r in period p of scenario ω (hm^3)
doh^{ω}_{t}, eoh^{ω}_{t}	Deficit of minimum yearly operation hours and surplus of maximum yearly operation hours of thermal unit t in scenario ω (h)
g^{ω}_{pr}	Release of reservoir r in period p of scenario ω (hm^3)
g^{ω}_{ph}	Release of storage hydro plant h in period p of scenario ω (hm^3)
pf^{ω}_{ph}	Production function of hydro plant h in period p of scenario ω (hWh/m^3)
tv^{ω}_{ph}	Tailrace volume of hydro plant h in period p of scenario ω (m)
rh^{ω}_{ph}	Reservoir elevation of reservoir r in period p of scenario ω (m)
wh^{ω}_{pr}	Headwater elevation of reservoir r in period p of scenario ω (m)

7.1 Introduction

Nowadays, under a deregulated framework in many countries electric companies manage their own generation resources and need detailed operation planning tools. In the next future, high penetration of renewable intermittent generation is going to change the electric system operation. Pumped storage hydro and storage hydro plants will play a much more important role due to their flexibility and complementary use with intermittent generation.

Operation planning models considering multiple interconnected cascaded hydroplants belonging to multiple basins can be classified into

- hydroelectric models that deal exclusively with hydropower plants and
- hydrothermal coordination models (HTCM) that manage the integrated operation planning of both hydropower and thermal power plants.

By nature, the later models are high-dimensional, dynamic, nonlinear, stochastic, and multiobjective. Solving these models is still a challenging task for large-scale systems. One key question for them is to obtain a feasible operation for each hydro plant, which is very difficult because the models require a huge amount of data, by the complexity of hydro subsystems and by the need to evaluate multiple hydrological scenarios. For these models no aggregation or disaggregation process for hydropower input and output is established. Besides, thermal power units are considered individually. Thus, rich marginal cost information is used for deciding hydro scheduling.

An HTCM determines the optimal yearly operation of all the thermal and hydropower plants taking into account multiple cascaded reservoirs in multiple basins. The objective function is based on cost minimization because the main goal is the medium-term hydro operation. However, the objective function can be easily modified to consider profit maximization if marginal prices are known (Stein-Erik and Trine Krogh 2008), which is a common assumption for price-taker companies.

This model is connected with other models within a hierarchical structure. At an upper level, a stochastic market equilibrium model (Cabero et al. 2005) with monthly periods is run to determine the hydro basin production. At a lower level, a stochastic simulation model (Latorre et al. 2007a) with daily periods details hydro plant power output. This later model analyzes for several scenarios the optimal operational policies proposed by the HTCM. In Fig. 7.1 it is represented the hierarchy of these three models. Adjustment feedbacks are allowed to assure the coherence among the output results.

The model presented in this chapter has two main uses. On one hand, Fig. 7.2 represents the typical horizon for yearly operation planning. It is a 2-year long scope beginning in October and ending in September, which corresponds to 2 consecutive hydrological years needed by the existence of multiannual reservoirs. This timeframe is used to avoid initial and terminal effects on the planning horizon because the natural planning period of interest is defined from January to December. On the other hand, Fig. 7.3 represents the second possible use of the model for obtaining optimal and "feasible" decisions under uncertainty in hydro inflows for the immediate future (for example, next 2 weeks). The operational decisions span for 2 years but only the first 2 weeks are actually implemented. Future decisions beyond these 2 weeks are not known with certainty. Once these 2-week decisions have been implemented, the model is reformulated with a new 2-year rolling horizon and solved again.

A recent review of the state of the art of hydro scheduling models is done in (Labadie 2004). According to stochasticity treatment models are classified into deterministic and stochastic ones.

Deterministic approaches are based on network flows, linear programming (LP), nonlinear programming (NLP) (Dembo et al. 1990), or mixed integer linear programming (MILP), where binary variables come from commitment decisions of thermal units or hydro plants or from piecewise linear approximation of nonlinear and nonconvex water head effects. For taking into account these nonlinear effects a successive LP solves are typically used. This process does not necessarily converge

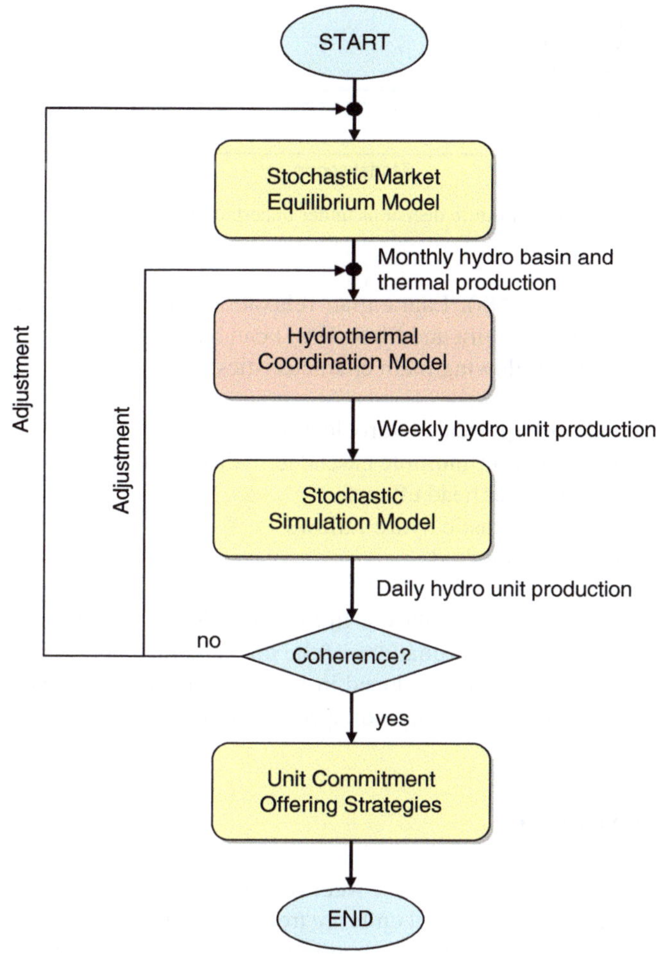

Fig. 7.1 Hierarchy of operation planning models

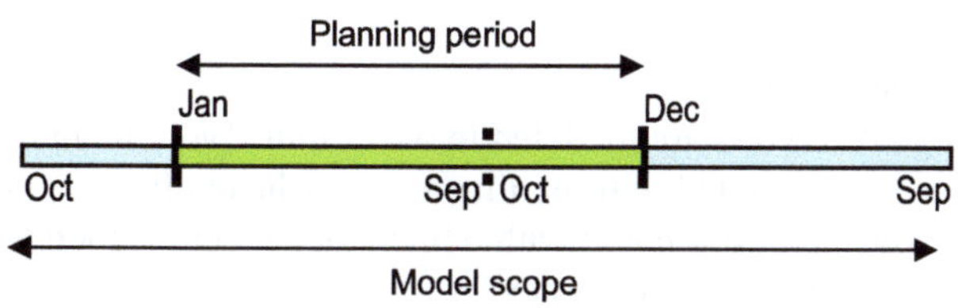

Fig. 7.2 Model scope for yearly operation planning

to the optimal solution; see (Bazaraa et al. 1993). This non-convergent behavior will also be tested with our model.

Stochastic approaches are represented by stochastic dynamic programming (SDP), stochastic linear programming (SLP) (Seifi and Hipel 2001), and stochastic nonlinear programming (SNLP). For SLP problems decomposition techniques like

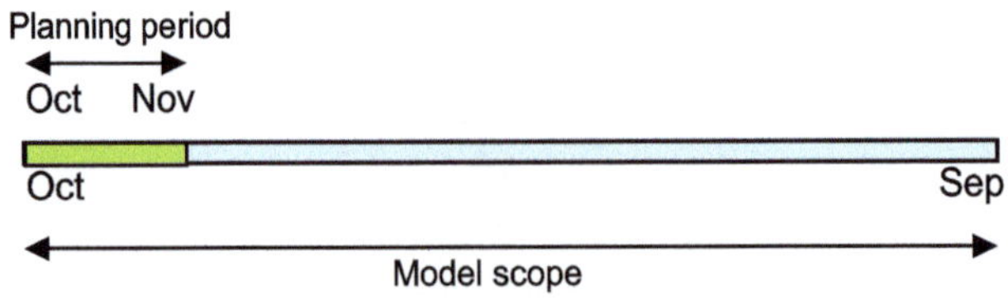

Fig. 7.3 Model scope for next future decisions under uncertain inflows

Benders (Jacobs et al. 1995), Lagrangian relaxation, or stochastic dual dynamic programming (SDDP) (Pereira and Pinto 1991) can be used.

This model has the following main characteristics:

- Specially suited for large-scale hydroelectric systems
- Deals with multireservoir, multiple cascaded hydro plants
- Consider nonlinear water head effects
- Takes into account stochastic hydro inflows
- Uses a robust solution method

The chapter is organized as follows. In Section 7.3 it is described the modeling of the system including all the equations of the mathematical optimization problem. The model implementation is introduced in Section 7.4. Then, the results for a real case study are presented and, finally, some conclusions are extracted.

7.2 System Modeling

The electric demand is modeled in a weekly basis with two constant load levels (peak and off-peak hours). Thermal units are treated individually. Commitment decisions of these units are considered as continuous variables given that the model is used for medium-term analysis. For hydro reservoirs a different modeling approach is followed depending on the following:

- Relevance of the reservoir
  Important large reservoirs are modeled in water units [volume in hm^3 and inflow in m^3/s]. They are modeled with nonlinear water head effects. On the contrary, smaller reservoirs are represented with a linear dependency; therefore, the model do not become unnecessarily complex.
- Owner company
  Hydropower plants belonging to other companies or state-operated reservoirs or the own small reservoirs are aggregated and modeled with one equivalent and independent reservoir each one, given that the reservoir and plant characteristics of some of them are generally ignored. They use energy units [volume and inflow in GWh].

Unregulated hydro inflows are assumed to be the dominant source of uncertainty in current Spanish electric system. In this system, stochasticity in hydro inflows have produced a hydroenergy availability ranging from 33.2 TWh in 2003 to 12.9 TWh in 2005 and hydroenergy generated has accounted for 20% in 2003 to only 9% in 2005 of the total energy demand, see (REE, http://www.ree.es).

Temporal changes in reservoir reserves are significant because of

- stochasticity in hydro inflows,
- highly seasonal pattern of inflows, and
- capacity of each reservoir with respect to its own inflow.

Stochasticity in hydro inflows is represented for the optimization problem by means of a multivariate scenario tree. This tree is generated by a neural gas clustering technique (Latorre et al. 2007b) that simultaneously takes into account the main stochastic inflow series and their spatial and temporal dependencies. The algorithm can take historical or synthetic series of hydro inflows as input data. Very extreme scenarios can be artificially introduced with a very low probability. The number of scenarios generated is enough for medium-term hydrothermal operation planning.

In Fig. 7.4 it is represented a scenario tree with eight scenarios. They correspond to the knee point of the quantization error function that measures the distance between the inflow series and the scenario tree versus the number of scenarios.

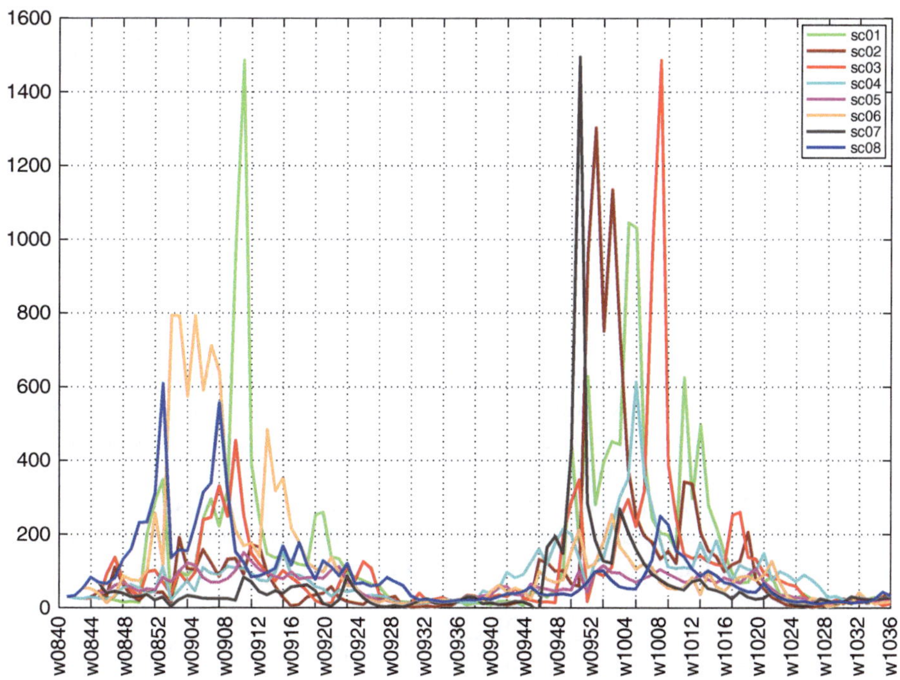

Fig. 7.4 Scenario tree with eight hydro inflows' scenarios from week 40 of year 2008 to week 39 of year 2010 expressed in (m^3/s)

7.3 Model Formulation

This HTCM is formulated as a stochastic nonlinear optimization problem as described in the following sections. The corresponding notation can be found at the beginning of the chapter. For clarity, uppercase letters are used for parameters and lower-case letters for variables.

The main results for each subperiod or load level (e.g., peak and off-peak) of each period (e.g., week) and scenario are storage hydro and pumped storage hydro plants operation and thermal unit operation, reservoir management, basin production, and marginal costs. As a byproduct the optimal water release tables for different stochastic inflows and reservoir volumes are obtained. They are computed by stochastic nested Benders' decomposition technique (Birge and Louveaux 1997) of a linear approximation of the stochastic nonlinear optimization problem. These release tables are used by the lower level daily stochastic simulation model, as seen in Fig. 7.1.

7.3.1 Constraints

The constraints introduced into the model are the following:

- Balance between generation and demand including pumping (MW)
  Generation of thermal units and storage hydro plants, $tp^{\omega}_{pp't}$ and $hp^{\omega}_{pp'h}$, respectively, minus consumption of pumped storage hydro plants, $pp^{\omega}_{pp'h}$, is equal to the demand $D_{pp'}$ for each scenario ω, period (week) p, and subperiod (load level) p':

$$\sum_t tp^{\omega}_{pp't} + \sum_h hp^{\omega}_{pp'h} - \sum_h \left(pp^{\omega}_{pp'h}/\eta_h\right) = D_{pp'} \qquad \forall p, p', \omega, \qquad (7.1)$$

  where η_h is the efficiency of pumped storage hydro plant h.
- Minimum and maximum yearly operation hours for each thermal unit in each scenario (h)
  These constraints are relaxed by introducing deficit and surplus variables, doh^{ω}_t and eoh^{ω}_t, respectively, that are penalized in the objective function; see Section 7.3.2. Those slack variables can be strictly necessary in the case of many scenarios of stochasticity where the larger the variability of hydro inflows the larger the change in a subset of thermal units.
  This type of constraints are introduced to account for some aspects that are not explicitly modeled into this model like unavailability of thermal units, domestic coal subsidies, CO_2 emission allowances, long-term capacity payments, etc.

$$\underline{H}_t - doh^{\omega}_t \le \frac{\sum_{pp'} \mathrm{DU}_{pp'} tp^{\omega}_{pp't}}{\overline{TP}_t} \le \overline{H}_t + eoh^{\omega}_t \qquad \forall t, \omega \qquad (7.2)$$

  being $\mathrm{DU}_{pp'}$ the duration of subperiod p' of period p.

- Minimum and maximum yearly average operation hours for each thermal unit (h)
Observe that this constraint does not have deficit and surplus variables because it corresponds to average generating hours:

$$\underline{H}_t \leq \frac{\sum_{pp'\omega} P_p^{\omega}\,\mathrm{DU}_{pp'}tp_{pp't}^{\omega}}{\overline{TP}_t} \leq \overline{H}_t \qquad \forall t, \tag{7.3}$$

where P_p^{ω} is the probability of scenario ω in period p.

- Water inventory balance for large reservoirs modeled in water units (hm^3)
Reservoir volume at the beginning of the period $r_{p-1,r}^{a(\omega)}$ plus unregulated inflows I_{pr}^{ω} plus spills from upstream reservoirs $\sum_{r'\in rr(r)} s_{pr'}^{\omega}$ minus spills from this reservoir s_{pr}^{ω} plus turbined water from upstream storage hydro plants $\sum_{r'\in rr(r)} g_{pr'}^{\omega}$ plus pumped water from downstream pumped storage hydro plants $\sum_{r'\in rr(r)} p_{pr'}^{\omega}$ minus turbined g_{pr}^{ω} and pumped water from this reservoir p_{pr}^{ω} is equal to reservoir volume at the end of the period r_{pr}^{ω}.
An artificial inflow $\mathrm{ar}_{pr}^{\omega}$ is allowed and penalized in the objective function; see Section 7.3.2. Hydro plant h that takes water from reservoir r is $rh(r)$ or releases it to reservoir r, $hr(r)$. The initial value of reservoir volume is assumed known. No lags are considered in water releases because 1 week is the time period unit:

$$r_{p-1,r}^{a(\omega)} + \mathrm{ar}_{pr}^{\omega} + I_{pr}^{\omega} + \sum_{r'\in rr(r)} \left(s_{pr'}^{\omega} + g_{pr'}^{\omega} + p_{pr'}^{\omega}\right) - s_{pr}^{\omega} - g_{pr}^{\omega} - p_{pr}^{\omega} = r_{pr}^{\omega} \quad \forall p, r, \omega, \tag{7.4}$$

where $a(\omega)$ is the ancestor scenario of scenario ω in previous period.

- Energy inventory balance for reservoirs modeled in energy (GWh)
Reservoir volume at the beginning of the period $r_{p-1,r}^{a(\omega)}$ plus unregulated inflows I_{pr}^{ω} minus spills from this reservoir s_{pr}^{ω} minus turbined water from this reservoir g_{pr}^{ω} is equal to reservoir volume at the end of the period r_{pr}^{ω}. An artificial inflow $\mathrm{ar}_{pr}^{\omega}$ is allowed and penalized in the objective function. The initial value of reservoir volume is assumed known:

$$r_{p-1,r}^{a(\omega)} + ar_{pr}^{\omega} + I_{pr}^{\omega} - s_{pr}^{\omega} - g_{pr}^{\omega} = r_{pr}^{\omega} \quad \forall p, r, \omega. \tag{7.5}$$

- Hydro plant generation (GWh) as a function of the water release
The hydro output can be expressed as the product of the water release g_{ph}^{ω}, the head of the plant ph_{ph}^{ω}, the gravity acceleration g, the efficiency of the turbine η, and of the generator η' and the water density ρ:

$$\sum_{p'} \mathrm{DU}_{pp'}hp_{pp'h}^{\omega} = g_{ph}^{\omega} \cdot ph_{ph}^{\omega} \cdot g \cdot \eta \cdot \eta' \cdot \rho \quad \forall p, h, \omega. \tag{7.6}$$

The last four terms can be approximated by the production function variable $\mathrm{pf}_{ph}^{\omega}$ (also called efficiency)

$$\mathrm{pf}_{ph}^{\omega} = ph_{ph}^{\omega} \cdot g \cdot \eta \cdot \eta' \cdot \rho \tag{7.7}$$

and, therefore, the energy produced by the plant is the product of the water release and the production function.

The production function is usually given by level curves that relate the power output of a plant with the net head for an amount of water released through the turbine. Figure 7.5 shows a typical hill diagram, similar to another found in (Diniz et al. 2007). It may be observed that given a net head (vertical dashed line) for the reservoir, there is an optimum water outflow (thick line).

Equation (7.6) is a nonlinear nonconvex constraint that considers the long-term effects of reservoir management and can be rewritten as

$$\sum_{p'} \mathrm{DU}_{pp'} hp_{pp'h}^{\omega} = g_{ph}^{\omega} pf_{ph}^{\omega} \qquad \forall p, h, \omega. \tag{7.8}$$

- Total reservoir release g_{pr}^{ω} is equal to the sum of reservoir releases from all the downstream hydro plants (hm^3)

$$g_{pr}^{\omega} = \sum_{h \in hr(r)} g_{ph}^{\omega} \qquad \forall p, r, \omega. \tag{7.9}$$

- Pumped water p_{pr}^{ω} in (hm^3) is equal to the pumped storage hydro plant consumption $pp_{pp'h}^{\omega}$ in (GWh) divided by the production function PF_h:

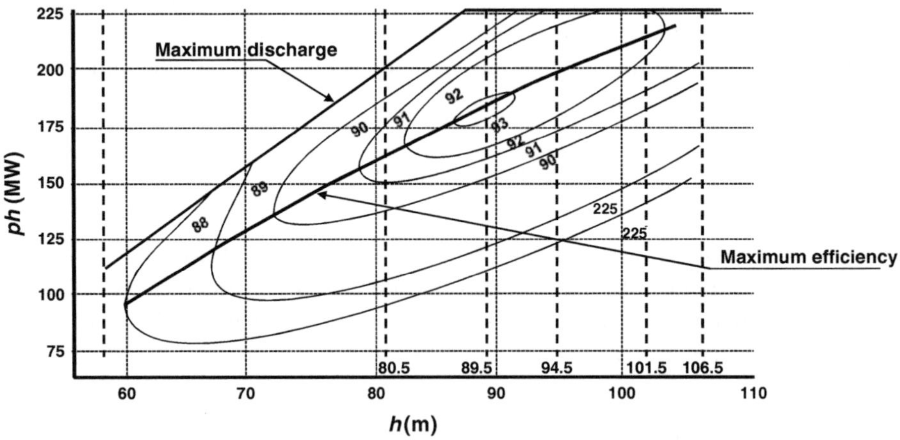

Fig. 7.5 Hill diagram of a hydro plant

$$p^{\omega}_{pr} = \sum_{p'} \sum_{h \in hr(r)} \mathrm{DU}_{pp'} \frac{pp^{\omega}_{pp'h}}{PF_h} \qquad \forall p, r, \omega. \tag{7.10}$$

- Achievement of a preestablished final reservoir volume FR_r with deficit dfr^{ω}_r and surplus variables efr^{ω}_r (hm^3)(GWh)
  This final reserve is determined by running a medium-term market equilibrium model, as seen in Fig. 7.1:

$$r^{\omega}_{pr} + dfr^{\omega}_r - efr^{\omega}_r = \mathrm{FR}_r \qquad \forall r, \omega. \tag{7.11}$$

- Minimum and maximum reservoir volume per period with deficit dmr^{ω}_{pr} and surplus variables emr^{ω}_{pr} (hm^3) (GWh)
  Those bounds are included to consider flood control curve, dead storage, and other plant operation concerns. The deficit variables will be strictly necessary in the case of many scenarios where inflow variety is higher:

$$\underline{R}_{pr} - dmr^{\omega}_{pr} \leq r^{\omega}_{pr} \leq \overline{R}_{pr} + emr^{\omega}_{pr} \qquad \forall p, r, \omega. \tag{7.12}$$

- Computation of the plant water head (m) and the production function variable ($\mathrm{hWh/m}^3$) as a linear function of it.
  Production function variable pf^{ω}_{ph} is a linear function of the water head of the plant that is determined as the forebay elevation of the reservoir rh^{ω}_{pr} minus the tailrace elevation of the plant th^{ω}_{ph}. Tailrace elevation of the plant is the maximum of the forebay elevation of downstream reservoir rh^{ω}_{pr} and the tailrace elevation of the plant TH_h. This value depends on the outflow through the power plant. However, in this medium-term model it has been assumed as constant:

$$\begin{aligned} pf^{\omega}_{ph} &= A_h + A'_h \left(rh^{\omega}_{pr} - th^{\omega}_{ph} \right) && \forall p, h, \omega, \\ tv^{\omega}_{ph} &\geq \max\left(rh^{\omega}_{pr}, TH_h \right) && \forall p, h, \omega. \end{aligned} \tag{7.13}$$

- Computation of the reservoir headwater elevation (m) and the reservoir volume (hm^3) as a nonlinear function of it.
  Reservoir headwater elevation wh^{ω}_{pr} is determined as the forebay elevation rh^{ω}_{pr} minus the reference elevation. Reserve volume r^{ω}_{pr} is a quadratic function of the reservoir headwater elevation wh^{ω}_{pr}:

$$\begin{aligned} wh^{\omega}_{pr} &= rh^{\omega}_{pr} - \mathrm{RH}_r && \forall p, r, \omega, \\ r^{\omega}_{pr} &= B_r + B'_r \left(wh^{\omega}_{pr} \right) + B''_r \left(wh^{\omega}_{pr} \right)^2 && \forall p, r, \omega. \end{aligned} \tag{7.14}$$

- Variable bounds, i.e., reservoir volumes between limits for each hydro reservoir and power operation between limits for each unit

$$
\begin{array}{ll}
0 \le \underline{TP}_{pt} \le tp^{\omega}_{tp't} \le \overline{TP}_{pt} & \forall p, p', t, \omega, \\
0 \le \text{HP}_{ph} \le hp^{\omega}_{hp'h} \le \overline{\text{HP}}_{ph} & \forall p, p', h, \omega, \\
0 \le \underline{PP}_{ph} \le pp^{\omega}_{pp'h} \le \overline{PP}_{ph} & \forall p, p', h, \omega, \\
0 \le s^{\omega}_{pr} & \forall p, r, \omega, \\
0 \le ar^{\omega}_{pr} & \forall p, r, \omega, \\
0 \le pf^{\omega}_{ph} & \forall p, h, \omega, \\
0 \le g^{\omega}_{prh} & \forall p, r, h, \omega, \\
\underline{G}_{pr} \le g^{\omega}_{pr} & \forall p, r, \omega, \\
0 \le pp^{\omega}_{pr} & \forall p, r, \omega, \\
0 \le dfr^{\omega}_{r} \le \text{FR}_r & \forall r, \omega, \\
0 \le efr^{\omega}_{r} & \forall r, \omega, \\
0 \le dmr^{\omega}_{pr}, emr^{\omega}_{pr} & \forall p, r, \omega, \\
0 \le doh^{\omega}_{t}, eoh^{\omega}_{t} \le 8760 & \forall t, \omega. \\
r^{\omega}_{0r} = \text{IR}_r &
\end{array}
\tag{7.15}
$$

7.3.2 Objective Function

The multiobjective function in [€] minimizes

- thermal variable costs plus,
- some penalty terms for deviations from ideal reservoir levels, i.e., deficit or surplus of final reservoir volumes, exceeding minimum and maximum operational rule curves, artificial inflows, and
- penalty terms for relaxing constraints like minimum and maximum yearly operation hours of thermal units.

It is important to notice the difficulties of finding a feasible solution for all the scenarios, so the penalties introduced into the objective function just accommodate these deviations in the best possible way. Different solutions and trade-offs can be obtained by changing these penalties and analyzing the stochastic optimization problem in a multicriteria decision-making framework:

$$
\begin{aligned}
\min \sum_{pp't\omega} & \; P^{\omega}_{p} \text{DU}_{pp'} \text{VC}_t tp^{\omega}_{pp't} \\
+ \sum_{r\omega} & \; P^{\omega}_{p} \left(\text{PDFR}_r dfr^{\omega}_{r} + \text{PEFR}_r efr^{\omega}_{r} \right) \\
+ \sum_{pr\omega} & \; P^{\omega}_{p} \left(\text{PDFR}_r dmr^{\omega}_{pr} + \text{PEFR}_r emr^{\omega}_{pr} \right) \\
+ \sum_{pr\omega} & \; P^{\omega}_{p} \left(\text{PDFR}_r ar^{\omega}_{pr} \right) \\
+ \sum_{pt\omega} & \; P^{\omega}_{p} \left(\text{PDH}_t doh^{\omega}_{t} + \text{PEH}_t eoh^{\omega}_{t} \right).
\end{aligned}
\tag{7.16}
$$

7.4 Model Implementation

According to (Labadie 2004) "the keys to success in implementation of reservoir system optimization models are (1) improving the levels of trust by more interactive of decision makers in system development; (2) better 'packaging' of these systems; and (3) improved linkage with simulation models which operators more readily accept." Following guideline (2) this model has been implemented with a spreadsheet-based graphical user interface that improves easiness and usability. It is able to represent any general reservoir system topology, given that it is not customized. The optimization problem is written in GAMS 23.3, see (Brooke et al. 2008), and automatically executed from the interface. The scenario tree generator is also embedded into the hydrothermal coordination model.

As guideline (3) suggests the optimal decisions obtained with this model are passed to another stochastic simulation model (Latorre et al. 2007a) to evaluate decisions at a daily level; see Fig. 7.1.

7.5 Case Study

The case study represents the Spanish electric system with 118 thermal units, 5 main basins with 49 hydro reservoirs, 56 hydro plants, and 2 pumped storage hydro plants. The hydro subsystem is very diverse. Hydro reservoir volumes range from 0.15 to 2433 hm^3 and hydro plant capacities go from 1.5 to 934 MW. We consider different number of scenarios of unregulated hydro inflows.

In the following sections we have done different runs to analyze the electric system and some modeling issues.

7.5.1 Computational Results

In this section we show the use of different nonlinear solvers for different case studies. For avoiding numerical problems a careful natural scaling of variables around 1 has been done and simple expressions are used in the nonlinear constraints, which are very efficiently managed by nonlinear solvers. The nonlinear problem is solved providing initial values and bounds for all the variables from the solution given by the linear solver CPLEX 12.1 (ILOG-CPLEX, http://www.ilog.com/products/cplex/) by an interior point method. Several nonlinear solvers have been tested with different cases to check their robustness and solution time. The tested solvers have been CONOPT3 3.14 based on a generalized reduced gradient method (Drud 1994), IPOPT 3.7 based on a primal–dual interior point filter line search algorithm (Wchter and Biegler 2006), KNITRO 5.1.2 using an interior point (Byrd et al. 2006), MINOS 5.51 based on a project Lagrangian algorithm (Murtagh and Saunders 1987). The default options and algorithms have been used for all the solvers.

The model has been run in a PC with a processor running at 1.83 GHz and with 1 GB of RAM memory. The problem with eight scenarios has been the biggest one

Table 7.1 Size of linear and nonlinear problems

	R	*V*	*E*	*R*	*V*	*E*	*NE*
1 scen	30,952	57,883	157,705	36,984	55,283	157,705	1248
4 scen	120,269	224,925	612,883	143,701	214,825	612,883	4848
8 scen	234,641	438,837	1,195,803	280,345	419,137	1,195,803	9456

Table 7.2 Solutions provided by linear and nonlinear solvers

	CPLEX		CONOPT		IPOPT		KNITRO		MINOS	
	O.F. (M€)	time (s)	O.F. (M€)	time (s)	O.F. (M€)	time (s)	O.F. (M€)	time (s)	O.F. (M€)	time (s)
1 scen	15,717.452	6	15,689.029	275	15,689.086	600[3]	15,689.103	601	15,689.875	78
4 scen	15,750.979	43	15,730.440	3202	15,730.502	4200	15,728.997	2309	15,730.796	2557
8 scen	15,764.817	132	15,750.062	6513	15,746.309	9010	15,754.388	5600	15,747.004	7628

that has fitted into the PC memory. Table 7.1 summarizes the sizes of the problems where *R* is the number of constraints, *V* number of variables, *E* nonzero elements, and *NE* nonlinear nonzero elements, and Table 7.2 the objective functions in (M€) and the solution times in seconds.

The solution of the nonlinear nonconvex problem by a linear approximation has been tested. The linearization is made by fixing the value of the production function obtained in previous iteration. The results are shown in Fig. 7.6. It can be concluded that the linear iterations do not converge necessarily to the NLP solution.

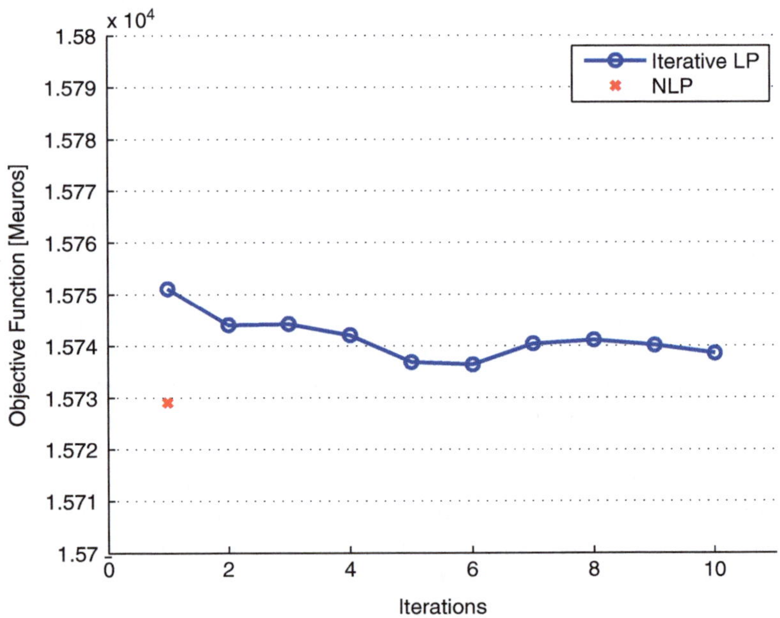

Fig. 7.6 Iterations of LP problem and the NLP solution

The relevant conclusion to extract from this analysis is that large-scale stochastic hydrothermal coordination problems can nowadays be solved by several general-purpose NLP solvers.

7.5.2 Hydro Reservoir Operation Planning

As it has been seen from the previous tables the impact of the nonlinear approximation is not very important regarding the objective function (less than a 0.2%). However, the operation of the hydro plants makes a crucial difference between them. The following figures show the different operation of a large hydro reservoir (with a maximum volume of approximately 900 hm^3) due to use of the linear or nonlinear modeling of water head effects in the four-scenario case. The curves represented in each graph correspond to minimum and maximum volume level, lower and upper operating rule curves, mode (most probable scenario), and five quantiles.

Although in a stochastic framework the only relevant decisions are those corresponding to the first stage, given that these here and now decisions will be implemented, it can be observed that the operation of the reservoir is smoother in the nonlinear approach than in the linear one, see, for example, curves in the weeks from s0852 to s0926 of Fig. 7.7. Besides, there is a strong difference in the optimal volume of the reservoir in the weeks from s0932 up to s1004. In the linear case, reservoir volume is remarkably lower because the production function does not depend on the water head and therefore the linear model is indifferent to this volume. This rational and realistic operation of the reservoir in the nonlinear model fully justifies the importance of using the nonlinear approximation and shows that feasibility is as important as optimality in stochastic hydrothermal planning models.

7.5.3 Scenario Analysis and Stochastic Measures

Figure 7.8 represents on the left y-axis the value of the objective function for the different scenarios with anticipative decisions and on the right y-axis the relative natural inflows (value 1 corresponds to the mean value). Changes in the objective function are around 10% among scenarios. Figure 7.9 plots the quadratic regression function of both variables, determining the impact of hydro inflows in the objective function.

Additionally, we have conducted a scenario analysis and determined some stochastic measures; see (Birge and Louveaux 1997) for their definition, whose values appear in Table 7.3. In this case it can be seen that the expected value of perfect information (EVPI) that measures the impact of the non-anticipative decisions and the value of the stochastic solution (VSS) are very small. The reason is that the branching process of the tree is done in early stages (at the end of the first month) and the scenarios are almost independent among them.

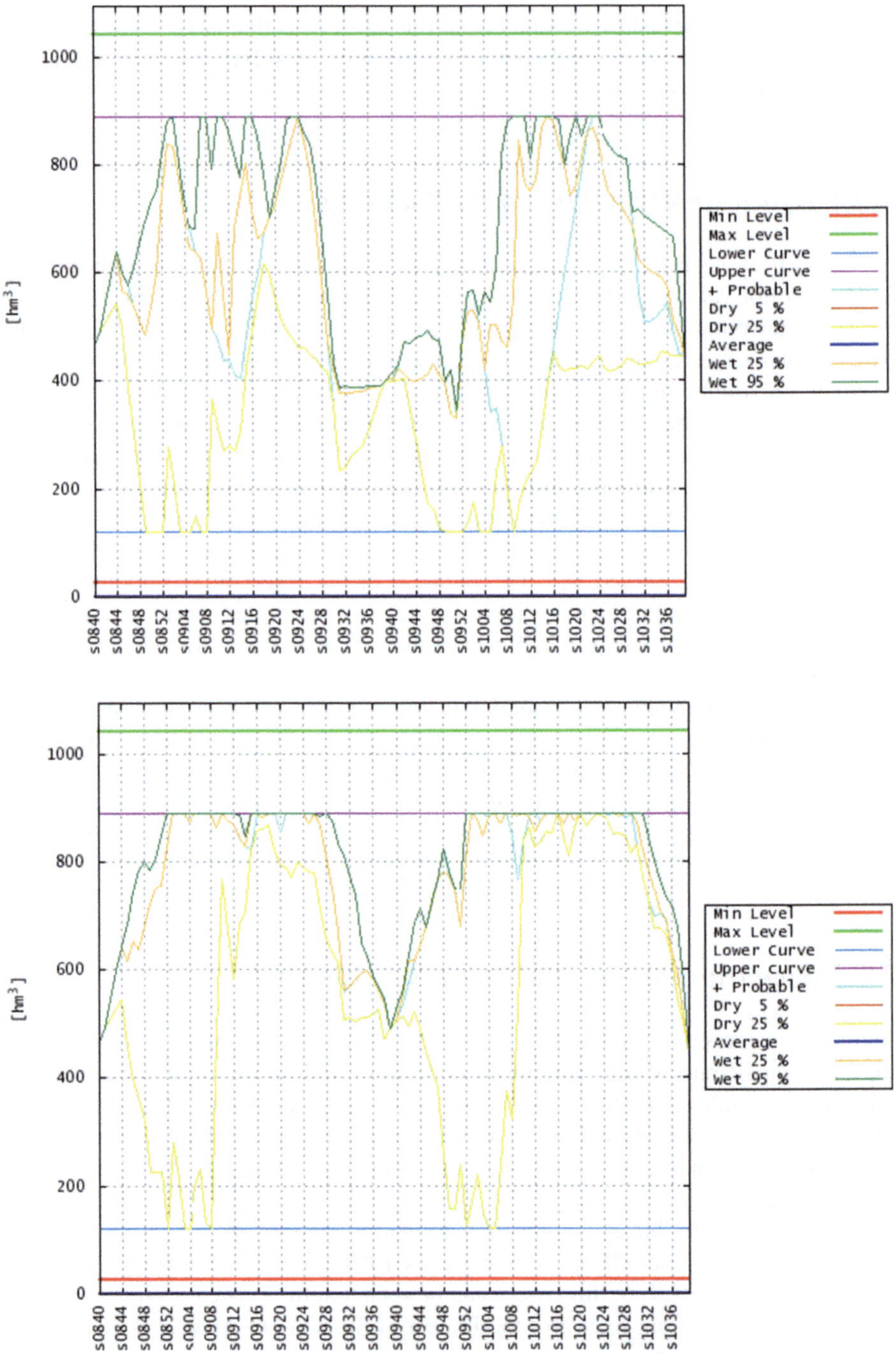

Fig. 7.7 Reserve volume for the planning horizon under stochastic hydro inflows with the linear (*above*) and nonlinear (*below*) approximation

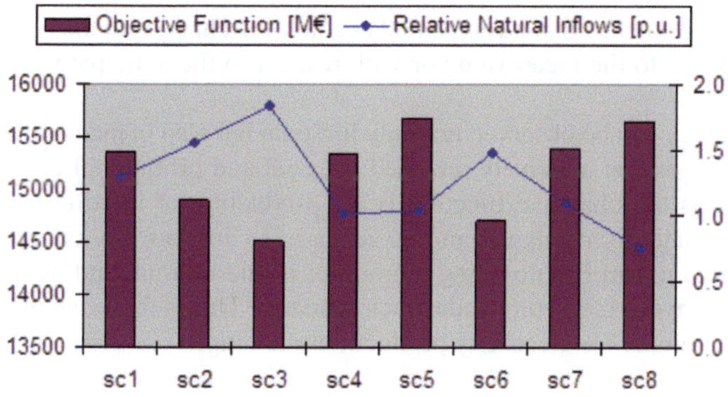

Fig. 7.8 Scenario analysis

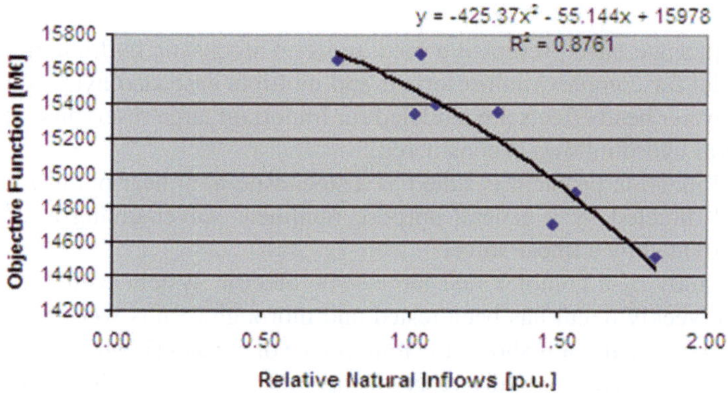

Fig. 7.9 Relation between natural inflows and total variable costs

Table 7.3 Stochastic measures

Expected value with perfect information	(EVWPI)	15,764.709
Expected value solution	(EV)	15,716.937
Stochastic solution	(SS)	15,764.817
Expected result of the expected value solution	(EEV)	15,764.755
Expected result of the stochastic solution	(ESS)	15,764.848
Value of the stochastic solution	(VSS)	0.062
Expected value of perfect information	(EVPI)	0.108

Another reason for that comes from the electric system. Since the last 5 years the Spanish electric system has had a strong investment in CCGTs. As a result nowadays there are only three thermal technologies: nuclear, coal, and natural gas, being natural gas the less competitive from a variable cost point of view. Stochasticity in hydro inflows may represent a variation of approximately 20 TWh of energy from a dry to a wet year and it is fully replaced by electricity produced by CCGTs. Those units are relatively new and therefore have similar heat rates. As the total variable cost of the system behaves linearly with respect to the stochasticity in hydro inflows, see Fig. 7.9, the stochastic measures, that account for changes in the objective function with respect to the mean value or with respect to the anticipative solution, are negligible.

This tendency can be observed not only in Spain but also in many other countries where natural gas has massively replaced old coal and oil thermal units. However, as important as the objective function is the production of the different units and this may substantially change from one scenario to another. So, the model results are much more useful by providing the output of the thermal and hydro units and the spillage of hydro reservoirs under each scenario. This is the value of a stochastic programming model for electricity production planning.

7.6 Conclusions

In this chapter we have presented a medium-term stochastic hydrothermal coordination model for complex multireservoir and multiple cascaded hydro subsystems. Nonlinear water head effects are modeled for important large reservoirs. Stochasticity of natural hydro inflows is considered.

The optimization problem is stated as a stochastic nonlinear optimization problem solved directed by a general-purpose nonlinear solver giving a close initial solution provided by a linear solver.

A case study of a complex and large-scale electric system with a 2-year time scope with weekly detail has been tested and thorough results were presented and discussed. In particular it is shown the importance of considering the nonlinear modeling in obtaining realistic hydro reservoir operations and the value of the stochasticity for this case.

References

M. S. Bazaraa, H. D. Sherali, and C. M. Shetty, *Nonlinear Programming. Theory and Algorithms*, 2nd ed. New York, NY: Wiley, 1993.

J. R. Birge and F. Louveaux, *Introduction to Stochastic Programming*, 1st ed. New York, NY: Springer, 1997.

A. Brooke, D. Kendrick, A. Meeraus, and R. Raman, *GAMS: A User's Guide*. Whashington, DC: GAMS Development Corporation, 2008.

R. H. Byrd, J. Nocedal, and R. A. Waltz, KNITRO: An Integrated Package for Nonlinear Optimization, in *Large-Scale Nonlinear Optimization*, G. di Pillo and M. Roma, Ed., Berlin: Springer, 2006, pp. 35–59.

J. Cabero, A. Baillo, S. Cerisola, M. Ventosa, A. Garcia-Alcalde, F. Peran, and G. Relano, A Medium-Term Integrated Risk Management Model for a Hydrothermal Generation Company, *IEEE Transactions on Power Systems*, **20**: 1379–1388, 2005.

R. S. Dembo, A. Chiarri, J. G. Martin, and L. Paradinas, Managing Hidroelectrica-Espanola Hydroelectric Power-System, *Interfaces*, **20**: 115–135, 1990.

A. L. Diniz, P. P. I. Esteves, and C. A. Sagastizabal, A Mathematical Model for the Efficiency Curves of Hydroelectric Units, *Ieee Power Engineering Society General Meeting*, **1–10**: 200–206, 2007.

A. Drud, CONOPT: A Large-Scale GRG Code, *INFORMS Journal on Computing*, **6**: 207–216, 1994.

J. Jacobs, G. Freeman, J. Grygier, D. Morton, G. Schultz, K. Staschus, and J. Stedinger, Socrates – A System for Scheduling Hydroelectric Generation under Uncertainty, *Annals of Operations Research*, **59**: 99–133, 1995.

J. W. Labadie, Optimal Operation of Multireservoir Systems: State-of-the-Art Review, *Journal of Water Resources Planning and Management-Asce*, **130**: 93–111, 2004.

J. M. Latorre, S. Cerisola, A. Ramos, R. Bellido, and A. Perea, Creation of Hydroelectric System Scheduling by Simulation, in *Complex Decision Making: Theory and Practice*, H. Qudrat-Ullah, J.M., Spector, P., Davidsen, Ed., New York, NY: Springer, 2007a, pp. 83–96.

J. M. Latorre, S. Cerisola, and A. Ramos, Clustering Algorithms for Scenario Tree Generation: Application to Natural Hydro Inflows, *European Journal Of Operational Research*, **181**: 1339–1353, 2007b.

B. A. Murtagh and M. A. Saunders, MINOS 5.1 User's Guide, *Stanford University Report* SOL 83-20R, 1987.

M. V. F. Pereira and L. M. V. G. Pinto, Multistage Stochastic Optimization Applied to Energy Planning, *Mathematical Programming*, **52**: 359–375, 1991.

A. Seifi and K. W. Hipel, Interior-Point Method for Reservoir Operation with Stochastic Inflows, *Journal of Water Resources Planning and Management-Asce*, **127**: 48–57, 2001.

F. Stein-Erik and K. Trine Krogh, Short-Term Hydropower Production Planning by Stochastic Programming, *Computers Operations Research*, **35**: 2656–2671, 2008.

A. Wchter and L. T. Biegler, On the Implementation of a Primal-Dual Interior Point Filter Line Search Algorithm for Large-Scale Nonlinear Programming, *Mathematical Programming*, **106**: 25–57, 2006.

Chapter 8
Hedging Electricity Portfolio for a Hydro-energy Producer via Stochastic Programming

Rosella Giacometti, Maria Teresa Vespucci, Marida Bertocchi, and Giovanni Barone Adesi

Abstract A stochastic multistage portfolio model for a hydropower producer operating in a competitive electricity market is proposed. The producer's portfolio includes the produced energy and a set of contracts for power delivery or purchase, including contracts of a financial nature such as forwards to be able to hedge against risks. The goal of using such a model is to maximise the profit of the producer and reduce the economic risks connected to the fact that energy spot and forward prices are highly volatile. We devise two scenario generation procedures: the first drives forward prices by the spot dynamics and the second models the forward curve dynamics directly. Our results, as expected, show that forward contracts are effective for hedging purposes and the optimal solution of the stochastic model with random coefficients generated by the second scenario procedure leads to the highest optimal value. Beyond financial gains, the advantage of using financial contracts consists in more efficient use of the hydroplant, allowing the possibility of pumping water and ending up with a higher final level of the reservoir.

Keywords Regime switching model · Hydro plants · Electricity forward contracts · Stochastic programming

8.1 Introduction

As a result of the liberalization of energy markets, generation companies are exposed to higher uncertainties. Risk management becomes a more pressing issue for electricity consumers and producers: contracts for future delivery of electricity (i.e. forward contracts) become a tool for hedging risk. The hydropower producer's portfolio includes the produced energy and a set of power contracts for delivery or purchase, including such contracts of a financial nature as forwards, to be able to hedge against various types of risks. In this chapter we develop a stochastic portfolio model for a hydropower producer operating in a competitive electricity market.

R. Giacometti (✉)
Department of Mathematics, Statistics, Computing and Applications, University of Bergamo, Bergamo, Italy
e-mail: rosella.giacometti@unibg.it

M. Bertocchi et al. (eds.), *Stochastic Optimization Methods in Finance and Energy*, International Series in Operations Research & Management Science 163,
DOI 10.1007/978-1-4419-9586-5_8,

The goal of using such a model is to reduce the economic risks connected to the fact that energy spot price may be highly volatile for various different, unpredictable reasons (i.e. a very cold winter) or to the possibility of a period of low rainfall or snowmelt. See Wallace et al. (2002), Conejo et al. (2008) and Nolde et al. (2008) for a discussion of the opportunity of using stochastic programming for such a problem. The basis risk factors include the wholesale spot and forward price of electrical energy, which are supposed to be unaffected by the decisions of the utility manager, and the uncertain inflow of hydro reservoirs; see also Fleten and Wallace (2009), Vitoriano et al. (2000), and Latorre et al. (2007). The model we propose differs from the one discussed in Fleten and Wallace (2009) since we want to concentrate on the advantage of using financial contracts and therefore we use both electricity spot prices and forward prices as a source of uncertainty, considering inflows as deterministic. We leave for future research the inclusion of stochastic inflows. Our approach combines both electricity production generation and the trading in the forward markets in order to hedge the risks described above. Our approach differs from the usual producer behaviour that keeps production planning separated from hedging strategies. Section 8.2 introduces the daily producer scheduling problem describing the main variables involved in such a problem. Section 8.3 discusses the Italian electricity forward market while Section 8.4 presents a stochastic model for hedging risks in order to match a scheduled production need. Section 8.5 is the core of the chapter and it describes the two models we adopt for the scenario generation involving spot and forward prices. Finally, Section 8.6 presents numerical results for the two models and shows the superiority of the stochastic approach versus the deterministic one. Conclusions on the importance of the model follow.

8.2 The Producer Daily Scheduling Model

Electricity generation is modeled relying on 1-day time space granularity, as common in medium-long term models, neglecting start-up and shut-down costs regarded as irrelevant.

We start by introducing the model of the hydroelectric system with hourly periods. The hydroelectric system consists of a number of cascades, i.e. sets of hydraulically interconnected hydro plants, pumped storage hydro plants and reservoirs. It is mathematically represented by a directed multi-graph, where nodes represent water storages (reservoirs) and arcs represent water flows (either power generation, or pumping, or spillage). Let J denote the set of nodes and I denote the set of arcs. The arc–node incidence matrix, whose (i, j)-entry is denoted by $A_{i,j}$, represents the interconnections among water storages and water flows in the hydroelectric system ($A_{i,j} = -1$, if arc i leaves node j; $A_{i,j} = 1$, if arc i enters node j; $A_{i,j} = 0$, otherwise). For every arc $i \in I$ or for every node $j \in J$ the following data are relevant:

- k_i (MWh/10^3 m^3): energy coefficient ($k_i > 0$, if arc i represents generation; $k_i < 0$, if arc i represents pumping; $k_i = 0$, if arc i represents spillage);
- $\overline{q}_i$ (10^3 m^3/h): maximum water flow in arc i;

- $\overline{v}_j$ (10^3 m^3): maximum storage volume in reservoir j;
- $v_{j,0}$ (10^3 m^3): initial storage volume in reservoir j;
- $\underline{v}_{j,T}$ (10^3 m^3): minimum storage volume required in reservoir j at the end of the planning period;
- $f_{j,t}$ (10^3 m^3/h): natural inflow in reservoir j in hour t.

The power producer must schedule the production of each hydro plant, which is expressed as the product of the hydro plant energy coefficient times the turbined volume in hour t, as well as the hourly pumped and spilled volumes. The decision variables of the hydro scheduling problem are

- $q_{i,t}$ (10^3 m^3/h): water flow on arc i in hour t (turbined volume, if arc i represents generation; pumped volume, if arc i represents pumping; spilled volume, if arc i represents spillage);
- $v_{j,t}$ (10^3 m^3): storage volume in reservoir j at the end of hour t.

The values assigned to the decision variables must satisfy the following constraints that describe the hydroelectric system:

- Flow on arc i in hour t is nonnegative and bounded above by the maximum volume that can be either turbined, or pumped, or spilled

$$0 \leq q_{i,t} \leq \overline{q}_i \qquad i \in I, \quad 1 \leq t \leq T. \tag{8.1}$$

- The storage volume in reservoir j at the end of hour t is nonnegative and bounded above by the maximum storage volume

$$0 \leq v_{j,t} \leq \overline{v}_j \qquad j \in J, \quad 1 \leq t \leq T. \tag{8.2}$$

- At the end of hour T, the last hour in the current planning period, the storage volume in reservoir j is bounded below by the minimum storage volume required at the end of the planning period, which will also represent an initial constraint on the period immediately following:

$$\underline{v}_{j,T} \leq v_{j,T}, \qquad j \in J. \tag{8.3}$$

- The storage volume in reservoir j at the end of hour t must be equal to the reservoir storage volume at the end of hour $t-1$ plus the sum of inflows in hour t minus the sum of outflows in hour t

$$v_{j,t} = v_{j,t-1} + f_{j,t} + \sum_{i \in I} A_{i,j} \cdot q_{i,t}, \quad j \in J, \quad 1 \leq t \leq T, \tag{8.4}$$

where $v_{j,0}$ is a data representing the initial storage volume in reservoir j. Reservoir inflows are natural inflows, turbine discharge from upstream hydro plants, pumped volumes from downstream hydro plants and spilled volumes

from upstream reservoirs. Reservoir outflows are turbine discharge to downstream hydro plants, pumped volumes to upstream hydro plants and spilled volumes to downstream reservoirs. In this chapter, the values of natural inflows $f_{j,t}$, $j \in J$, $1 \leq t \leq T$, are assumed to be known with certainty.
- The value of the reservoir at the end of the horizon $V(v_{j,T})$ is a function of reservoir level that has to be specified to avoid end effects.

This is the general model used by the producer for daily scheduling where the time unit is the hour. When we want to include financial contracts we will assume working on a daily basis. Thus, all the previous equations will be transformed to fit the new time period and t will represent the day. From now on, T represents the time horizon expressed in multiple of days.

8.3 The Use of Forward Contracts

An electricity forward contract is the obligation to buy or sell a specified amount of power −1 megawatt (MW) during every hour (i.e. base load) – at a predetermined delivery price, the forward price, during a delivery period fixed at the issue time of the contract. Additional to these so-called base-load contracts, there are peak-load contracts, which deliver 1 MW every hour during working days from 8 am to 8 pm, in the delivery period only. In this chapter we will not consider them.

The forward contracts are standardized by the following characteristics: volume, delivery period and settlement. The volume is the number of MWh underlying the contract; for contracts with a fixed rate (energy amount per hour) of 1 MW, it is equivalent to the number of hours in the delivery period. As an example, for an April contract with monthly delivery period (in the following we will specify the days in the delivery period as DP, hence DP $= 30$ days in our example), this means a total of 1 MW $\times$ 30 days $\times$ 24 h/day $=$ 720 MWh. The quoted forward price is the price at which the owner of the contract will buy/sell energy during the delivery period per 1 MWh. The value of the contract is the product of the quotation and the volume. For each buying or selling of the contract, we consider a transaction fee of 0.01 per MW and an estimated bid-ask spread of 3% of the forward price. In our example the transaction fee is 0.01 EUR $\times$ 30 days $\times$ 24 h/day $=$ 7.2 EUR.

Delivery periods may be monthly (M1, M2, . . . , M12), quarterly (Q1, Q2, Q3, Q4) or annual within the current calendar year. Shorter delivery periods such as daily, weekly or during week-ends may also be considered within specific standardized forward markets. For each contract we can distinguish between a trading period and a delivery period. The trading in a given contract stops when it enters the delivery period. Another relevant characteristic is the settlement. We can distinguish between financial contracts and physical contracts. The former requires a cash settlement of forward price against the realized spot prices during the delivery period. The latter requires energy's delivery at the delivery price during the delivery period.

Let us consider, for modelling reasons, the accounting of a forward contract. This specification is relevant to understand the optimization model we propose in the

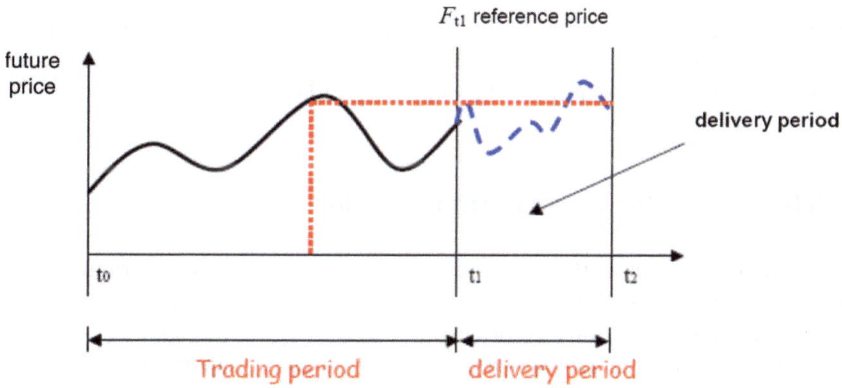

Fig. 8.1 The evolution of forward price

following section, where we want to select which contract holds and which contract closes before the delivery period. Assume that we enter in a long forward position at time T_b and that we maintain the contract till the delivery period which starts in T_e and lasts DP days. See Fig. 8.1 for an illustration of all these quantities where $t_1 = T_e$. We can decompose the loss/gain on the contracts into two components: the mark to market, during the trading period, and the settlement, during the delivery period. The first mechanism implies that from the purchase day till the last trading day, we close the position every day and immediately we reopen it at the new forward price. The daily gain/loss is given by the price variations. At the end of the trading period the holder of the long position has a gain/loss of $24 \cdot \text{DP} \cdot (F_{T_e} - F_{T_b})$. The last position of the trading period is a forward with the quote F_{T_e}, called the reference price. During the delivery period we distinguish between cash settlement and physical delivery. The former is computed considering the daily variations between the reference price and the spot price without a physical exchange of electricity. The latter consists in the payment of the reference price against the physical delivery of electricity. In (8.5) we show that this mechanism is equivalent to the classical settlement where the exchange is between the forward price fixed at the purchase of the contract and the spot price during the delivery period:

$$\sum_{t \in [T_e, T_e + 24 \cdot DP]} (S_t - F_{T_b}) = 24 \cdot \text{DP} \cdot \sum_{t \in [T_b, T_e]} (F_t - F_{t-1}) + \\ + \sum_{t \in (T_e, T_e + 24 \cdot DP]} (S_t - F_{T_e}). \quad (8.5)$$

If we consider physical settlement, the exchange is between the forward price fixed at the purchase of the contract and the physical delivery of energy during the delivery period.

In this work we consider base-load contracts with a physical settlement, since we decide to model the daily movements of the spot process. The introduction of peak-load contracts would require the modelling of hourly prices.

8.4 The Stochastic Programming Model

In this chapter we assume that the natural inflows are known with certainty and we concentrate our attention on the financial aspects, i.e. the uncertainty in the electricity spot prices and in the forward prices.

A scenario tree (see Dupačová et al. (2000) and Chapters 14, 15 and 16) represents the information on the daily energy spot price and contract forward price, where each path from the root to a leaf of the tree corresponds to one scenario. The stochastic model is written in terms of the nodes $\{1, \ldots, n, \ldots, N\}$ of the scenario tree and the tree structure is described by giving each node n the probability P_n, $1 \leq n \leq N$, and a pointer to its parent pred(n), $2 \leq n \leq N$ (i.e. except the root of the tree). The planning horizon is divided into K stages, where each stage k, $1 \leq k \leq K$, is associated with the number of days T_k and to the set of nodes N_k, where $k(n)$ is the stage associated with node n. The model can be extended to any time length.

The power producer is assumed to be a price taker, i.e. not able to influence the electricity market price, which is therefore exogenous to the model. We suppose to be able to hedge the producer's portfolio by buying and selling forward base-load contracts; the variable $\text{buy}_{l,t,n}$ ($\text{sell}_{l,t,n}$) is used to denote the number of positions at time t and l represents the type of the contract to be bought (sold) on the forward market in node n, with $n \in N_k$ if $t \in T_k$, $1 \leq k \leq K$; $x_{l,t,n}$ denotes the open positions in forward contracts at time t for contract l at node n.

Generally, the producer can buy or sell forward contracts with different delivery period, i.e. forward contracts with weekly, monthly, quarterly and yearly delivery. If the producer has a planning horizon T of 1 year (expressed in days), he can decide to hedge his risk by buying or selling forward contracts with different delivery periods in the different days of his planning horizon. For contract l, $T_e(l)$ indicates the maturity of the contract, $T_d(l)$ the end of the delivery period, $T_s(l)$ indicates the time when the contract is first traded on the market. We define the length of the contract $\text{DP}(l) = T_d(l) - T_e(l)$. The market price of contract l is the forward price indicated by $F_{l,t,n}$ while the spot price at time t in node n is $S_{t,n}$.

The decision variables $q_{i,t,n}$ and $v_{j,t,n}$ represent, respectively, the flow on arc i in day t and the storage volume in reservoir j at the end of day t in node n.

We indicate with r the risk-free interest rate, $W_{t,n}$ the cumulative wealth at time t and node n, $U(W)$ an increasing concave utility function of wealth describing the producer risk aversion. We set $W_{0,0} = 0$ and $F_{l,0,1} = 0 \quad \forall l$.

The stochastic program finds values of the decision variables $q_{i,t,n}$, $v_{j,t,n}$, $x_{l,t,n}$, $\text{buy}_{l,t,n}$, $\text{sell}_{l,t,n}$, for $1 \leq k \leq K$, $n \in N_k$, $t \in T_k$, $i \in I$ and $j \in J$, which maximize the objective function representing the producer's profit: it consists of various parts taking into account net sales in the spot market, selling and buying forward contracts, and the value of end reservoir, $V(v_{j,T,n})$, which represents the monetary value of the water volume of reservoir j at the end of period T in node n:

$$\max \sum_{n \in N_K} P_n U(W_{T,n} + \sum_j V(v_{j,T,n})) \tag{8.6}$$

with respect to the decision variables $q_{i,t,n}$, $v_{j,t,n}$, $x_{l,t,n}$, $\text{buy}_{l,t,n}$, $\text{sell}_{l,t,n}$, for $1 \leq k \leq K, n \in N_k, t \in T_k, i \in I$ and $j \in J$, subject to

$$\begin{aligned} W_{t,n} = {} & W_{t-1,\nu}(1+r) + S_{t,n}\left(\sum_i k_i q_{i,t,n} + \sum_l x_{l,t,n} \cdot 1_{t \in [T_e(l), T_d(l)]}\right) + \\ & - \sum_l F_{l,T_e(l),n} \cdot x_{l,t,n} \cdot 1_{t \in [T_e(l), T_d(l)]} + \\ & + \sum_l \text{DP}(l)(F_{l,t,n} - F_{l,t-1,n}) \cdot x_{l,t,n} \cdot 1_{t \in [T_s(l), T_e(l))} + \\ & - \sum_l (\text{buy}_{l,t,n} + \text{sell}_{l,t,n})(tc + ba \cdot F_{l,t,n}) \cdot 1_{t \in [T_s(l), T_e(l))}, \end{aligned} \tag{8.7}$$

$$x_{l,t,n} = x_{l,t-1,\nu} + (\text{buy}_{l,t,n} - sell_{l,t,n}) 1_{t \in [T_s(l), T_e(l))}, \qquad t \in T_k, \quad n \in N_k, \tag{8.8}$$

$$\text{buy}_{l,t,n} \geq 0\,, \qquad \text{sell}_{l,t,n} \geq 0, \qquad t > t_{k(n)}, \qquad n \in N_k, \tag{8.9}$$

where

$$\sum_i k_i q_{i,t,n} + \sum_l x_{l,t,n} \cdot 1_{t \in [T_e(l), T_d(l)]} = D_t, \tag{8.10}$$

where D_t is the exogenous production scheduling at time t which has to be met;

Constraints similar to (8.1), (8.2), (8.3) and (8.4) are added, where the time unit t is the day:

$$v_{j,t,n} = v_{j,t-1,\nu} + f_{j,t,n} + \sum_{i \in I} A_{i,j} \cdot q_{i,t,n} j \in J, \ t \in T_k, \ n \in N_k, \ 1 \leq k \leq K, \tag{8.11}$$

$$0 \leq q_{i,t,n} \leq \overline{q}_i, \qquad i \in I, \ t \in T_k, \ n \in N_k, \ 1 \leq k \leq K, \tag{8.12}$$

$$0 \leq v_{j,t,n} \leq \overline{v}_j, \qquad j \in J, \ t \in T_k, \ n \in N_k, \ 1 \leq k \leq K, \tag{8.13}$$

$$\underline{v}_{j,T} \leq v_{j,T,n}, \qquad j \in J, \quad n \in N_K. \tag{8.14}$$

In the wealth equation (8.7), in the mass balance equations (8.11) of the hydro system model and in the financial balance equation (8.8) $\nu = n$, if $t-1, t \in T_k$, and $\nu = \text{pred}(n)$, if $t-1 \in T_{k-1}$ and $t \in T_k$. The objective function in (8.6) represents the expected utility of the final wealth plus the monetary value of water in reservoirs at the horizon. Constraints (8.7), (8.8), (8.10), (8.11) describe, respectively, the cash flow balance, the open position balance, the volume balance between stages and the matching between produced and bought energy and the scheduled production. Constraints (8.11), (8.12), (8.13), (8.14) are the equivalent nodal formulation of constraints (8.4), (8.1), (8.2), (8.3).

8.5 Scenario Generation for Spot and Forward Electricity Prices

While the Italian electricity forward market is pretty recent – opened at the end of 2008 – with a limited number of transaction, the electricity spot market was opened in 2003, and with its increased activity during recent years, it can be considered a liquid market with many daily transactions. In our analysis we consider the daily base-load spot prices time series from 1/1/2008 to 9/9/2009. After removing the daily and weekly seasonal components, we analyse the log prices data and we find stationarity but no strong presence of spikes: only four observations are larger than three times the standard deviation for the whole period. The log spot price exhibits autocorrelation, heteroscedasticity but not a dramatic kurtosis. In line with recent researches (see De Jong 2006; Serati et al. 2006), we fit a regime switching model able to capture different market conditions, in terms of changing mean and volatilities. We assume that y_t, the log price process, follows an AR(1) model depending on the state variables s_t:

$$y_t = \mu_{s_t} + \Phi_{s_t} y_t + \epsilon_t, \quad \text{where} \quad \epsilon_t \sim \text{i.i.d } N(0, \sigma^2_{s_t}), \tag{8.15}$$

where s_t changes through time and takes values $j = 1, \ldots, J$. The changes in s_t are described by a Markov chain $P(s_{t+1} = j | s_t = i) = p_{ij}$. We do not observe s_t directly, but we only infer it through the observed behaviour of y_t.

On the complete data set we find evidence of two regimes. The parameters necessary to fully describe the probability law governing y_t are then the volatility of the Gaussian innovation, the autoregressive coefficients, the two intercepts and the two state transition probabilities, p_{11}, p_{12}, p_{21} and p_{22}. In Tables 8.1 and 8.2 we present the estimated parameters with the t-statistics in parenthesis.

Table 8.1 Table related to Fig. 8.2

	μ	Φ_{s_t}	σ
State 1	0.0140	0.1510	0.0760
	(2.4140)	(1.9610)	(18.5366)
State 2	−0.0210	−0.0507	0.1409
	(−1.8584)	(0.6225)	(19.8451)

Table 8.2 Transition probabilities between the two regimes in Fig. 8.2

Transition	State 1	State 2
State 1	0.96 (0.05)	0.04 (0.02)
State 2	0.05 (0.03)	0.95 (0.04)

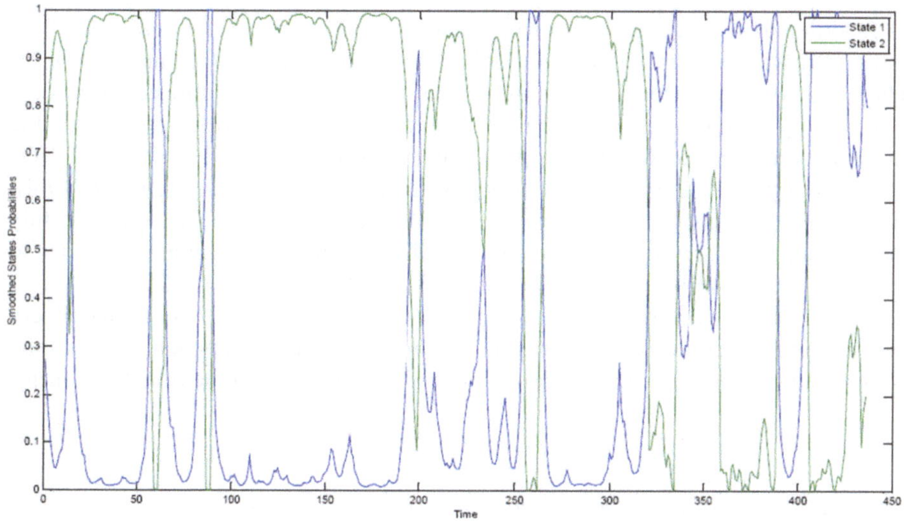

Fig. 8.2 The spot switching model

For a multistage stochastic programming model a scenario tree is needed where information is revealed. Hence, we generate 100 independent scenarios, which describe the spot price evolution over 1 year, and then aggregate them into a three-stage scenario tree applying the scenario reduction scheme proposed by Pflug (2001) (see Chapter 15).

The theoretical dynamic of the forward prices can be derived, alternatively, from the spot dynamic or explicitly modelling the forward curves. The second alternative is motivated by the empirical evidence that the electricity spot and forward prices are not closely related, as it is typical for other commodities, such as crude oil. Electricity spot and forward prices can be very far from each other. Often, the spot and futures markets are so dissimilar that the relationship between spot and futures prices breaks down; see Borovkova and Geman (2006a, b).

For this reason we decide to enrich our analysis and to model explicitly the dynamic of the forward curve with the condition that the scenarios generated for the spot and forward contracts are correlated and consistent with present observed contract quotations; see Giacometti et al. (2007) for further details. We restrict our analysis to one segment of the term structure, the quarterly contracts and leave a more comprehensive analysis to future research. Here, we apply the model to the forward prices of quarterly contracts in the period 20/10/08 to 9/09/09 on Italian data of OTC contracts quoted at the TFS, a private platform which at present is more liquid than the standardized market, the IDEX (Italian Derivatives Energy Exchange) just started at the beginning of 2009.

The idea is to compute from the daily quotations of the forward contracts, the forward term structure for fixed key rates and analyse the dynamics of the term structure.

In order to derive the forward curve we have to remove the seasonal effect associated with forward contracts. We follow Borovkova and Geman (2006a, b). They observe that seasonal effects in the spot price and in the futures contracts are significantly different and that the main feature of electricity forward curves is the seasonality attached to the delivery period, not to the trading day. Let $F(t, T, T+Q)$ be the day-t price of the forward contract expiring in T with $T = (T1, T2, T3, T4)$. We fix the beginning of the delivery period at four dates (January 1st, April 1st, July 1st and September 1st) and the length of delivery period to 3 months in order to represent the quarterly contracts $Q1$, $Q2$, $Q3$, and $Q4$.

We estimate the deterministic seasonal forward premium, $\pi(T)$, for each delivery date assuming that the forward price is the product of two components, a seasonal component (the premium) and a component which depends on the time to maturity

$$F(t, T, T + Q) = \bar{F}(t)e^{\pi(T)-\gamma(t,T-t)(T-t)}, \tag{8.16}$$

where the deseasonalized forward price $F^{\mathrm{DS}}(t, T, T + Q)$ is

$$F^{\mathrm{DS}}(t, T, T + Q) = F(t, T, T + Q)e^{-\pi(T)}. \tag{8.17}$$

The seasonal premium is defined as the average deviation from the mean value of the log forward quotations

$$\hat{\pi}(T) = \frac{1}{n}\sum_t (\ln(F(t, T, T + Q)) - \ln(\bar{F}(t))), \quad T = T1, \ldots, T4, \tag{8.18}$$

where

$$\ln(\bar{F}(t)) = \frac{1}{4}\sum_{T=T1,\ldots,T4} \ln(F(t, T, T + Q)). \tag{8.19}$$

The estimated seasonal premia for quarterly electricity forwards are shown in Fig. 8.3.

As expected, forwards expiring in winter are at a premium with respect to the average price level and summer forwards at a discount. For electricity the January premium is the highest, at 15%, while that for April is −12%.

Once we have removed the seasonal premium, we can derive the term structure for quarterly contracts. More precisely, we concentrate our attention on four key rates relative to the four quarterly maturities. The procedure used in the forecasting approach for scenario generation basically involves two distinct steps. By principal component analysis (PCA) on the daily deseasonalized historical returns of forward key rates, we find the orthogonal factors. The first three factors explain 92% of the forward curve and correspond to a parallel shift (first factor which explains 61%), a tilting (second factor which explains 17%) and a curvature effect (third factor explains 14%); see Fig. 8.4. We consider as relevant the first two factors that have an explanatory power of about 80% of the variability.

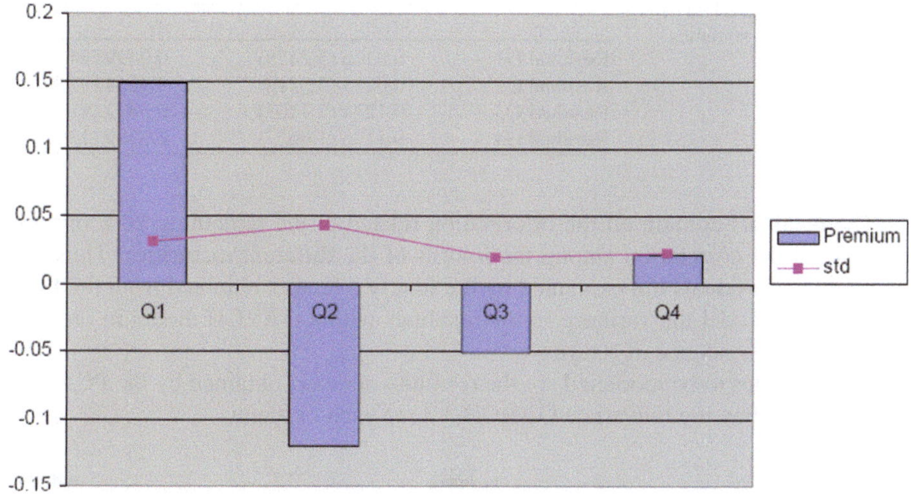

Fig. 8.3 The forward seasonal premia

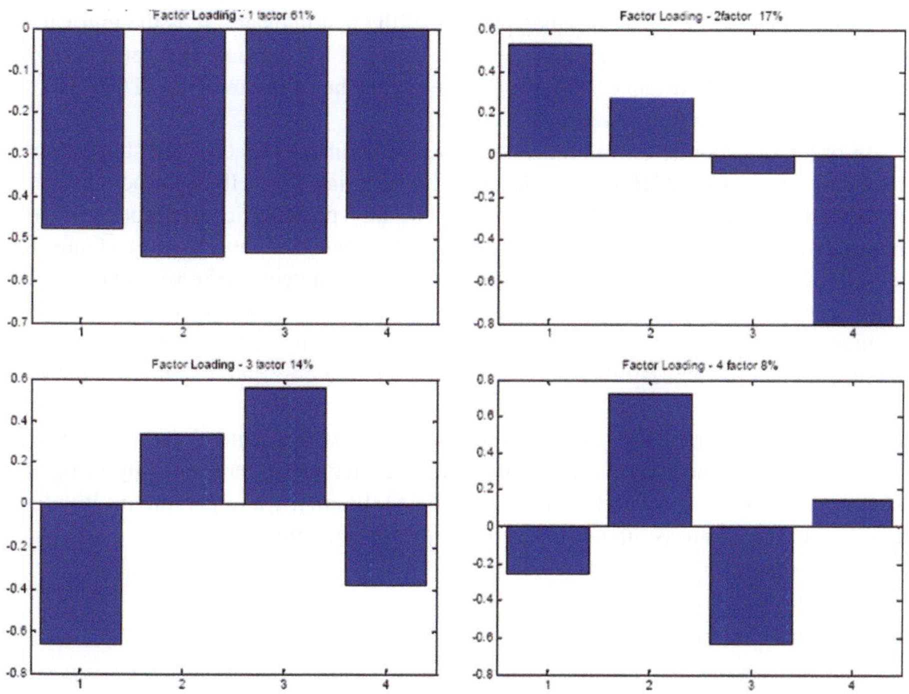

Fig. 8.4 Factor loadings

Table 8.3 GARCH(1,1) parameters and t-statistics

	α_1	α_2
Residual Q1	0.0376 (3.2478)	0.9499 (54.0736)
Residual Q2	0.0233 (3.3218)	0.9654 (97.9361)
Residual Q3	0.1274 (2.7203)	0.8492 (21.4536)
Residual Q4	0.1125 (1.6657)	0.8589 (13.5745)

The residuals contain all the information related to the remaining 20% of variability and the correlation among the returns of the different maturities. Thus, we compute the residuals not explained by the first two factors obtained from the PCA and we then model the variance of the residuals with a GARCH model in order to capture the dependence of returns.

The variables to be modelled are the residuals series r_t obtained by the PCA. We applied the following univariate GARCH(1,1) to each of them:

$$r_t = \epsilon_t, \tag{8.20}$$

$$h_t^2 = \alpha_1 \epsilon_{t-1}^2 + \alpha_2 h_{t-1}^2, \tag{8.21}$$

where h_t^2 is the conditional variance process of the residuals and ϵ_t is the innovation of the time series process, with $\epsilon_t = z_t h_t$ and z_t is a Gaussian i.i.d. process with zero mean and unit variance. In Table 8.3 we report the estimate of the GARCH(1,1) models and asymptotic t-statistics.

In order to generate correlated scenarios, we combine together the standardized residuals of the GARCH(1,1) model and the residuals from the regime switching model for the same days. We do not impose any parametric assumption on the marginal distributions and use the empirical cumulative distribution to fit a Gaussian copula of the historical residual vectors. We simulate a vector of correlated innovations from the Gaussian copula and reconstruct the forecasted scenarios using the estimated principal factors for the forward price scenario and the regime switching for the log spot price. By this procedure we generate correlated scenarios for spot and forward prices.

Hence, we generate 100 correlated scenarios and we aggregated them in a recombining tree using the backward scenario reduction technique proposed by Pflug and Hochreiter (2007) (see also Chapters 14 and 15). Finally, we adjust the multivariate tree, as described above, in order to guarantee market consistency.

8.6 Numerical Results

In this section we discuss the numerical results obtained by solving the stochastic programming problem in two case studies. The simulation framework is based on MATLAB release 12 and on GAMS release 21.5, for modelling and solving the optimization problem by nonlinear optimization package (MINOS). The hydro system is made up of one cascade with three basins and three hydro plants, one of these

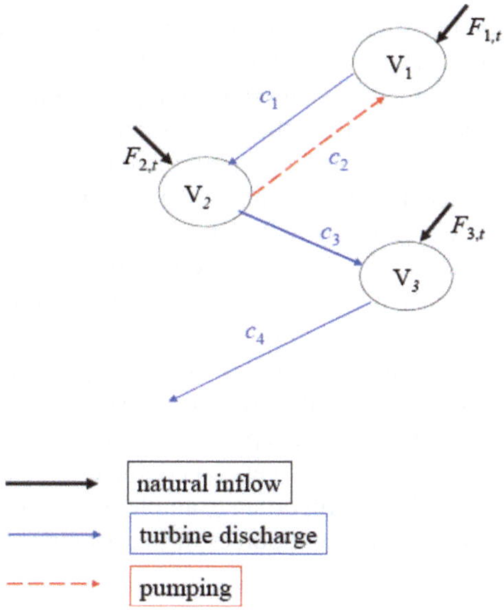

Fig. 8.5 The hydro system

Table 8.4 Hydro basin data: capacity, initial and minimum final storage volumes

Basin	$\overline{v}_j$	$v_{j,0}$	$\underline{v}_{j,T}$
v_1	1000	100	0
v_2	2000	1000	500
v_3	2000	1000	500

Table 8.5 Hydro arc data: energy coefficients and capacities

Arc	k_i	$\overline{q}_i$
c_1	1.0	100
c_2	−1.7	50
c_3	1.1	150
c_4	0.9	120

being a pumped storage hydro plant as shown in Fig. 8.5 (see Tables 8.4 and 8.5 for input data of the hydro system). In order to represent the scenarios we introduce the following notation, see also Vespucci et al. (2010), $T_1 = 1$, $T_2 = \{t : 2 \leq t \leq 20\}$, $T_3 = \{t : 21 \leq t \leq 191\}$, where the subscripts represent stage numbers. We have considered 25 scenarios represented by means of a scenario tree where the nodes are as follows: $N_1 = 1$, $N_2 = \{2, \ldots, 6\}$, $N_3 = \{7, \ldots, 31\}$.

As first step, we solve the model considering only one source of variability, the spot price. We note that the objective function is the sum of two components, the final wealth and the monetary value of water volumes in the reservoirs at the end of period. If we consider the first component in the objective function, i.e. the final cumulated wealth obtained by buying and selling energy on the spot and forward markets, we notice a decrease both in its expected value and variance. This result is not surprising since the contracts are hedging instruments able to reduce the risk related to movements of the spot prices. Moreover, if we compare the optimal solutions with and without financial contracts, we observe a different behaviour in the management of the plant. The introduction of financial contracts leads to a more efficient use of pumping, ending up with a higher expected value of water volumes in the reservoirs at the time horizon T. This is due to a more efficient use of the plant, in particular of the pumping feature. Overall, the effect of using forwards leads to an increase of the value of the objective function, expressed as certainty equivalent, from 232,068.61 to 247,709.52.

As second step, we introduce the second source of variability, the forward prices explicitly modelled as described in the previous section. As before, we notice that the use of forward contracts reduces the variability of the final wealth and, again, allows a more efficient use of water pumping with a higher final level of water in the basins. Overall, the effect is an increase in the objective function, from 232,068.61 to 251,308.03. We report the certainty equivalent obtained by using a power utility function with risk aversion coefficient −0.5.

Finally, we report an example of optimal buying and selling in forward contracts in the case of two sources of uncertainty; see Tables 8.6 and 8.7.

Table 8.6 Results for one and two sources of stochasticity

Profit value (Euro)	One source Certainty equivalent	Two sources Certainty equivalent
Stochastic model	247,709.52	251,308.03
Modified expected mean value model	240,813.01	244,509.78
Modified VSS	6696.51	6798.25

Table 8.7 Buying and selling in forward contracts with two sources of stochasticity

Optimal values of variables 'buy' Time	Maturity	Node	Quantity
1	Q1	1	3.2
2	Q1	2	16.97
2	Q1	3	15.7
2	Q1	4	0.95
2	Q1	5	2.66
3	Q2	3	6.15
3	Q2	6	9.1
10	Q2	5	51.3
21	Q2	2	8.03
21	Q2	2	0.45
21	Q3	3	1.97
21	Q4	4	9.0
Optimal values of variables 'sell'			
2	Q4	6	49.6
21	Q2	2	1.86
21	Q2	4	1.46
21	Q4	2	50.1
21	Q4	3	49.7
21	Q4	4	51.3

In order to assess the value of modelling uncertainty, we follow the procedure introduced in the literature by Escudero et al. (2007) and Vespucci et al. (2010) for evaluating the value of the stochastic solution for three-stage problem.

The procedure is based on the idea of reproducing the decision process as the uncertainty reveals: this procedure is suitable for multistage problems and is not prone to infeasibility.

The optimal objective value obtained in stage 3 is called *modified EEV (MEEV)*. Technically, this is computed as follows:

1. Scenario tree $\mathscr{T}_{1,\text{mean}}$ (see Fig. 8.6a) is defined by considering the expected value of the uncertain parameters (spot and forward prices); stochastic program with scenario tree $T_{1,\text{mean}}$ is solved and the optimal values of the first-stage variables are stored. In this way the optimal solution of the EV problem is computed.
2. Scenario tree $\mathscr{T}_{2,\text{mean}}$ (see Fig. 8.6b) is the expected value of the spot and forward prices on nodes belonging to N_3. The stochastic program with scenario tree $T_{2,\text{mean}}$ is solved having assigned the value stored at step 1 to the first-stage decision variables. The optimal value of second-stage variables are stored.
3. The stochastic program on benchmark tree $\mathscr{T}$ (see Fig. 8.6c) is solved, assigning to the first-stage decision variables the values stored at step 1 and to the second-stage decision variables the values stored at step 2 (see Fig. 8.6c).

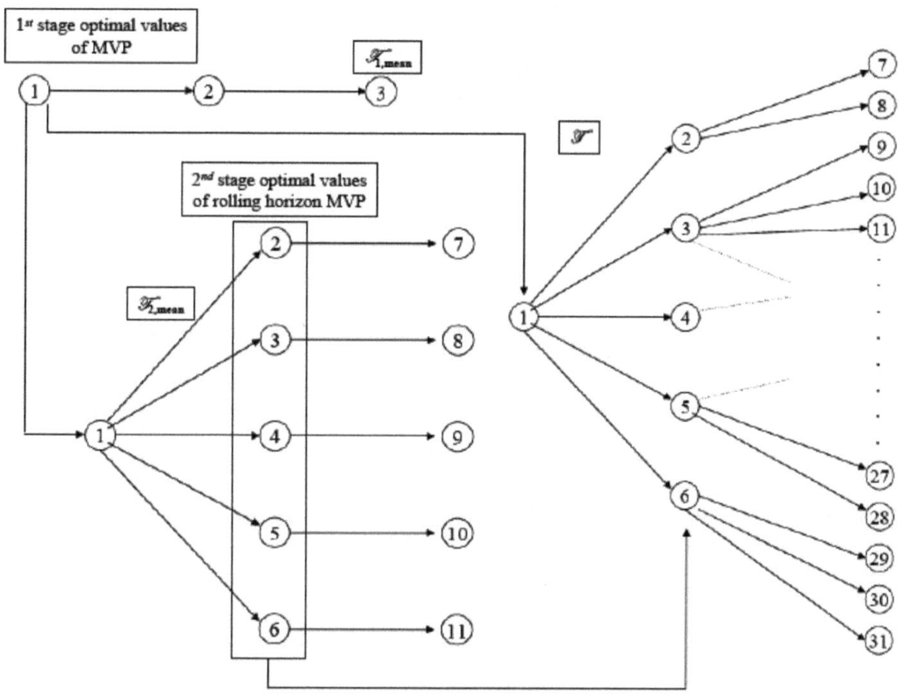

Fig. 8.6 Scheme for the computation of MEEV

We compute the modified EEV (respectively, MEEV = 240,813.01 and MEEV = 244,509.78). The value of the stochastic solution for the one and two sources of variability are, respectively, 6696.51 and 6798.25, giving us the goodness of the expected solution value when the expected values are replaced by the random values for the input variables. The introduction of the second source of variability gives a slightly higher value for the certainty equivalent and a better value for the stochastic solution.

8.7 Conclusions

In this chapter we have introduced a model for the daily hydropower system scheduling problem. The model is stochastic multistage nonlinear in which profit comes from direct production and financial operations in the forward energy market. We consider as stochastic parameters spot energy price and forward prices. The introduction of forward contracts reduces the variability of final wealth, confirming their hedging properties, with an overall increase in the objective function. Our results show that, apart from the financial gains, the advantage of using financial contracts is a more efficient use of the hydroplant, allowing the possibility of pumping water and ending up with a higher final level of the reservoir.

Acknowledgements The authors greatly acknowledge the support of Bergamo University grant 2009–2010 (coordinator: M. Bertocchi), grant 2010 (coordinator: R. Giacometti) and grant 'Metodi di integrazione delle fonti energetiche rinnovabile e monitoraggio satellitare dell'impatto ambientale', Regione Lombardia.

References

Borovkova, S., Geman, H. (2006a) 'Analysis and modelling of electricity futures prices studies', *Studies in Nonlinear Dynamics and Econometrics*, Vol. 10(3), pp. 1–14.

Borovkova, S., Geman, H. (2006b) 'Seasonal and stochastic effects of commodity forward curve', *Review of Derivatives Research*, Vol. 9(2), pp. 167–186.

Conejo, A.J., Garcia-Bertrand, R., Carrion, M., Caballero, A., de Andres, A. (2008) 'Optimal involvement in future markets of a power producer', *IEEE Transactions on Power Systems*, Vol. 23(2), pp. 703–711.

De Jong, C. (2006) 'The nature of power spikes: A regime-switch approach', *Studies in Nonlinear Dynamics & Econometrics*, Vol. 10(3).

Dupačová, J., Consigli, G., Wallace, S.W. (2000) 'Scenarios for multistage stochastic programs', *Annals of Operations Research*, Vol. 100(1–4), pp. 25–53.

Escudero, L.F., Garin, M., Merino, M., Perez, G. (2007) 'The value of the stochastic solution in multistage problems,' *Top*, Vol. 15, pp. 48–64.

Fleten, S.E., Wallace, S.W. (2009) 'Delta-hedging a hydropower plant using stochastic programming', *In* J. Kallrath, P.M. Pardalos, S. Rebennack, M. Scheidt (eds.), *Optimization in the Energy Industry*, Berlin, Heidelberg: Springer, pp. 507–524.

Fleten, S.E., Wallace, S.W. (1998) 'Power scheduling in forward contracts', *Proceedings of the Nordic MPS*, Molde, May 9–10.

Giacometti, R., Vespucci, M.T., Bertocchi, M., Barone Adesi, G. (2007) 'A stochastic model for hedging electricity portfolio for an hydro-energy producer', *DSMIA Tecnichal report*, 2, pp. 1–24.

Latorre, J., Cerisola, S., Ramos, A. (2007) 'Clustering algorithms for scenario tree generation: application to natural hydro inflows', *European Journal of Operational Research*, Vol. 181(3), pp. 1339–1353.

Nolde, K., Uhr, M., Morari, M. (2008) 'Medium term scheduling of a hydro-thermal system using stochastic model predictive control', *Automatica*, Vol. 44, pp. 1585–1594.

Pflug, G. (2001) 'Scenario tree generation for multiperiod financial optimization by optimal discretization', *Mathematical Programming B*, Vol. 89, pp. 251–271.

Pflug, G., Hochreiter, R. (2007) 'Financial scenario generation for stochastic multi-stage decision processes as facility location problem', *Annals of Operations Research*, Vol. 152(1), pp. 257–272.

Serati, M., Manera, M., Plotegher, M. (2008) 'Modeling electricity prices: From the State of the Art to a Draft of a New Proposal', *Fondazione Eni Enrico Mattei.*

Vespucci, M.T., Maggioni, F., Bertocchi, M., Innorta, M. (2010) 'Hedging electricity portfolios via stochastic programming', *In* G. Pflug (ed.), *Special Issue of APMOD Conference 2008, Annals of Operations Research*, to appear.

Vitoriano, B., Cerisola, S., Ramos, A. (2000) 'Generating scenario trees for hydro inflows', *6th International Conference on Probabilistic Methods Applied to Power Systems PSP3-106*, Madeira, Portugal.

Wallace, S.W., Fleten, S.E., Ziemba, W.T. (2002) 'Hedging electricity portfolios via stochastic programming', *In* C. Greengard, A. Ruszczyński (eds.), *Decision Making under Uncertainty: Energy and Power, IMA Volumes in Mathematics and Its Applications*, New York, NY: Springer, Vol. 128, pp. 71–93.

Chapter 9
Short-Term Trading for Electricity Producers

Chefi Triki, Antonio J. Conejo, and Lina P. Garcés

Abstract This chapter considers a price-taker power producer that trades in an electricity pool and provides models for weekly scheduling, contracting, and daily offering. On a weekly basis, a stochastic programming model is formulated to derive the on–off schedule of the production units, the contracting for the entire week and the offering curves for Monday day-ahead market. On a daily basis, a different stochastic programming model is formulated to derive the offering curve in the day-ahead markets of weekdays other than Monday. As a spinoff of the daily model, offering curves for adjustment markets within each day are also derived. Two illustrative examples clarify the models proposed.

Keywords Price-taker power producer · Weekly scheduling · Weekly contracting · Daily offering · Multi-stage stochastic programming · Risk · CVaR

9.1 Short-Term Trading

This chapter provides optimization-based decision support tools for a price-taker power producer operating under uncertainty in a restructured electricity market. We focus on short-term trading and formulate two stochastic programming models that describe the producer's trading on two different but linked time horizons. For the weekly horizon, a three-stage stochastic programming model is formulated to derive the on–off schedule of the production units, the contracting for the entire week, and the offering curves for Monday day-ahead auction. For the daily horizon, additional trading auctions are considered and a three-stage stochastic programming model is formulated to derive the offering curves for all the available auctions (e.g., day-ahead and adjustment) run in the pool for each day of the week. Stochastic programming background can be found in Birge and Louveaux (1997).

C. Triki (✉)
University of Salento, Lecce, Italy
e-mail: chefi.triki@unisalento.it

M. Bertocchi et al. (eds.), *Stochastic Optimization Methods in Finance and Energy*, International Series in Operations Research & Management Science 163,
DOI 10.1007/978-1-4419-9586-5_9, © Springer Science+Business Media, LLC 2011

9.1.1 Market Auctions

On a weekly basis it is a common industry practice that a power producer determines not only the best start-up and shut-down schedule of its production units but also which weekly contracts to sign. Moreover, the producer should derive its pool offering strategy for all the days in the considered week. As a consequence, an appropriate decision support tool for the short-term trading of a power producer should embody a two-step decision framework, weekly and daily:

1. Weekly step: the producer decides the on/off status of each generating unit and which contracts to sign for the whole week. Additionally, a detailed offering strategy for the Monday day-ahead auction and coarse ones for the other days of the week need to be determined.
2. Daily step: the producer determines its daily offering strategy for each auction organized within the electricity pool of the considered day.

The pool is often organized in several successive energy auctions that run independently. Typically, the market includes one day-ahead auction, several adjustment auctions, one regulation (automatic generation control) auction, and hourly balancing sessions. Reserve auctions are also held on a daily basis but for the sake of simplicity such auctions are not considered in this chapter. For each of the energy auctions, offers can be submitted by the producers until the closing hour, when the auction is cleared by the market operator and the results are published before the offering deadline of the next auction.

For example, a typical pool timing for day d is the following:

1. The day-ahead auction (DAA) is closed and then immediately cleared during the morning of the day $d-1$ (e.g., 11 am).
2. The first adjustment auction (AA) (for simplicity only one is considered in this chapter) is cleared in the afternoon of the same day $d-1$ (e.g., 4 pm).
3. The regulation auction (RGA) is cleared in the late afternoon of day $d-1$ (e.g., 6 pm).
4. Balancing auctions (BA) are organized on an hourly basis throughout day d, not day $d-1$.

Weekly and daily decision-making steps are illustrated in Figs. 9.1 and 9.2, respectively.

9.1.2 Uncertainty

In addressing the aforementioned decision-making problems, the producer has to take into account the complexity introduced by uncertainty. Indeed, all the decisions should be made before knowing the actual realizations of the pool prices. While only day-ahead prices are considered for solving the weekly problem, it becomes necessary to take into account the clearing prices of all pool auctions for dealing with the daily trading.

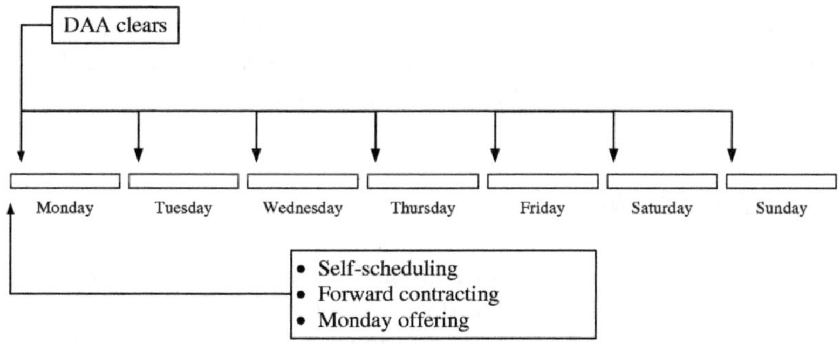

Fig. 9.1 Time sequence for the weekly problem (DAA: day-ahead auction)

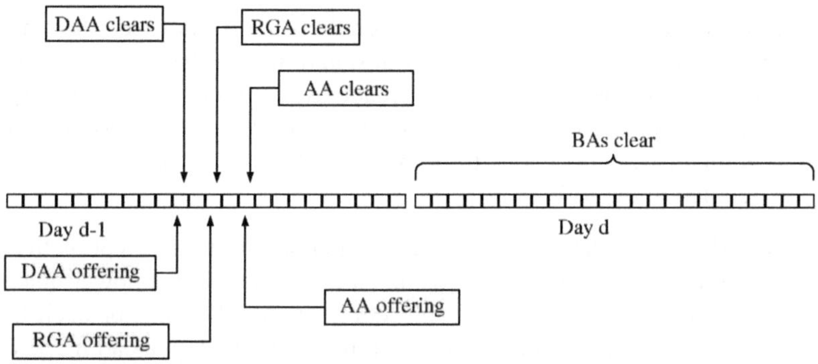

Fig. 9.2 Time sequence for the daily problem (DAA: day-ahead auction; AA: adjustment auction; RGA: regulation auction; BA: balancing auction)

Uncertain prices are characterized by using stochastic processes and historical information and are incorporated into the multi-stage models by means of a scenario set arranged in a tree (an example of such a tree is shown in Fig. 9.3). In practice, such scenario tree can be constructed by using, for example, a seasonal ARIMA model, which requires to tune up properly the corresponding time series parameters on the basis of the available historical data, as for example in Conejo et al. (2005).

To enrich the presentation and for tutorial purposes, we adopt a scenario-wise representation in the case of the weekly model and a node-wise representation in the case of the daily model. More specifically, the weekly model uses explicitly only the scenario probabilities (leaves of the tree) that can be calculated as the products of the probabilities of occurrences of the stochastic process along a path from the root node to the leaves. On the other hand, the daily model uses the information at each node within the sequential decision process, so that we keep track in each node of the previous decisions and of the scenario information observed so far.

In both cases, an accurate representation of the probability distributions for real-world applications leads usually to the generation of a large number of scenarios that may affect the tractability of the problems considered. In order to overcome this difficulty while keeping as intact as possible the stochastic information that the scenario sets embody, scenario-reduction techniques can be used. Interested readers can find detailed information on these techniques in Dupačová et al. (2003) and on their use for electricity trading in Morales et al. (2009).

9.1.3 Decision Making

The weekly self-scheduling problem consists in solving a unit commitment-type problem that in addition to the on–off decisions of the units includes decisions on contract selection and on the building of offering curves for Monday trading as well. Such decisions are made far sighted by considering into the model other decisions related to weekdays other than Monday, which are revised afterward through the successive daily models.

On the other hand, daily models consider the on–off status of the units and the contract commitments as known quantities, and tune up the day-ahead offering curves (on the basis of recent information made available) of other weekdays, and define the offering curves for other auctions and every day of the considered week.

More specifically, the decisions to be made within the weekly model and their timing are detailed in the following:

1. First-stage decisions: determine, before the opening of Monday's pool, the on/off schedule for each production unit for the entire week, and the power quantity to be sold through each available weekly contract. In addition, for each hour of Monday derive an offering curve to be submitted to the day-ahead market. Note that these are *here and now* decisions.
2. Second-stage decisions: on the basis of the prices so far observed (Monday day-ahead prices), determine, for weekdays other than Monday, the offering curves to be submitted to the day-ahead auctions pertaining to weekdays other than Monday. Note that these are *wait and see* decisions for Monday's prices and *here and now* for prices of weekdays other than Monday.
3. Third stage: the third stage represents the trading in the day-ahead market for every scenario of prices spanning the seven days of the week.

Regarding the daily offering problem, the decision framework is as follows:

1. First-stage decisions: these decisions involve determining the offering curves to be submitted to the day-ahead auctions for each hour of day d (Tuesday to Sunday). Note that these are *here and now* decisions.
2. Second-stage decisions: these decisions are the offering curves to be submitted to the subsequent adjustment auctions during day d. Note that these are *wait and see* decisions for day-ahead prices and *here and now* for prices of the adjustment markets.
3. Third stage: the third stage represents the trading in the pool for every scenario for prices spanning the whole day.

Note that Monday daily problem only involves offering in the adjustment market since Monday day-ahead offer curves are derived in the weekly problem. Note also that day-ahead offering curves for weekdays other than Monday are derived in both the weekly and daily problems. On a weekly basis we make up coarse-grained decisions for day-ahead offering of weekdays other than Monday that we tune up later through the repetitive solution of the daily model. Indeed, on Monday we make farseeing decisions that involve scenarios related to the pool prices covering all weekdays. While time moves toward the dispatching day more precise information on the demand, and thus on the price scenarios, becomes available and day-by-day updated offering curves for the day-ahead auction are necessary. The new decisions, having a daily basis, include also detailed scenarios related to other auctions of the pool.

9.1.4 Multi-stage Stochastic Models

The considered models should reflect the two salient characteristics of electricity trading: the fact that markets are dynamic (since the decision process is time dependent) and stochastic (since the market prices are not known in advance).

As a consequence, a multi-stage stochastic framework is the most appropriate modeling framework to deal with both problems under scrutiny (Birge and Louveaux 1997).

9.1.5 Notation

For clarity, the notation used throughout the chapter is stated below.

9.1.5.1 Constants

c_m	Linear generation cost of unit m (\$/MWh).
$\overline{G_m}$	Capacity of unit m (MW).
$\underline{G_m}$	Minimum power output of unit m (MW).
R_m^{d}	Ramp-down limit of unit m (MW/h).
R_m^{u}	Ramp-up limit of unit m (MW/h).
D_c	Time periods spanned by contract c (h).
$\overline{P_c}$	Maximum power that can be sold through contract c (MW).
π_c	Energy price pertaining to contract c (\$/MWh).
T_t	Time duration of period t (h).
α	Per unit confidence level to be used to compute the conditional value at risk.
β	Weighting positive factor to achieve an appropriate trade-off profit versus risk.
$A(k,s)$	Binary constant which is equal to 1 if scenarios k and $k+1$ are equal up to stage s, and 0 otherwise.
ϕ_k	Probability of occurrence of scenario k.
$S(t)$	Stage in which day-ahead offering happens in period t.

9.1.5.2 Random Variable Realizations

π_{tk}^{D} Price in the day-ahead auction, during period t and under scenario k (\$/MWh).

π_{tk}^{A} Price in the adjustment auction, during period t and under scenario k (\$/MWh).

9.1.5.3 Decision Variables

G_{mtk}^{D} Power generated by unit m for the day-ahead auction in period t under scenario k(MW).

G_{mtk}^{A} Power generated by unit m for the adjustment auction in period t under scenario k(MW).

P_c Power sold through contract c(MW).

P_{tk}^{D} Power sold in the day-ahead auction during period t under scenario k(MW).

P_{tk}^{A} Power sold in the adjustment auction during period t under scenario k(MW).

x_{mt} Binary variable which is equal to 1 if unit m is online during period t, and 0 otherwise.

ζ Variable to calculate the value at risk.

η_k Auxiliary variable used to compute the conditional value at risk.

9.1.5.4 Sets and Numbers

C_t Set of contract available during period t.

N_K Number of realizations of the stochastic process.

9.1.6 Chapter Organization

Once introduced the short-term trading problem of a power producer, the core of this chapter includes two sections. Section 9.2 is devoted to the description of the three-stage stochastic model that covers the weekly horizon, whereas Section 9.3 focuses on the offering strategies on a daily basis. Both sections include simple illustrating examples that help understanding the models and the results. The chapter concludes stating some remarks and conclusions.

9.2 Weekly Scheduling, Contracting, and Monday Day-Ahead Offering

Within a weekly horizon, we consider a producer that maximizes its profit from selling electric energy both in the pool and through contracts, while controlling the risk of variability of that profit. The sales in the day-ahead auction may result in high profit volatility since day-ahead prices vary substantially. On the other hand, selling through contracts at fixed prices results in no volatility of the profit. However, this last option prevents the producer from taking advantage of periods where the day-ahead prices are high. The level of profit variability is controlled using the conditional value at risk (CVaR) technique, which is often used in problems modeling uncertainty through scenarios (Rockafellar and Uryasev 2000, 2002).

The proposed model is a three-stage stochastic programming model. This model allows the power producer, before the beginning of each week, to optimally schedule the production of its units, to decide which weekly contracts to sign in order to sell electric energy in the futures market, as well as its offering strategy in Monday's day-ahead auction.

9.2.1 Weekly Model

We propose a stochastic programming model that provides an optimal solution to the weekly decision-making problem of a producer. This optimal solution defines the weekly self-scheduling (on/off status) of the production units, as well as the weekly contracts to be signed, and the producer offering curves for each period of Monday's day-ahead auction. Considering an average profit for any given risk level, the optimal solution obtained represents the best strategy the producer can make.

The objective function to be maximized is the expected profit of the producer plus a risk controlling CVaR term. The expected profit is the sum of the revenue from selling through contracts plus the expected pool profit for the whole week minus the expected production cost. That is,

$$\begin{aligned}\text{maximize} & \sum_c \pi_c P_c D_c \\ & + \sum_k \left[\sum_t (\pi^{\text{D}}_{tk} P^{\text{D}}_{tk} - \sum_m c_m G^{\text{D}}_{mtk}) T_t \right] \phi_k \\ & + \beta \left[\zeta - \frac{1}{(1-\alpha)} \sum_k \phi_k \eta_k \right]. \end{aligned} \tag{9.1}$$

Constraints are

$$\sum_m G^{\text{D}}_{mtk} = \sum_{c \in C_t} P_c + P^{\text{D}}_{tk} \quad \forall t, \forall k, \tag{9.2}$$

$$P^{\text{D}}_{tk} = P^{\text{D}}_{t(k+1)} \quad \forall t, k = 1, \ldots, N_K - 1; \quad \text{if} \quad A(k, S(t)) = 1, \tag{9.3}$$

$$P^{\text{D}}_{tk} \geq P^{\text{D}}_{t\tilde{k}} \quad \forall t \text{ and } \forall k, \tilde{k} : [\pi^{\text{D}}_{tk} \geq \pi^{\text{D}}_{t\tilde{k}}], \tag{9.4}$$

$$0 \leq P_c \leq \overline{P_c} \quad \forall c, \tag{9.5}$$

$$\underline{G_m}\, x_{mt} \leq G^{\mathrm{D}}_{mtk} \leq \overline{G_m}\;\; x_{mt} \quad \forall m, \forall t, \forall k, \tag{9.6}$$

$$G^{\mathrm{D}}_{m(t-1)k} - R^{\mathrm{d}}_{m} \leq G^{\mathrm{D}}_{mtk} \leq G^{\mathrm{D}}_{m(t-1)k} + R^{\mathrm{u}}_{m} \quad \forall m,\;\; \forall t,\;\; \forall k, \tag{9.7}$$

$$P^{\mathrm{D}}_{tk} \geq 0 \quad \forall t,\;\; \forall k, \tag{9.8}$$

$$-\sum_{c} \pi_c P_c D_c - \sum_{t} \left(\pi^{\mathrm{D}}_{tk} P^{\mathrm{D}}_{tk} - \sum_{m} c_m G^{\mathrm{D}}_{mtk} \right) T_t + \zeta - \eta_k \leq 0 \quad \forall k, \tag{9.9}$$

$$\eta_k \geq 0 \quad \forall k. \tag{9.10}$$

Constraints (9.2) enforce energy balance for each period and each scenario. The non-anticipativity constraints are given by (9.3). Constraints (9.4) ensure higher production at higher prices (offer condition). Constraints (9.5) impose lower and upper bounds on the power sold through contracts. Power production limits are given by (9.6). Ramping limits for production units are imposed by (9.7). The non-negativeness of the energy offered in the pool is enforced in (9.8). Finally, constraints (9.9) and (9.10) enforce conditions pertaining to the risk-metric CVaR.

Non-anticipativity and increasing offer constraints (9.3) and (9.4) deserve some additional explanations. Constraints (9.3) enforce partially non-anticipativity conditions. Since the desired output is an hourly offer curve (price vs. power) and not a single hourly power quantity, the production in each hour is made dependent on the price in that hour, which involves a relaxation of the non-anticipativity constraints. This relevant idea was first stated in Baíllo et al. (2004). Constraints (9.4) impose that a higher hourly price implies a higher hourly production level, which is needed to construct a monotonically increasing offer curve, a must in most pool markets. Note finally that constraints (9.3) and (9.4) together allow building optimal offering curves.

The above formulation is a mixed-integer linear model that can be solved by using general-purpose solvers. Relevant references addressing weekly scheduling and contracting include Garcés and Conejo (2010), Conejo et al. (2008), Gedra (1994), Kaye et al. (1990), Niu et al. (2005), Pineda et al. (2008), Shrestha et al. (2005), and Tanlapco et al. (2002).

9.2.2 Weekly Decision Making: Illustrative Example

For the sake of simplicity, the time framework considered for the week in this example is six periods, four of them representing Monday and two representing the remaining days of the week (Tuesday to Sunday). Note that a realistic case study requires 24 hourly period within Monday and 144 hourly period representing the remaining days of the week (Tuesday to Sunday).

All simulations have been carried out using CPLEX under GAMS (Rosenthal 2008). Solution times are negligible.

The pool prices are described using a tree of six scenarios, three for Monday and two for weekdays other than Monday and each realization of Monday prices ($6 = 3 \times 2$). A given scenario includes four prices corresponding with a realization of Monday pool prices, plus two values representing prices for days Tuesday to Sunday. Table 9.1 provides the pool prices for each period and each scenario. The scenario tree of this example is shown in Fig. 9.3. Note that a realistic case study may require several hundred scenarios to represent Monday prices (each one including 24 price values) plus several hundred scenarios representing price in weekdays other than Monday (each one including 144 price values).

The considered producer owns one coal unit and one gas unit whose technical data are provided in Table 9.2 and has the possibility of signing the two contracts described in Table 9.3. Signing a contract implies the selling of a single block of energy spanning the six periods at a fixed price.

The considered confidence level to obtain CVaR is $\alpha = 0.95$.

Table 9.1 Weekly problem: price scenarios for the day-ahead auction. Prices in $/MWh

	Scenario					
Period	1	2	3	4	5	6
1	39	39	33	33	22	22
2	45	45	36	36	29	29
3	43	43	31	31	24	24
4	41	41	34	34	27	27
5	47	21	47	21	47	21
6	42	25	42	25	42	25
Probability	0.1670	0.1670	0.1665	0.1665	0.1665	0.1665

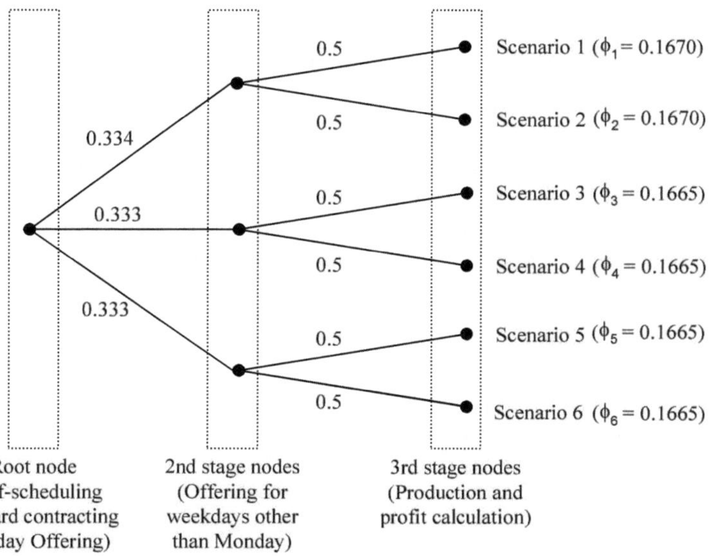

Fig. 9.3 Weekly problem: scenario tree example

Table 9.2 Technical data of the generation units

Type	$\underline{G_m}$ (MW)	$\overline{G_m}$ (MW)	R_m^{u} (MW/h)	R_m^{d} (MW/h)	P_{0m} (MW)	c_m ($)
Coal	50	150	50	50	0	10
Gas	100	300	50	50	100	24

Table 9.3 contract data

Contract type	$\overline{P_c}$ (MW)	π_c ($/MWh)
A	75	31
B	50	28

To illustrate and clarify the non-anticipativity constraints that appear in the general formulation, we write below those corresponding to this example.

Offers to be submitted to Monday's pool are decided at the beginning of the week (first node or root node in Fig. 9.3). Variables $P_{11}^{\mathrm{D}}, P_{12}^{\mathrm{D}}, P_{13}^{\mathrm{D}}, P_{14}^{\mathrm{D}}, P_{15}^{\mathrm{D}}, P_{16}^{\mathrm{D}}, P_{21}^{\mathrm{D}}, P_{22}^{\mathrm{D}}, P_{23}^{\mathrm{D}}, P_{24}^{\mathrm{D}}, P_{25}^{\mathrm{D}}, P_{26}^{\mathrm{D}}, P_{31}^{\mathrm{D}}, P_{32}^{\mathrm{D}}, P_{33}^{\mathrm{D}}, P_{34}^{\mathrm{D}}, P_{35}^{\mathrm{D}}, P_{36}^{\mathrm{D}}, P_{41}^{\mathrm{D}}, P_{42}^{\mathrm{D}}, P_{43}^{\mathrm{D}}, P_{44}^{\mathrm{D}}, P_{45}^{\mathrm{D}}$, and P_{46}^{D} are the offers for each Monday period ($t = 1, 2, 3$, and 4) and for each price scenario considered ($k = 1, \ldots, 6$).

As depicted in Fig. 9.3, leaves 1 and 2, corresponding to scenarios 1 and 2, have the same predecessor node. Offering strategies for Monday day-ahead auction made in this predecessor node must comply with the conditions below:

$$P_{11}^{\mathrm{D}} = P_{12}^{\mathrm{D}},\ P_{21}^{\mathrm{D}} = P_{22}^{\mathrm{D}},\ P_{31}^{\mathrm{D}} = P_{32}^{\mathrm{D}},\ P_{41}^{\mathrm{D}} = P_{42}^{\mathrm{D}}.$$

In the same way, other non-anticipativity constraints are

$$P_{13}^{\mathrm{D}} = P_{14}^{\mathrm{D}},\ P_{23}^{\mathrm{D}} = P_{24}^{\mathrm{D}},\ P_{33}^{\mathrm{D}} = P_{34}^{\mathrm{D}},\ P_{43}^{\mathrm{D}} = P_{44}^{\mathrm{D}},$$
$$P_{15}^{\mathrm{D}} = P_{16}^{\mathrm{D}},\ P_{25}^{\mathrm{D}} = P_{26}^{\mathrm{D}},\ P_{35}^{\mathrm{D}} = P_{36}^{\mathrm{D}},\ P_{45}^{\mathrm{D}} = P_{46}^{\mathrm{D}}.$$

This example is solved considering five different values of the risk parameter, $\beta = \{0, 1, 2, 3, 5\}$. For each value of β, Table 9.4 provides the expected revenue from Monday day-ahead auction, the expected pool revenue for weekdays other than Monday, the revenue from selling through contracts, the total revenue, and the

Table 9.4 Weekly problem: expected revenues and costs

β	Monday's auction (million $)	Other weekdays' auction (million $)	Contracts (million $)	Total revenue (million $)	Total production cost (million $)
0	0.2653	2.0940	0	2.3593	1.3014
1	0.2047	1.7295	0.3906	2.3248	1.3014
2	0.1643	1.4865	0.6258	2.2766	1.3014
3	0.1592	1.4307	0.6258	2.2157	1.2534
5	0.1565	1.3497	0.6258	2.1320	1.1934

Table 9.5 Weekly problem: expected profit and CVaR

β	Expected profit (million $)	CVaR (million $)
0	1.0579	0.3072
1	1.0234	0.4035
2	0.9752	0.4425
3	0.9623	0.4488
5	0.9386	0.4551

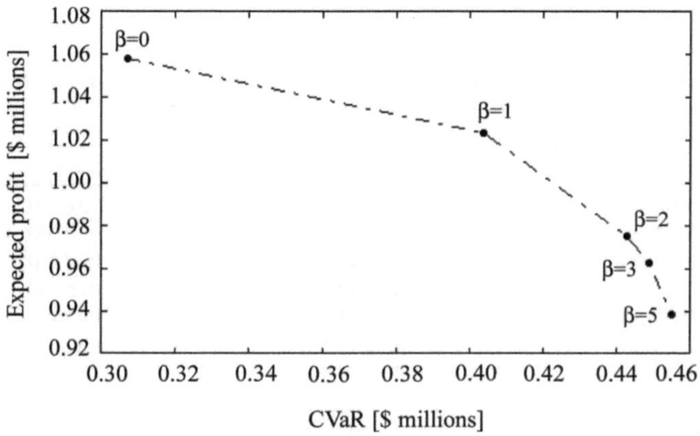

Fig. 9.4 Weekly problem: expected profit versus CVaR

expected total production cost. If β changes from 0 to 5, the expected pool revenue of the entire week (columns 2 and 3) decreases while the expected revenue from selling through contracts (column 4) increases. These results are consistent because a risk-concerned producer avoids a high variability of the profit at the cost of a reduction on its total expected profit.

Table 9.5 provides the expected total profit and the CVaR while Fig. 9.4 depicts the efficient frontier, i.e., the expected profit versus the CVaR for different values of β. As expected, Table 9.5 and Fig. 9.4 show that the higher the expected profit the lower its CVaR.

For the risk-neutral case ($\beta = 0$), the expected profit is $1.0579 million while the CVaR is $0.3072 million. On the other hand, the risk-averse case ($\beta = 5$) results in an expected profit of $0.9386 million and a CVaR of $0.4551 million. Consequently, if β is increased from 0 to 5, the expected profit decreases 11.28% while the CVaR increases 48.14%.

Table 9.6 provides the power sold through contracts in each period. Note that for the risk-neutral case, the power sold through contracts is zero, i.e., no contracts are signed, while the most risk-averse cases ($\beta = 2$–5) involve signing all contracts available. For a value of $\beta = 1$, the producer only signs contract A that implies the selling of 75 MW throughout the week.

The power generated and the power sold in the pool by the producer for each scenario and period is provided in Table 9.7. For the risk-neutral case, all the power

Table 9.6 Weekly problem: power sold through contracts in MW

Period	$\beta = 0$	$\beta = 1$	$\beta = 2$	$\beta = 3$	$\beta = 5$
1	0	75	125	125	125
2	0	75	125	125	125
3	0	75	125	125	125
4	0	75	125	125	125
5	0	75	125	125	125
6	0	75	125	125	125

Table 9.7 Weekly problem: power generated/power sold in the pool in MW

	$\beta = 0$					
Period	$k = 1$	$k = 2$	$k = 3$	$k = 4$	$k = 5$	$k = 6$
1	200/200	200/200	200/200	200/200	200/200	200/200
2	300/300	300/300	300/300	300/300	300/300	300/300
3	400/400	400/400	400/400	400/400	350/350	350/350
4	450/450	450/450	400/400	400/400	400/400	400/400
5	450/450	400/400	450/450	350/350	450/450	350/350
6	450/450	450/450	450/450	400/400	450/450	400/400
	$\beta = 5$					
Period	$k = 1$	$k = 2$	$k = 3$	$k = 4$	$k = 5$	$k = 6$
1	200/75	200/75	200/75	200/75	200/75	200/75
2	300/175	300/175	300/175	300/175	300/175	300/175
3	400/275	400/275	400/275	400/275	300/175	300/175
4	450/325	450/325	400/275	400/275	300/175	300/175
5	450/325	400/275	450/325	350/225	350/225	250/125
6	450/325	450/325	450/325	400/275	400/275	300/175

produced in each period is sold in the pool while for the risk-averse cases, only one part of this power is sold in the pool. In some periods and scenarios, the relation power generated/power sold in the pool changes significantly from the risk-neutral case to the risk-averse one ($\beta = 5$), e.g., in period 4 and scenario 5, this relation changes from 400/400 to 300/175.

Considering both the risk-neutral and the risk-averse cases, Fig. 9.5 depicts the average production of the two generation units in each period. In both cases, Fig. 9.5 indicates that units 1 and 2 are up in all periods. The power quantity produced by unit 2 changes from the risk-neutral case to the risk-averse one, while unit 1 is scheduled to produce the same quantities in both cases.

Figure 9.6 depicts, for the risk-neutral and risk-averse cases, the offers to be submitted to the day-ahead auction for each Monday period. Note that Monday's pool offering in each period of the day-ahead auction for the risk-neutral case involves higher quantities than for the risk-averse case; this is so because in the risk-neutral case the producer offers the greatest possible amount of power, taking the maximum possible advantage from high pool prices.

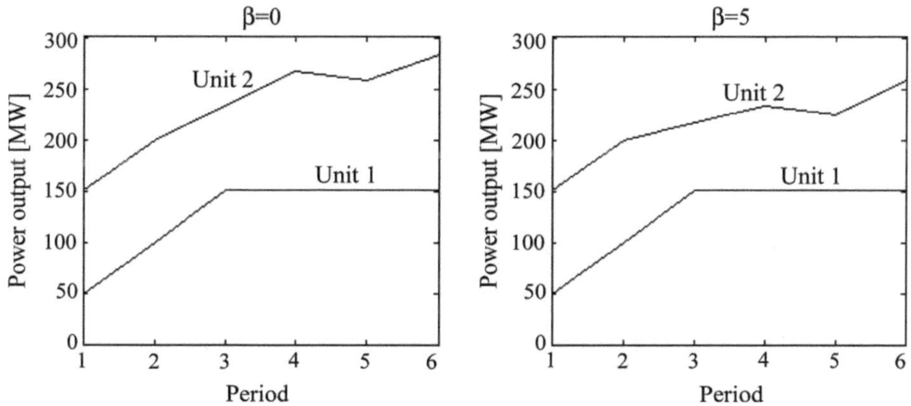

Fig. 9.5 Weekly problem: average power output in MW of the generation units

9.3 Daily Trading

We consider below the problem of a producer seeking to maximize its profit from selling in an electricity pool within a daily time framework. For simplicity, we just consider two auctions as representative of the spectrum of all possible trading floors, namely, the day-ahead auction and one adjustment auction (including additional auctions is simple). As a result, the corresponding model has three stages, two corresponding to the two auctions and a third one representing dispatching output and profit calculation.

It is worth noting that, even though the volume traded in the adjustment market is limited (typically below 10% of the amount traded in the day-ahead auction), it is advantageous participating in such an auction since it represents an opportunity not only for profit making but also for reacting against undesirable events such as operation infeasibilities.

9.3.1 Day-Ahead Model

To develop a decision-support tool for the producer's participation in the pool each day we propose a stochastic programming model that defines, for every hour, optimal nondecreasing offering curves for both the day-ahead and adjustment auctions. In order to achieve this, day-ahead decisions (P_{tk}^{D} and G_{mtk}^{D}) are made dependent on the observations of the second-stage scenarios and similarly adjustment auction decisions (P_{tk}^{A} and G_{mtk}^{A}) are defined such that they depend on the outcomes of the third-stage scenarios. Thus, non-anticipativity conditions are partly relaxed as proposed in Baíllo et al. (2004).

Several features can be included into the model to allow the producer not only to express its general strategy but also to comply with specific bidding rules. However, for the sake of simplicity, we restrict our attention to those features that ensure the

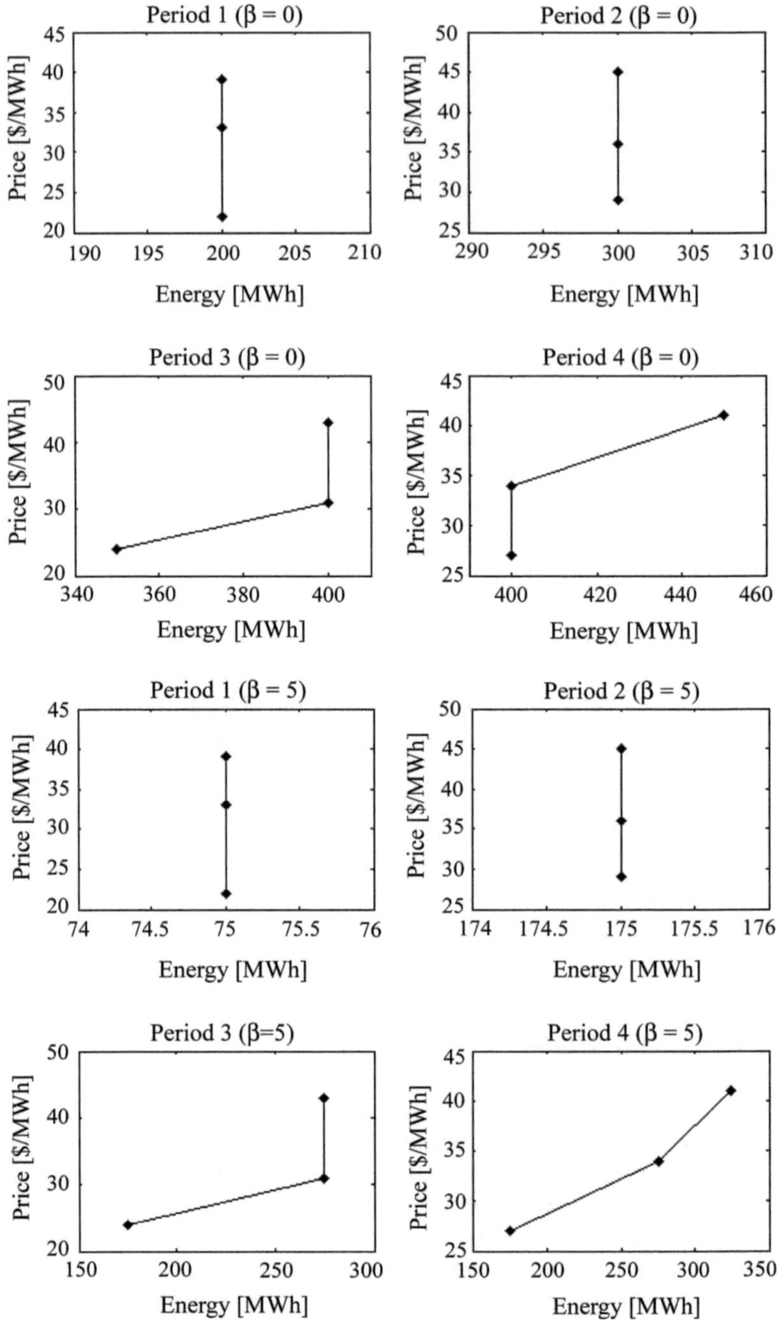

Fig. 9.6 Weekly problem: pool offering for Monday day-ahead auction with $\beta = 0$ and $\beta = 5$

definition of a basic offering strategy. The proposed multi-stage stochastic formulation is presented below as a deterministic equivalent model.

The objective function to be maximized is the sum of the expected profits from the two auctions minus the expected production cost. The production cost is represented linearly through linear variable costs c_m. The extension to consider a more complex cost function is simple in the modeling phase but may require the employment of sophisticated solution techniques. Thus, the objective is to

$$\text{maximize} \sum_t \left[\sum_k \left(\pi_{tk}^{\mathrm{D}} P_{tk}^{\mathrm{D}} - \sum_m c_m G_{mtk}^{\mathrm{D}} \right) \phi_k + \sum_k \left(\pi_{tk}^{\mathrm{A}} P_{tk}^{\mathrm{A}} - \sum_m c_m G_{mtk}^{\mathrm{A}} \right) \phi_k \right]. \tag{9.11}$$

Constraints are

$$\sum_m G_{mtk}^{\mathrm{D}} = P_{tk}^{\mathrm{D}} \quad \forall t, \ \forall k, \tag{9.12}$$

$$\sum_m G_{mtk}^{\mathrm{A}} = P_{tk}^{\mathrm{A}} \quad \forall t, \ \forall k, \tag{9.13}$$

$$P_{tk}^{\mathrm{D}} = P_{t(k+1)}^{\mathrm{D}} \quad \forall t, \ k = 1, \ldots, N_K - 1; \quad \text{if} \quad A(k, S(t)) = 1, \tag{9.14}$$

$$G_{mtk}^{\mathrm{D}} \geq G_{mt\tilde{k}}^{\mathrm{D}} \quad \forall m, \forall t \quad \text{and} \quad \forall k, \tilde{k} : [\pi_{tk}^{\mathrm{D}} \geq \pi_{t\tilde{k}}^{\mathrm{D}}], \tag{9.15}$$

$$G_{mtk}^{\mathrm{A}} \geq G_{mt\tilde{k}}^{\mathrm{A}} \quad \forall m, \forall t \quad \text{and} \quad \forall k, \tilde{k} : [\pi_{tk}^{\mathrm{A}} \geq \pi_{t\tilde{k}}^{\mathrm{A}}], \tag{9.16}$$

$$\underline{G_m} \leq G_{mtk}^{\mathrm{D}} + G_{mtk}^{\mathrm{A}} \leq \overline{G_m} \quad \forall m, \ \forall t, \ \forall k, \tag{9.17}$$

$$\begin{aligned} &(G_{m(t-1)k}^{\mathrm{D}} + G_{m(t-1)k}^{\mathrm{A}}) - R_m^{\mathrm{d}} \leq G_{mtk}^{\mathrm{D}} + G_{mtk}^{\mathrm{A}}, \\ &\quad \leq (G_{m(t-1)k}^{\mathrm{D}} + G_{m(t-1)k}^{\mathrm{A}}) + R_m^{u} \quad \forall m, \ \forall t, \ \forall k. \end{aligned} \tag{9.18}$$

Constraints (9.12) and (9.13) ensure the energy balance in both the day-ahead and the adjustment auctions, respectively. The non-anticipativity constraints (partly relaxed) on the day-ahead decisions are expressed by means of (9.14). The monotonously increasing nature of offering stacks for both auctions is ensured through constraints (9.15) and (9.16). All technical limits that guarantee the correct functioning of the generating units should be expressed as constraints of this mathematical formulation. Specifically, constraints (9.17) ensure for each unit the maximum/minimum power outputs and constraints (9.18) enforce the up/down-ramping limits.

This formulation of the multi-stage stochastic program is a linear model that can be solved by using any general-purpose linear software package. However, extending the model to consider other auctions may require the use of binary variables, which, in turn, may require the use of specialized solution procedures to tackle large-scale instances of this model.

Relevant references addressing offering strategies for producers in a daily pool include Baíllo et al. (2004), Arroyo and Conejo (2000), Conejo et al. (2002), Triki et al. (2005), Plazas et al. (2005), and Musmanno et al. (2010).

9.3.2 Day-Ahead Offering: Illustrative Example

We consider in this example the same data provided in Section 9.2.2 with a time framework of six periods spanning 4 h each. For simplicity, We assume that all the production capacity is available for pool trading and that no power is committed through contracts.

The pool price realizations are represented by means of the scenario tree depicted in Fig. 9.7. The day-ahead and adjustment prices for each period together with the corresponding probability values are provided in Table 9.8.

For the sake of clarity, it is convenient to write explicitly the non-anticipativity constraints that can be expressed as follows:

$$\begin{gathered}
P^{\mathrm{D}}_{11} = P^{\mathrm{D}}_{12},\, P^{\mathrm{D}}_{21} = P^{\mathrm{D}}_{22},\, P^{\mathrm{D}}_{31} = P^{\mathrm{D}}_{32},\, P^{\mathrm{D}}_{41} = P^{\mathrm{D}}_{42},\, P^{\mathrm{D}}_{51} = P^{\mathrm{D}}_{52},\, P^{\mathrm{D}}_{61} = P^{\mathrm{D}}_{62},\\
P^{\mathrm{D}}_{13} = P^{\mathrm{D}}_{14},\, P^{\mathrm{D}}_{23} = P^{\mathrm{D}}_{24},\, P^{\mathrm{D}}_{33} = P^{\mathrm{D}}_{34},\, P^{\mathrm{D}}_{43} = P^{\mathrm{D}}_{44},\, P^{\mathrm{D}}_{53} = P^{\mathrm{D}}_{54},\, P^{\mathrm{D}}_{63} = P^{\mathrm{D}}_{64},\\
P^{\mathrm{D}}_{15} = P^{\mathrm{D}}_{16},\, P^{\mathrm{D}}_{25} = P^{\mathrm{D}}_{26},\, P^{\mathrm{D}}_{35} = P^{\mathrm{D}}_{36},\, P^{\mathrm{D}}_{45} = P^{\mathrm{D}}_{46},\, P^{\mathrm{D}}_{55} = P^{\mathrm{D}}_{56},\, P^{\mathrm{D}}_{65} = P^{\mathrm{D}}_{66}.
\end{gathered}$$

The offering curves for the day-ahead and adjustment auctions, for both production units and for all the periods, are depicted in Figs. 9.8 and 9.9, respectively. As

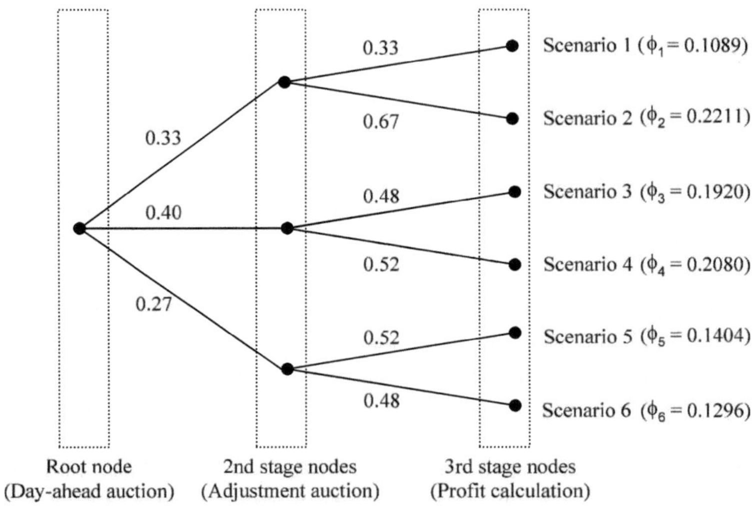

Fig. 9.7 Scenario tree for the daily offering example

Table 9.8 Daily problem: pool prices in \$/MWh and probability values for each scenario

	Day-ahead			Adjustment					
Period	1	2	3	1	2	3	4	5	6
1	27	43	22	23	32	36	44	19	28
2	28	41	23	26	31	37	43	20	24
3	26	45	21	21	29	40	45	19	29
4	25	40	23	23	27	38	41	18	26
5	26	39	20	24	28	36	40	18	23
6	27	43	22	23	31	39	44	21	23
Probability	0.33	0.40	0.27	0.33	0.67	0.48	0.52	0.52	0.48

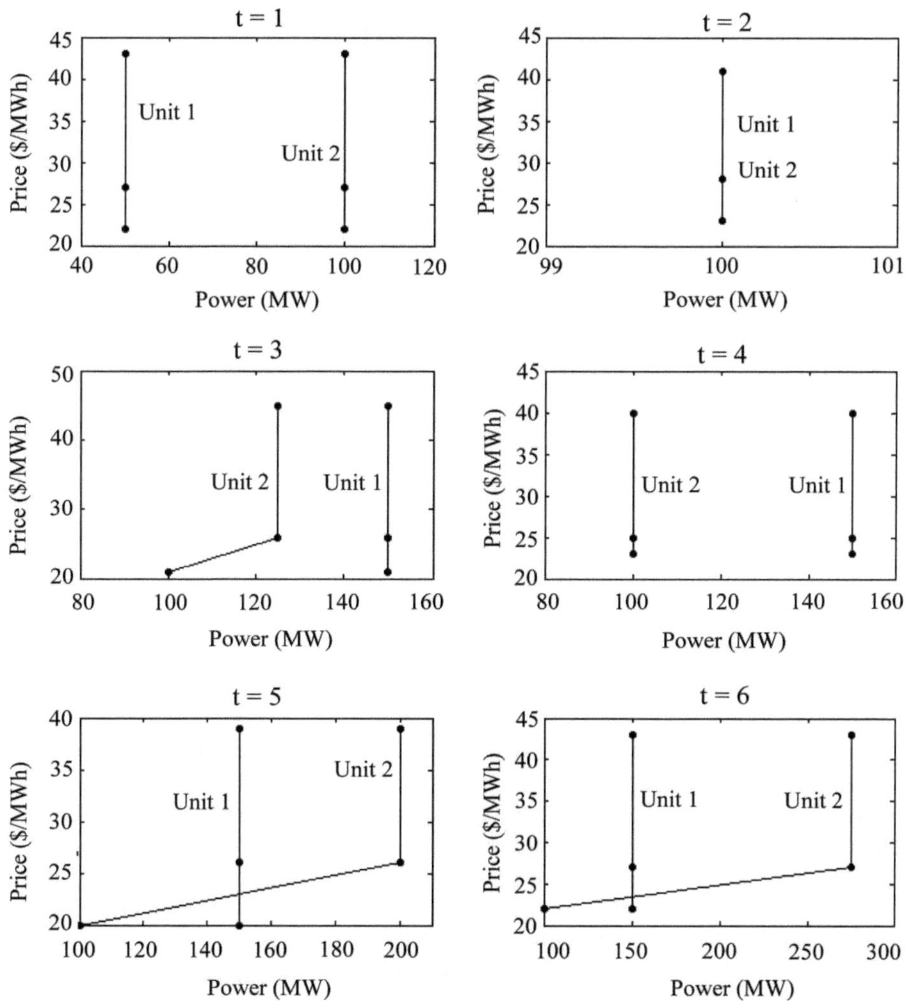

Fig. 9.8 Daily problem: offering curves for the day-ahead auction

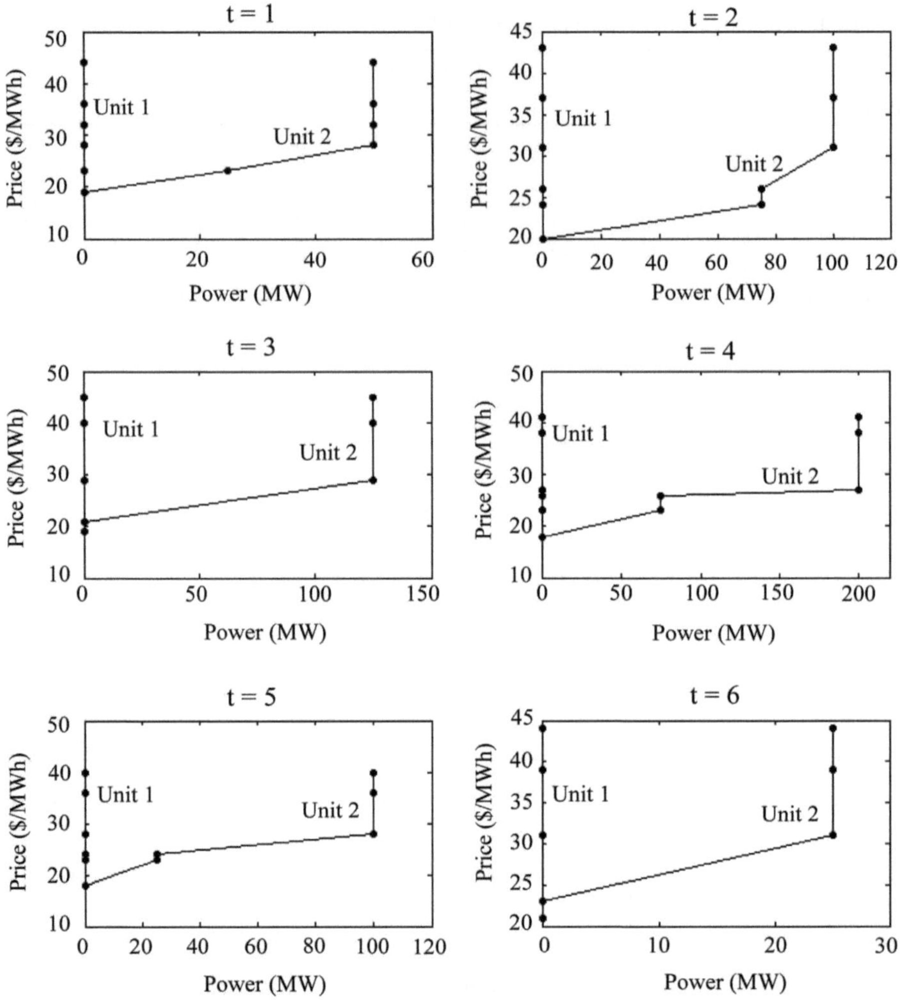

Fig. 9.9 Daily problem: offering curves for the adjustment auction

non-anticipativity constraints are partly relaxed and pool prices in each scenario are known, Figs. 9.8 and 9.9 are formed by a set of values which represent the power offered in the day-ahead or adjustment auction, respectively, for each price scenario. Thus, the power–price pairs are plotted in such a way as to generate nondecreasing offering curves. Specifically, each curve in Fig. 9.8 represents a stack of three power–price pairs to be offered in the day-ahead auction. Clearly, each power–price pair corresponds to a possible realization in the second stage of the scenario tree but only one of the three outcomes will be observed after the auction clearing. Similarly, each curve in Fig. 9.9 represents a stack of six power–price pairs that anticipate the possible outcomes of the third-stage scenarios.

We can also note in Figs. 9.8 and 9.9 that while both production units are considered for offering in the day-ahead auction, only unit 2 is used to offer in the adjustment auction.

In Table 9.9 we provide not only expected values but also scenario values for the revenues, cost, and profit. It can be noted, for example, that the profit corresponding to scenario 4 is higher than that of the other scenarios. This is not surprising since scenario 4 is characterized by both high clearing prices and high probability values (see Table 9.8).

It should be also observed that at the first stage, i.e., while defining the offering strategy under uncertainty, the decision maker can observe only the expected values of the revenues, cost, and profit (last row in Table 9.9) through the daily optimization model. Only at the third stage, i.e., after the clearing of all the auctions, it is possible to know the actual scenario observed and, thus, which one among the possible profits provided in Table 9.9 has actually materialized.

Finally, we provide in Table 9.10 the quantity of power sold in both the day-ahead and adjustment auctions. It is possible to observe that, for each period, while the quantity sold in the adjustment auction may vary from one scenario to another, at most three different power values can be observed in the day-ahead auction. Indeed, these last values are always the same for scenarios 1 and 2, for scenarios 3 and 4, and for scenarios 5 and 6. This result is consistent with the scenario tree considered and also with the non-anticipativity conditions imposed on the day-ahead decisions.

Table 9.9 Daily problem: revenues and profits

Scenario	Day-ahead revenue (million $)	Adjustment revenue (million $)	Total revenue (million $)	Production cost (million $)	Total profit (million $)
1	0.1745	0.0194	0.1939	0.1356	0.0583
2	0.1745	0.0692	0.2437	0.1740	0.0697
3	0.2758	0.0907	0.3665	0.1740	0.1925
4	0.2758	0.1017	0.3775	0.1740	0.2035
5	0.1176	0	0.1176	0.0876	0.0300
6	0.1176	0.0374	0.1550	0.1212	0.0338
Expected	0.1997	0.0608	0.2605	0.1508	0.1097

Table 9.10 Power sold in day-ahead auction/power sold in adjustment auction in MW

Period	Scenario 1	Scenario 2	Scenario 3	Scenario 4	Scenario 5	Scenario 6
1	150/25	150/50	150/50	150/50	150/0	150/50
2	200/75	200/100	200/100	200/100	200/0	200/75
3	275/0	275/125	275/125	275/125	250/0	250/125
4	250/75	250/200	250/200	250/200	250/0	250/75
5	350/25	350/100	350/100	350/100	250/0	250/25
6	425/0	425/25	425/25	425/25	250/0	250/0

9.4 Summary and Conclusions

In this chapter we provide multi-stage stochastic models to build short-term trading strategies for a price-taker producer operating in an electricity pool. We define a decision-support tool consisting on a two-step decision framework involving the weekly and the daily horizons. Decisions concerning the weekly horizon involve the production units status, contract selection, and Monday day-ahead offering. In the daily horizon we derive offering strategies for day-ahead auctions and extend the trading opportunities to consider also the adjustment auctions. In this context, each offer is not defined by simply a single power–price pair but rather through an optimal offering curve constituted by several nondecreasing power–price pairs, which entails the partial relaxation of non-anticipativity constraints.

Uncertainty related to the pool prices are modeled by using an intuitive scenario tree representation based on historical information.

The optimization models developed are expressed in terms of deterministic equivalent formulations that can be solved by using general-purpose algorithms. However, the dynamic and stochastic natures of the application may make the resulting problems intractable, specially if large-scale instances are considered.

In order to improve the clarity of the description of the optimization models related to the weekly and daily planning problems we provide illustrative examples with realistic data. Moreover, the weekly example provides a detailed discussion of the effect of the risk on the decision process.

The following conclusions are in order:

1. Stochastic programming is an appropriate tool to address the weekly self-scheduling and contracting problems faced by a power producer.
2. On a daily basis, stochastic programming allows building models to derive offering curves for the different auctions of the pool.
3. If the uncertainty involved requires considering a large number of scenarios, appropriate scenario reduction techniques need to be used.
4. To derive optimal offering curves and not single hourly optimal quantities, non-anticipativity constraints need to be partially relaxed.
5. The resulting models can be formulated linearly, avoiding generally a large number of binary variables. Thus, they are generally tractable even for large producers.
6. Risk control on the decision-making process can be easily and efficiently incorporated into the model through the CVaR.
7. Extensive simulations carried out considering different markets and producers allow concluding that the proposed models constitute relevant tools for power producers.

References

Arroyo J. M., Conejo A. J. (2000) Optimal response of a thermal unit to an electricity spot market. IEEE Trans Power Syst 15(3):1098–1104.

Baíllo A., Ventosa M., Rivier M., Ramos A. (2004) Optimal offering strategies for generation companies operating in electricity spot markets. IEEE Trans Power Syst 19(2):745–753.

Birge J. R., Louveaux F. (1997) Introduction to stochastic programming. Springer, New York, NY.

Conejo A. J., Nogales F. J., Arroyo J. M. (2002) Price-taker bidding strategy under price uncertainty. IEEE Trans Power Syst 17(4):1081–1088.

Conejo A. J., Contreras J., Espínola R., Plazas M. A. (2005) Forecasting electricity prices for a day-ahead pool-based electric energy market. Int J Forecasting 21(3): 435–462.

Conejo A. J., García-Bertrand R., Carrión M., Caballero A., de Andrés A. (2008) Optimal involvement in futures markets of a power producer. IEEE Trans Power Syst 23:703–711.

Dupačová J., Gröwe-Kuska N., and Römisch W. (2003) Scenario reduction in stochastic programming: An approach using probability metrics. Math Program Ser A 95(3):493–511.

Garcés L. P., Conejo A. J. (2010) Weekly self-scheduling, forward contracting and offering strategy for a producer. IEEE Trans Power Syst 25(2):657–666.

Gedra T. W. (1994) Optional forward contracts for electric power markets. IEEE Trans Power Syst 9:1766–1773.

Kaye R. J., Outhred H. R., Bannister C. H. (1990) Forward contracts for the operation of an electricity industry under spot pricing. IEEE Trans Power Syst 5:46–52.

Morales J. M., Pineda S., Conejo A. J., Carrión M. (2009) A scenario reduction for futures market trading in electricity markets. IEEE Trans Power Syst 24(2):878–888.

Musmanno R., Scordino N., Triki C., Violi A. (2010) A multistage formulation for generation companies in a multi-auction electricity market. IMA J Manage Math 21(2):165–181.

Niu N., Baldick R., Guidong Z. (2005) Supply function equilibrium bidding strategies with fixed forward contracts. IEEE Trans Power Syst 20:1859–1867.

Pineda S., Conejo A. J., Carrión M. (2008) Impact of unit failure on forward contracting. IEEE Trans Power Syst 23(4):1768–1775.

Plazas M. A., Conejo A. J., Prieto F. J. (2005) Multi-market optimal bidding for a power producer. IEEE Trans Power Syst 20(4):2041–2050.

Rockafellar R. T., Uryasev S. (2000) Optimization of conditional value-at-risk. J Risk 2:21–41.

Rockafellar R. T., Uryasev S. (2002) Conditional value-at-risk for general loss distributions. J Bank Financ 26:1443–1471.

Rosenthal R. E. (2008) GAMS, a user's guide. GAMS Development Corporation, Washington, DC.

Shrestha G. B., Pokharel B. K., Lie T. T., Fleten S.-E. (2005) Medium term power planning with bilateral contracts. IEEE Trans Power Syst 20:627–633.

Tanlapco E., Lawarree J., Liu C. C. (2002) Hedging with futures contracts in a deregulated electricity industry. IEEE Trans Power Syst 17:577–582.

Triki C., Beraldi P., Gross G. (2005) Optimal capacity allocation in multi-auction electricity markets under uncertainty. Comput Oper Res 32:201–217.

Chapter 10
Structuring Bilateral Energy Contract Portfolios in Competitive Markets

Antonio Alonso-Ayuso, Nico di Domenica, Laureano F. Escudero, and Celeste Pizarro

Abstract A multistage complete recourse model for structuring energy contract portfolios in competitive markets is presented for price-taker operators. The main uncertain parameters are spot price, exogenous water inflow to the hydro system and fuel-oil and gas cost. A mean-risk objective function is considered as a composite function of the expected trading profit and the weighted probability of reaching a given profit target. The expected profit is given by the bilateral contract profit and the spot market trading profit along the time horizon over the scenarios. The uncertainty is represented by a set of scenarios. The problem is formulated as a mixed 0–1 deterministic equivalent model. Only 0–1 variables have nonzero coefficients in the first-stage constraint system, such that the continuous variables only show up in the formulation of the later stages. A problem-solving approach based on a splitting variable mathematical representation of the scenario clusters is considered. The approach uses the twin node family concept within the algorithmic framework presented in the chapter. The Kyoto protocol-based regulations for the pollutant emission are considered.

Keywords Energy trading contracts portfolio · Stochastic programming · Mean risk · Mixed 0–1 models · Splitting variable · Branch-and-fix coordination

10.1 Introduction

Given a power generation system and a set of committed as well as candidate energy trading contracts, the *Energy Contract Portfolio Problem* (*ECP2*) is concerned with selecting the bilateral contracts for energy purchasing and selling to be delivered along a given time horizon. It is one of the main problems faced today by the power generation companies and the energy service providers. The main uncertain parameters are spot market price, exogenous water inflow to the hydro system, and fuel-oil and gas cost. The maximization of the weighted *reaching probability (RP)* is included as a risk measure in the objective function, together with the expected

L.F. Escudero (✉)
Departamento de Estadística e Investigación Operativa, Universidad Rey Juan Carlos, Móstoles (Madrid), Spain
e-mail: laureano.escudero@urjc.es

M. Bertocchi et al. (eds.), *Stochastic Optimization Methods in Finance and Energy*, International Series in Operations Research & Management Science 163,
DOI 10.1007/978-1-4419-9586-5_10,

energy trading profit along the time horizon over the scenarios, and the economic value of satisfying the Kyoto protocol-based regulations for the pollutant emissions by the thermal generators. The constraint system comprises the power generation constraints for the thermal and hydro generators, the logistic constraints related to the energy contracts, and the power load already committed to purchasing and selling. See Tanlapco et al. (2002) for different types of financial hedging contracts. See also Fleten et al. (2001) for an approach to hedge electricity portfolios via stochastic optimization. An approach in Pritchard et al. (2005) uses dynamic programming for problem solving. Shrestha et al. (2005) present a medium-term hydropower planning with bilateral contracts. Sen et al. (2006) integrate the unit commitment model with financial decision-making and spot market considerations in a non-hydropower generation environment. Fleten and Kristoffsen (2007) present a model for determining optimal bidding strategies taking uncertainty into account by a price-taking hydropower producer.

Nowadays, most of the traded energy (around 75%) is done via bilateral contracts (e.g. the daily market APX in the Netherlands has only between 10 and 15% of the total energy of the country and the daily market EEX in Germany sells something less than 15% of the total energy). Moreover, there are some other countries (Spain is a good example, see Perez Arriaga (2005)), with a very small bilateral contracts development due, mainly, to disruption of market regulations.

The deterministic version of the problem can be represented as a mixed 0–1 model. Given today's state-of-the-art optimization tools, no major difficulties should arise for problem solving of moderate size instances, at least. However, given the uncertainty of the main parameters, the modelling and algorithmic approach to the problem makes *ECP2* an interesting application case of stochastic integer programming.

The aim of the model that we present in the chapter consists of helping a better decision making on the energy selling/purchasing (so-called first-stage) policy. For this purpose, the future uncertainties in the spot market and generation availability are considered. (Since the model assumes price-taking decision maker, no restrictions are considered for satisfying the uncertain demand.)

Some of the stochastic approaches related to power generation only consider continuous variables. Moreover, there are different schemes to address the energy power generation planning under uncertainty, by modelling it with 0–1 and continuous variables; see Carpentier et al. (1996), Caroe and Schultz (1998, 1999), Dentcheva et al. (1997), Dentcheva and Römisch (1998), Gollmer et al. (1999), Hemmecke and Schultz (2001), Klein Haneveld and van der Vlerk (2001), Ni et al. (2004), Nürnberg and Römisch (2002), Schultz et al. (2005), Takriti and Birge (2000), and Triki et al. (2005), among others. Most of these approaches only consider mean (expected) objective functions and Lagrangian decomposition schemes, some of which propose bidding strategies for 24 h horizons.

Alternatively, very few approaches in general deal with the mean-risk measures by considering semi-deviations (Ogryczak and Ruszczynski 1999), excess probabilities (Alonso-Ayuso et al. 2009; Schultz and Tiedemann 2004), value-at-risk (VaR), conditional VaR (Rockafellar and Uryasev 2000, Schultz and Tiedemann 2006) and stochastic dominance constraint strategies (Gollmer et al. (2011) and references therein); see also Dert (1998); Lulli and Sen (2004); Schultz (2003) and Valente (2002). These approaches are more amenable for large-scale problem solving than the classical mean – variance schemes, particularly, in the presence of 0–1 variables.

In this chapter we present a mixed 0–1 *deterministic equivalent model* (*DEM*) for the multistage stochastic *ECP2* with complete recourse, where the parameters' uncertainty is represented by a set of scenarios. Only 0–1 variables (that represent strategic trading decisions) have nonzero coefficients in the first-stage constraint system, such that the continuous variables (that represent operational decisions) only show up in the formulation of the later stages. The values of the variables do consider all scenarios without being subordinated to any of them. The *reaching probability (RP)* is considered as a risk measure in contrast with approaches where only mean functions appear in the objective function to optimize.

The model has no direct relationship with the enterprise-wide risk management function or the bidding to each submarket function. However, the max mean-risk function to consider assumes this interrelation with the enterprise-wide risk management function. On the other hand, the elements of the spot market in the model assume the interrelation with the bidding function in an aggregated way. After structuring the portfolio, the model can be run again, in a rolling horizon approach, once new opportunities for energy selling/purchasing, restructuring and cancellation appear in the horizon.

One of the main contributions of the chapter consists of considering a coupling of a generic electricity generation planning problem and the contract management problem, since in our opinion the structuring of the bilateral contract portfolio must take into account the spot market availability and the energy generation restrictions albeit in an aggregate way.

We present a *branch-and-fix coordination (BFC)* scheme to exploit the structure of the *splitting variable* representation of the mixed 0–1 *DEM* and, specifically, the structure of the *nonanticipativity* constraints for the 0–1 variables. The algorithm makes use of the *twin node family (TNF)* concept introduced in Alonso-Ayuso et al. (2003a). It is specially designed for coordinating and reinforcing the branching node and the branching variable selections at each *branch-and-fix (BF)* tree. The trees result from the relaxation of the *nonanticipativity* constraints in the *splitting variable* representation of the *DEM* for the scenario *clusters*.

Additionally, the proposed approach considers the *DEM* at each *TNF integer set*, i.e. each set of integer nodes, one for each *BF* tree, where the *nonanticipativity* constraints for the 0–1 variables are satisfied. By fixing those variables to the nodes' values, the *DEM* has only continuous variables. The approach ensures the satisfaction of the *nonanticipativity* constraints for the continuous variables associated with each scenario group in the scenario tree and so, provides the *LP* optimal solution for the given *TNF integer set*.

The remainder of the chapter is organized as follows. Section 10.2 states the *ECP2* and introduces the mixed 0–1 *DEM* for the multistage stochastic model, compact representation. Section 10.3 introduces the mean-risk function and the splitting variable representation of the *DEM*. Section 10.4 presents our approach for problem solving. Section 10.5 concludes. The Appendix gives more algorithmic details of the approach presented in this chapter.

10.2 Problem Description

10.2.1 Problem Statement

Consider a set of thermal power generators, and a set of hydropower generators distributed along river basins whose reservoirs can store water from one period to the next one in a given time horizon. Consider also a set of scenarios for the uncertain parameters, mainly, spot market price, exogenous water inflow to the hydro system, and fuel-oil and gas cost. Finally, let a set of candidate energy bilateral trading (purchasing/selling) contracts be considered. Each contract is characterized by the energy price depending on the scenario to occur and the lower and upper bounds of the purchasing/selling load to deliver in each period of the time horizon. The load in the selling contracts will be assumed to be dependent on the scenario to occur. It is assumed that the energy generator/trading company is a price taker when operating in the spot market. Note that the contract portfolio can combine contracts in different ways to reduce risk and profit less or, vice versa, increase profit and increase risk.

The problem consists of deciding which candidate purchasing/selling contracts to accept, such that some logistic constraints related to the combination of the contracts are satisfied, the power to generate satisfies the thermal and hydro generation constraints, the balance equations of the water flow along the periods in the river basins are satisfied, the Kyoto protocol-based pollutant emission constraints are satisfied by the thermal units, and the purchasing/selling loads already committed are also satisfied. See also Conejo and Prieto (2001) and Escudero (2001). The objective function to maximize is a composite function of the estimated profit along the time horizon and the weighted *RP* of having a profit greater than or equal to a given threshold over the scenarios.

We present below a mixed 0–1 model for structuring the energy contract portfolio, where the uncertainty is treated such that the occurrence of the events is represented by a multistage scenario tree.

10.2.2 Scenario Tree

Let the tree shown in Fig. 10.1 represent the stochasticity of the problem to be dealt with. Each node in the figure is associated with a point in time where a decision can be made. Once a decision is made, some contingencies can happen (in this example, the number of contingencies is 3 for time period $t = 2$), and information related to them is available at the beginning of the period. Each root-to-leaf path in the tree represents one specific scenario and corresponds to one realization of

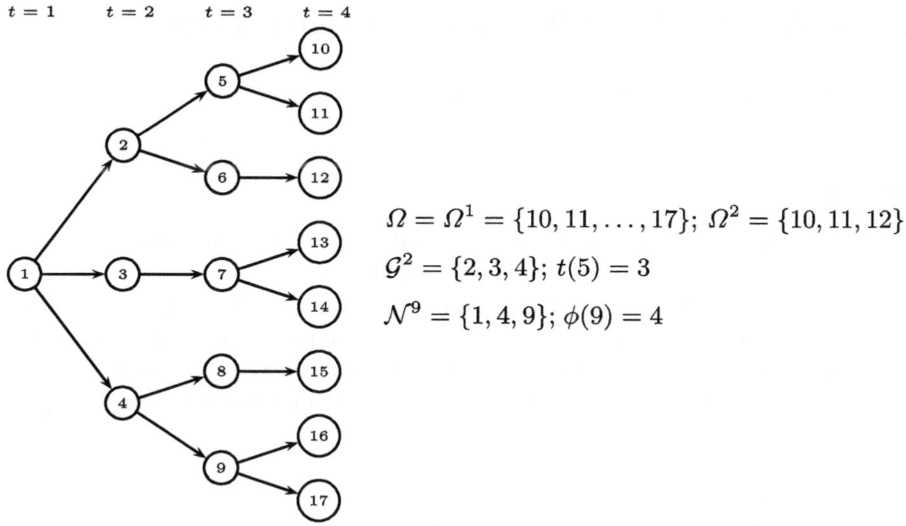

Fig. 10.1 Scenario tree

the whole set of the uncertain parameters along a time horizon. Each node in the tree can be associated with a scenario group, such that two scenarios belong to the same group for a given time period, provided that they have the same realizations of the uncertain parameters up to that period. The *nonanticipativity* principle stated by Wets (1974) and restated in Rockafellar and Wets (1991), see Birge and Louveaux (1997), among many others, requires that both scenarios have the same values for the related variables up to the given period.

The following notation related to the scenario tree will be used throughout the chapter:

$\mathcal{T}$, set of stages (periods, in our case). $\mathcal{T}^- = \mathcal{T} - \{|\mathcal{T}|\}$
Ω, set of (consecutively numbered) scenarios
Γ, set of multiperiods for pollutant emission regulation. Note: The regulation is performed at the multiperiod level (e.g. quarter, year), instead of the period level (e.g. week)
$\mathcal{M}_\gamma \subseteq \mathcal{T}$, set of (consecutive) time periods for multiperiod γ, for $\gamma \in \Gamma$.
τ_γ, last period for multiperiod γ, for $\gamma \in \Gamma$
$\mathcal{G}$, set of (consecutively numbered) scenario groups
$\mathcal{G}^t$, set of scenario groups in period t, for $t \in \mathcal{T}$, such that the scenarios that belong to the same group are identical in all realizations of the uncertain parameters up to period t ($\mathcal{G}^t \subseteq \mathcal{G}$)
$t(g)$, period for scenario group g, for $g \in \mathcal{G}$. Note: $g \in \mathcal{G}^{t(g)}$
Ω^g, set of scenarios in group g, for $g \in \mathcal{G}$ ($\Omega^g \subseteq \Omega$)
$\mathcal{N}^g$, set of scenario groups $\{k\}$ such that $\Omega^g \subseteq \Omega^k$, for $g \in \mathcal{G}$ ($\mathcal{N}^g \subseteq \mathcal{G}$), that is, set of ancestor nodes of node g
$\phi(g)$, immediate ancestor node of node g, for $g \in \mathcal{G}$. Note: $\phi(g) \in \mathcal{N}^g$

10.2.3 Mixed 0–1 DEM: Mean Function Maximization

The following is additional notation used in *ECP2*.

10.2.3.1 Sets

$\mathcal{D}$, set of thermal-power generator types (coal, fuel-oil, gas, combined cycle, etc.)
$\mathcal{I}$, set of thermal-power generators in the problem
$d(i)$, type of generator i, such that $d(i) \in \mathcal{D}$, for $i \in \mathcal{I}$. Note: Two generators that belong to the same type share the allowed pollutant emission regulations
$\mathcal{J}$, set of hydro generation units in the river basins under consideration
$\mathcal{U}_j$, set of immediate up-stream hydro units to hydro unit j to represent the topology of the river basins, for $j \in \mathcal{J}$
$\mathcal{F}^{\text{cu}}$, current energy bilateral selling contracts
$\mathcal{K}^{\text{cu}}$, current energy bilateral purchasing contracts
$\mathcal{F}^{\text{ca}}$, candidate energy bilateral selling contracts
$\mathcal{K}^{\text{ca}}$, candidate energy bilateral purchasing contracts
$\mathcal{F} = \mathcal{F}^{\text{cu}} \bigcup \mathcal{F}^{\text{ca}}$ and $\mathcal{K} = \mathcal{K}^{\text{cu}} \bigcup \mathcal{K}^{\text{ca}}$

10.2.3.2 Deterministic Function and Parameters

$p_j\left(x_j^g, \hat{z}_j^g\right)$, power output of hydro unit j for a given water release, say, x_j^g and storage level, say, z_j^g at period $t(g)$ under scenario group g, for $g \in \mathcal{G}$, $j \in \mathcal{J}$. Although the formulation in terms of the variables z_j would provide a better representation of the influence of the water-related variables in the hydropower generation function, it would result in a significant increase in problem complexity. As a compromise, a representative value for the stored water, say $\hat{z}_j^g$, is used in the formulation to linearize the resulting expression. It is represented by a concave piecewise function on x_j^g.
$\overline{z}_j$, upper bound of the stored water in the reservoir associated with hydro unit j at any period, for $j \in \mathcal{J}$
$\overline{xy}_j$, upper bound of the (released and spilled) water flow downstream from the reservoir associated with hydro unit j at any time period, for $j \in \mathcal{J}, t \in \mathcal{T}$
$\overline{p}_j$, upper bound of the power to generate by hydro unit j at any time period, for $j \in \mathcal{J}$
$\overline{p}_i$, upper bound of the power to generate by thermal unit i, for $i \in \mathcal{I}$
$\underline{\theta}_{kt}, \overline{\theta}_{kt}$, lower and upper bounds for energy bilateral purchasing contract k during period t, respectively, for $k \in \mathcal{K}, t \in \mathcal{T}$. This instrument is evidently a physical instrument, but it can be used as a financial one and, in particular, can be used as a load factor contract, see its definition in Mo et al. (2001)
υ_d, unit pollutant emission from thermal generator type d, for $d \in \mathcal{D}$
$\hat{\varrho}_d$, threshold for the non-penalized pollutant emission for the Kyoto protocol regulation in the thermal generation type d, for $d \in \mathcal{D}$

ζ, unit penalty (res., reward) for Kyoto protocol-based pollutant emission excess (res., deficit) from the allowed threshold. It is constant for the time horizon, at least

a, unit trading fee for energy purchasing/selling from the spot market

10.2.3.3 Feasibility Spaces

Φ_i, operating feasible region for the production of thermal generator i at any time period, for $i \in \mathcal{I}$. There is ample bibliography in the open literature describing the operating-constraint system to define the feasible space; see Arroyo and Conejo (2000), Dentcheva et al. (1997), Conejo et al. (2002), Dentcheva and Römisch (1998), Groewe-Kuska et al. (2001), Hobbs et al. (2001), Jimenez Redondo and Conejo (1997), Klein Haneveld and van der Vlerk (2001), Schultz et al. (2005), Nürnberg and Römisch (2002), and Wood and Wollenberg (1996), among many others. However, since the problem treated in this chapter considers a long time horizon, the set of operating constraints has not been considered.

Δ, feasible region for the set of energy bilateral selling contracts to be chosen, such as maximum percentage for each contract, regional considerations, exclusivity constraints mainly in the financial contracts, etc.

Ψ, feasible region for the set of energy bilateral purchasing contracts to be chosen, such as maximum allowed budget, maximum percentage for each contract, regional considerations, and exclusivity constraints mainly in the financial contracts.

10.2.3.4 Stochastic Function and Parameters for Period $t(g)$ Under Scenario Group g, for $g \in \mathcal{G}$

$c_i^g(p_i^g)$, production cost for power output, say, p_i^g of thermal generator i, for $i \in \mathcal{I}$. It can be represented by a convex piecewise linear function, where the stochasticity is mainly due to the cost of the fuel-oil and gas along the time horizon.

$\underline{z}_j^g$, lower bound of the stored water in reservoir j, for $j \in \mathcal{J}$. Special attention should be given to the lower bounds of the stored water during the last period of the time horizon.

$\underline{xy}_j^g$, lower bound of the (released and spilled) water flow downstream from reservoir j, for $j \in \mathcal{J}$.

e_j^g, net exogenous water inflow to reservoir j, for $j \in \mathcal{J}$.

ι_f^g, energy requirement from bilateral selling contract f, for $f \in \mathcal{F}$.

μ_f^g, unit payment received from bilateral selling contract f, for $f \in \mathcal{F}$.

λ_k^g, unit payment due for bilateral purchasing contract k, for $k \in \mathcal{K}$.

η^g, unit energy price in the spot market.

Let ω^g denote the weight assigned to scenario group g, for $g \in \mathcal{G}$.

10.2.3.5 Generation and Trading Variables for Period $t(g)$ Under Scenario Group g, for $g \in \mathcal{G}$

θ_k^g, energy requirement from bilateral purchasing contract k, for $k \in \mathcal{K}$.
p_i^g, power output of thermal generator i, for $i \in \mathcal{I}$.
ϱ_d^{+g}, ϱ_d^{-g}, excess and deficit pollutant emissions by the thermal generators of type d from the allowed target for satisfying the Kyoto protocol-based regulations, respectively, for $d \in \mathcal{D}$, $g \in \mathcal{G}^{\tau_\gamma}$, $\gamma \in \Gamma$.
x_j^g, released water from reservoir j, for $j \in \mathcal{J}$.
y_j^g, spilled water from reservoir j, for $j \in \mathcal{J}$.
z_j^g, stored water at (the end of) period $t(g)$ in reservoir j, for $j \in \mathcal{J}$.
s^g, power output to be sold to the spot market by the generation/energy service company.
b^g, power to be purchased from the spot market by the generation/energy service company.

10.2.3.6 First-Stage Variables: Candidate Contracts

δ_f, 0–1 variable having the value 1 if the candidate energy bilateral selling contract f is chosen and, otherwise, its value is 0, for $f \in \mathcal{F}^{ca}$.
β_k, 0–1 variable having the value 1 if the candidate energy bilateral purchasing contract k is chosen and, otherwise, its value is 0, for $k \in \mathcal{K}^{ca}$.

The following is a so-called *compact* representation of the mixed 0–1 *DEM* for the *multistage* stochastic problem with complete recourse.

Objective: Determine the portfolio of the energy bilateral trading contracts to maximize the mean (expected) profit from the contract portfolio exploitation and spot market trading along the time horizon over the scenarios, subject to constraints (10.2)–(10.14):

$$Q_E = \sum_{g \in \mathcal{G}} \sum_{f \in \mathcal{F}^{cu}} w^g \mu_f^g \iota_f^g + \max \sum_{g \in \mathcal{G}} w^g \Bigg[\sum_{f \in \mathcal{F}^{ca}} \mu_f^g \iota_f^g \delta_f + (\eta^g - a) s^g \\ - \Bigg(\sum_{k \in \mathcal{K}} \lambda_k^g \theta_k^g + \sum_{i \in \mathcal{I}} c_i^g (p_i^g) + (\eta^g + a) b^g + \sum_{d \in \mathcal{D} | g \in \mathcal{G}^{\tau_\gamma}, \gamma \in \Gamma} \zeta (\varrho_d^{+g} - \varrho_d^{-g}) \Bigg) \Bigg]. \tag{10.1}$$

Constraints:

$$\sum_{i \in \mathcal{I}} p_i^g + \sum_{j \in \mathcal{J}} p_j(x_j^g, \hat{z}_j^g) + \sum_{k \in \mathcal{K}} \theta_k^g + b^g = \sum_{f \in \mathcal{F}^{cu}} \iota_f^g + \sum_{f \in \mathcal{F}^{ca}} \iota_f^g \delta_f + s^g \quad \forall g \in \mathcal{G}, \tag{10.2}$$

$$0 \leq p_i^g \leq \overline{p}_i \quad \forall i \in \mathcal{I},\ g \in \mathcal{G}, \tag{10.3}$$

$$v_d \sum_{k \in \mathcal{N}^g | t(k) \in \mathcal{M}_\gamma} \sum_{i \in \mathcal{I} | d(i) = d,} p_i^k + \varrho_d^{-g} = \hat{\varrho}_d + \varrho_d^{+g} \quad d \in \mathcal{D},\ g \in \mathcal{G}^{\tau_\gamma},\ \gamma \in \Gamma, \tag{10.4}$$

$$p_j(x_j^g, \hat{z}_j^g) \leq \overline{p}_j \quad \forall j \in \mathcal{J},\ g \in \mathcal{G}, \tag{10.5}$$

$$-z_j^{\phi(g)} - \sum_{u \in \mathcal{U}_j} (x_u^g + y_u^g) + x_j^g + y_j^g + z_j^g = e_j^g \quad \forall j \in \mathcal{J},\ g \in \mathcal{G}, \tag{10.6}$$

$$\underline{z}_j^g \leq z_j^g \leq \overline{z}_j \quad \forall j \in \mathcal{J},\ g \in \mathcal{G}, \tag{10.7}$$

$$\underline{xy}_j^g \leq x_j^g + y_j^g \leq \overline{xy}_j \quad \forall j \in \mathcal{J},\ g \in \mathcal{G}, \tag{10.8}$$

$$\underline{\theta}_{k,t(g)} \leq \theta_k^g \leq \overline{\theta}_{k,t(g)} \quad \forall k \in \mathcal{K}^{\text{cu}},\ g \in \mathcal{G}, \tag{10.9}$$

$$\underline{\theta}_{k,t(g)} \beta_k \leq \theta_k^g \leq \overline{\theta}_{k,t(g)} \beta_k \quad \forall k \in \mathcal{K}^{\text{ca}},\ g \in \mathcal{G}, \tag{10.10}$$

$$\delta_f \in \Delta \quad \forall f \in \mathcal{F}, \tag{10.11}$$

$$\beta_k \in \Psi \quad \forall k \in \mathcal{K}, \tag{10.12}$$

$$x_j^g, y_j^g, s^g, b^g \geq 0 \quad \forall i \in \mathcal{I},\ j \in \mathcal{J},\ g \in \mathcal{G}, \tag{10.13}$$

$$\varrho_d^{+g},\quad \varrho_d^{-g} \geq 0 \quad \forall d \in \mathcal{D},\quad g \in \mathcal{G}^{\tau_\gamma},\ \gamma \in \Gamma. \tag{10.14}$$

The model determines the selling/purchasing contract (so-called *first-stage*) policy to be followed when restructuring the energy bilateral contracts portfolio. The restructuring aims to maximize the expected profit given by the exploitation of the portfolio as well as the spot market trading as represented in (10.1). The balance equations (10.2) equate the power availability (i.e., thermal and hydro generation plus purchased power already committed plus candidate power to purchase in bilateral trading contracts and power to purchase on the spot market) on one side and the power needs (i.e., sold power already committed plus candidate power to sell in bilateral trading contracts and power to sell on the spot market) on the other side. Bounds (10.3) ensure the feasibility of the thermal power to generate. Equations (10.4) define the positive and negative output from the pollutant thermal system to satisfy the regulation based on the Kyoto protocol. Constraints (10.5) force upper bounds on the hydropower generation. Balance equations (10.6) ensure the water availability along the time horizon, given the topology of the river basins under the control of the generation company (see in Cervellera et al. (2006) an approach for optimizing large-scale water reservoir network by stochastic dynamic programming).

Bounds (10.7) and (10.8) ensure the bounding of the stored water in the reservoirs and the lower and upper bounds of the water flow (i.e., released and spilled water) in the segments of the river basins. Bounds (10.9) ensure that the generation company/energy service provider will purchase the required energy from the current bilateral contracts satisfying the requirements stated in the contracts. Constraints (10.10) ensure that the delivery of the energy from the candidate bilateral purchasing contracts will satisfy the requirements of the contracts, if any. Memberships (10.11) and (10.12) ensure that the logistic constraints are satisfied for the set of bilateral energy selling and purchasing contracts to be chosen.

It is worth pointing out the different mechanism of the energy bilateral selling and purchasing contracts. The model decides the energy to purchase, within a given set of lower and upper bounds. On the other hand, the decisions on the energy to be sold in a given period are exogenous to the model (i.e., they are made by the buyer within committed lower and upper bounds); they are represented by uncertain energy requirements which are modeled under the scenario groups for both types of contracts, current, and candidate.

Notice that the decision on the acceptance of the contracts considers the scenario-related system, but it is not subordinated to any scenario.

Remark: The bilateral contracts enter the constraint system in the power balance equation restriction (10.2), but they also appear in the constraints that define the feasible spaces (10.11) and (10.12) besides the objective function (10.1) (and function (10.15)). It is enough for deciding the selling/purchasing (so-called first-stage) policy. Notice also that the physical contracts are the main contracts in the model; however, the financial ones are also allowed. The differences between the two types of contracts are considered in the spaces defined by (10.11) and (10.12). It is up to the contract manager to selling or purchasing a financial contract but (s)he should take into account according to the model that there is a cost for energy purchasing (including fees) and an income for energy selling (subtracting the related fees), such that the solution must also satisfy constraints (10.2), (10.11), and (10.12).

10.3 Mean-Risk Function Maximization and *VaR*

As an alternative to maximizing the risk neutral mean function (10.2), we propose to consider a risk averse mean-risk objective function, defined as a composite function of the expected trading profit and RP, the probability of reaching a given profit threshold, say, τ. It can be formulated as follows:

$$\max Q_E + \rho Q_P, \tag{10.15}$$

where

$$Q_E = \sum_{\omega \in \Omega} w_\omega \mathcal{O}^\omega, \tag{10.16}$$

w_ω gives the weight assigned to scenario ω, $\mathcal{O}^\omega$ gives the profit to obtain under scenario ω for $\omega \in \Omega$, ρ is a nonnegative weighting parameter, and Q_P is the *reaching probability,* RP. The profit $\mathcal{O}^\omega$ can be expressed as

$$
\begin{aligned}
\mathcal{O}^{\omega} = & \sum_{g \in \mathcal{N}^n} \left[\sum_{f \in \mathcal{F}^{\text{cu}}} \mu_f^g \iota_f^g + \sum_{f \in \mathcal{F}^{\text{ca}}} \mu_f^g \iota_f^g \delta_f + (\eta^g - a) s^g \right] \\
& - \sum_{g \in \mathcal{N}^n} \left[\sum_{k \in \mathcal{K}} \lambda_k^g \theta_k^g + \sum_{i \in \mathcal{I}} c_i^g (p_i^g) + (\eta^g + a) b^g + \zeta \sum_{d \in \mathcal{D} | g \in \mathcal{G}^{\tau_\gamma}, \gamma \in \Gamma} (\varrho_d^{+g} - \varrho_d^{-g}) \right],
\end{aligned} \tag{10.17}
$$

where $n \in \mathcal{G}^{|\mathcal{T}|}$ such that $\omega \in \Omega^n$, and RP can be expressed as

$$
Q_P = P\left(\omega \in \Omega : \mathcal{O}^{\omega} \geq \tau\right). \tag{10.18}
$$

Based on an approach for modeling *RP* given in Schultz and Tiedemann (2004), a more amenable expression of (10.15) for computational purposes, at least, is given by

$$
\begin{aligned}
& \max \sum_{\omega \in \Omega} w_{\omega} \left(\mathcal{O}^{\omega} + \rho \nu^{\omega}\right), \\
& \text{s.t. } \mathcal{O}^{\omega} + M(1 - \nu^{\omega}) \geq \tau \quad \forall \omega \in \Omega, \\
& \nu^{\omega} \in \{0, 1\} \quad \forall \omega \in \Omega,
\end{aligned} \tag{10.19}
$$

where M is a parameter whose value is chosen as the smallest value which does not eliminate any feasible solution under any scenario, and ν^{ω} is a 0–1 variable such that

$$
\nu^{\omega} = \begin{cases} 1, & \text{if the trading profit } \mathcal{O}^{\omega} \text{ under scenario } \omega \\ & \text{is greater or equal than threshold } \tau \\ 0, & \text{otherwise} \end{cases} \qquad \forall \omega \in \Omega.
$$

Note: The term $M(1 - \nu^{\omega})$ in (10.19) allows the benefit $\mathcal{O}^{\omega}$ to be negative for some scenario $\omega \in \Omega$.

As alternatives to the mean-risk optimization, the *value at risk* (VaR) and *conditional VaR* functionals to maximize for a given α-risk, where $0 \leq \alpha < 1$, can be expressed as

$$
\begin{aligned}
& \max \text{ VaR} \\
& \text{s.t. } \mathcal{O}^{\omega} + M(1 - \vartheta^{\omega}) \geq \text{VaR } \forall \omega \in \Omega \\
& \sum_{\omega \in \Omega} w_{\omega} \vartheta^{\omega} \geq 1 - \alpha \\
& \vartheta^{\omega} \in \{0, 1\} \quad \forall \omega \in \Omega,
\end{aligned} \tag{10.20}
$$

for the *value at risk* (VaR), where ϑ^{ω} is a 0–1 variable, and

$$\max \text{CVaR}$$
$$\text{s.t. } \kappa + \frac{1}{1-\alpha} \sum_{\omega \in \Omega} \left[w_\omega \mathcal{O}^\omega - \kappa \right]^+ \geq \text{CVaR}, \tag{10.21}$$

for the *conditional VaR*, where $[x]^+$ gives the positive value of x and κ gives the *VaR* value. Note: The Kyoto-based pollutant emission regulation has not been considered.

Notice that the replacement of the expected trading profit (10.2) by the mean-risk system (10.19) with (10.17) does not change the structure of the model. On the contrary, the maximization of VaR (10.20) and the maximization of *conditional VAR* (10.21) destroy the structure of the model, since the related constraints are non-separable.

The instances of the mixed 0–1 DEM (10.1)–(10.14) can have such large dimensions that the plain use of state-of-the-art optimization engines becomes unaffordable.

Additionally, a *splitting variable* representation can be introduced by replacing the variables with their siblings, such that θ_k^g, p_i^g, x_j^g, y_j^g, z_j^g, s^g and b^g are replaced with θ_{kt}^ω, p_{it}^ω, x_{jt}^ω, y_{jt}^ω, z_{jt}^ω, s_t^ω and b_t^ω, respectively, for $t = t(g)$, $\omega \in \Omega^g$, $g \in \mathcal{G}$; ϱ_d^{+g} and ϱ_d^{-g} are replaced with $\varrho_{dt}^{+\omega}$ and $\varrho_{dt}^{-\omega}$, respectively, for $\omega \in \Omega^g$, $g \in \mathcal{G}^t$, $t = \tau_\gamma$, $\gamma \in \Gamma$; and δ_f and β_k are replaced with δ_f^ω and β_k^ω, respectively, for $\omega \in \Omega$. Additionally, the *nonanticipativity* constraints (10.22)–(10.32) are appended to the model, $\forall \omega \in \Omega^g$, $g \in \mathcal{G}^t$, $t \in \mathcal{T}^-$, where from now on $\omega + 1$ is assumed to be any scenario in set Ω^g, but the last one, whenever ω is the last scenario in the set:

$$\theta_{kt}^\omega - \theta_{kt}^{\omega+1} = 0 \quad \forall k \in \mathcal{K}, \tag{10.22}$$
$$p_{it}^\omega - p_{it}^{\omega+1} = 0 \quad \forall i \in \mathcal{I}, \tag{10.23}$$
$$\varrho_{dt}^{+\omega} - \varrho_{dt}^{+\omega+1} = 0 \quad \forall d \in \mathcal{D}, \gamma \in \Gamma, \text{ where } t = \tau_\gamma, \tag{10.24}$$
$$\varrho_{dt}^{-\omega} - \varrho_{dt}^{-\omega+1} = 0 \quad \forall d \in \mathcal{D}, \gamma \in \Gamma, \text{ where } t = \tau_\gamma, \tag{10.25}$$
$$x_{jt}^\omega - x_{jt}^{\omega+1} = 0 \quad \forall j \in \mathcal{J}, \tag{10.26}$$
$$y_{jt}^\omega - y_{jt}^{\omega+1} = 0 \quad \forall j \in \mathcal{J}, \tag{10.27}$$
$$z_{jt}^\omega - z_{jt}^{\omega+1} = 0 \quad \forall j \in \mathcal{J}, \tag{10.28}$$
$$s_t^\omega - s_t^{\omega+1} = 0, \tag{10.29}$$
$$b_t^\omega - b_t^{\omega+1} = 0, \tag{10.30}$$
$$\delta_f^\omega - \delta_f^{\omega+1} = 0 \quad \forall f \in \mathcal{F}^{\text{ca}}, \tag{10.31}$$
$$\beta_k^\omega - \beta_k^{\omega+1} = 0 \quad \forall k \in \mathcal{K}^{\text{ca}}. \tag{10.32}$$

By considering the *nonanticipativity* constraints (10.22)–(10.32) as the "difficult" ones, Benders (1962) decomposition schemes can be used; see Andrade et al. (2006), Birge and Louveaux (1997), Escudero et al. (2007), and Laporte and Louveaux (1993), among others. Alternatively, by dualizing these constraints at each so-called supernode of a branch-and-bound phase, Lagrangian decomposition schemes can be used for solving the model given by the function (10.19) with (10.17) and the constraint system (10.2)–(10.14); see Caroe and Schultz (1999), Hemmecke and Schultz (2001), Klein Haneveld and van der Vlerk (1999, 2001), Nürnberg and Römisch (2002), Römisch and Schultz (2001), Schultz (2003), Schultz et al. (2005), Schultz and Tiedemann (2004), and Takriti and Birge (2000), among others. However, some heuristics could be needed for obtaining solutions to the problem, such that the *nonanticipativity* constraints are also satisfied. Independently of the constraints' dualization, a *Branch-and-Fix Coordination* approach can be used, see below.

10.4 A Branch-and-Fix Coordination (BFC) Approach

10.4.1 BFC Methodology

By slightly abusing the notation, consider the following *splitting variable* representation of the DEM: maximize (10.19) with (10.17) subject to (10.2)–(10.14) and (10.22)–(10.32).

$$Z^{\mathrm{IP}} = \max \sum_{\omega \in \Omega} w_\omega \left(\sum_{r \in \mathcal{R}} e_r^\omega \gamma_r^\omega + \sum_{t \in \mathcal{T}} c_t^\omega v_t^\omega + \rho \nu^\omega \right)$$

$$\begin{aligned}
\text{s.t.} \sum_{r \in \mathcal{R}} e_r^\omega \gamma_r^\omega + \sum_{t \in \mathcal{T}} c_t^\omega v_t^\omega + M(1 - \nu^\omega) \geq \tau \;\; \forall \omega \in \Omega, \\
& A_t^\omega \gamma^\omega + B_t'^\omega v_{t-1}^\omega + B_t^\omega v_t^\omega = b_t^\omega && \forall \omega \in \Omega, t \in \mathcal{T}, \\
& \gamma_r^\omega - \gamma_r^{\omega+1} = 0 && \forall \omega \in \Omega, r \in \mathcal{R}, \\
& v_{ht}^\omega - v_{ht}^{\omega+1} = 0 && \forall \omega \in \Omega^g, g \in \mathcal{G}^t, t \in \mathcal{T}^-, h \in \mathcal{H}, \\
& v_{ht}^\omega \geq 0 && \forall \omega \in \Omega, t \in \mathcal{T}, h \in \mathcal{H}, \\
& \gamma_r^\omega, \nu^\omega \in \{0, 1\} && \forall \omega \in \Omega, r \in \mathcal{R},
\end{aligned} \tag{10.33}$$

where v_t^ω denotes the vector of the continuous variables $\{v_{ht}^\omega, \forall h \in \mathcal{H}\}$ for $\omega \in \Omega$, $t \in \mathcal{T}$, representing the variables θ_{kt}^ω, p_{it}^ω, $\varrho_{dt}^{+\omega}$, $\varrho_{dt}^{-\omega}$, x_{jt}^ω, y_{jt}^ω, z_{jt}^ω, s_t^ω and b_t^ω in *ECP2*; $\mathcal{H}$ is the set of indices of the variables in v_t^ω; γ^ω denotes the vector of the 0–1 variables $\{\gamma_r^\omega, \forall r \in \mathcal{R}\}$ for $\omega \in \Omega$ representing the variables δ_f^ω $\forall f \in \mathcal{F}$ and β_k^ω $\forall k \in \mathcal{K}$; $\mathcal{R}$ is the set of indices of the variables in γ^ω; c_t^ω is the row vector of the objective function coefficients for the vector v_t^ω and e_r^ω is the objective function coefficient for the variable γ_r^ω; A_t^ω, $B_t'^\omega$, and B_t^ω are the constraint matrices for the vectors γ^ω, υ_{t-1}^ω, and υ_t^ω, respectively; b_t^ω is from now on the right-hand side (*rhs*) vector for $\omega \in \Omega$, $t \in \mathcal{T}$; and τ and ν are as defined above, all of them with conformable dimensions.

Notice that the relaxation of the *nonanticipativity* constraints

$$\gamma_r^\omega - \gamma_r^{\omega+1} = 0 \quad \forall \omega \in \Omega, r \in \mathcal{R} \tag{10.34}$$

$$\upsilon_{ht}^\omega - \upsilon_{ht}^{\omega+1} = 0 \quad \forall \omega \in \Omega^g, g \in \mathcal{G}^t, t \in \mathcal{T}^-, h \in \mathcal{H} \tag{10.35}$$

in model (10.33) results in a set of $|\Omega|$ independent mixed 0–1 models, where (10.36) is the model for scenario $\omega \in \Omega$:

$$\begin{aligned}
&\max \sum_{r\in\mathcal{R}} e_r^\omega \gamma_r^\omega + \sum_{t\in\mathcal{T}} c_t^\omega \upsilon_t^\omega + \rho\nu^\omega \\
&\text{s.t.} \sum_{r\in\mathcal{R}} e_r^\omega \gamma_r^\omega + \sum_{t\in\mathcal{T}} c_t^\omega \upsilon_t^\omega + M(1-\nu^\omega) \geq \tau \\
&A_t^\omega \gamma^\omega + B_t'^\omega \upsilon_{t-1}^\omega + B_t^\omega \upsilon_t^\omega = b_t^\omega \qquad \forall t \in \mathcal{T}, \\
&\upsilon_{ht}^\omega \geq 0 \qquad \forall t \in \mathcal{T}, h \in \mathcal{H}, \\
&\gamma_r^\omega, \nu^\omega \in \{0,1\} \qquad \forall r \in \mathcal{R}.
\end{aligned} \tag{10.36}$$

To improve computational efficiency, it is not necessary to relax all constraints of the form (10.34) and (10.35). The number of constraints to relax in a given model will depend on the dimensions of the scenario-related model (10.36) (i.e., the parameters $|\mathcal{T}|$, $|\mathcal{I}|$, $|\mathcal{J}|$, $|\mathcal{D}|$, $|\mathcal{F}^{\text{ca}}|$, and $|\mathcal{K}|$ in model (10.2)–(10.14)). Let q denote the number of scenario *clusters* that results after taking into account the size considerations mentioned above. Let Ω_p denote the set of scenarios that belong to *cluster* p. The criterion for the choice of scenario *clusters* for the sets $\Omega_1, \ldots, \Omega_q$, such that $\Omega_p \cap \Omega_{p'} = \emptyset$, $p, p' = 1, \ldots, q : p \neq p'$ and $\Omega = \cup_{p=1}^q \Omega_p$ could alternatively be based on the smallest internal deviation of the uncertain parameters, the greatest deviation, etc. We favour the assignation of a higher priority to include in the same *cluster* those scenarios with greater number of common ancestor nodes in the associated scenario tree.

By slightly continuing to abuse the notation, the model to consider for scenario *cluster* $p = 1, \ldots, q$ can be expressed by the *compact* representation (10.37), where ω for $n \in G^{|\mathcal{T}|}$ is such that ω is the unique scenario in Ω^n and, on the other hand, $G_p = \{g \in \mathcal{G} : \Omega^g \bigcap \Omega_p \neq \emptyset\}$:

$$Z_p^{IP} = \max \sum_{n\in\mathcal{G}^{|\mathcal{T}|}\bigcap\mathcal{G}_p} w_\omega \left[\sum_{g\in\mathcal{N}^n} \left(e^g\gamma^p + c^g\upsilon^g\right) + \rho\nu^\omega \right]$$

subject to

$$
\begin{aligned}
&\sum_{g\in\mathcal{N}^n} \left(e^g\gamma^p + c^g v^g\right) + M(1-\nu^\omega) \geq \tau \ \forall n \in \mathcal{G}^{|\mathcal{T}|}\bigcap \mathcal{G}_p, \\
&A^g\gamma^p + B'^g v^{\phi(g)} + B^g v^g = b^g \qquad \forall g \in \mathcal{G}_p, \\
&v_h^g \geq 0 \qquad \forall g \in \mathcal{G}_p, h \in \mathcal{H}, \\
&\gamma_r^p,\ \nu^\omega \in \{0,1\} \qquad \forall \omega \in \Omega_p, r \in \mathcal{R},
\end{aligned}
\tag{10.37}
$$

where v^g is the vector of the continuous variables for the constraint system related to period $t(g)$ under scenario group g, γ^p is the vector of the 0–1 variables, e^g and c^g are the row vectors of the objective function coefficients for the γ^p and v^g vectors, respectively, A^g, B'^g, and B^g are the constraint matrices, b^g is the *rhs* under scenario group g, and the other parameters are as above, all with conformable dimensions. Note that model (10.37) for $q = 1$ is the *compact* representation equivalent to the *splitting variable* representation (10.33), and it is the model (10.36) for $q = |\Omega|$.

The q models (10.37) are linked by the *nonanticipativity* constraints:

$$\gamma^p - \gamma^{p'} = 0, \tag{10.38}$$

$$v^{g^p} - v^{g^{p'}} = 0, \tag{10.39}$$

$\forall g \in \mathcal{G}_p \bigcap \mathcal{G}_{p'}$, $p, p' = 1, \ldots, q : p \neq p'$, and γ^p and $\gamma^{p'}$ are the γ vectors and v^{g^p} and $v^{g^{p'}}$ are the v vectors for submodels (10.37) related to the *clusters* p and p', respectively.

We could apply a branch-and-bound procedure to ensure the integrality condition in models (10.37), for $p = 1, \ldots, q$. However, instead of obtaining independently the optimal solution for each one, elsewhere (Alonso-Ayuso et al. 2003a, b) we propose an approach, the so-called *Branch-and-Fix Coordination* (BFC). Our specialization in this chapter is designed to coordinate the selection of the branching node and branching variable for each scenario *cluster*-related so-called *Branch-and-Fix* (BF) tree, such that the relaxed constraints (10.38) are satisfied when fixing the appropriate γ variables to either 1 or 0. The proposed approach also coordinates and reinforces the scenario *cluster*-related *BF* node pruning, the variable fixing, and the objective function bounding for the submodels attached to the nodes. See also Alonso-Ayuso et al. (2009), Escudero (2009), Escudero et al. (2007, 2009a, b, 2010)

Let $\mathcal{C}_p$ denote the *BF* tree associated with scenario *cluster* p and $\mathcal{A}_p$ the set of active nodes in $\mathcal{C}_p$ for $p = 1, \ldots, q$. Any two nodes, say, $a \in \mathcal{A}_p$ and $a' \in \mathcal{A}_{p'}$, are said to be *twin* nodes if either they are the *root* nodes or the paths from their *root* nodes to each of them in their own *BF* trees $\mathcal{C}_p$ and $\mathcal{C}_{p'}$, respectively, have branched on/fixed at the same values of some or all their (*first-stage*) variables γ_r^p and $\gamma_r^{p'}$, for $p, p' = 1, \ldots, q : p \neq p'$. Note that in order to satisfy the *nonanticipativity* constraints (10.38), the γ variables must be branched on/fixed at the same 0–1 value for the *twin* nodes. A *Twin Node Family* (TNF), say, $\mathcal{L}_n$ is a set of active nodes, such that any one is a *twin* node to all the other members of the family, for $n \in \Upsilon$, where Υ is the set of *TNFs*.

10.4.2 On Obtaining LP Optimal Solutions for TNF Integer Sets

Recall that a TNF *integer set* is each set of integer nodes, one for each *BF* tree, where the *nonanticipativity* constraints (10.38) for the 0–1 variables are satisfied.

Consider the following *LP* model, obtained after fixing in model (10.33) the γ and ν variables to their 0–1 values, say, $\hat{\gamma}_r\ \forall r \in \mathcal{R}$ and $\hat{\nu}^\omega\ \forall \omega \in \Omega$ for a given TNF *integer set*, respectively,

$$Z_{\text{TNF}}^{\text{LP}} = \sum_{\omega\in\Omega} w_\omega \left(e^\omega \hat{\gamma} + \rho \hat{\nu^\omega}\right) + \max \sum_{\omega\in\Omega}\sum_{t\in\mathcal{T}} w_\omega c_t^\omega v_t^\omega$$

subject to

$$\begin{aligned}
&\sum_{t\in\mathcal{T}} c_t^\omega v_t^\omega \geq \tau - e^\omega\hat{\gamma} - M(1-\hat{\nu}^\omega)\ \forall\omega\in\Omega, \\
&B_t'^\omega v_{t-1}^\omega + B_t^\omega v_t^\omega = b_t^\omega - A_t^\omega \hat{\gamma} && \forall \omega\in\Omega, t\in\mathcal{T}, \\
&v_{ht}^\omega - v_{ht}^{\omega+1} = 0 && \forall\omega\in\Omega^g, g\in\mathcal{G}^t, t\in\mathcal{T}^-, h\in\mathcal{H}, \\
&v_{ht}^\omega \geq 0 && \forall\omega\in\Omega, t\in\mathcal{T}, h\in\mathcal{H}.
\end{aligned} \tag{10.40}$$

Notice that the *RP*-related constraint for scenario ω is redundant for $\hat{\nu^\omega} = 0$, $\omega \in \Omega$.

10.4.3 On TNF Integer Set Bounding

Let $\mathcal{R}f$ denote the set of non-yet branched on/fixed at γ variables in a given TNF *integer set*, such that $\mathcal{R}01 = \mathcal{R} - \mathcal{R}f$. Let also $A_t^\omega = (Af_t^\omega, A01_t^\omega)\ \forall\omega\in\Omega$, where Af_t^ω and $A01_t^\omega$ are the constraint submatrices related to the sets $\mathcal{R}f$ and $\mathcal{R}01$, respectively, for $t\in\mathcal{T}, \omega\in\Omega$. Similarly, let $\Omega = \Omega f \bigcup \Omega 01$, such that Ωf and $\Omega 01$ are the set of scenarios where the ν variables have not yet been branched on/fixed at 0–1 and the set where these variables have already been branched on/fixed at integer 0–1 values, respectively.

Model (10.41) enforces the satisfaction of the *nonanticipativity* constraints (10.34) and (10.35) but, on the other hand, it allows the γ and ν variables from the sets $\mathcal{R}f$ and Ωf, respectively, to take continuous values for the given *TNF integer set*.

$$\begin{aligned}
Z_{\text{CON}}^{\text{LP}} = &\sum_{\omega\in\Omega}\sum_{r\in\mathcal{R}01} w_\omega e^r \gamma^r + \rho \sum_{\omega\in\Omega 01} w_\omega \nu^\omega \\
&+ \max \sum_{\omega\in\Omega}\sum_{r\in\mathcal{R}f} w_\omega e^r\gamma^r + \sum_{\omega\in\Omega} w_\omega c^\omega v^\omega + \rho\sum_{\omega\in\Omega f} w_\omega \nu^\omega
\end{aligned}$$

subject to

$$
\begin{aligned}
& e^{\omega f}\gamma^{\omega f} + c^{\omega}\upsilon^{\omega} + M^{\omega f}(1-\nu^{\omega}) \geq \\
& \tau - e^{\omega 01}\gamma^{\omega 01} - M^{\omega 01}(1-\nu^{\omega}) && \forall \omega \in \Omega, \\
& Af_t^{\omega}\gamma^{\omega f} + B_t^{\prime\omega}\upsilon_{t-1}^{\omega} + B_t^{\omega}\upsilon_t^{\omega} = b_t^{\omega} - A01_t^{\omega}\gamma^{\omega 01} && \forall \omega \in \Omega, t \in \mathcal{T}, \\
& \gamma^{\omega f} - \gamma^{\omega f+1} = 0 && \forall \omega \in \Omega, \\
& \upsilon_{ht}^{\omega} - \upsilon_{ht}^{\omega+1} = 0 && \forall \omega \in \Omega^{g}, g \in \mathcal{G}^{t}, t \in \mathcal{T}^{-}, h \in \mathcal{H}, \\
& \upsilon_{ht}^{\omega} \geq 0 && \forall \omega \in \Omega^{g}, g \in \mathcal{G}^{t}, t \in \mathcal{T}, h \in \mathcal{H}, \\
& 0 \leq \gamma^{\omega f} \leq 1 && \forall \omega \in \Omega, \\
& 0 \leq \nu^{\omega} \leq 1, && \forall \omega \in \Omega f,
\end{aligned}
\tag{10.41}
$$

where $\gamma^{\omega 01}$ and $\gamma^{\omega f}$ give the subvectors of the vector γ^{ω} for the sets $\mathcal{R}01$ and $\mathcal{R}f$, respectively, similarly for the subvectors $e^{\omega 01}$ and $e^{\omega f}$, and the subvectors $M^{f\omega} = M$ and $M^{01\omega} = M$ for $\omega \in \Omega f$ and $\omega \in \Omega 01$, respectively, and, otherwise, they are zero.

10.4.4 BFC Algorithm

Consider the following algorithm for solving model (10.33) by using the scenario *cluster*-related submodels (10.37).

Step 1: Solve the LP relaxations of the q models (10.37). Each model is attached to the *root* node in the trees $\mathcal{C}_p\ \forall p = 1, \ldots, q$. If the integrality constraints and the constraints (10.38) and (10.39) are satisfied then stop, the optimal solution to the original mixed 0–1 model (10.33) has been obtained.

Step 2: The following parameters are saved in a centralized device (CD): the values of the variables and the solution value (i.e., the optimal objective function value) of the LP models attached to the nodes in $\mathcal{A}_p\ \forall p = 1, \ldots, q$, as well as the appropriate information for branching on the 0–1 γ and ν variables in the TNFs $\mathcal{L}_n\ \forall n \in \Upsilon$. A decision is made in *CD* for the selection of the branching TNF and the branching variable. The decision is made available for the execution of each scenario *cluster*-related BF phase.

Step 3: Optimization of the LP models attached to the newly created TNF after branching on the chosen 0–1 variable. Prune the TNF if its related *LP* model is infeasible or its solution value (i.e., the weighted sum of the solution values of its node members) is not greater than the value of the incumbent solution and, then, go to step 8.

Step 4: If the optimal solution that has been obtained in step 3 has 0–1 values for all the γ and ν variables and satisfies constraints (10.38) (i.e. it is a TNF *integer set*), either of the two following situations has happened:

(a) The *nonanticipativity* constraints (10.39) have been satisfied and, then, a new solution has been found for the original mixed 0–1 model (10.33). The related *incumbent* solution can be updated and, additionally, the updating of the sets $\mathcal{A}_p$ at the trees $\mathcal{C}_p$ $\forall p = 1, \ldots, q$ can also be performed. In any case, the TNF is pruned. Go to step 8.

(b) The *nonanticipativity* constraints (10.39) have not been satisfied. Go to step 5.

Otherwise, go to step 2.

Step 5: Optimizing the *LP* model (10.40) that results from fixing the γ and ν variables in model (10.33) to the 0–1 values given in the TNF *integer set* whose associated models (10.37) have been optimized in step 3. See Section 10.4.2. If the LP model is feasible (and, thus, its optimal solution has been obtained), then the updating of the *incumbent* solution and active node sets can be performed.

Step 6: Solve the LP model (10.41) that results from fixing the branched on/fixed at γ and ν variables in model (10.33) to their integer values and allowing the other 0–1 variables to take continuous values. See Section 10.4.3.

Step 7: If the solution value of model (10.41) for the given TNF is not greater than the *incumbent* solution value, or the LP models to be optimized in steps 5 and 6 yield the same solution value, then the family is pruned.

Step 8: If the sets of active nodes are empty then stop, since the optimality of the *incumbent* solution has been proved, if any. Otherwise, go to step 2.

10.5 Conclusions

In this chapter we have introduced a multistage stochastic mixed 0–1 model with complete recourse for structuring portfolios of bilateral energy trading contracts in competitive electricity markets. The treatment of the uncertainty plays a central role in the approach that we have presented. Scenario treatment has proved to be a useful mechanism to represent the uncertainty. The risk measure given by the composite function of the expected trading profit and the *reaching probability* functional is considered, instead of optimizing mean values alone. The modelling approach to select candidate purchasing and selling contracts has been proved to be successful for maximizing the mean-risk function and, thus, balancing the net revenue from the contracts exploitation and the trading in the spot market under each scenario group along the time horizon. It is worth noting that the approach can be used for portfolio structuring by energy service providers in addition to energy generation companies. It can also be easily adapted to the existence of fixed costs associated with the contracts and to the consideration of contract cancellations. The model considers the influence of the constraints imposed by the Kyoto protocol-based pollutant emission regulations. A specialization of the *Branch-and-Fix Coordination* algorithm has been proposed for optimizing multistage stochastic problems with 0–1 variables in the first stage and continuous variables in the other stages.

Appendix

Different types of implementations can be considered within the algorithmic framework presented in Section 10.4. This appendix presents the version that we favour.

We use the *depth-first* strategy for selecting the branching *TNF*. The branching criterion followed for the energy selling contracts is to perform first a "branching on the ones" and, afterwards, a "branching on the zeros". However, the branching criterion for the energy purchasing contracts performs first a "branching on the zeros" and, afterwards, a "branching on the ones".

Given the significance of the ν variables, they have the highest priority for branching purposes. The priority is given according to the non-increasing scenario weight criterion. It is branched first on the ones and then on the zeros.

Another topic of interest is the branching ordering for the energy trading contracts. We have considered a *static* ordering as follows: The highest priority is given to the selling contracts over the purchasing contracts. Within each category, the priority is given according to the non-increasing expected income criterion for the selling contracts

$$\sum_{g \in \mathcal{G}} w^g \mu_f^g \iota_f^g \quad \forall f \in \mathcal{F}^{\text{ca}},$$

and the non-decreasing expected cost criterion for the purchasing contracts,

$$\sum_{g \in \mathcal{G}} w^g \lambda_k^g \frac{1}{2} \left(\overline{\theta}_{k,t(g)} + \underline{\theta}_{k,t(g)} \right) \quad \forall k \in \mathcal{K}^{\text{ca}}.$$

Notice that a TNF can be pruned for any of the following reasons: (a) the LP relaxation of the scenario *cluster* model (10.37) attached to any node member is infeasible, (b) there is no guarantee that a better solution than the *incumbent* one can be obtained from the best descendant TNF *integer set* (currently, it is based on the LP objective function value), (c) the LP model (10.40) attached to the TNF *integer set* is infeasible or its solution value is not better than the value of the *incumbent* solution, in case that all γ and ν variables have already been branched on/fixed at for the family, and (d) see below the case where some integer-valued γ or ν variable has not yet been branched on/fixed at.

Once a TNF has been pruned, the same branching criterion allows us to perform either a "branching on the zeros (resp., "on the ones") in the case that the TNF has already been "branched on the ones" for the energy selling contracts (resp., "on the zeros" for the energy purchasing contracts), or a *backtracking* to the previous branched TNF if it has already been branched "on the ones" and "on the zeros".

It is worth noting, see also Escudero et al. (2007), that if $Z_{\text{TNF}}^{\text{LP}}$ (10.40) = $Z_{\text{CON}}^{\text{LP}}$ (10.41), then the related TNF can be pruned. Notice that the solution space defined by model (10.40) is included in the space defined by model (10.41) and, so, in this case, there is not a greater solution value than $Z_{\text{TNF}}^{\text{LP}}$ from the descendant nodes to be obtained by branching on the non-yet branched on/fixed at γ and ν variables. For the same reason, the family is also pruned if $Z_{\mathrm{CON}}^{\mathrm{LP}}$ is not greater than the *incumbent* solution value.

To gain computational efficiency, the optimization of model (10.41) should not be performed for a small number of the γ and ν variables already branched on/fixed at in the given TNF. Let λ denote the minimum fraction of the branched on/fixed at variables that is required for optimizing the model, where $0 \leq \lambda \leq 1$.

For presenting the detailed BFC algorithm to solve model (10.33), let the following additional notation:

LP_p, LP relaxation of the scenario *cluster*-related model (10.37) attached to a node member from the *BF* tree $\mathcal{C}_p$ in the given TNF, for $p = 1, \ldots, q$.

Z_p^{LP}, solution value of the LP model LP_p, for $p = 1, \ldots, q$.

$\overline{Z}^{\mathrm{IP}}$, upper bound of the solution value of the original model (10.33) to be obtained from the best descendant *TNF integer set* for a given family. It is computed as $\overline{Z}^{\mathrm{IP}} = \sum_{p=1,\ldots,q} Z_p^{\mathrm{LP}}$.

$\underline{Z}^{\mathrm{IP}}$, lower bound of the solution value of the original model (10.33). It is the *incumbent* solution value, and it ends up with the value of the optimal solution.

By convention, $Z_*^{\mathrm{LP}} = -\infty$ for any infeasible LP_* model.

Procedure

Step 0: Initialize $\underline{Z}^{\mathrm{IP}} := -\infty$.
Assign $\sigma_r := 1$ where r represents a scenario or an energy selling contract, and $\sigma_r := 0$ if it is an energy purchasing contract, for $r \in \Omega \bigcup \mathcal{R}$.

Step 1: Solve the q independent LP models LP_p, $\forall p = 1, \ldots, q$, and compute $\overline{Z}^{\mathrm{IP}}$. If the γ and ν variables take integer values and the *nonanticipativity* constraints (10.38) and (10.39) are satisfied, then the optimal solution to the original model (10.33) has been found and, so, $\underline{Z}^{\mathrm{IP}} := \overline{Z}^{\mathrm{IP}}$ and stop.

Step 2: If there is any γ or ν variable that takes continuous values or there is a γ variable that takes different values for some of the q scenario *clusters* then go to step 3.
Solve the LP model (10.40) to satisfy constraints (10.35) for the v variables in the given TNF *integer set* (in this case, the set of *root* nodes in the BF trees). Note that the solution value is denoted by $Z_{\mathrm{TNF}}^{\mathrm{LP}}$.
Update $\underline{Z}^{\mathrm{IP}} := Z_{\mathrm{TNF}}^{\mathrm{LP}}$. If $\underline{Z}^{\mathrm{IP}} = \overline{Z}^{\mathrm{IP}}$ then the optimal solution to the original model has been found and stop.

Step 3: Initialize $r := 1$ and go to step 5.

Step 4: Reset $r := r + 1$.
If $r = |\Omega \bigcup \mathcal{R}| + 1$ then go to step 9.

Step 5: Branch $\gamma_r^p := \sigma_r$, $\forall p = 1, \ldots, q$.

Step 6: Solve the linear models LP_p, $\forall p = 1, \ldots, q$ and compute $\overline{Z}^{\mathrm{IP}}$.

If $\overline{Z}^{\text{IP}} \leq \underline{Z}^{\text{IP}}$ then go to step 8.
If there is any γ or ν variable that takes continuous values or there is a γ variable that takes different values for some of the q scenario *clusters* then go to step 4.
If all the υ variables take the same value for all scenario *clusters* $p = 1, \ldots, q$ then update $\underline{Z}^{\text{IP}} := \overline{Z}^{\text{IP}}$ and go to step 8.
Solve the LP model (10.40) to satisfy constraints (10.35) for the υ variables in the given *TNF integer set*.
Update $\underline{Z}^{\text{IP}} := \max\{Z^{\text{LP}}_{\text{TNF}}, \underline{Z}^{\text{IP}}\}$. If $r = |\Omega \bigcup \mathcal{R}|$ then go to step 8.
If $r < \lambda|\Omega \bigcup \mathcal{R}|$ then go to step 4.

Step 7: Solve the LP model (10.41), where the 0–1 continuous variables are the non-yet been branched on/fixed at γ and ν variables in the solution of the current TNF. Notice that the solution value is denoted by $Z^{\text{LP}}_{\text{CON}}$. If $Z^{\text{LP}}_{\text{TNF}} < Z^{\text{LP}}_{\text{CON}}$ and $Z^{\text{LP}}_{\text{CON}} > \underline{Z}^{\text{IP}}$ then go to step 4.

Step 8: Prune the branch.
If $\gamma_r^p = \sigma_r, \forall p = 1, \ldots, q$ then go to step 11.

Step 9: Reset $r := r - 1$.
If $r = 0$ then stop, since the optimal solution $\underline{Z}^{\text{IP}}$ has been found, if any.

Step 10: If $\gamma_r = 1 - \sigma_r, \forall p = 1, \ldots, q$ then go to step 9.

Step 11: Branch $\gamma_r := 1 - \sigma_r \ \forall p = 1, \ldots, q$.
Go to step 6.

Acknowledgements This research has been supported by the grants RM URJC-CM-2008-CET-3703 and RIESGOS CM from Comunidad de Madrid, and OPTIMOS2 MTM2009-14039-C06-03 and PLANIN MTM2009-14087-C04-01 from Ministry of Science and Innovation, Spain.

References

A. Alonso-Ayuso, L.F. Escudero and M.T. Ortuño. BFC, a Branch-and-Fix Coordination algorithmic framework for solving some types of stochastic pure and mixed 0-1 programs. *European Journal of Operational Research*, 151:503–519, 2003a.

A. Alonso-Ayuso, L.F. Escudero, A. Garín, M.T. Ortuño and G. Pérez. An approach for strategic supply chain planning based on stochastic 0–1 programming. *Journal of Global Optimization*, 26:97–124, 2003b.

A. Alonso-Ayuso, L.F. Escudero and C. Pizarro. On SIP algorithms for minimizing the mean-risk function in the Multi Period Single Source Problem under uncertainty. *Annals of Operations Research*, 166:223–242, 2009.

R. Andrade, A. Lisser, N. Maculan and G. Plateau. Enhancing a Branch-and-Bound algorithm for two-stage stochastic integer network design-based models. *Management Science*, 52:1450–1455, 2006.

J.M. Arroyo and A.J. Conejo. Optimal response of a thermal unit to an electricity spot market. *IEEE Transactions on Power Systems*, 15:1098–1104, 2000.

J.F. Benders. Partitioning procedures for solving mixed variables programming problems. *Numerische Mathematik*, 4:238–252, 1962.

J.R. Birge and F.V. Louveaux. *Introduction to Stochastic Programming*. Springer, New York, NY, 1997.

C.C. Caroe and R. Schultz. A two-stage stochastic program for unit commitment under uncertainty in a hydro-thermal power system. Report SC98-13, Konrad-Kuse für Informationstechnik Berlin (ZIB), Germany, 1998.

C.C. Caroe and R. Schultz. Dual decomposition in stochastic integer programming. *Operations Research Letters*, 24:37–45, 1999.

P. Carpentier, G. Cohen, J.C. Culioli and A. Renaud. Stochastic optimization of unit commitment: A new decomposition framework. *IEEE Transactions on Power Systems*, 11:1067–1073, 1996.

C. Cervellera, V.C.P. Chen and A. Wen. Optimization of a large scale water reservoir network by stochastic dynamic programming with efficient state space discretization. *European Journal of Operational Research*, 171:1139–1151, 2006.

A.J. Conejo and F.J. Prieto. Mathematical programming and electricity markets. *TOP*, 9:1–54, 2001

A.J. Conejo, F.J. Nogales and J.M. Arroyo. Price-taker bidding strategy under price uncertainty. *IEEE Transactions on Power Systems* 17:1081–1088, 2002.

D. Dentcheva, R. Gollmer, A. Möller, W. Römisch and R. Schultz. Solving the unit commitment problem in power generation by primal and dual methods. In M. Broens, M.P. Bendsoe and M.P. Soerensen, editors. *Progress in Industrial Mathematics at ECMI 96*, Teubner, Stuttgart, Germany, 332–339, 1997.

D. Dentcheva and W. Römisch. Optimal power generation under uncertainty via stochastic programming. In K. Marti and P. Kall, editors. *Stochastic Programming Methods and Technical Applications*. Springer, Berlin, 22–56, 1998.

C.K. Dert. A dynamic model for asset liability management for defined benefit pension funds. In W.T. Ziemba and J.M. Mulvey, editors, *Worldwide Asset and Liability Modeling*. Cambridge University Press, Cambridge, 501–536, 1998.

L.F. Escudero. Discussion on mathematical programming and electricity markets. *TOP*, 9:23–30, 2001.

L.F. Escudero. On a mixture of Fix-and-Relax Coordination and Lagrangean Substitution schemes for multistage stochastic mixed integer programming. *TOP*, 17:5–29, 2009.

L.F. Escudero, A. Garín, M. Merino and G. Pérez. A two-stage stochastic integer programming approach as a mixture of Branch-and-Fix Coordination and Benders Decomposition schemes. *Annals of Operations Research*, 152:395–420, 2007.

L.F. Escudero, A. Garín, M. Merino and G. Pérez. A general algorithm for solving two-stage stochastic mixed 0-1 first stage problems. *Computers & Operations Research*, 36:2590–2600, 2009a.

L.F. Escudero, A. Garín, M. Merino and G. Pérez. BFC-MSMIP: an exact Branch-and-Fix Coordination approach for solving multistage stochastic mixed 0–1 problems. *TOP*, 17:96–122, 2009b.

L.F. Escudero, A. Garín, M. Merino and G. Pérez. On BFC-MSMIP strategies for scenario clustering, and Twin Node Family branching selection and bounding for multistage stochastic mixed 0–1 problems. *Computers & Operations Research*, 37:738–753, 2010.

S.-E. Fleten and T.K. Kristoffsen. Stochastic programming for optimizing bidding strategies in a Nordic hydropower producer. *European Journal of Operational Research*, 181:916–928, 2007.

S.-E. Fleten, S.W. Wallace and W.T. Ziemba. Hedging electricity portfolios via stochastic programming. In C. Greengard and A. Ruszczynski, editors. *Decision making under uncertainty: Energy and Power*, *The IMA Volumes in Mathematics and Its Applications*, Vol. 128, pp. 71–93. Springer, 2001.

R. Gollmer, W. Gotzes and R. Schultz. A note on second-order stochastic dominance constraints induced by mixed-integer recourse. *Mathematical Programming Series A*, 126:179–190, 2011.

R. Gollmer, A. Möller, M.P. Nowak, W. Römisch and R. Schultz. Primal and dual methods for unit commitment in a hydro-thermal power system. *Proceedings 13th Power Systems Computational Conference*, Trondheim, Norway, 2:724–730, 1999.

N. Groewe-Kuska, K. Kiwiel, M.P. Nowak, W. Römisch and I. Wegner. Power management in a hydro-thermal system under uncertainty by Lagrangian relaxation. In C. Greengard and A.

Ruszczynski, editors. *Decision Making Under Uncertainty: Energy and Power, The IMA Volumes in Mathematics and Its Applications*, Vol. 128, pp. 39–70. Springer, 2001.

R. Hemmecke and R. Schultz. Decomposition methods for two-stage stochastic Integer Programs. In M. Grötschel, S.O. Krumke and J. Rambau, editors. *Online Optimization of Large Scale Systems*. Springer, Berlin, 601–622, 2001.

B.F. Hobbs, M.-H. Rothkopf, R.P. O'Neill and H. Chao, editors. *The Next Generation of Electric Power Unit Commitment Models*. Kluwer, Dordrecht, 2001.

N. Jimenez Redondo and A.J. Conejo. Short-term hydro-thermal coordination by Lagrangian relaxation: solution of the dual problem. *IEEE Transactions on Power Systems*, 14:89–95, 1997.

W.K. Klein Haneveld and M.H. van der Vlerk. Stochastic integer programming: General models and algorithms. *Annals of Operations Research*, 85:39–57, 1999.

W.K. Klein Haneveld and M.H. van der Vlerk. Optimizing electricity distribution using integer recourse models. In S. Uryasev and P.M. Pardalos, editors. *Stochastic Optimization: Algorithms and Applications*. Kluwer, Dordrecht, 137–154, 2001.

G. Laporte and F.V. Louveaux. The integer L-shaped method for stochastic integer programs with complete recourse. *Operations Research Letters*, 13:133–142, 1993.

G. Lulli and S. Sen. A branch-and-price algorithm for multistage stochastic integer programming with application to stochastic batch-sizing problems. *Management Science*, 50:786–796, 2004.

B. Mo, A. Gjelsvik and A. Grundt. Integrated risk management of hydro power scheduling and contract management. *IEEE Transactions on Power Systems*, 16:216–221, 2001.

E. Ni, P.B. Luh and S. Rourke. Optimal integrated generation bidding and scheduling with risk management under a deregulated power market. *IEEE Transactions on Power Systems* 19:600–609, 2004.

R. Nürnberg and W. Römisch. A two-stage planning model for power scheduling in a hydro-thermal system under uncertainty. *Optimization and Engineering*, 3:355–378, 2002.

W. Ogryczak and A. Ruszczynski. From stochastic dominance to mean-risk models: semi-deviations as risk measures. *European Journal of Operational Research*, 116:33–50, 1999.

J.I. Perez Arriaga. Libro blanco sobre la reforma del marco regulatorio de la generacion de energia electrica en España, www6.mityc.es/energia, p. 80, (2005).

G. Pritchard, A.B. Philpott and P.J. Neame. Hydroelectric reservoir optimization in a pool market. *Mathematical Programming, Series A*, 103:445–461, 2005.

R.T. Rockafellar and S. Uryasev. Optimization of conditional value-at-risk. *Journal of Risk*, 2:21–41, 2000.

R.T. Rockafellar and R.J-B Wets. Scenario and policy aggregation in optimization under uncertainty. *Mathematics of Operations Research*, 16:119–147, 1991.

W. Römisch and R. Schultz. Multistage stochastic integer programs: An introduction. In M. Grötschel, S.O. Krumke and J. Rambau, editors. *Online Optimization of Large Scale Systems*. Springer, Berlin, 581–600, 2001.

R. Schultz. Stochastic programming with integer variables. *Mathematical Programming, Series B*, 97:285–309, 2003.

R. Schultz and S. Tiedemann. Risk aversion via excess probabilities in stochastic programs with mixed-integer recourse. *SIAM Journal on Optimization*, 14:115–128, 2004.

R. Schultz and S. Tiedemann. Conditional Value-at-Risk in stochastic programs with mixed integer recourse. *Mathematical Programming, Series B*, 105:365–386, 2006.

R. Schultz, M.P. Nowak and M. Westphalen A Stochastic Integer Programming model for incorporating day-ahead trading of electricity into hydro-thermal unit commitment. *Optimization and Engineering*, 6:163–175, 2005.

S. Sen, L. Yu and T. Genc. A stochastic programming approach to power portfolio optimization. *Operations Research*, 54:55–72, 2006.

G.B. Shrestha, B.K. Pokharel, T.T. Lie and S.-E. Fleten. Medium term power planning with bilateral contracts. *IEEE Transactions on Power Systems* 20:627–633, 2005.

S. Takriti and J.R. Birge. Lagrangean solution techniques and bounds for loosely coupled mixed-integer stochastic programs. *Operations Research*, 48:91–98, 2000.

E. Tanlapco, J. Lawarree and Ch.-Ch. Liu. Hedging with futures contracts in a deregulated electricity industry. *IEEE Transactions on Power Systems*, 17:577–582, 2002.

Ch. Triki, P. Beraldi and G. Gross. Optimal capacity allocation in multi-auction electricity markets under uncertainty. *Computers & Operations Research*, 32:201–217, 2005.

P. Valente. Software tools for the investigation of stochastic programming problems. PhD thesis, Dept. of Mathematics and Computation, Brunel University, UK, 2002.

R.J.-B. Wets. Stochastic programs with fixed recourse: The equivalent deterministic program. *SIAM Review*, 16:309–339, 1974.

A.J. Wood and B.F. Wollenberg. *Power Generation, Operation and Control*. Wiley, New York, NY, 1996.

Chapter 11
Tactical Portfolio Planning in the Natural Gas Supply Chain

Marte Fodstad, Kjetil T. Midthun, Frode Rømo, and Asgeir Tomasgard

Abstract We present a decision support tool for tactical planning in the natural gas supply chain. Our perspective is that of a large producer with a portfolio of production fields. The tool takes a global view of the supply chain, including elements such as production fields, booking of transportation capacity, bilateral contracts and spot markets. The bilateral contracts are typically take-or-pay contracts where the buyer's nomination and the prices are uncertain parameters. Also the spot prices in the market nodes are uncertain. To handle the uncertain parameters, the tool is based on stochastic programming. The goal for the producer is to prioritize production over the planning period in a way that makes sure that both delivery obligations are satisfied and that profits are maximized. The flexibility provided by the short-term markets gives the producer a possibility to further increase his profits. Production and transportation booking decisions in the early periods are taken under the uncertainty of the coming obligations and prices which makes flexible and robust solutions important. There will be a trade-off between maximum profits and robustness with respect to delivery in long-term contracts.

Keywords Portfolio planning · Natural gas · Stochastic programming

Notation

Sets

$\mathcal{N}$	The nodes in the transportation network
$\mathcal{B}$	Booking nodes, $\mathcal{B} \subseteq \mathcal{N}$
$\mathcal{G}$	Production fields, $\mathcal{G} \subseteq \mathcal{B}$
$\mathcal{D}$	Delivery nodes for the contracts, $\mathcal{D} \subseteq \mathcal{B}$
$\mathcal{M}$	Spot markets in the network, $\mathcal{M} \subseteq \mathcal{B}$ and $\mathcal{M} \cap \mathcal{D} = \emptyset$
$\mathcal{I}(n)$	Nodes with outflow going to node n
$\mathcal{O}(n)$	Nodes with inflow coming from node n
$\mathcal{C}$	The contracts in the portfolio

A. Tomasgard (✉)
Department of Industrial Economics and Technology Management, Norwegian University of Science and Technology, Trondheim, Norway; SINTEF Technology and Society, NO-7491 Trondheim, Norway
e-mail: asgeir.tomasgard@sintef.no; asgeir.tomasgard@iot.ntnu.no

M. Bertocchi et al. (eds.), *Stochastic Optimization Methods in Finance and Energy*, International Series in Operations Research & Management Science 163,
DOI 10.1007/978-1-4419-9586-5_11,

$\mathcal{C}^{\text{split}}$	Contracts in the portfolio with multiple delivery nodes, $\mathcal{C}^{\text{split}} \subseteq \mathcal{C}$
$\mathcal{C}(d)$	The contracts in delivery node d, $\mathcal{C}(d) \subseteq \mathcal{C}$
$\mathcal{D}(c)$	The delivery nodes of contract c, $\mathcal{D}(c) \subseteq \mathcal{D}$
$\mathcal{Y}$	The years included in the optimization horizon
$\mathcal{T}$	The time periods in the optimization horizon
$\mathcal{T}^{\text{booking}}$	The time periods where booking decisions can be made
$\mathcal{T}(y)$	The time periods included in year y
$\mathcal{S}$	The scenarios
$\mathcal{Z}$	Event nodes in the scenario tree
$\mathcal{S}(z)$	Scenarios passing through event node z

Constants

K_g	The unit cost for production in field g
H_b	The per unit tariff in booking node b
X_{bt}	Booked firm capacity in booking node b for transportation in time t
A_{bt}	Volume available for booking in node b for transportation in time t
Q_m	The maximum trade in spot market m, time t and scenario s
$C_{cd}^{\max}$	The maximum fraction of nominated gas in contract c that can be delivered in delivery node d
$C_{cd}^{\min}$	The minimum fraction of nominated gas in contract c that can be delivered in delivery node d
γ_c	The fraction of gas that can be sourced freely for delivery in contract c
F_{ij}	The flow capacity between the downstream nodes i and j
$\overline{F}_{gt}$	The maximum daily production in field g and time t (aggregated to match period length)
$\underline{F}_{gt}$	The maximum daily production in field g and time t (aggregated to match period length)
F_{gt}^{year}	The maximum yearly production in field g and year y
T_z	The time period of event node z

Stochastic parameters

P_{mts}^{spot}	The spot price in market m in time t in scenario s
P_{cts}^{contr}	The price in contract c in time t in scenario s
V_{cts}	The demand in take-or-pay contract c in time t in scenario s
π_{ts}	The probability of scenario s

Variables

k_{gts}	Production in field g in time t in scenario s
q_{mts}	Spot sale in time t in scenario s. Negative values represent purchase
v_{cdts}	Volume delivered in take-or-pay contract c in delivery node d in time t in scenario s
v_{cdts}^{eq}	Equity gas delivered in split contract c in delivery node d in time t in scenario s

$a_{b\tau ts}$ The balance of transportation capacity booked from booking node i to booking node j at time τ for transportation in time t in scenario s

$h_{b\tau ts}$ The booking of transportation capacity from booking node i to booking node j in time τ for transportation in time t in scenario s

f_{ijts} Flow from nodes i to node j in time t and scenario s

11.1 Introduction

Portfolio optimization is commonly used to manage portfolios of financial assets (Mulvey 2001; Zenios 1993; Ziemba and Vickson 2006), but also physical asset portfolios can benefit from this methodology. We look at portfolio optimization applied for the natural gas supply chain, with a special focus on the subsea system on the Norwegian Continental Shelf (NCS) which is illustrated in Fig. 11.1. The basic components of this supply chain are production fields, intermediate nodes, storages, the contract delivery points and downstream spot markets, all connected with a grid of pipelines for transportation. Traditionally long-term contracts have been most common and some large producers have dominated in this supply chain.

The portfolio perspective is particularly interesting given the liberalization process which the European natural gas business is going through at the moment. The process is mainly driven by two EU directives (EU Commission 1998, 2003; Directorate-General for Energy European Commission and Transport 2002). This liberalization process has led to the emergence of new short-term market hubs, i.e. in Zeebrugge, and we also see developing derivative markets with natural gas as the underlying commodity, for instance, the International Petroleum Exchange (IPE). It could be noted that the evolution of the UK market, NBP, which is the most developed European market, was mainly market driven and started prior to the EU directives (Midthun 2007).

On the NCS, the main changes include the separation of transportation and production into separate companies. This is accompanied with third-party access to the infrastructure. The tariffs for transportation are regulated by the Norwegian Ministry of Petroleum and Energy with the objective that profits should be generated in production and sale, not in transportation. Further, each producer now sells their gas independently, not through the mutual Gas Negotiating Committee as before (Dahl 2001).

In Ulstein et al. (2007) planning of offshore petroleum production is studied on a tactical level. The model has a supply chain approach where production plans, network routing, processing of natural gas and sales in the markets are considered. In addition, quality restrictions in the markets and multi-commodity flows are considered. The pressure constraints in the network are, however, not included in the model. The non-linear splitting for chemical processing is linearized with binary variables. The resulting model is a mixed integer programming model.

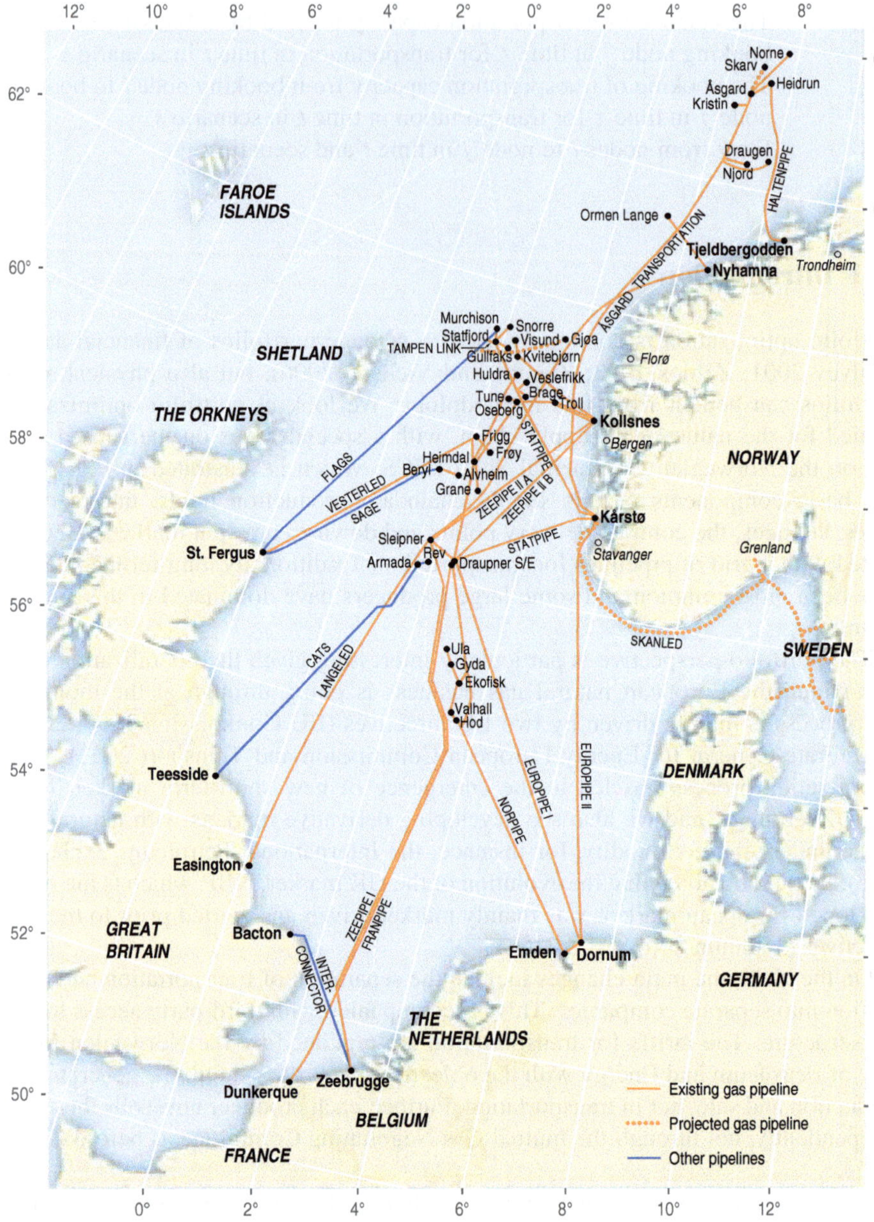

Fig. 11.1 The transport network on the Norwegian Continental Shelf (Ministry of Petroleum and Energy/Norwegian Petroleum Directorate 2009)

Selot et al. (2008) presents an operational model for production planning and routing in the natural gas supply chain. The model combines a detailed infrastructure model with a complex contractual model. There is no market for natural gas included in the model. The infrastructure model includes non-linear equations for relating pressure and flow in wells and pipelines, multi-commodity flows and contractual agreements in the market nodes (delivery pressure and quality of the gas). The contractual model is based on a set of logical conditions for production sharing and customer requirements. The combined model is a mixed integer non-linear programming model (MINLP). In addition, the model is non-convex due to the pressure-flow relationship and the modelling of multi-commodity flows.

A tactical portfolio optimization model with a focus on the physical properties of the natural gas transportation network is presented in Tomasgard et al. (2007). The paper provides a stochastic formulation, but do not include any numerical examples. Midthun et al. (2009) show how the properties of pressure and flow of gas in pipelines give system effects in a network that affects efficient utilization. Midthun (2007) presents an operational portfolio optimization model where decisions are taken under uncertainty in demand and spot prices. Especially the paper focuses on the commercial value of utilizing line-pack, which is excess storage capacity in the pipelines system.

In this chapter we focus on the business environment faced by a large natural gas producer: how can the portfolio of production rights, booking rights and market opportunities be handled in an optimal way? The physical network and routing are not included since these decisions are made by an independent system operator and are out of the producers' control. This means that the most important decisions made by the producer are the booking of transportation capacity, distribution of production over the planning period, sales in spot markets and delivery in contracts. We present a multi-stage stochastic optimization model and provide numerical examples to illustrate the value of portfolio optimization in the natural gas supply chain.

In Section 11.2 we present the portfolio perspective in our model. The mathematical formulation is given in Section 11.3 before we discuss scenario generation for multi-stage stochastic models in Section 11.4. In Section 11.5 we provide some results and numerical examples before we give some conclusions in Section 11.5.

11.2 Portfolio Optimization

Even though the natural gas producers do not control the routing in the network they still face bottlenecks that make the portfolio perspective valuable:

- Limited liquidity in the market nodes
- Equity gas requirements in the contracts
- Booking capacity
- Production capacity

The limited liquidity in the market nodes makes it challenging to match production plans from uncoordinated fields with the delivery obligations downstream. For some of the delivery contracts there may be requirements regarding equity gas. This means that a ratio of the total gas delivered should come from the producer's own production (and not from the spot markets). Lastly, limited booking and production capacity make the coordination between markets favourable with respect to prioritizing between the fields.

11.2.1 Planning Perspectives

As the operational framework and market structure evolve, also the producer's activities and organization may change. This is reflected in a evolution line of different planning perspectives illustrated in Fig. 11.2. Traditional production planning has the focus on balancing the production portfolio with the contract portfolio. With the access to short-term and derivative markets, the possibility of combined production and market optimization is opened. At this level emphasis is on using the market flexibility to avoid physical bottlenecks and thereby maximize the total profit. This can evolve further into a trading level where transactions in the financial markets to maximize profits are done independently of the physical operations. Finally, for a risk-averse company portfolio management can be integrated with risk management. Whether or not such integration is advisable depends among other on the completeness of the markets, existence of market power and organizational costs. This is discussed further in Bjørkvoll et al. (2001) and Fleten et al. (2002) related to the electricity market. In this chapter we will focus on a model for the production- and market optimization level, but with the greedy nature of a optimization model the border to trading is not as clear in the model as in the organizational structure and strategies of a company.

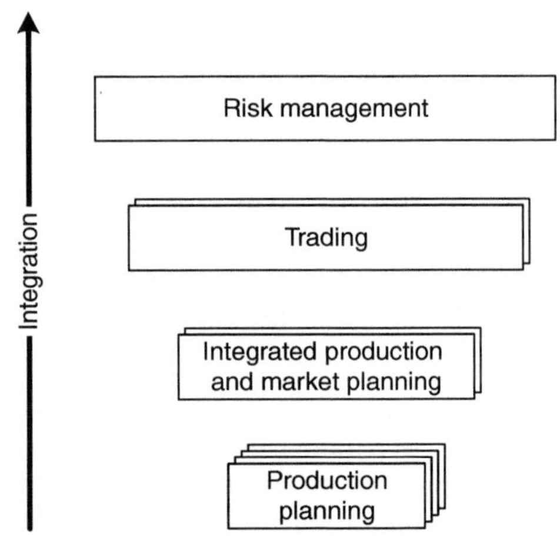

Fig. 11.2 Evolution in planning perspectives

Anthony (1965) suggests to classify planning and control activities in three classes that are often named *strategic*, *tactical* or *operational*. This classification is frequently used in *hierarchical planning* (see, e.g., Hax and Meal (1975) and Bitran and Tirupati (1993)) that can also be applied on the natural gas supply chain. A producer's planning can be seen as a hierarchy of strategic, tactical and operational planning where the more long-term plans give limitations and guidelines for the more short-term plans. The strategic planning has several years horizon with a focus on investments, long-term contracts and energy allocation between the years. Tactical planning typically covers up to 3 years with a focus on energy allocation in a seasonal perspective, transportation capacity booking and positioning in the short-term markets. Operational planning relates to daily or weekly planning with short-term production planning based on market possibilities, secondary market transportation booking and physical constraints. All the hierarchical levels can utilize portfolio planning to facilitate a global view of the available resources. In this chapter the focus is on the tactical level.

A tactical portfolio optimization model gives decision support in several areas. It optimizes how to employ the production capacities of different fields and thereby helps on establishing production plans. Further it finds preferable transactions to make in the natural gas markets that can be used as input to tactical energy allocation. Similarly, the model illustrates the need for transportation booking that can be input when booking decisions are to be made. Suggestions on booking and market transactions are useful both to initiate actions and as guidance for the operational planning. Besides the operational planning a tactical portfolio optimization model can be used to evaluate possible strategic decisions and for valuation of assets in the supply chain.

11.3 Model Description

The model presented here is a multi-period multi-stage linear programming problem. The period length can be chosen freely, but to support readability we assume all periods to have equal length in this presentation. The uncertainty is represented discretely by scenarios with outcomes for all the uncertain variables in each period.

The network that forms the basic structure of the model consists of fields, contract delivery nodes, spot markets and intermediate nodes. Any of these can be entry or exit nodes of the transportation market, here denoted as booking nodes. The possible flows are given by directed transportation links. Fields are sources that cannot have any inflow, whereas all other nodes can have both inflow and outflow. Market and delivery nodes can only have outflows going to other market or delivery nodes. This comes from the fact that there are no upstream markets and the direction of flow is determined in all the export pipelines from NCS. An example of a network is given in Fig. 11.3.

We use a steady-state representation where the natural gas flows through this network without any time lag. In reality the time for a production rate change to be

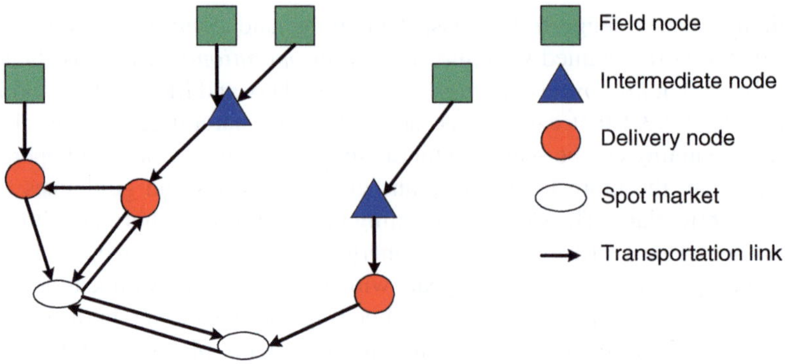

Fig. 11.3 Example of a valid network

observable in the downstream markets can be several days, but this is still assumed to be neglectable in a tactical horizon.

11.3.1 Constraints and Objective

11.3.1.1 Mass Balance

In all nodes we have to make sure that the volumes entering the node correspond to the volumes leaving the node. Since fields are assumed to have no inflow this implies production should equal total outflow in each field:

$$k_{gts} = \sum_{i \in \mathcal{O}(g)} f_{gits}, \quad g \in \mathcal{G}, t \in \mathcal{T}, \ s \in \mathcal{S}. \tag{11.1}$$

In other nodes than fields we require total inflow to equal total outflow minus what is sold or delivered to contracts within the node. Note that v_{cnts} and q_{nts} only exist for $n \in \mathcal{D}$ and $n \in \mathcal{M}$, respectively:

$$\sum_{i \in \mathcal{I}(n)} f_{ints} = \sum_{c \in \mathcal{C}(n)} v_{cnts} + q_{nts} + \sum_{i \in \mathcal{O}(n)} f_{nits} \tag{11.2}$$
$$n \in \mathcal{N} \setminus \mathcal{G}, \ t \in \mathcal{T}, \ s \in \mathcal{S}.$$

11.3.1.2 Transportation Capacity

We model a transportation market similar to the one existing on the NCS. The operations of the transportation network are unbundled from the production and marketing of natural gas, so physical requirements like pressure and gas blending are taken care of by a independent system operator (ISO). The network modelled here is a commercial network where booking nodes and transportation links are included according to how transportation capacities are made available by the ISO.

Because of the ISO's flexibility of swapping different producers' gas the commercial network is typically more flexible than the underlying physical one.

The booking system on the NCS is closely related to a zonal system with entry and exit booking in each zone. It consists of a primary and a secondary market. In the primary market producers can buy capacity from the ISO within booking time windows at a fixed price. Each producer has an upper booking limit in each booking node that is calculated by the ISO based on the network capacity and the producer's production capacity and long-term obligations. The secondary market is a bilateral market where a producer can resell booked capacity to another producer. The secondary market is not included in the model presented here because this market has a very limited liquidity which makes it unreasonable to base tactical planning on the ability to trade in the market.

To model the transportation market we use two sets of variables, transaction variables $h_{b\tau ts}$ representing the booking decisions and balance variables $a_{b\tau ts}$ representing the amount that is booked so far. In each booking period the balance is updated according to the booking decisions and the balance from the previous period. The balance is initiated with the amount booked prior to the model horizon:

$$a_{b\tau ts} = a_{b,\tau-1,ts} + h_{b\tau ts}, \quad b \in \mathcal{B}, t \in \mathcal{T}^{\text{booking}}, \tau \in \mathcal{T}, t \geq \tau, s \in \mathcal{S}, \tag{11.3}$$

$$a_{b0ts} = X_{bt}, \quad b \in \mathcal{B}, t \in \mathcal{T}, s \in \mathcal{S}. \tag{11.4}$$

The booking should not be allowed to exceed the upper booking limit described above. Since selling transportation capacity is not included in the model, it is sufficient to make sure the balance in the period of transportation does not exceed the upper limit:

$$a_{btts} \leq A_{bt}, \quad b \in \mathcal{B}, t \in \mathcal{T}, s \in \mathcal{S}. \tag{11.5}$$

At last we restrict the total flow into a booking node from exceeding the booked capacity. Since fields do not have any inflow this constraint relates to total outflow for fields.

$$\sum_{i \in \mathcal{O}(g)} f_{gits} \leq a_{gtts}, \quad g \in \mathcal{G}, \ t \in \mathcal{T}, \ s \in \mathcal{S}, \tag{11.6}$$

$$\sum_{i \in \mathcal{I}(b)} f_{ibts} \leq a_{btts}, \quad b \in \mathcal{B} \setminus \mathcal{G}, t \in \mathcal{T}, s \in \mathcal{S}. \tag{11.7}$$

11.3.1.3 Fields

The production in the field nodes is restricted by the minimum and maximum daily level. Some fields also have maximum yearly production limits which are concessions from the authorities. The flexibility of a field is reflected in how these levels relate. Many fields produce both gas, condensate and oil simultaneously, and the producer has very limited possibility to affect the ratio between the products. Since gas is the least valuable of these products typically the difference between daily minimum and maximum production is small for a field having a low gas-to-oil ratio. Fields mainly producing gas typically have a wider daily flexibility. The concessions are tighter than what could be achieved within the daily maximum production limits, which gives flexibility in how to allocate the gas within the year.

The daily and yearly limits are modelled below. The constants of the daily limits are aggregated to match the length of the periods in the model:

$$\underline{F}_{gt} \leq k_{gts} \leq \overline{F}_{gt}, \quad g \in \mathcal{G},\ t \in \mathcal{T},\ s \in \mathcal{S}, \tag{11.8}$$

$$\sum_{t \in \mathcal{T}(y)} k_{gts} \leq F_{gy}^{\text{year}}, \quad g \in \mathcal{G},\ y \in \mathcal{Y},\ s \in \mathcal{S}. \tag{11.9}$$

11.3.1.4 Markets

Trade Limits: The market does not have perfect competition, but a constant price within a interval as modelled below. To approximate how large volumes would influence the price several price intervals can be used to give a piecewise linear convex function. Alternatively the price elasticity could be expressed in a quadratic objective.

$$q_{mts} \leq Q_m, \quad m \in \mathcal{M},\ t \in \mathcal{T},\ s \in \mathcal{S}, \tag{11.10}$$

$$q_{mts} \geq -Q_m, \quad m \in \mathcal{M},\ t \in \mathcal{T},\ s \in \mathcal{S}. \tag{11.11}$$

Split Contracts: Some of the contracts give the producers the flexibility to choose which delivery point to send the gas to. This is constrained by upper and lower limits on the fraction of the delivery that can be sent to each delivery node:

$$v_{cdts} \leq C_{cd}^{\max} V_{cts}, \quad d \in \mathcal{D},\ c \in \mathcal{C}(d),\ t \in \mathcal{T},\ s \in \mathcal{S}, \tag{11.12}$$

$$v_{cdts} \geq C_{cd}^{\min} V_{cts}, \quad d \in \mathcal{D},\ c \in \mathcal{C}(d),\ t \in \mathcal{T},\ s \in \mathcal{S}. \tag{11.13}$$

Meet Demand: The demand in each contract should always be met (either by equity gas or by utilizing the spot markets):

$$\sum_{d \in \mathcal{D}(c)} v_{cdts} = V_{cts}, \quad c \in \mathcal{C},\ t \in \mathcal{T},\ s \in \mathcal{S}. \tag{11.14}$$

11.3.1.5 Equity Gas

For some of the contracts there is a requirement that parts of the deliveries should be equity gas. This means that a fraction (γ_c) can be sourced freely, while $1 - \gamma_c$ must come from the producer's own production. The alternative to own production is gas bought in a spot market. According to the network definition spot gas can appear in market nodes and delivery nodes only. This means gas arriving the delivery node from fields or intermediate nodes but not from other delivery nodes or market nodes is defined as equity gas.

The source requirement is modelled with two constraints in order to also take care of both the contracts with single and multiple delivery nodes. We start with a formulation for the contracts with single delivery nodes:

$$\sum_{n \in \mathcal{I}(d) \backslash \mathcal{M} \backslash \mathcal{D}} f_{ndts} - \sum_{c \in \mathcal{C}^{\text{split}}(d)} v^{eq}_{cdts} - \sum_{n \in \mathcal{O}(d)} f_{dnts} \geq \sum_{c \in \mathcal{C}(d) \backslash \mathcal{C}^{\text{split}}} (1 - \gamma_c)\, v_{cdts},$$

$$d \in \mathcal{D},\ t \in \mathcal{T},\ s \in \mathcal{S}. \tag{11.15}$$

The equity gas available for delivery in the delivery node d is given by the inflows f_{ndts}. This gas can be used in two different ways: it can be delivered in a long-term contract or it can be transported to a connected downstream node. The contract deliveries can be divided further in single and multiple delivery node contracts. The right-hand side in constraint (11.15) gives the total deliveries in contracts with one delivery node multiplied with the equity gas requirement. The second and third term on the left-hand side then gives the gas used for other purposes, the gas delivered in contracts with multiple delivery points, and the gas transported out of the node, respectively. In sum, the constraint specifies that the delivery of equity gas in contracts with one delivery point has to satisfy the equity gas requirement in the contracts.

It then remains to take care of the contracts with multiple delivery points. Since v^{eq}_{cdts} is known to represent equity gas by the previous equations, we add up these equity gas deliveries from all the possible delivery nodes and require this sum to at least correspond to the required amount of equity gas:

$$\sum_{d \in \mathcal{D}(c)} v^{\text{eq}}_{cdts} \geq (1 - \gamma_c) \sum_{d \in \mathcal{D}(c)} v_{cdts},\quad c \in \mathcal{C}^{\text{split}},\ t \in \mathcal{T},\ s \in \mathcal{S}. \tag{11.16}$$

11.3.1.6 Non-anticipativity

We use a scenario tree, as illustrated in the upper part of Fig. 11.4, to represent the information structure and possible outcomes of the stochastic variables. The information structure is the sequence of decision points and information flow telling what will be known and what will be uncertain at the time a decision should be taken. Branches in the scenario tree represent points where new information becomes available and some stochastic variables become certain. Decisions are taken in each node in the tree. A stage starts with a decision and ends with a branching.

Our model formulation corresponds to the scenario representation given in the lower part of Fig. 11.4. To make sure the decisions taken in the early stages do not depend on foresight we need to add non-anticipativity constraints for all nodes for which the history of information is equal. These constraints force all decisions taken in one node to be equal for all scenarios containing that node (see Rockafellar and Wets 1991):

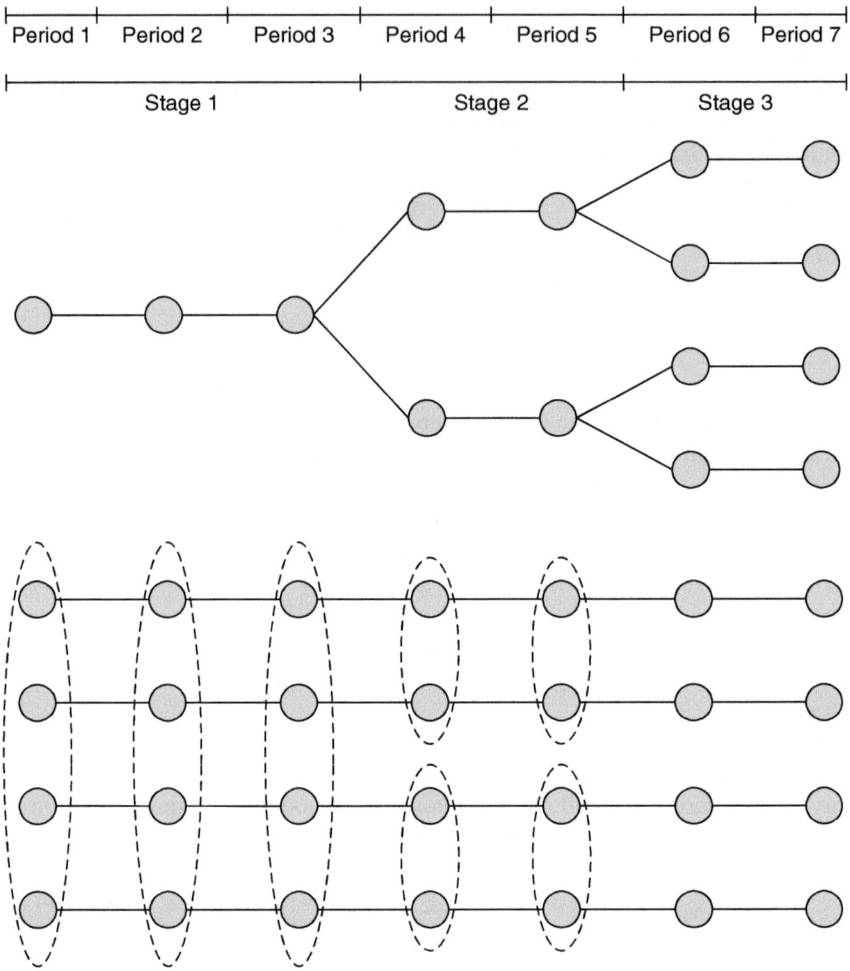

Fig. 11.4 Representation of uncertainty. The *upper part* as a scenario tree and the *lower part* as single scenarios with non-anticipativity constraints

$$\frac{1}{|\mathcal{S}(z)|}\sum_{s'\in\mathcal{S}(z)}\left(k_{gts'}, q_{mts'}, f_{ijts'}, v_{cmts'}, v^{\text{eq}}_{c'mts'}, a_{b\tau ts'}, h_{b\tau ts'}\right) \tag{11.17}$$
$$= \left(k_{gts}, q_{mts}, f_{ijts}, v_{cmts}, v^{\text{eq}}_{cmts}, a_{b\tau ts}, h_{b\tau ts}\right), \; z\in\mathcal{Z}, \; s\in S(z), \; t\in T(z).$$

11.3.1.7 Objective Function

The objective is to maximize profits from contract sales, spot market trades, production and booking decisions. This leads to the following mathematical formulation:

$$\max \sum_{t\in\mathcal{T}} \sum_{s\in\mathcal{S}} \pi_{ts} \left(\sum_{c\in\mathcal{C}^{split}} P_{cts}^{\text{contr}} \sum_{d\in\mathcal{D}(c)} v_{cdts} + \sum_{m\in\mathcal{M}} P_{mts}^{\text{spot}} q_{mts} \right.$$
$$\left. - \sum_{g\in\mathcal{G}} K_g k_{gts} - \sum_{b\in\mathcal{B}} \sum_{\tau\in\mathcal{T}:\tau\leq t} H_b h_{b\tau ts} \right). \qquad (11.18)$$

The first term gives the income from deliveries in the long-term contracts; the second term gives the income from trades in the spot market while the third term is the production costs and the fourth the costs from additional capacity booking. Only income from contracts with multiple delivery nodes is included in the objective, since there are no decision flexibility in the other contracts and the contract prices are given. Similarly the cost of booking decisions taken prior to the model horizon is left out. When the model is used for asset valuation these constant terms might be added after the optimization.

11.4 Scenario Generation

The uncertain parameters in our portfolio optimization model are the natural gas spot price in the markets, the demand in the bilateral contracts and the price in the bilateral contracts. The price in the contracts will typically depend on an underlying commodity such as the spot price of natural gas or of a competing fuel. In order to represent the uncertainty in our model, we construct scenario trees.

The structure of the scenario tree depends on the flow of information in our decision problem. In periods where we receive new information, we should include more than one branch. The size of the scenario tree will, however, directly influence the size of the optimization model. This means that we will have to keep the scenario trees at a reasonable size, and thus a trade-off between accurately describing the information flow and the total model size is important. In addition, the properties of the stochastic parameters will influence how many branches we need to add for each stage. In the following, we give an introduction to how scenario generation can be performed for multi-stage models. A nice discussion and overview on scenario generation for multi-stage models are given in Dupačová et al. (2000), and an evaluation of different methods can be found in Kaut and Wallace (2007).

There are two important elements to consider when building scenario trees for multi-stage problems: a good representation of the properties of the stochastic parameters in each branching and the linking of time periods. In a scenario generation tool developed at SINTEF, the linking of time periods is done with a prediction function while the branching is done by moment matching. The moment matching technique is based on finding a discrete representation of continuous distributions, where the first four moments (expectation, variance, skewness and kurtosis) as well as the correlation between the different stochastic parameters are kept. The moment-matching procedure is based on Høyland et al. (2003). The scenario generation is done in four steps:

1. Estimate prediction function
2. Find prediction errors
3. Build scenario tree for the prediction errors
4. Use the prediction function on the scenario tree

The procedure is independent of the chosen prediction function. This gives the user flexibility when it comes to representing the uncertainty in the given decision problem. For presentational purposes, we will here focus on an Ornstein–Uhlenbeck price process (Uhlenbeck and Ornstein 1930). The Ornstein–Uhlenbeck process is a mean reverting process that is given by the following stochastic differential equation:

$$dp_t = \eta\left(\overline{p} - p_t\right)dt + \sigma dW_t, \tag{11.19}$$

where p_t is the price in time t, the long-run equilibrium level is given by $\overline{p}$, the volatility by σ and the rate of mean reversion by η. W_t denotes the Wiener process.

11.4.1 Example of a Scenario Generation Procedure

In the following example, we focus on the uncertainty in spot prices. In each node in the scenario tree, there will then be a spot price for each of the market hubs in the network. We use the Ornstein–Uhlenbeck models to represent the spot price in all market nodes. A similar procedure can also be used for other price processes/forecasting methods. The Ornstein–Uhlenbeck model can be discretized in the following manner:

$$p_t = e^{\ln\left[p_{t-1}\right]e^{-\eta\Delta t}+\left(1-e^{-\eta\Delta t}\right)\left(\ln\left[\overline{p}\right]-\frac{\sigma^2}{4\eta}\right)+\sigma\sqrt{\frac{1-e^{-2\eta\Delta t}}{2\eta}}N\left(0,\sqrt{\Delta t}\right)}, \tag{11.20}$$

where the last term, $N\left(0, \sqrt{t}\right)$, represents sample from a normal distribution. It is this last term that forms the basis for our scenario generation. We use the general scenario generation procedure presented in Section 11.4 to generate scenarios for the normal distribution. The moment matching is performed with the given values for mean and variance, as well as the standard values for skewness and kurtosis. Figure 11.5 shows an example of a scenario tree for the standard normal distribution. The indexes f_1 and f_2 give the number of branches in the first and second stages, respectively. The value of $\epsilon_t^{stage,\ branch}$ is zero in all nodes in the scenario tree, except for the nodes in the first period in a new stage (corresponding to period $t+1$ and $t+6$ in Fig. 11.5). We generate S multivariate scenarios for the prediction error with the correct correlation between the markets and with correct moments for the individual error terms.

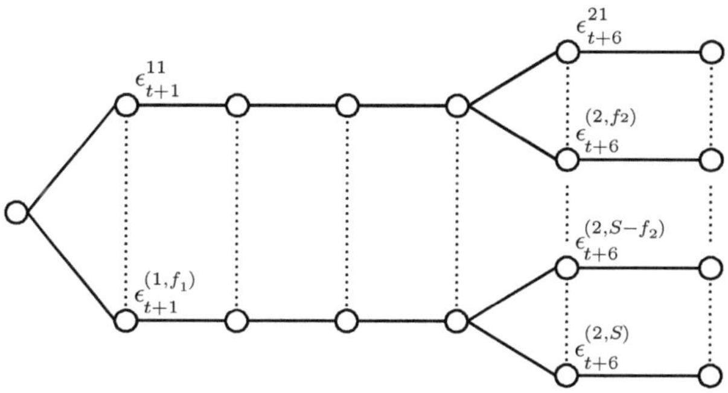

Fig. 11.5 The scenario tree for the prediction errors

Finally, we combine the discretization of the Ornstein–Uhlenbeck process (11.20) with the scenario tree for the prediction errors to one scenario tree. Each scenario presents a path from the root node to the leaf node (there are S unique paths through the tree). The value in each node in a path through the scenario tree can then easily be found by applying (11.20). Hence we use the forecasting method to predict the expected price, and scenario generation to describe the variation (error) around this price.

11.4.1.1 Uncertain Demand in the Bilateral Contracts

Traditionally, the bilateral contracts in the North Sea have a take-or-pay structure where the yearly off-take is given. However, some of these contracts give the customers substantial flexibility. This is true both with respect to the yearly volume and the daily volume. The yearly and daily off-take must be within given limits. For instance, the daily off-take can be within 50 and 110% of a daily average contracted level. This means that for the producers, the volume uncertainty represents a challenge with respect to production and portfolio planning.

We model the uncertainty in the bilateral contracts is modelled by assuming that the customers in the contracts treat the contracts as real options. When the spot price is higher than the contract price, the customers will nominate a large volume of gas in the contract. On the other hand, when the spot price is lower than the contract price, the customer will nominate a small volume in the contract. Since some of the customers will have limited flexibility with respect to drastically changing their nominations based on the spread between spot price and contract price (due to limited liquidity in the markets and supply from their own customers), we include two different customer groups (this is a similar approach to the one used in Midthun (2007)). The two customer groups are illustrated in Fig. 11.6.

11.4.1.2 Risk Aversion

Risk can be handled in several ways in the portfolio optimization model. In the version we present in this chapter we have focused on the risk of not being able to deliver according to the obligations in the bilateral contracts. Since the market

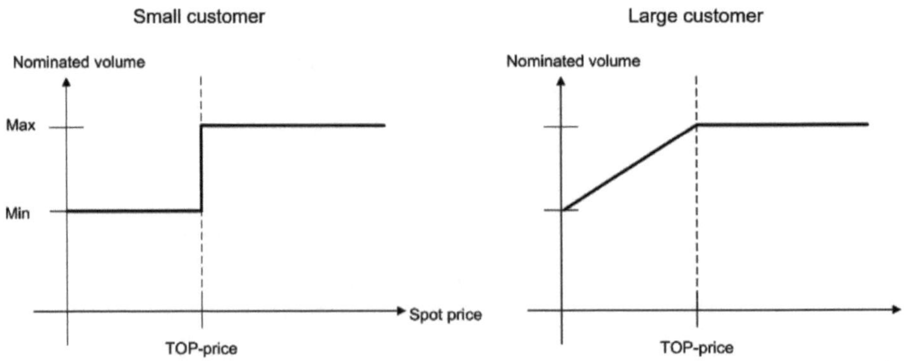

Fig. 11.6 An illustration of two different customer types

liquidity is limited, this means that we have to be able to supply the customers mainly from our own production. This also means that we must distribute the production concessions over the planning period to be able to deliver in the contracts. In this perspective it may be interesting to evaluate the situation where the scenario tree does not represent the real underlying uncertainty well in the tails of the distributions. If the real demand outcome follows the spot price in the manner expected during scenario generation, the decisions in the model will always make it possible to deliver according to the obligations. If, however, the demand turns out to be higher than we have anticipated in our scenario tree, we may risk having insufficient production concessions compared to the demand in the contracts. In order to handle this risk, we can add extreme scenarios to our scenario tree. This means that we introduce highly unlikely scenario (scenario with a zero probability of occurring) with maximum demand over the planning period. By including these scenarios we constrain our feasible region and thus we will also get a lower (or equally good) solution as before these scenarios were added. Since the probability of them occurring is zero, the profit in the scenarios will not be included in the objective function of the model. By running the model with and without these extreme scenarios, we can also find the cost of maintaining a high security of supply.

11.5 Numerical Examples

In this section we provide numerical examples to illustrate the importance of portfolio optimization and the use of a stochastic model to handle the uncertainty faced by the decision maker. We start with a simplified setting to show how the availability of short-term markets provides the producer with the flexibility to do profitable time swaps and geographical swaps. Then we use a realistic data set to recognize these effects in a large-scale setting.

11.5.1 Time Swap

We will illustrate how a producer can gain valuable flexibility through coordinated optimization of physical balancing and transactions in the spot market. To make the effects as visible as possible, we use a simplified example with one field, one market and no transportation limitations or costs. The producer has daily production capacity limitations and a concession limiting the yearly production. Further, the producer has one take-or-pay contract where the buyer each morning decides on the daily volumes to be delivered. Let us assume for simplicity that the contract price is known with a seasonal variation, but the delivery obligation is a stochastic parameter since the producer does not know the buyer's nomination in advance. On the other hand, the producer and buyer have had a long-lasting business relation, so the producer is very confident with its delivery obligation forecast. Based on this we use only two scenarios in each stage, the forecast and an extreme scenario with a infinitely small probability and a volume given by the maximum contracted obligation. Both price and the two scenarios are plotted in Fig. 11.7.

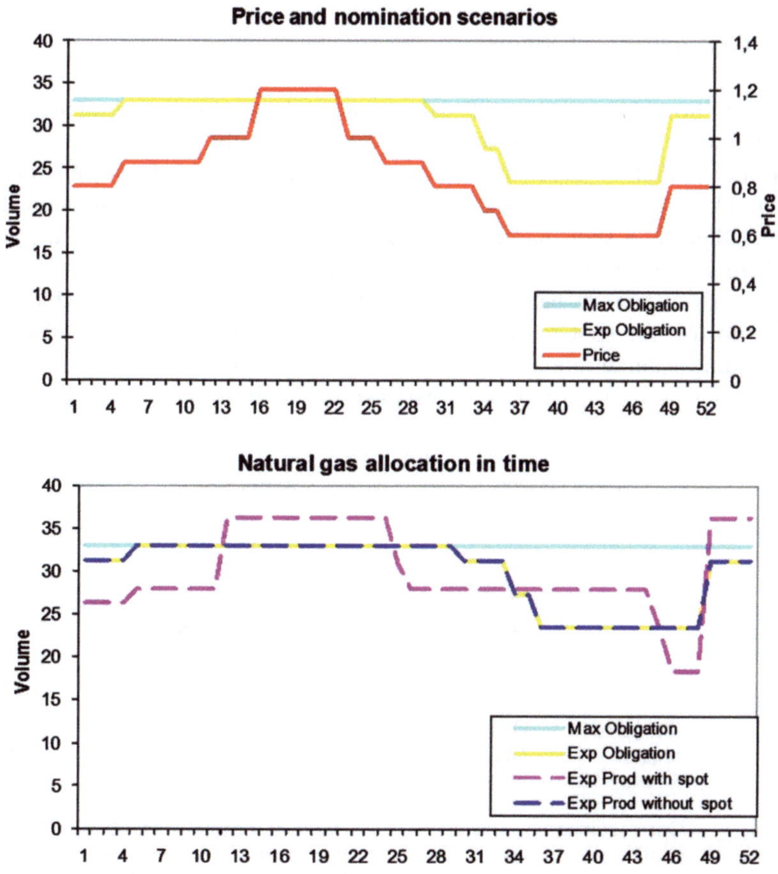

Fig. 11.7 Illustration of the case study with and without a spot market. The figure on the *top* shows the price and demand scenarios, while the figure on the *bottom* shows the production results in the model version with and without the availability of a spot market

Let us analyse the producer's planning problem in the situation with and without a spot market. Without a spot market the producer has no real decisions to make. Each day he will have to produce and deliver the volume defined by the buyer which gives an expected production equal to the expected obligation. We include a spot market with a spot price equal to the contract price and a limitation on the volumes assumed available for that price. Figure 11.7 shows the expected production together with the two scenarios for obligations. When the production is below the expected obligation it means spot purchase and the other way around. As can be seen the producer will use the spot market to move the production capacity from periods with low price to periods with better price. The volumes to swap is limited by the assumed spot liquidity and the extreme scenario forcing the producer to hold back enough gas to fulfil the obligation if the unlikely extreme scenario is realized.

11.5.2 Geographical Swap

To illustrate how the existence of short-term markets makes geographical swaps profitable, we use a similar setting to the one used for time swaps. This time we consider a slightly larger network with one field node and two market nodes (see Fig. 11.8). In market node A, the producer has a contract obligation. In the first case, there is a spot market available only in market node B. In the second case, there is a spot market available also in market node A. When there is a positive price difference between markets B and A, it will be profitable for the producer to use the spot market in node A to fulfil his commitments and sell a corresponding volume in market node B. The example is summarized in Fig. 11.8. In networks where the producer is responsible for the routing in addition to the

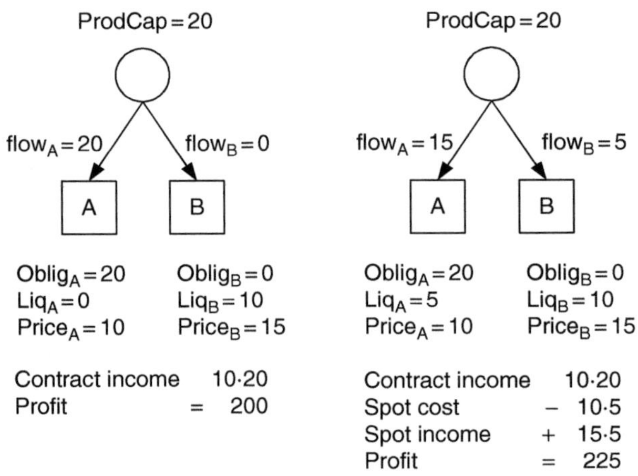

Fig. 11.8 Illustration of the case study with varying liquidity in the spot markets

production/booking decisions, the effect of geographical swaps will be even larger. The existence of booking limits will, nevertheless, give the possibility to perform geographical swaps a potential high value also in the setting that we present in this chapter.

11.5.3 Comparison of Stochastic and Deterministic Models

We will illustrate the difference in plans suggested by our multi-stage stochastic model and a corresponding deterministic model. For our tests we assume the scenario tree is an exact representation of the future. The deterministic model uses a single scenario with the expected values from the scenario tree. This corresponds to letting the deterministic model use the same forecasting method as the stochastic model. The deterministic model is run in a dynamic way where the forecast is recalculated and the plan is reoptimized every time new information becomes available according to the scenario tree. In order to highlight the effects from including the stochasticity in the model, we use small cases. The structures will, however, appear (and often be enhanced) in real data sets.

When comparing stochastic and deterministic models the notion of *expected value of expected solution* (EEV) is frequently used. The expected solution is the solution of a deterministic model where all uncertain data are replaced with their expected values. EEV is defined for two-stage models as the expected objective value of the stochastic model if the first-stage decisions are fixed according to the expected solution (Birge and Louveaux 1997). On the other hand the stochastic model (recourse problem) gives the stochastic solution and the objective value denoted *RP*. The *value of the stochastic solution* (VSS), defined as $\mathrm{VSS} = \mathrm{RP} - \mathrm{EEV}$ for maximization problems, is a measure of the expected objective value gain from using the stochastic model instead of the deterministic for the given description of the stochastic future. VSS is non-negative, since RP is optimizing its solution on the scenario tree while EEV is just evaluating the given expected solution on the same scenario tree.

Escudero et al. (2007) extend these concepts into a multi-stage setting through a dynamic way of defining the expected solution. For every node i in the scenario tree a solution is calculated for a problem where the future is described by the average value of the descendents of i and the decisions of the ancestors of i are fixed according to the previously calculated dynamic expected solutions of those nodes. Based on this procedure the expected value of the dynamic expected solution for a period t (EDEV_t) can be calculated as the weighted average objective values of the scenario tree nodes of that period. The *dynamic value of the stochastic solution* is defined as $(\mathrm{VSS}^{\mathrm{D}}) = \mathrm{RP} - \mathrm{EDEV}_T$ where T is the last time period. Analogous to VSS, $\mathrm{VSS}^{\mathrm{D}}$ is non-negative. We use this dynamic procedure to represent the deterministic model in our comparison.

We use a test case with three time periods, each with duration of 120 days. The network consists of one field, one contract and one spot hub. The contract can be supplied from both the field and the spot market in the hub. The production has a

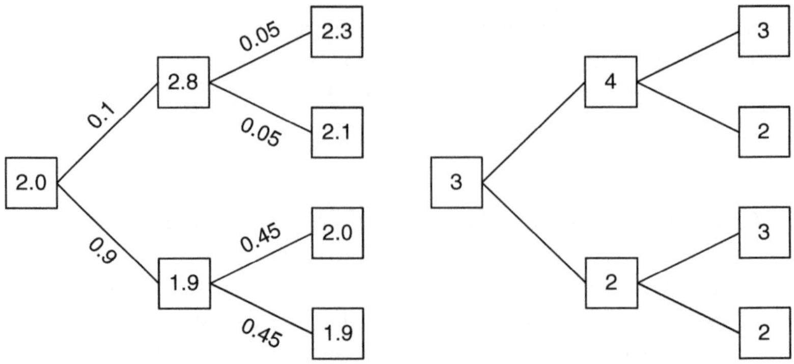

Fig. 11.9 Input data for small test case. *Left part*: spot prices [MNOK/MSm^3] and probabilities. *Right part*: Delivery obligations [MSm^3/day]

constant daily limitation of 10 MSm^3/day and a yearly limitation of 1200 MSm^3. No production cost is included. Transportation capacity booking is required for the exit from the field at a fixed price of 0.01 MNOK/MSm^3. Firm capacity equals the daily production capacity in the first period and is zero the two last periods. Until 10 MSm^3/day of capacity for each of the two last periods can be booked in the first period. The trade limit in the spot market is 5 MSm^3/day. The contract obligation and spot prices are uncertain and given in Fig. 11.9.

Note that the scenario tree is not balanced in this case, but rather has an 'upside' scenario with high price and obligation at a low probability. The expected spot price is falling throughout the model horizon. Further, note that the field is very flexible in the sense that the daily production capacity is high compared to the yearly capacity.

Let us look at how the yearly production capacity is allocated by the two models. The production and spot trade decisions are given in Fig. 11.10. The deterministic model uses the production capacity as early as possible and utilizes the whole sales trade capacity the first period. This is reasonable, since the model takes its decisions based on the constantly falling expected spot price curve. On the contrary, the stochastic model saves capacity in the first period; to be able to utilize the high price in the second period if the 'upside' scenario is realized. If the 'upside' scenario is not realized the gas is sold in the last period since this gives a better expected price than the middle period. The value of using the stochastic model instead of the deterministic (VSS^D) is for this test case a 3% addition to the expected profit achieved through exploiting the volatility of the spot prices. This might seem like a small payoff, but in a business where the profits are very large, the values can be substantial.

Now, let us change the trade limit in the spot market to 2 MSm^3/day and otherwise keep the test case unchanged. The new production and spot trade decisions are given in Fig. 11.11. In this new situation the contract can no longer be fully supplied by the spot market which implies the model has no longer relative complete recourse. This causes the deterministic model to become infeasible in two of the four scenarios in the last stage. There are two decisions in the early stages that

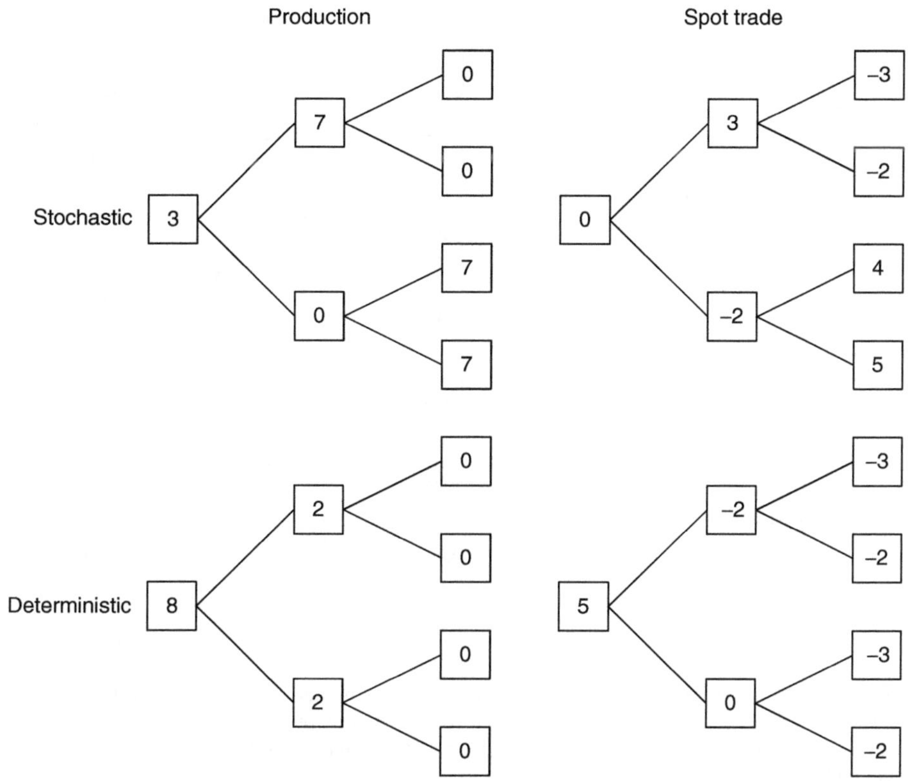

Fig. 11.10 Result from the stochastic and deterministic models on the small test case with 5 MSm^3/day as trade limit (all results are given as MSm^3/day). *Left part*: production decisions. *Right part*: Spot trade decisions (positive means sale). *Upper part*: Stochastic model. *Lower part*: Deterministic model

cause these infeasibilities, too little production capacity saved for the last period and too little transportation capacity booked for the last period.

To fulfil the obligation in the last period at least 1 MSm^3/day of the production capacity must be available, but as the deterministic model bases the decisions on expected values it only sees the need for saving 0.5 MSm^3/day. In the 'upside' scenario it can clearly be seen how the deterministic model saves less than 0.5 MSm^3/day of the yearly production capacity for the last period and thereby becomes infeasible.

The first period is the only possible booking period in this test case. Table 11.1 contains the transportation booking decisions made by the two models. The deterministic model prefers early deliveries to late deliveries because of the falling expected spot price, which gives a similar pattern for the transportation booking. This implies booking enough transportation capacity to both fullfil the expected delivery obligation and fully exhausts the spot trade limit in the middle period. Booking for the last period corresponds to the remaining yearly production

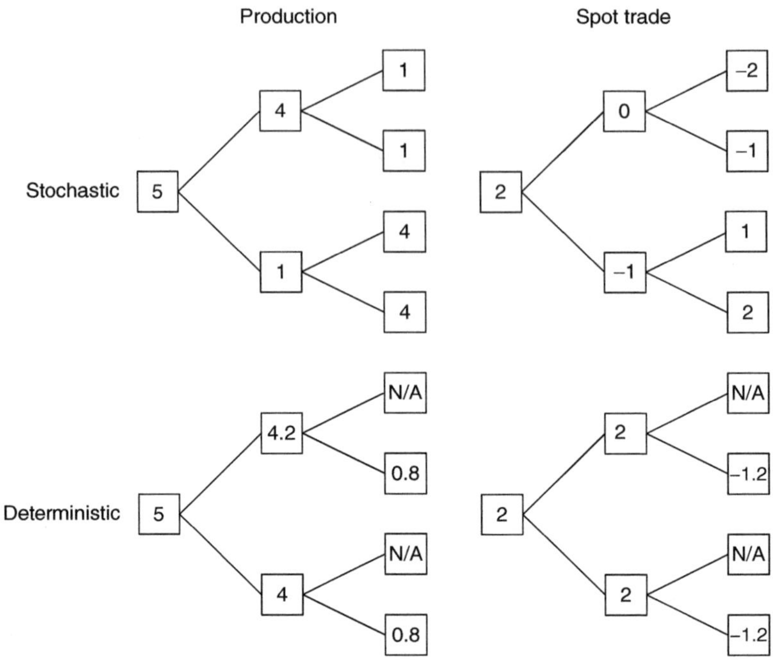

Fig. 11.11 Result from the stochastic and deterministic models on the small test case with 2 MSm^3/day as trade limit. All results in MSm^3/day. *Left part*: production decisions. *Right part*: Spot trade decisions (positive means sale). *Upper part*: Stochastic model. *Lower part*: Deterministic model

Table 11.1 Booking decisions (MSm^3/day)

Model	Middle period	Last period
Deterministic	4.2	0.8
Stochastic	4	4

capacity that cannot be delivered the two first periods. Since this remainder is less than the 1 MSm^3/day needed to fulfil the last period obligation in two scenarios this transportation booking decision makes the deterministic model infeasible in these two scenarios.

Theoretically we could argue that these infeasibilities mean the VSS^D is infinite in this situation. In real business, there typically are more instruments available to treat an 'infeasible' state, but these can be very expensive. Examples are buying replacement gas beyond the trade limit at a very high price or failing to fulfil an obligation with penalty fees and weakened reputation as a consequence.

In general, what we have seen in these two situations is how the stochastic model sees a value of making robust decisions in the early stages by making capacity (production and transportation) available till more information is available.

11.5.4 Experiences with Large-Scale Realistic Data

In this section we use a large-scale example with realistic data for the NCS to show the same effects as illustrated in the previous sections. The data set represents a gas year with 6 time periods of 2 months each. The scenario tree is symmetric with 4 stages, 14 branches from each stage and 2744 scenarios. The example has 112 nodes, of which 35 are fields, 6 are spot markets and 7 have delivery obligations. We use real data describing fields, the spot prices are based on historical data while the data on contracts and transportation booking rights are sensitive data in the business and therefore substituted by synthetic but realistic data. In this section the model built from this large-scale example with the model description given in Section 11.3 is defined as the base case. All results reported are expected values.

The model is implemented in Mosel and solved by Xpress version 7.0.0 (www.fico.com). The base case has approximately 386,000 variables and 205,000 constraints after presolve and is solved in 71 s on a computer with 2.33 GHz CPU and 3 GB RAM.

We will first analyse the value of coordinating market and production planning and thereby being able to use time and geographical swaps. The model is run with and without the possibility of buying spot gas in the markets, since this is a condition to be able to make swaps. The results reported in Table 11.2 show a 10% decrease in the spot income. The differences in total volume traded are only marginal, so the profit decrease is mainly a result of achieving lower prices for the gas. Totally the decrease in profit is 4% which in absolute values is in the order of 200 million Euro (Note that only the decision-dependent profit is included in the calculations. That includes transportation cost, production cost, spot sales income and income from contracts with optional delivery nodes.)

Further we look into the robustness and risk profile of different model structures. We use three models, the first iteration of the dynamic deterministic model (deterministic) described in Section 11.5.3, the stochastic model (base) and the stochastic model with a extended scenario tree (stochastic extreme). The deterministic model uses a single scenario given by the expected values from the scenario tree of the base model. The extended scenario tree has a new extreme scenario after each node except the leaf nodes. These extreme scenarios have zero probability and contract obligations equal to the maximum level the customer can nominate within each contract. Adding extreme scenarios in the stochastic model corresponds to a risk-averse policy where the probability of not being able to fulfilling an obligation is zero.

Table 11.3 shows how the expected value of the objective function components deviates from the base model. The deterministic model achieves better expected values and the stochastic extreme model achieves worse than the base model. There

Table 11.2 Effect of removing spot purchase possibility

Model	Profit	Spot income	Transportation cost
Without spot purchase	−4%	−10%	14%

Table 11.3 Deviations from base case with different models tested on the large-scale example

Model	Profit	Spot income	Transportation cost
Deterministic	1%	1%	−29%
Stochastic extreme	−3%	−7%	177%

are no differences in contract income and only marginal differences in volumes traded spot in the three models. The extreme scenarios reduce the flexibility to make spot trade decisions based on preferable prices which limits the spot income in the stochastic extreme model. Further the capacity booking increases with 177% compared to the base model to be able to cover the obligations in the extreme scenarios, but because of low tariffs the effect on the total expected profit is limited.

Note that these results order the models according to the spread of the scenario tree optimized over, which can be interpreted as ordering according to the level of risk aversion. The deterministic model utilizes the resources very efficiently but the solution is less robust for deviations from the expected scenario. The stochastic extreme model on the other hand can handle any outcome in the support of the stochastic parameters, but for this robustness a risk premium is paid.

The true values of these solutions are not realized until the actual outcome of the stochastic parameters is known. This is recognized in the procedure calculating EDEV (Section 11.5.3) where the solution of the deterministic model is evaluated on another tree than the single scenario tree it is optimized over. For the large-scale example the dynamic deterministic solution turns out to be infeasible in the last iteration, partly because of too aggressive spot sales in the spring and partly because of lacking transportation booking, which corresponds to the effect pointed out for the small examples. On the other hand, if a less than maximum off-take in the contracts is realized the stochastic extreme model will prove unnecessarily cautious. The choice on which model to use should be taken based on the risk policy in the company and a judgement of the possibility and cost of handling high off-takes with means not included in the model.

11.6 Conclusions

We have presented a tactical portfolio optimization model for a natural gas producer. The model includes concession on fields, short-term markets, booking of transportation capacity, handling of contract commitments and uncertain parameters (price and demand in long-term contracts and price in the short-term markets). The model has been used for different analyses concerning portfolio optimization by a large natural gas producer for several years.

The numerical examples illustrate the potentially high value of utilizing short-term markets for geographical swaps and time swaps. In order to utilize this flexibility in an optimal manner, the portfolio view on the set of concessions, contracts and market opportunities is vital. We have also included a large-scale example based on realistic data (market data, network data and field data), where we show a substantial economic potential of using stochastic programming and a portfolio perspective with coordinated production and trade decisions in planning.

Acknowledgments This project has been supported by Statoil and the Research Council of Norway (project number 144217/212), for which we are grateful.

References

Robert N. Anthony. *Planning and Control Systems: A Framework for Analysis*. Studies in Management Control. Division of Research, Graduate School of Business Administration, Harvard University, Boston, 1965.

J.R. Birge and F. Louveaux. *Introduction to Stochastic Programming*. Springer, New York, NY, 1997.

Gabriel R. Bitran and Devanath Tirupati. Hierarchical production planning. In S. C. Graves, A. H. G. Rinnooy Kan, and Paul H. Zipkin, editors, *Logistics of Production and Inventory*, volume 4 of *Handbooks in Operations Research and Management Science*, chapter 10, pages 523–568. Elsevier Science, Amsterdam, 1993.

Thor Bjørkvoll, Stein-Erik Fleten, Matthias Peter Nowak, Asgeir Tomasgard, and Stein W. Wallace. Power generation planning and risk management in a liberalised market. In *2001 IEEE Porto Power Tech Proceedings*, pages 426–431, 2001.

Hans Jørgen Dahl. *Norwegian Natural Gas Transportation Systems. Operations in a Liberalized European Gas Market*. PhD thesis, Norwegian University of Science and Technology, 2001.

Directorate-General for Energy European Commission and Transport. Opening up to choice: Launching the single european gas market, 2002.

Jitka Dupačová, Giorgio Consigli, and Stein W. Wallace. Scenarios for multistage stochastic programs. *Annals of Operations Research*, 100:25–53, December 2000.

Laureano F. Escudero, Araceli Garín, María Merino, and Gloria Pérez. The value of the stochastic solution in multistage problems. *TOP*, 15:48–64, 2007.

EU Commission. Directive 98/30/EC of the European Parliament and of the Council of 22 June 1998 concerning common rules for the internal market in natural gas, 1998.

EU Commission. Directive 2003/55/EC of the European Parliament and of the Council of 26 June 2003 concerning common rules for the internal market in natural gas and repealing Directive 98/30/EC, 2003.

Stein-Erik Fleten, Stein W. Wallace, and William T. Ziemba. Hedging electricity portfolios via stochastic programming. In Claude Greengard and Andrzei Ruszczyński, editors, *Decision Making under Uncertainty: Energy and Power*, volume 128 of *IMA Volumes on Mathematics and its Applications*, pages 71–93. Springer, New York, NY, 2002.

Arnoldo C. Hax and Harlan C. Meal. Hierarchical intergration of product planning and scheduling. In M. A. Geisler, editor, *Studies in Management Sciences, Vol. 1: Logistics*. Elsevier, New York, NY, 1975.

Kjetil Høyland, Michal Kaut, and Stein W. Wallace. A heuristic for moment-matching scenario generation. *Computational Optimization and Applications*, 24(2-3):169–185, February–March 2003.

Michal Kaut and Stein W. Wallace. Evaluation of scenario generation methods for stochastic programming. *Pacific Journal of Optimization*, 3:257–271, 2007.

Kjetil Trovik Midthun. *Optimization Models for Liberalized Natural Gas Markets*. PhD thesis, Department of Industrial Economics and Technology Management, Norwegian University of Science and Technology, Trondheim, Norway, 2007. Theses at NTNU, 2007:205.

Kjetil T. Midthun, Mette Bjørndal, and Asgeir Tomasgard. Modeling optimal economic dispatch and system effects in natural gas networks. *Energy Journal*, 30:155–180, 2009.

Ministry of Petroleum and Energy/Norwegian Petroleum Directorate. Facts 2009 – the Norwegian petroleum sector. http://www.npd.no/en/Publications/Facts/, 05 2009.

John M. Mulvey. Introduction to financial optimization: Mathematical programming special issue. *Mathematical Programming, Series B*, 89:205–216, 2001.

R. T. Rockafellar and Roger J-B. Wets. Scenarios and policy aggregation in optimization under uncertainty. *Mathematics of Operations Research*, 16(1):119–147, February 1991.

A. Selot, L.K. Kuok, M. Robinson, T.L. Mason, and P.I. Barton. A short-term operational planning model for natural gas production systems. *AIChE Journal*, 54(2):495–515, 2008.

A. Tomasgard, F. Rømo, M. Fodstad, and K.T. Midthun. Optimization models for the natural gas value chain. In G. Hasle, K.-A. Lie, and E. Quak, editors, *Geometric Modelling, Numerical Simulation and Optimization*, pages 521–558. Springer, Berlin, 2007.

G. E. Uhlenbeck and L.S. Ornstein. On the theory of Brownian motion. Physical Review, 36:823–841, 1930.

Nina Linn Ulstein, Bjørn Nygreen, and Jan Richard Sagli. Tactical planning of offshore petroleum production. *European Journal of Operational Research*, 176(1):550–564, January 2007.

S. A. Zenios, editor. *Financial Optimization*. Cambridge University Press, Cambridge, 1993.

William T. Ziemba and Raymond G. Vickson. *Stochastic Optimization Models in Finance*. World Scientific Publishing Company, Singapore, 2006.

Chapter 12
Risk Management with Stochastic Dominance Models in Energy Systems with Dispersed Generation

Dimitri Drapkin, Ralf Gollmer, Uwe Gotzes, Frederike Neise, and Rüdiger Schultz

Abstract Dispersed power generation is the source of many challenging optimization problems with uncertain data. We review algorithmic approaches to risk aversion with stochastic dominance constraints. Dispersed power generation provides the practical background for illustration and comparison of the methods.

Keywords Stochastic programming · Dispersed power generation · Risk aversion · Stochastic dominance constraints · Decomposition algorithms

12.1 Introduction

Dispersed generation systems are innovative structures in power generation. They consist of small distributed units, often involving renewables (wind and photovoltaics), and usually located in the vicinity of consumers. This promotes sustainability by reduction of transmission losses and by a more efficient exploitation of heat as a coupled product (cogeneration units for power and heat). Sometimes these systems are called virtual power plants, since they do not represent a physical power unit but rather a virtual one by their controlled interaction.

Design and operation of power systems with dispersed generation pose new challenges for the handling of uncertainty and risk in an optimization context. This is due to the multitude of random influences on the system (load, prices, and infeed from renewables) and to the more complex coordination requirements when operating the dispersed generation units as a virtual power plant.

Stochastic programming offers flexible methodology for handling uncertainty in optimization problems, (Birge and Louveaux 1997; Ruszczyński and Shapiro 2003; Shapiro et al. 2009). In this chapter, we will pick up ideas from two-stage stochastic integer programming (Louveaux and Schultz 2003; Schultz 2003), for risk modeling and optimization. Motivated by dispersed power generation, we will consider optimization problems under uncertainty where, in a first stage, nonanticipative

R. Schultz (✉)
Faculty of Mathematics, University of Duisburg-Essen, D-47057 Duisburg, Germany
e-mail: ruediger.schultz@uni-due.de; schultz@mail.math.uni-duisburg.de

M. Bertocchi et al. (eds.), *Stochastic Optimization Methods in Finance and Energy*, International Series in Operations Research & Management Science 163,
DOI 10.1007/978-1-4419-9586-5_12,

here-and-now decisions have to be taken which must not foresee future information. After uncertainty has been unveiled, second-stage decisions are taken, with an aspiration to optimize, given the first-stage decisions taken before and the random data observed.

In power systems with dispersed generation this approach to optimization under uncertainty is fruitful in many respects. As an example take the design of these systems: Decisions on which and how many generation units to install must be taken before knowing data such as electrical and thermal load, fuel and power prices, or infeed from wind and solar energy that are crucial for optimization of system operation. So these design decisions would form the first stage, while operation decisions after knowing the mentioned uncertain data are accumulated into the second stage.

Conceptually, a similar case arises in retailer problems where forward contracting and customer prices must be fixed in the first stage, while customer supply under uncertainty of load and pool prices makes up the second stage.

A third example refers to planning the operation of a power system over some time horizon. Typically, information about uncertain data such as load, pool prices, or infeed from renewables is available with some certainty at best for some initial section of the planning horizon. So operation decisions belonging to this time section are put into the first stage, while decisions belonging to the rest of the planning horizon form the second stage.

In the outlined two-stage stochastic optimization context, the issue of risk is conventionally addressed by minimizing statistical parameters reflecting different perceptions of risk (risk measures) over the set of all feasible first-stage solutions; see Ruszczyński and Shapiro (2003), Schultz (2003), and Shapiro et al. (2009) for general methodology and Schultz and Neise (2007) for applications to power systems with dispersed generation.

Here we adopt an alternative view at risk modeling in two-stage stochastic programming. With a stochastic benchmark reflecting a cost profile just acceptable to the decision maker, the set of all first-stage decisions leading to overall costs preferable to this benchmark is identified, and an objective function is minimized over this set. Preference to the benchmark is expressed using concepts of stochastic dominance (Dentcheva and Ruszczyński 2003, 2004; Fishburn 1970; Müller and Stoyan 2002). This leads to a new class of two-stage stochastic programming models for which basic results were derived in Drapkin and Schultz (2010) and Gollmer et al. (2008, 2010).

In this chapter, we review algorithmic approaches to these models and point to suitable applications in power system optimization. In Section 12.2, the dominance relations and resulting two-stage stochastic programs are made mathematically rigorous. Section 12.3 deals with mixed-integer linear programming equivalents in case the underlying probability spaces are finite. A Lagrangean decomposition method working for models with integer variables in both stages is presented in Section 12.4. Numerical experiments for this method with test instances from dispersed power generation are reported in Section 12.5. An enhancement of Lagrangean decomposition by cutting planes is possible if integer variables are missing in the second stage of the two-stage stochastic program. This issue is elaborated in Section 12.6. Finally, we have a conclusions section.

12.2 Stochastic Integer Programs with Dominance Constraints

Consider the (mixed-integer linear) optimization problem under uncertainty

$$\min\left\{c^\top x + q^\top y \ :\ Tx + Wy = z(\omega),\ x \in X,\ y \in Y\right\}, \tag{12.1}$$

together with the requirement that x must be selected without anticipating the realization of the random data $z(\omega)$.

This leads to a two-stage scheme of alternating decision and observation: The first-stage decision x is followed by observing $z(\omega)$ and then the second-stage decision y is taken, thus depending on x and ω. For ease of presentation we have restricted uncertainty in (12.1) to the right-hand side z. Uncertainty may be present in c, q, T, W as well. By X, Y we denote solution sets to systems of linear (in)equalities possibly involving integer requirements to components of x, y. Hence, (12.1) indeed is a mixed-integer linear program.

The two-stage dynamics mentioned in the introduction becomes explicit by a reformulation of (12.1):

$$\begin{aligned}&\min_x \left\{ c^\top x + \underbrace{\min_y\{q^\top y \ :\ Wy = z(\omega) - Tx,\ y \in Y\}}_{\Phi(x,z(\omega))} \ :\ x \in X \right\}\\ &= \min_x\{c^\top x \ +\ \Phi(x, z(\omega)) \ :\ x \in X\}.\end{aligned} \tag{12.2}$$

Let us take the following view at the above model: Each feasible first-stage decision $x \in X$ induces a random variable $f(x, z(\omega)) := c^\top x + \Phi(x, z(\omega))$, and the model suggests to look for a "minimal" among those random variables. In conventional stochastic programming, see, e.g., Birge and Louveaux (1997), Ruszczyński and Shapiro (2003), Schultz (2003), Schultz and Tiedemann (2003, 2006), and Shapiro et al. (2009) for appropriate assumptions and statements; these random variables are ranked according to a weighted sum of statistical parameters reflecting mean and risk:

$$\min\{(\mathbb{E} + \rho\mathcal{R})\left[f(x, z(\omega))\right] \ :\ x \in X\}. \tag{12.3}$$

Here, $\mathbb{E}$ denotes the expected value, $\rho \geq 0$ is a fixed weight factor, and $\mathcal{R}$ is a risk measure.

In Märkert and Schultz (2005) and Schultz and Tiedemann (2003, 2006), specifications of $\mathcal{R}$ in terms of excess probability, conditional value-at-risk, expected excess, and semideviation have been considered. It was shown that, for $z(\omega)$ living in a finite probability space, the nonlinear, nonconvex minimization problem (12.3) turns into a large-scale, block-structured, mixed-integer linear program. With taylored decomposition algorithms, these mixed-integer linear programs become numerically tractable for instances where standard software such as ILOG-Cplex (2005) fails; see Schultz and Neise (2007) for experiments with instances derived from optimization in power systems with dispersed generation.

Rather than to look for "best possible" random variables $f(x, z(\omega)), x \in X$, the aim of this chapter is to identify "acceptable" members in this family. To this end, we introduce a benchmark random variable $a(\omega)$ describing a random profile of the total costs which is just acceptable in the context of the application at hand. It indicates different cost levels that are acceptable with certain probabilities, i.e., in certain ratios of the total number of outcomes. Of course, a deterministic benchmark a, acceptable in 100% of the outcomes, arises as a special case. The random benchmark $a(\omega)$, however, is a more appropriate reflection of the stochastic nature inherent to (12.1) and allows for more flexibility in determining what is acceptable to decision makers.

Stochastic dominance (Dentcheva and Ruszczyński 2003, 2004; Fishburn 1970; Müller and Stoyan 2002) offers concepts to make the acceptance outlined above mathematically rigorous. Employed to the random variables $f(x, z(\omega))$ and the benchmarks $a(\omega)$ it leads to the following:

Denoting with μ, ν the probability distributions of $z(\omega), a(\omega)$, we say that $f(x, z(\omega))$ is stochastically smaller than $a(\omega)$ (or dominates $a(\omega)$ to first order), denoted $f(x, z(\omega)) \preceq_1 a(\omega)$, if and only if

$$\mu\left(\{z : f(x, z) \leq \eta\}\right) \geq \nu\left(\{a : a \leq \eta\}\right) \quad \text{for all } \eta \in I\!R. \tag{12.4}$$

This means that, for any real-valued cost level η, the probability that $f(x, z(\omega))$ does not exceed η is greater than or equal to the probability that the benchmark costs $a(\omega)$ do not exceed η.

We say that $f(x, z(\omega))$ is stochastically smaller than $a(\omega)$ in increasing convex order (or dominates $a(\omega)$ to second order), denoted $f(x, z(\omega)) \preceq_2 a(\omega)$, if and only if

$$\int_{I\!R^s} \max\{f(x, z) - \eta, 0\}\, \mu(dz) \leq \int_{I\!R} \max\{a - \eta, 0\}\, \nu(da) \quad \text{for all } \eta \in I\!R. \tag{12.5}$$

This means that, for any real-valued cost level η, the expected excess of $f(x, z(\omega))$ over η is less than or equal to the expected excess of $a(\omega)$ over η.

Our notions of stochastic dominance are based on preferring small outcomes of random variables to big ones, since the random variables $f(x, z(\omega))$ arise in a minimization context. Historically, stochastic dominance has been developed for preference of big outcomes (Müller and Stoyan 2002). Of course, both settings transform directly into each other. In this way, the above relation "stochastically smaller" corresponds to traditional first-order dominance; and likewise "stochastically smaller in increasing convex order" to second-order dominance; see Müller and Stoyan (2002). This explains our slight abuse of notation when referring to $\preceq_1$ and $\preceq_2$ as "dominance" relations although being dominant means being smaller rather than bigger.

Both $\preceq_1$ and $\preceq_2$ define partial orders on the space of random variables. It can be shown that first-order stochastic dominance implies second-order stochastic dominance to hold, but not vice versa. Therefore, a bigger class of random variables stands in relation $\preceq_2$ than this is the case for $\preceq_1$. From the modeling perspective, $\preceq_1$ is relevant when comparison to the benchmark is based on probabilities of certain level sets of $f(x, z(\omega))$. Second-order dominance refers to comparison via function values of $f(x, z(\omega))$, more precisely, expected values of excesses over certain targets.

With these ingredients, and starting out from (12.1), we formulate the following stochastic integer programs with dominance constraints

$$\min\{g^\top x \; : \; f(x, z(\omega)) \preceq_i a(\omega), \; x \in X\} \quad (i = 1, 2). \tag{12.6}$$

In contrast with (12.3), problems (12.6) do not aim at finding first-stage solutions minimizing (a weighted sum of expectation and) risk. Rather, they start out from the sets $\{x \; : \; f(x, z(\omega)) \preceq_i a(\omega), \; x \in X\}$ of first-stage solutions which are acceptable in terms of the risk benchmark $a(\omega)$, and over this set the objective function $g^\top x$ is minimized.

The specification of $g^\top x$ allows to include additional features into the optimization. In Section 12.5 we will consider specifications where the random optimization problem (12.1) corresponds to minimizing operation costs of a dispersed generation system under load uncertainty. The set $\{x \; : \; f(x, z(\omega)) \preceq_i a(\omega), \; x \in X\}$ reflects all nonanticipative generation policies x which are economically and technologically feasible, and whose (random) costs are stochastically preferable to a prescribed benchmark. The objective function $g^\top x$ is counting the number of start-ups of units, thus (12.6) is minimizing abrasion over all acceptable generation policies.

In another application, see Gollmer et al. (2010) and Gotzes (2009), problem (12.1) corresponds to an investment planning problem for electricity generation. In the first stage, decisions on capacity expansions for different generation technologies under budget constraints and supply guarantee are made. The second stage concerns the minimization of production costs for electricity under capacity constraints and load uncertainty. Proper specifications of $g^\top x$ in (12.6) then allow to minimize expansion of certain technologies (environmentally hazardous ones, for instance) over all nonanticipative investment policies which are feasible and whose costs are preferable to the benchmark $a(\omega)$.

12.3 Equivalent Block-Structured Mixed-Integer Linear Programs

In view of definitions (12.4) and (12.5), the conditions $f(x, z(\omega)) \preceq_i a(\omega),\ i = 1, 2$, in (12.6) amount to semi-infinite constraints, i.e., involving a decision variable x of finite dimension but an infinite number, in fact a continuum, of constraints. Numerical tractability of these objects, however, improves in case $z(\omega)$ and $a(\omega)$ are following finite discrete probability distributions.

Assume that $z(\omega)$ and $a(\omega)$ have discrete distributions with realizations z_l, $l = 1, \ldots, L$, and $a_k,\ k = 1, \ldots, K$, as well as probabilities $\pi_l,\ l = 1, \ldots, L$, and $p_k,\ k = 1, \ldots, K$, respectively. Then the following equivalences hold (Müller and Stoyan 2002; Noyan et al. 2006): The first-order stochastic dominance relation (12.4) is valid if and only if

$$\mu\left(\{z : f(x,z) \le a_k\}\right) \ \ge\ \nu\left(\{a : a \le a_k\}\right), \quad k = 1, \ldots, K.$$

The second-order stochastic dominance relation (12.5) holds if and only if

$$\int_{\mathbb{R}^s} \max\{f(x,z) - a_k, 0\}\, \mu(dz) \ \le\ \int_{\mathbb{R}} \max\{a - a_k, 0\}\, \nu(da), \quad k = 1, \ldots, K.$$

So the continuum of constraints turns into a finite number. For the above equivalences it was only essential that $a(\omega)$ is finite discrete. They hold for general random variables $z(\omega)$. If in addition $z(\omega)$ is finite discrete then the optimization problems (12.6) turn into large-scale mixed-integer linear programs, see Gollmer et al. (2008) and Gotzes (2009) for proofs. These optimization problems look as follows.

For first-order stochastic dominance ($i = 1$), problem (12.6) is equivalent to

$$\min\left\{ \begin{array}{llll} g^\top x : c^\top x + q^\top y_{lk} - a_k & \le \mathsf{M}\theta_{lk} & \forall l\ \forall k \\ Tx + W y_{lk} & = z_l & \forall l\ \forall k \\ \sum_{l=1}^{L} \pi_l \theta_{lk} & \le \bar{a}_k & \forall k \\ x \in X,\ y_{lk} \in Y,\ \theta_{lk} \in \{0,1\} & & \forall l\ \forall k \end{array} \right\}. \tag{12.7}$$

Here, M denotes a sufficiently big constant ("Big M"), and $\bar{a}_k := 1 - \nu(\{a : a \le a_k\}),\ k = 1, \ldots, K$.

For second-order stochastic dominance ($i = 2$), problem (12.6) is equivalent to

$$\min\left\{ \begin{array}{llll} g^\top x : c^\top x + q^\top y_{lk} - a_k & \le v_{lk} & \forall l\ \forall k \\ Tx + W y_{lk} & = z_l & \forall l\ \forall k \\ \sum_{l=1}^{L} \pi_l v_{lk} & \le \hat{a}_k & \forall k \\ x \in X,\ y_{lk} \in Y,\ v_{lk} \ge 0 & & \forall l\ \forall k \end{array} \right\}, \tag{12.8}$$

with $\hat{a}_k := \int_{\mathbb{R}} \max\{a - a_k, 0\}\, \nu(da),\ k = 1, \ldots, K$.

When compared to the mixed-integer linear programs resulting from the mean-risk minimization problem (12.3), see Ruszczyński and Shapiro (2003), Schultz (2003), Schultz and Neise (2007), and Schultz and Tiedemann (2003, 2006), the mixed-integer linear programs (12.7) and (12.8) bear some similarities as well as essential differences.

The constraints

$$\sum_{l=1}^{L} \pi_l \theta_{lk} \leq \bar{a}_k \quad \forall k \tag{12.9}$$

and

$$\sum_{l=1}^{L} \pi_l v_{lk} \leq \hat{a}_k \quad \forall k \tag{12.10}$$

have a special role in (12.7) and (12.8). Apart from them all other constraints are requirements to individual pairs $(l, k) \in \{1, \ldots, L\} \times \{1, \ldots, K\}$. These other constraints are merely coupled by the occurrence of the common variable x. The mentioned mixed-integer linear programming equivalents to (12.3), with the exception of $\mathcal{R}$ specified as the semideviation (Märkert and Schultz 2005), all have the property that coupling among their constraints is caused by joint occurrence of x only.

So the essential qualitative difference between the mixed-integer linear programming equivalents to (12.3) and to (12.6) is that in (12.7), (12.8) the constraints (12.9), (12.10) induce further coupling. This will be of crucial importance for the design of the decomposition algorithms we are going to address subsequently.

12.4 Lagrangean Decomposition Algorithms

The starting point of our algorithmic investigations is to recall that (12.7) and (12.8) are representations of the nonconvex global minimization problems (12.6). A proven tool for global minimization is branch-and-bound. With our optimization problem in mind, its essence can be described as follows:

The set X is partitioned with increasing granularity. Linear inequalities are used for this partitioning to maintain the (mixed-integer) *linear* description. On the current elements of the partition upper and lower bounds for the optimal objective function value are sought. This is embedded into a coordination procedure to guide the partitioning and to prune elements due to infeasibility, optimality, or inferiority. Altogether, tighter and tighter lower and upper bounds for the global optimal values are generated this way. The feature of decomposition will come up in the upper- and lower-bounding parts of the algorithm.

To formulate a generic branch-and-bound algorithm for (12.7) and (12.8), let $\mathbf{P}$ denote a list of problems, and $\varphi_{\text{LB}}(P)$ be a lower bound for the optimal value of $P \in \mathbf{P}$. Moreover, $\bar{\varphi}$ denotes the currently best upper bound to the globally optimal value, and $X(P)$ is the element in the partition of X belonging to P. The algorithm then proceeds as follows:

Algorithm 1

STEP 1 (INITIALIZATION):
Let $\mathbf{P} := \{(12.7)\}$ *for* $i = 1$ *or* $\mathbf{P} := \{(12.8)\}$ *for* $i = 2$. *Put* $\bar{\varphi} := +\infty$.

STEP 2 (TERMINATION):
If $\mathbf{P} = \emptyset$ *then the* $\bar{x}$ *that yielded* $\bar{\varphi} = g^{\top}\bar{x}$ *is optimal.*

STEP 3 (BOUNDING):
Select and delete a problem P *from* $\mathbf{P}$. *Compute a lower bound* $\varphi_{LB}(P)$ *and apply a feasibility heuristics to find a feasible point* $\bar{x}$ *of* P.

STEP 4 (PRUNING):
If $\varphi_{LB}(P) = +\infty$ *(infeasibility of a subproblem) or* $\varphi_{LB}(P) > \bar{\varphi}$ *(inferiority of* P*), then go to step 2.*
If $\varphi_{LB}(P) = g^{\top}\bar{x}$ *(optimality for* P*), then check whether* $g^{\top}\bar{x} < \bar{\varphi}$. *If yes, then* $\bar{\varphi} := g^{\top}\bar{x}$. *Go to step 2.*
If $g^{\top}\bar{x} < \bar{\varphi}$, *then* $\bar{\varphi} := g^{\top}\bar{x}$.

STEP 5 (BRANCHING):
Create two new subproblems by partitioning the set $X(P)$ *by means of linear inequalities. Add these subproblems to* $\mathbf{P}$ *and go to step 2.*

This generic algorithm is turned into specific solution methods for the first- and second-order models ($i = 1, 2$) by different specifications of the lower- and upper-bounding procedures of step 3; see Gollmer et al. (2008, 2010) for detailed expositions. In what follows we describe these specifications for the first-order model. For the second-order model an analogous analysis applies.

First, we model the nonanticipativity of x explicitly, i.e., we introduce copies $x_l, l = 1, \ldots, L$, of x and add the requirement $x_1 = x_2 = \cdots = x_L$. Coupling in the model now is provided by these nonanticipativity constraints and by (12.9). We obtain a lower bound to the global optimal value when relaxing these constraints. More specifically, we perform Lagrangean relaxation of (12.9) and just omit the nonanticipativity constraints $x_1 = x_2 = \cdots = x_L$. The reason for not including the latter into Lagrangean relaxation is a trade-off between tightness of lower bounds and numerical effort: Lagrangean relaxation of nonanticipativity would not only improve the lower bounds but also increase considerably the dimension of the Lagrangean dual (see (12.12)). The numerical tests to be reported in Section 12.5 have confirmed that the loss in precision of the lower bounds is tolerable for the problem instances handled there.

These considerations lead to the following Lagrangean function:

$$\mathcal{L}(x, \theta, \lambda) := \sum_{l=1}^{L} \pi_l g^{\top} x_l + \sum_{k=1}^{K} \lambda_k \left(\sum_{l=1}^{L} \pi_l \theta_{lk} - \bar{a}_k \right)$$

which can be written as

$$\mathcal{L}(x,\theta,\lambda) \;:=\; \sum_{l=1}^{L} \mathcal{L}_l(x_l,\theta_l,\lambda), \tag{12.11}$$

where

$$\mathcal{L}_l(x_l,\theta_l,\lambda) \;:=\; \pi_l g^\top x_l \;+\; \pi_l \sum_{k=1}^{K} \lambda_k(\theta_{lk} - \bar{a}_k).$$

The Lagrangean dual now reads

$$\max\{D(\lambda) \;:\; \lambda \geq 0\} \tag{12.12}$$

where $D(\lambda)$ is the optimal value to

$$\min\left\{\begin{array}{lll}\mathcal{L}(x,\theta,\lambda) \;: c^\top x_l + q^\top y_{lk} - a_k & \leq M\theta_{lk} & \forall l\ \forall k \\ \qquad\qquad\qquad T x_l + W y_{lk} & = z_l & \forall l\ \forall k \\ \qquad\qquad\qquad x_l \in X,\ \ y_{lk} \in Y, & \theta_{lk} \in \{0,1\} & \forall l\ \forall k\end{array}\right\}.$$

Taking (12.11) into account, this optimal value is obtained by summing up for $l = 1, \ldots, L$ the optimal values of the single-scenario problems

$$\min\left\{\begin{array}{lll}\mathcal{L}_l(x_l,\theta_l,\lambda) \;: c^\top x_l + q^\top y_{lk} - a_k & \leq M\theta_{lk} & \forall k \\ \qquad\qquad\qquad T x_l + W y_{lk} & = z_l & \forall k \\ \qquad\qquad\qquad x_l \in X,\ \ y_{lk} \in Y, & \theta_{lk} \in \{0,1\} & \forall k\end{array}\right\}. \tag{12.13}$$

This is the announced decomposition feature. Instead of working with the full-size mixed-integer linear program behind $D(\lambda)$, much smaller mixed-integer linear programs corresponding to the individual realizations $z_l, l = 1, \ldots, L$, can be employed. The Lagrangean dual is a nonsmooth concave maximization (or convex minimization) problem which is tackled with bundle subgradient methods from nondifferentiable optimization. In our implementation we did resort to Helmberg and Kiwiel (2002).

This concludes the description of the lower bounding procedure which is applied with $X(P)$ instead of X in every loop of the above branch-and-bound algorithm.

Upper bounding is accomplished by the following heuristics that aims at finding a feasible point to (12.7). The input to the heuristics consists of the x_l-parts $\tilde{x}_l$ of optimal solutions to the above single-scenario problems for optimal or nearly optimal λ.

Algorithm 2

STEP 1:
Understand $\tilde{x}_l$, $l = 1, \ldots, L$, *as proposals for* x *and pick a "reasonable candidate"* $\bar{x}$, *for instance, one arising most frequently, or one with minimal* $\mathcal{L}_l(x_l, \theta_l, \lambda)$, *or average the* $\tilde{x}_l$, $l = 1, \ldots, L$, *and round to integers if necessary.*

STEP 2:
Solve for each $l = 1, \ldots, L$*:*

$$\min \left\{ \sum_{k=1}^{K} \theta_{lk} \; : \begin{array}{ll} c^\top \bar{x} + q^\top y_{lk} - a_k & \leq \mathrm{M}\theta_{lk} \\ T\bar{x} + W y_{lk} & = z_l \\ y_{lk} \in Y, \theta_{lk} \in \{0,1\}, \; k = 1, \ldots, K & \end{array} \right\}.$$

STEP 3:
Check whether the θ_{lk} *found in step 2 fulfill*

$$\sum_{l=1}^{L} \pi_l \theta_{lk} \leq \bar{a}_k, \qquad k = 1, \ldots, K.$$

If so, then a feasible solution to (12.7) is found. The heuristics stops with the upper bound $g^\top \bar{x}$*. Otherwise, the heuristics stops without a feasible solution to (12.7) and assigns the formal upper bound* $+\infty$*.*

This completes the description of the algorithm for the first-order model. For the second-order model (12.8) an analogous analysis applies. The model even is slightly more accessible, since the dominance relation itself does not lead to further Boolean variables, as is the case with $\theta_{lk} \in \{0, 1\}$ for the first-order model (12.7). Regarding algorithmic performance the main difference is in the single-scenario subproblems. Here, (12.13) involves the Boolean variables $\theta_{lk} \in \{0, 1\}$ where its second-order counterpart has continuous variables $v_{lk} \geq 0$ (for fixed l in both cases).

As will be seen in the next section, models (12.7) and (12.8) may become huge mixed-integer linear programs, too huge even for most advanced general-purpose mixed-integer linear programming software such as ILOG-Cplex (2005). The outlined branch-and-bound algorithm with its decomposition features in the bounding parts provides a viable alternative in that it is able to solve problem instances to optimality where the mentioned standard software not even provides feasible solutions.

12.5 Numerical Experiments with a Dispersed Generation System

We illustrate the computational performance of the methodologies developed above at the optimization of operation planning in a dispersed generation system under load uncertainty.

The system consists of 5 engine-based cogeneration (CG) stations, producing power and heat simultaneously, 12 wind turbines, and 1 hydroelectric power plant. The CG stations include eight gas boilers, nine gas motors, and one gas turbine, and each is equipped with a thermal storage and a cooling device. While the heat is distributed locally, the electricity is fed into the global distribution network. The cost minimal operation of this system (over a time horizon of 24 h discretized into quarter-hourly intervals) such that load is met and relevant technical constraints are fulfilled can be formulated as a model fitting into (12.1). This random mixed-integer linear program has about 17,500 variables (9000 Boolean and 8500 continuous) and 22,000 constraints; see Handschin et al. (2006) and Neise (2008) for detailed model descriptions.

The problem dimensions of the resulting first-order stochastic dominance model (12.7) for four benchmark scenarios and different numbers of data scenarios are displayed in Table 12.1.

Tables 12.2, 12.3, and 12.4 report comparative numerical experiments for first-order models (12.7) with our implementation ddsip.vSD of the decomposition method from Section 12.4 and the standard solver ILOG-Cplex (2005). Computations were done on a Linux-PC with a 3.2 GHz pentium processor and 2 GB ram.

We report instances with $L = 10, 30, 50$ data and $K = 4$ benchmark scenarios. In all tables, the benchmark costs increase successively from Instance 1 on. So the dominance constraints become easier to fulfill. Recall that the objective function $g^{\top}x$ counts the number of start-ups. It is nicely seen how optimal values decrease in each table from Instance 1 on.

In each table, the status of the optimization for different points in time is displayed. Usually the first two points show when either the decomposition method or Ilog-Cplex find the first feasible solution (values in UB – upper bound – columns). Also for the timelimit of 8 h (28,800 s) the objective values and the best lower bounds (LB) are given for each solver, unless optimality was proven earlier.

Advantages of the decomposition algorithm are evident, in particular as the number of scenarios is getting more substantial ($L = 30, 50$). The standard solver then no longer could find feasible points (upper bounds) and ran out of memory

Table 12.1 Problem dimensions of the first-order model for $K = 4$

Number of	10 scenarios	30 scenarios	50 scenarios
Boolean variables	299,159	894,439	1,489,719
Continuous variables	283,013	846,213	1,409,413
Constraints	742,648	2,220,488	3,698,328

Table 12.2 Results for $L = 10$ data and $K = 4$ benchmark scenarios

		Ilog-Cplex		ddsip.vSD	
Inst.	Time (s)	UB	LB	UB	LB
1	430.43	–	29	29	15
	899.16	–	29	29	29
	15,325.75	29	29	29	29
2	192.48	–	27	28	15
	418.90	28	28	28	15
	802.94	28	28	28	28
3	144.63	–	21	21	12
	428.61	21	21	21	18
	678.79	21	21	21	21
4	164.34	–	11	13	10
	818.26	–	12	13	13
	28,800.00	13	12	13	13
5	171.52	–	7	8	8
	3304.02	8	8	8	8

Table 12.3 Results for $L = 30$ data and $K = 4$ benchmark scenarios

		Ilog-Cplex		ddsip.vSD	
Inst.	Time (s)	UB	LB	UB	LB
1	473.27	–	28	29	12
	1658.02	–	29	29	29
	3255.99	–	29 m.	29	29
2	1001.53	–	26	28	18
	2694.93	–	27	28	28
	3372.24	–	27 m.	28	28
3	469.93	–	17	23	10
	3681.15	–	18 m.	21	20
	28,800.00	–	–	21	20
4	618.21	–	10	14	8
	3095.02	–	11 m.	14	10
	28,800.00	–	–	14	13
5	672.73	–	7	8	8
	8504.88	–	8 m.	8	8

(marked by "m."). For the instances with 50 data scenarios, the standard solver not even could supply lower bounds since the problem instances of (12.7) are so large that the available memory is insufficient to build up the model (lp-) file for Ilog-Cplex. It can also be seen that the lower bounds provided by the decomposition method are pretty weak in early stages of the computation. This is due to the rather harsh relaxation of nonanticipativity mentioned in Section 12.4.

Compared to the above results for the first-order model (12.7), computations with the second-order model (12.8) show a similar tendency; see Gollmer et al. (2010) and Gotzes (2009) for a detailed exposition. The model sizes are comparable to those in Table 12.1. For instances with 10 data scenarios the standard solver still is competitive, but for instances with 30 and 50 scenarios it runs into the same trouble

Table 12.4 Results for $L = 50$ data and $K = 4$ benchmark scenarios

		Ilog-Cplex		ddsip.vSD	
Inst.	Time (s)	UB	LB	UB	LB
1	745.87	–	–	29	11
	2534.21	–	–	29	29
2	1549.22	–	–	28	18
	4168.89	–	–	28	28
3	756.06	–	–	23	10
	28800.00	–	–	21	20
4	975.20	–	–	15	8
	28800.00	–	–	13	12
5	1150.95	–	–	8	8

as above. The decomposition method, on the other hand, is able to solve these instances to optimality within computing times similar to those reported above.

Another application of our stochastic dominance models is a two-stage investment planning problem for electricity generation as described in the Introduction and at the end of Section 12.2, see also Louveaux and Smeers (1988) for an earlier multistage model that provided some inspiration.

Here the dimensions of the initial random optimization model (12.1) are smaller such that computations with up to $L = 500$ data scenarios and $K = 20$ benchmark scenarios were possible. The biggest instances of (12.7) then have up to 1 million variables and 300,000 constraints. Similar to the results for the operation planning instances, the standard solver has trouble with finding feasible points. The decomposition method performs better in that it always provides feasible points and yields lower bounds that are tighter than the ones obtained with the standard solver. Further details can be found in Gollmer et al. (2010) and Gotzes (2009).

12.6 Cutting Plane Decomposition Algorithms

A major structural complication in the models discussed in Sections 12.2 and 12.3 is due to the integer requirements for the second-stage decisions y. As a consequence, the second-stage value function $\Phi(x, z(\omega))$ in (12.2) may become discontinuous in x. Algorithmic options are quite restricted then. In particular, powerful concepts and instruments such as convexity or duality cannot be employed.

The situation improves considerably if all second-stage decisions y in (12.1) are real valued, i.e., if Y is a polyhedron. Under mild conditions, see Birge and Louveaux (1997), Ruszczyński and Shapiro (2003), and Shapiro et al. (2009), the value function Φ then is convex in x, enabling powerful cutting plane algorithms.

In the energy sector, the assumption that second-stage variables are continuous may be restrictive in that it excludes second-stage switching of units, to name just one prominent feature. On the other hand, there are important problems where continuity of second-stage variables is less debatable.

As an example let us consider the retailer problem addressed in Carrión et al. (2009): An electricity retailer has to make forward contracting purchases and fix selling prices offered to potential clients at the beginning of the year. In the course of the year, the retailer supplies the customers' demands by either resorting to the forward contracts or by participating in a pool market. Selling to the pool excessive energy previously contracted is an additional option. At the beginning of the year, future pool prices and client demands are available to the retailer in terms of probability distributions derived from statistical data. The selling prices the retailer offers to potential customers shall be as attractive (= low) as possible, provided the retailer's profit "compares favourably to a pre-specified benchmark".

This corresponds to a dominance-constrained two-stage stochastic program as in (12.6) if the text in quotes is made mathematically rigorous by a benchmark probability distribution and a dominance constraint for the comparison. While first-stage variables usually include integralities (commitments to contracts), the second-stage decisions (mainly pool transactions) usually do not.

The dominance constraint says that the profit random variable dominates to second order a random benchmark. Selecting the benchmark increasingly risk averse, i.e., reducing or even forbidding losses, or decreasing dispersion, increases selling prices and shifts bigger and bigger portions of electricity from the pool market to the forward contracts. By shaping the benchmark the retailer can explore the trade-off between profit aspiration and the need to have attractive selling prices for customers; see Carrión et al. (2009) for further details.

To start the more formal discussion let us consider the following counterparts to (12.1)

$$\min\left\{c^\top x + q^\top y \;:\; Tx + Wy \geq z(\omega),\; x \in X,\; y \in I\!R_+^{m_2}\right\},$$

and (12.2)

$$\begin{aligned}
&\min_x \left\{ c^\top x + \underbrace{\min_y \{q^\top y \;:\; Wy \geq z(\omega) - Tx,\; y \geq 0\}}_{\Phi(x,z(\omega))} \;:\; x \in X \right\} \\
&= \min_x \{c^\top x + \Phi(x, z(\omega)) \;:\; x \in X\}.
\end{aligned}$$

With $f(x, z(\omega)) := c^\top x + \Phi(x, z(\omega))$, the dominance-constrained stochastic program (12.6) transfers verbatim:

$$\min\{g^\top x \;:\; f(x, z(\omega)) \preceq_1 a(\omega),\; x \in X\}$$

where we have confined ourselves to the first-order model, i.e.,

$$\min\left\{g^\top x \ : \ x \in X, \ \mu\left(\{z \ : \ f(x,z) \leq \eta\}\right) \geq \nu\left(\{a \ : \ a \leq \eta\}\right) \forall \eta \in I\!R\right\} \tag{12.14}$$

with μ, ν again denoting the probability distributions of $z(\omega) \in I\!R^s$, $a(\omega) \in I\!R$.

Now Φ is given by a linear program which enables the utilization of linear programming duality. This leads to the following equivalence (Drapkin and Schultz 2010):

$$f(x,z) = c^\top x + \Phi(x,z) \quad \leq \quad \eta$$

if and only if

$$(z - Tx)^\top \delta_i \ + \ (c^\top x - \eta)\delta_{io} \quad \leq \quad 0 \quad \text{for all} \quad i = 1, \ldots, I. \tag{12.15}$$

where $(\delta_i, \delta_{io}) \in I\!R^{s+1}, i = 1, \ldots, I$, denote the vertices of

$$\Delta = \left\{(u, u_o) \in I\!R^{s+1} \ : \ 0 \leq u \leq \mathbf{1}, \ 0 \leq u_o \leq 1, \ W^\top u - q u_o \leq 0\right\}.$$

Algorithmically, it will be crucial that Δ is the constraint set of the linear programming dual to the following feasibility problem:

$$\min_{y,\tau,\tau_o}\left\{\mathbf{1}^\top \tau + \tau_o \ : \ Wy + \tau \geq z - Tx, \ c^\top x + q^\top y - \tau_o \leq \eta, \ y \geq 0, \ \tau \geq 0, \ \tau_o \geq 0\right\}$$

whose optimal value is zero if and only if $c^\top x + \Phi(x,z) \leq \eta$.

As in Section 12.3, let $z(\omega)$ and $a(\omega)$ follow discrete distributions with realizations z_l, $l = 1, \ldots, L$, and a_k, $k = 1, \ldots, K$, as well as probabilities π_l, $l = 1, \ldots, L$, and p_k, $k = 1, \ldots, K$, respectively. Then problem (12.14) is equivalent to the following mixed-integer linear program (Drapkin and Schultz 2010),

$$\min\left\{\begin{array}{lll} g^\top x \ : \ (z_l - Tx)^\top \delta_i \ + \ (c^\top x - a_k)\delta_{io} \leq \mathsf{M}\theta_{lk} & \forall l \ \forall k \ \forall i \\ \sum_{l=1}^{L} \pi_l \theta_{lk} \leq \bar{a}_k & \forall k \\ x \in X, \ \theta_{lk} \in \{0,1\} & \forall l \ \forall k \end{array}\right\}, \tag{12.16}$$

where $\bar{a}_k := 1 - \nu[a \leq a_k]$, $k = 1, \ldots, K$, and M denotes a sufficiently big constant.

Compared with (12.7), the above model has $L \times K \times dim\ y$ variables less which may be quite a number. On the other hand, the first group of constraints in (12.16) is huge in general, since it may involve an exponential number of vertices of Δ. From the stochastic programming point of view, however, the latter is less dramatic. In analogy to traditional stochastic programming with linear recourse (Birge and Louveaux 1997, Ruszczyński and Shapiro 2003; Shapiro et al. 2009), inequalities

(12.15) can be seen as feasibility cuts. Adapting the principle of L-shaped decomposition (Van Slyke and Wets 1969), the numerical solution of (12.16) is accomplished by generating only those feasibility cuts which are needed for the progress of the algorithm.

Indeed, the constraints

$$(z_l - Tx)^\top \delta_i + (c^\top x - a_k)\delta_{io} \le \mathsf{M}\theta_{lk} \quad \forall l \ \forall k \ \forall i \tag{12.17}$$

are given only implicitly, since the vertices $(\delta_i, \delta_{io}), i = 1, \ldots, I$, of Δ are not available a priorily. Therefore, the algorithm works with master problems

$$\left.\begin{aligned} \min \{g^\top x : (z_l - Tx)^\top \delta_i + (c^\top x - a_k)\delta_{io} &\le \mathsf{M}\theta_{lk} \ (l,k,i) \in \mathcal{I}_n \\ \textstyle\sum_{l=1}^L \pi_l \theta_{lk} &\le \bar{a}_k \qquad \forall k \\ x \in X, \ \theta_{lk} \in \{0,1\} & \qquad\qquad \forall l \ \forall k \end{aligned}\right\}, \tag{12.18}$$

where $\mathcal{I}_n \subseteq \{(l,k,i) : \forall l \ \forall k \ \forall i\}$. Unless optimality has been reached, violated cuts, meaning inequalities from (12.17), are added to the current master problem in each loop. These cuts are derived via optimal dual solutions to subproblems

$$\left.\begin{aligned} \min_{y,\tau,\tau_o} \{\mathbf{1}^\top \tau + \tau_o : Wy + \tau &\ge z_l - Tx, \\ c^\top x + q^\top y - \tau_o &\le a_k, \\ y \ge 0, \ \tau \ge 0, \ \tau_o &\ge 0 \end{aligned}\right\}. \tag{12.19}$$

Altogether, we obtain the following cutting plane decomposition method, whose correctness is shown in Drapkin and Schultz (2010).

Algorithm 3

STEP 1 (INITIALIZATION):
Set $n := 0$ *and* $\mathcal{I}_n := \emptyset$.

STEP 2 (MASTER PROBLEM):
Solve the current master problem (12.18) yielding an optimal solution (x^n, θ^n).

STEP 3 (SUBPROBLEMS):
Solve, with $x := x^n$, *subproblems (12.19) for all* (l,k) *such that* $\theta^n_{lk} = 0$.
If all these subproblems have optimal value zero, then stop, x^n *is optimal for (12.16).*
If some of these subproblems, say for $(l,k) \in \mathcal{J}_n$, *have optimal value greater than zero, then the optimal solutions to their duals yield a number of vertices* $(\delta_i, \delta_{io}), i \in \mathcal{V}_n$, *of* Δ. *The resulting cuts from (12.17) with* $(l,k,i) \in \hat{\mathcal{I}}_n \subseteq \mathcal{J}_n \times \mathcal{V}_n$ *are added to the master problem. Set* $n := n+1$ *and* $\mathcal{I}_{n+1} := \mathcal{I}_n \cup \hat{\mathcal{I}}_n$; *go to step 2.*

Table 12.5 Dimensions of mixed-integer linear programming equivalents (12.7)

K	L	Boolean variables	Continuous variables	constraints
10	20	200	145,801	81,411
	50	500	364,501	203,511
	100	1000	729,001	407,011
	200	2000	1,458,001	814,011
	300	3000	2,187,001	1,221,011
	500	5000	3,645,001	2,035,011

Table 12.6 CPU times in seconds for solving the problems to optimality

K	L	Ilog-Cplex	Algorithm 3	Algorithm 1
10	20	217.24	40.85	187.47
	50	1658.21	153.70	728.79
	100	6624.46	303.68	2281.94
	200	29,570.15	731.51	5378.27
	300	63,861.25	954.19	9668.09
	500	–	1665.75	13,324.63

The method achieves solution of the full problem (12.16) by solving tractable mixed-integer linear master problems which are updated by the addition of cuts. For the generation of the latter, suitable linear programs (12.19) corresponding to individual realizations of the underlying probability distributions must be solved. Here, decomposition becomes effective.

This is much more efficient than solving with standard software such as ILOG-Cplex (2005) the mixed-integer linear programming equivalent (12.7) which of course is valid for continuous second-stage variables y, too.

It also turns out more efficient than solving (12.7) with the suitable specification of Algorithm 1. These principal rankings are illustrated in Tables 12.5 and 12.6 reporting sizes and scaling of runtime with problem size for academic test instances without direct relation to real-life industrial problems.

Cutting plane methods in general are receiving increased attention in current research on dominance-constrained stochastic programs. This includes second-order dominance models as well; see Fábián et al. (2010), Noyan and Ruszczyński (2008), and Rudolf and Ruszczyński (2008) for recent contributions.

12.7 Conclusions

Stochastic programs with dominance constraints, as pioneered in Dentcheva and Ruszczyński (2003, 2004), open up new flexibility for addressing risk aversion in an optimization context. We have incorporated stochastic dominance into the framework of two-stage stochastic integer programming.

Within the latter, numerous decision problems in power optimization under uncertainty can be captured. This begins with the design and operation of dispersed generation systems, continues with energy planning and trading, and goes up to complex models crossing over technological, economical, and environmental aspects.

For finite probability spaces, dominance-constrained stochastic integer programs are becoming huge mixed-integer linear programs, quickly exceeding the capabilities of even the most advanced general-purpose solvers. To overcome this, we propose decomposition methods for these models.

If there are integer variables in both stages of the stochastic program, then Lagrangean relaxation of dominance-related constraints together with branch-and-bound in the space of first-stage variables can be developed into a decomposition algorithm.

Without integer variables in the second stage, this method is enhanced by appealing to structural similarities with L-shaped or regularized decomposition (Ruszczyński and Shapiro 2003; Van Slyke and Wets 1969). This leads to a Benders-type cutting plane decomposition algorithm.

Numerical experiments confirm the superiority of both decomposition algorithms over direct application of general-purpose mixed-integer linear programming solvers. Real-life models from power optimization are brought into the reach of practical computation this way.

References

J.R. Birge and F.V. Louveaux. *Introduction to Stochastic Programming*. Springer, New York, NY, 1997.

M. Carrión, U. Gotzes, and R. Schultz. Risk aversion for an electricity retailer with second-order stochastic dominance constraints. *Computational Management Science*, 6:233Ű–250, 2009.

D. Dentcheva and A. Ruszczyński. Optimization with stochastic dominance constraints. *SIAM Journal on Optimization*, 14:548–566, 2003.

D. Dentcheva and A. Ruszczyński. Optimality and duality theory for stochastic optimization problems with nonlinear dominance constraints. *Mathematical Programming*, 99:329–350, 2004.

D. Drapkin and R. Schultz. An algorithm for stochastic programs with first-order dominance constraints induced by linear recourse. *Discrete Applied Mathematics*, 158/4:291–297, 2010.

C.I. Fábián, G. Mitra, and D. Roman. Processing second-order stochastic dominance models using cutting-plane representations. *Mathematical Programming*, to appear, DOI:10.1007/s10107-009-0326-1, 2010.

P.C. Fishburn. *Utility Theory for Decision Making*. Wiley, New York, NY, 1970.

R. Gollmer, F. Neise, and R. Schultz. Stochastic programs with first-order dominance constraints induced by mixed-integer linear recourse. *SIAM Journal on Optimization*, 19:552–571, 2008.

R. Gollmer, U. Gotzes, and R. Schultz. A note on second-order stochastic dominance constraints induced by mixed-integer linear recourse.*Mathematical Programming*, 126:179–190, 2011.

U. Gotzes. *Decision Making with Dominance Constraints in Two-Stage Stochastic Integer Programming*. Vieweg+Teubner, Wiesbaden, 2009.

E. Handschin, F. Neise, H. Neumann, and R. Schultz. Optimal operation of dispersed generation under uncertainty using mathematical programming. *International Journal of Electrical Power and Energy Systems*, 28:618–626, 2006.

C. Helmberg and K.C. Kiwiel. A spectral bundle method with bounds. *Mathematical Programming*, 93:173–194, 2002.

ILOG. CPLEX Callable Library 9.1.3, 2005.

F.V. Louveaux and R. Schultz. Stochastic integer programming. In A. Ruszczyński and A. Shapiro, editors, *Handbooks in Operations Research and Management Science, 10: Stochastic Programming*, pages 213–266. Elsevier, Amsterdam, 2003.

F.V. Louveaux and Y. Smeers. Optimal investments for electricity generation: A stochastic model and a test problem. In Yu. Ermoliev and J.-B. Wets, editors, *Numerical Techniques for Stochastic Optimization*, pages 445–453. Springer, Berlin, 1988.

A. Märkert and R. Schultz. On deviation measures in stochastic integer programming. *Operations Research Letters*, 33:441–449, 2005.

A. Müller and D. Stoyan. *Comparison Methods for Stochastic Models and Risks*. Wiley, Chichester, 2002.

F. Neise. *Risk Management in Stochastic Integer Programming, with Application to Dispersed Power Generation*. Vieweg+Teubner, Wiesbaden, 2008.

N. Noyan and A. Ruszczyński. Valid inequalities and restrictions for stochastic programming problems with first order stochastic dominance constraints. *Mathematical Programming*, 114: 249–275, 2008.

N. Noyan, G. Rudolf, and A. Ruszczyński. Relaxations of linear programming problems with first order stochastic dominance constraints. *Operations Research Letters*, 34:653–659, 2006.

G. Rudolf and A. Ruszczyński. Optimization problems with second order stochastic dominance constraints: Duality, compact formulations, and cut generation methods. *SIAM Journal on Optimization*, 19:1326–1343, 2008.

A. Ruszczyński and A. Shapiro, editors. *Handbooks in Operations Research and Management Science, 10: Stochastic Programming*. Elsevier, Amsterdam, 2003.

R. Schultz. Stochastic programming with integer variables. *Mathematical Programming*, 97: 285–309, 2003.

R. Schultz and F. Neise. Algorithms for mean-risk stochastic integer programs in energy. *Revista Investigación Operacional*, 28:4–16, 2007.

R. Schultz and S. Tiedemann. Risk aversion via excess probabilities in stochastic programs with mixed-integer recourse. *SIAM Journal on Optimization*, 14:115–138, 2003.

R. Schultz and S. Tiedemann. Conditional value-at-risk in stochastic programs with mixed-integer recourse. *Mathematical Programming*, 105:365–386, 2006.

A. Shapiro, D. Dentcheva, and A. Ruszczyński. *Lectures on Stochastic Programming, Modeling and Theory*. MPS-SIAM, Philadelphia, PA, 2009.

R. Van Slyke and R.J-B. Wets. L-shaped linear programs with applications to optimal control and stochastic programming. *SIAM Journal on Applied Mathematics*, 17:638–663, 1969.

Chapter 13
Stochastic Equilibrium Models for Generation Capacity Expansion

Andreas Ehrenmann and Yves Smeers

Abstract Capacity expansion models in the power sector were among the first applications of operations research to the industry. The models lost some of their appeal at the inception of restructuring even though they still offer a lot of possibilities and are in many respect irreplaceable provided they are adapted to the new environment. We introduce stochastic equilibrium versions of these models that we believe provide a relevant context for looking at the current very risky market where the power industry invests and operates. We then take up different questions raised by the new environment. Some are due to developments of the industry like demand side management: an optimization framework has difficulties accommodating them but the more general equilibrium paradigm offers additional possibilities. We then look at the insertion of risk-related investment practices that developed with the new environment and may not be easy to accommodate in an optimization context. Specifically we consider the use of plant-specific discount rates that we derive by including stochastic discount rates in the equilibrium model. Linear discount factors only price systematic risk. We therefore complete the discussion by inserting different risk functions (for different agents) in order to account for additional unpriced idiosyncratic risk in investments. These different models can be cast in a single mathematical representation but they do not have the same mathematical properties. We illustrate the impact of these phenomena on a small but realistic example.

Keywords Capacity adequacy · Risk functions · Stochastic equilibrium models · Stochastic discount factors

13.1 Introduction

The restructuring of the electricity industry led to an explosion of literature transposing and extending optimization models of short-term operations to equilibrium models of electricity markets with given infrastructure. The optimal dispatch and its

---

Disclaimer The views and opinions expressed herein do not necessarily state or reflect those of GDF Suez

A. Ehrenmann (✉)
Center of Expertise in Economic Modeling and Studies, GDF SUEZ,
Brussels, Belgium
e-mail: andreas.ehrenmann@gdfsuez.com

M. Bertocchi et al. (eds.), *Stochastic Optimization Methods in Finance and Energy*,
International Series in Operations Research & Management Science 163,
DOI 10.1007/978-1-4419-9586-5_13,

extension to the optimal power flow are the reference optimization paradigms at the origin of that literature. They cover energy and transmission and sometimes encompass other services such as reserve. These models were instrumental in analyzing market design. Variations of these models that encompass market power were also extensively developed to examine market structure.

Capacity expansion models are as old as the optimal dispatch models but the transition from optimization to equilibrium models has not yet taken place. The early optimization models of capacity expansion go back to the late 1950s when the industry was still regulated (Morlat and Bessière 1971). The problem was first formulated as a linear program but further developments quickly followed suit and extensions covered all types of optimization techniques. Capacity expansion, which was initially seen as a true planning exercise was easily reinterpreted in terms of equilibrium in a competitive energy economy in the early 1970s after the first energy crisis. The power industry of the 1970s was still regulated on a cost plus basis that largely protected it from risk. Deterministic models were thus satisfactory in the situation of the time. Restructuring removed that protection at the same time that various new policies and external events dramatically increased the risk surrounding the electricity sector. This emergence of risk in the investment process strongly suggests to move the analysis from a deterministic to a stochastic environment. The question is thus to transpose former optimization capacity expansion models to stochastic equilibrium models. This extension is the subject of this chapter.

The first analysis of a capacity expansion problem in terms of a stochastic equilibrium capacity expansion model in the energy area is probably found in Haurie et al. (1988). The model deals with gas developments and was formulated as an open-loop Cournot equilibrium under demand uncertainty. This model could be converted to an optimization model that was later used in Gürkan et al. (1999) to illustrate the method of "sample path" since elaborated by several authors. Lin and Fukushima (2009) recently reviewed different models of stochastic equilibrium, among them the one used by Gürkan et al. (1999) in their application of sample path to the investments in gas production. This model is stated as a stochastic variational inequality problem; we adopt the closely related formulation of stochastic complementarity problems as the modeling paradigm of the investment problem throughout this chapter.

Section 13.2 of the chapter introduces a very simple and standard two-stage version of a stochastic optimization capacity expansion model as could have been constructed in the regulated environment. We adopt a standard stochastic programming approach and present the model in terms of its first and second stages. We then immediately reformulate this problem in the stochastic equilibrium format that drives the whole chapter. Section 13.3 discusses the possibilities and limitations of stochastic equilibrium models to account for idiosyncrasies of restructured electricity markets.

The rest of the chapter analyzes different risk issues encountered in the investment process. The standard approach in investment problems is to reflect risk in the discount rate. The discount rate is normally regulated when the industry operates as a monopoly; this may have raised economic controversies but did not create modeling difficulties as the discount rate is just a single parameter of the model. The problem is quite different in a world where "project finance" drives the capacity expansion process and requires that plants are evaluated on the basis of different discount rates. The CAPM and the APT are the reference theories for finding these discount rates. Expositions of these theories can be found in any textbook of corporate finance and we take them for granted. The adoption of a project finance approach therefore requires the stochastic equilibrium model to accommodate plant-specific discount rates while maintaining the interpretation of a competitive economy that is the justification of the model. A first treatment of the question is given in Section 13.4 leading to a fixed point formulation. Section 13.5 adopts an alternative, probably more rigorous but also less usual representation of risk. Starting again from a CAPM-based formulation it assumes that the different risks affecting plants can be taken care of by modifying the payoff of the different plants using a linear stochastic discount rate. Discounting is then conducted at the risk-free rate but with risk-adjusted cash flows computed with CAPM-based stochastic discount rates. Section 13.6 considers an alternative version of the risk-neutral discounting where the adjustment to the cash flow is derived from risk functions. Risk functions were initially developed by Artzner et al. (1999) and have been recently cast in an optimization context (see the book by Shapiro et al. (2009) for a comprehensive treatment). We extend this view to an equilibrium context to construct alternative adjustments of the cash flows of the plants. We provide a simplified but realistic illustration of these notions in Section 13.7. Conclusion summarizes the chapter. In order to simplify the presentation the discussion is entirely conducted on a two- or three-stage models depending on our needs.

13.2 The Basic Capacity Expansion Model

13.2.1 The Optimization Model

Consider a two-stage setup where one invests in a mix of new technologies in stage 0 and operates them in stage 1. The objective is to satisfy a time-segmented, price-insensitive demand so as to minimize total (annual in this simple case) cost. The first versions of these models go back to the late 1950s. They were initially formulated as linear programs and later expanded to take advantage of essentially all optimization techniques. We introduce these models as follows.

Consider a set of capacity types K and a load duration curve decomposed in different time segments L as depicted in Fig. 13.1. The left figure gives a general decomposition and characterizes each time segment by its duration $\tau(\ell)$ and demand level $d(\ell)$. The right figure depicts the particular case of a decomposition into peak and off-peak segments.

Assume in order to simplify the presentation that there is no existing capacity. We introduce the following notation: $x(k)$ is the investment in capacity (in MW) of

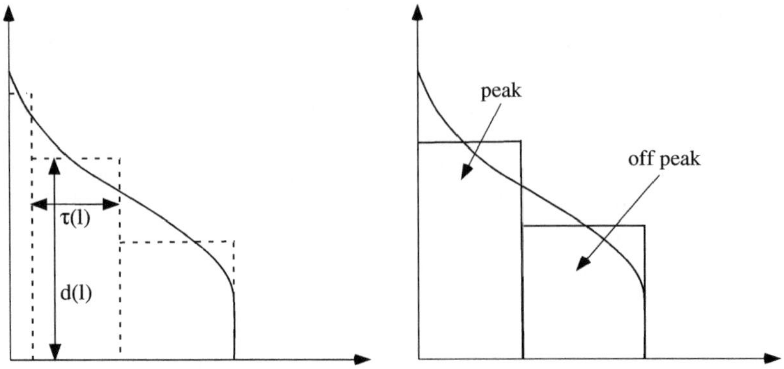

Fig. 13.1 Decomposition of the load–duration curve

technology $k \in K$ in the period; this capacity is operated at level $y(k, \ell)$ (in MW) in time segment ℓ of duration $\tau(\ell)$ (in hours) when demand level is $d(\ell)$ (in MW). $\sum_{\ell} \tau(\ell) = 8760$ is the number of hours in a year. The annual investment cost of technology k is $I(k)$ (in €/MW) and its (constant) marginal operating cost is $c(k)$ (in €/MWh). PC is the value of lost load (VOLL), that is, the economic value (in €/MWh) of unsatisfied electricity demand. The notion has been around since several decades but has so far escaped any precise evaluation. VOLL is thus often taken as a conventional value ranging between 1000 and 10,000 €/MWh. PC can alternatively be interpreted as a price cap, that is, an upper bound on the electricity price set by the regulator (figures ranging between 300 and 3000 €/MWh are then found in practice); $z(\ell)$ is the unsatisfied demand in time segment ℓ (in MW).

Adopting the standard two-stage approach of stochastic programming, we successively write the second- and first-stage optimization problems (dual variables are stated at the right of the equations) as follows. The (short-term) operations problem is stated as the following short-term cost minimization problem

$$Q(x) \equiv \min_{y,z} \sum_{\ell \in L} \tau(\ell) \left[\sum_{k \in K} c(k)\, y(k, \ell) + PC\, z(\ell) \right], \tag{13.1}$$

s.t.

$$0 \le x(k) - y(k, \ell), \qquad \tau(\ell)\mu(k, \ell), \tag{13.2}$$

$$0 \le \sum_{k \in K} y(k, \ell) + z(\ell) - d(\ell), \qquad \tau(\ell)\pi(\ell), \tag{13.3}$$

$$0 \le y(k, \ell). \tag{13.4}$$

The (long-term) investment problem is stated as a long-term cost minimization problem

$$\min_{x \geq 0} \sum_{k \in K} I(k)\, x(k) + Q(x). \tag{13.5}$$

In accordance with the units used for defining the parameters and primal variables, $\mu(k,\ell)$ and $\pi(\ell)$ are in €/MW. It is more convenient to refer to them as €/MWh.

13.2.2 The Equilibrium Version of the Optimization Model

The conversion of the optimization model to an equilibrium model is obtained by writing duality relations. The KKT conditions of the operations problem are stated as

$$0 \leq x(k) - y(k,\ell) \perp \mu(k,\ell) \geq 0, \tag{13.6}$$

$$0 \leq \sum_{k \in K} y(k,\ell) + z(\ell) - d(\ell) \perp \pi(\ell) \geq 0, \tag{13.7}$$

$$0 \leq c(k) + \mu(k,\ell) - \pi(\ell) \perp y(k,\ell) \geq 0, \tag{13.8}$$

$$0 \leq PC - \pi(\ell) \perp z(\ell) \geq 0. \tag{13.9}$$

Those of the investment problem are

$$0 \leq I(k) - \sum_{\ell \in L} \tau(\ell)\, \mu(k,\ell) \perp x(k) \geq 0. \tag{13.10}$$

Complementarity formulations of capacity expansion models have been used by Gürkan et al. (2009) to examine different investment incentive policies in restructured markets. Our focus in this chapter is different; we assume a single investment incentive mechanism throughout the chapter (as embedded in model (13.6) to (13.10) and usually referred to as "energy only"); we then concentrate on the investment cost $I(k)$ that we examine through different theories of corporate finance.

Relations (13.6) to (13.10) can easily be interpreted in terms of a perfect competition equilibrium. For the sake of brevity, we present this discussion in Appendix 1 and only report here the economic interpretation of the dual variables that play a key role in the rest of the discussion. Most important for our purpose, $\mu(k,\ell)$ is the hourly marginal value of capacity k in time segment ℓ. It is zero if the capacity is not fully utilized in that time segment; it is positive and $\tau(\ell)\,\mu(k,\ell)$ measures the hourly marginal decrease of the operating cost of the system in time segment ℓ if one adds a unit of capacity k. The hourly marginal generation cost in time segment ℓ is measured by $\pi(\ell)$. It is the sum of the operating cost $c(k,\ell)$ and of the hourly marginal value $\mu(k,\ell)$ of capacity k when it is operating. The price $\pi(\ell)$ is set to PC when load is curtailed.

The discussion developed in this chapter focuses on the investment criterion (13.10). Its interpretation is that one invests in technology k when the investment cost $I(k)$ is equal to the weighted (by $\tau(\ell)$) sum of the hourly marginal values of the capacity $\mu(k,\ell)$ computed over all time segments. We refer to this weighted sum $\sum_\ell \tau(\ell)\,\mu(k,\ell)$ as the gross margin of plant k and note it in abridged form $\mu(k)$. $\mu(K)$ is the vector of the $\mu(k)$ for $k \in K$. One does not invest in technology k when the gross margin is insufficient to cover $I(k)$. Because of the interactions between the operation and investment stages, $\mu(K)$ is an element of the subdifferential $\partial_x Q(x)$ of the operating cost with respect to x (see Appendix 2.1). This is a point to set mapping from $R_+^{|K|}$ into $R_+^{|K|}$ (Appendix 2.2). We can then restate $\mu(k,\ell)$ as $\mu(x,k,\ell)$ (and $\mu(k)$ as $\mu(x,k)$) in the above investment relation in order to express the dependence of the marginal value of capacity on the amount of capacity. The investment criterion is then written as

$$0 \leq I(k) - \sum_{\ell \in L} \tau(\ell)\,\mu(x,k,\ell) \perp x(k) \geq 0 \tag{13.11}$$

or in abridged form after writing $\mu(x,k) = \sum_{\ell \in L} \tau(\ell), \mu(x,k,\ell)$ and collecting these expressions (13.11) for all k

$$0 \leq I(K) - \mu(x,K) \perp x(K) \geq 0. \tag{13.12}$$

This complementarity relation summarizes the whole capacity investment model: assuming that one knows the mapping $\mu(x,K)$ for a particular problem, the capacity expansion problem is entirely summarized in (13.12). The properties of the mapping $\mu(x,K)$ intervening in this model have important implications on the existence and unicity of equilibria. Because of the expository nature of this work, we only briefly mention these properties in passing and leave their detailed analysis for a further paper. We now extend this model to the stochastic case.

13.2.3 Introducing Risk

The introduction of risk factors generalizes the above model to a stochastic environment. Fuel costs $c(k)$ and demand levels $d(\ell)$ are standard risk factors that immediately come to mind. Regulatory risks are new risks that take a prominent role in the current environment of the firm. They originate both from the regulation of the power sector and from related domains like environmental regulation. The cost PC, when interpreted as a price cap set by the regulator is an illustration of regulatory risk: the regulator sets the price during curtailment. We make no assumption on the dependence or independence of these risk factors. Let ω denote a scenario and Ω be the set of these scenarios. We note $c(k,\omega)$, $d(\ell,\omega)$, $\mathrm{PC}(\omega)$ as the exogenous fuel cost, demand, and PC scenario. We also note $y(k,\ell,\omega)$, $\mu(k,\omega)$, $\pi(\ell,\omega)$ as the endogenous realization of the primal and dual variables of the operation model in that scenario. The formulation of the optimization and complementarity problems extends to this more general problem by assuming risk-neutral generators (investors) that see the same scenarios and share the same beliefs (probability $p(\omega)$) about their occurrence. This is stated as follows.

13.2.3.1 The Stochastic Capacity Expansion Optimization Model

The second-stage operations model in scenario ω becomes

$$Q(x,\omega) \equiv \min_{y,z} \sum_{\ell \in L} \tau(\ell) \left[\sum_{k \in K} c(k,\omega)\, y(k,\ell,\omega) + PC(\omega)\, z(\ell,\omega) \right], \tag{13.13}$$

s.t.

$$0 \leq x(k) - y(k,\ell,\omega), \qquad \tau(\ell)\mu(k,\ell,\omega), \tag{13.14}$$

$$0 \leq \sum_{k \in K} y(k,\ell,\omega) + z(\ell,\omega) - d(\ell,\omega), \quad \tau(\ell)\pi(\ell,\omega), \tag{13.15}$$

$$0 \leq y(k,\ell,\omega). \tag{13.16}$$

The first-stage investment part of the model is stated as

$$\min_{x \geq 0} \sum_{k \in K} I(k)\, x(k) + E_p Q(x,\omega), \tag{13.17}$$

where E_p denotes the expectation under the p measure. For notational convenience, we sometimes also refer to $E_p Q(x,\omega)$ as $Q(x)$.

13.2.3.2 The Stochastic Capacity Expansion Equilibrium Model

As before, the equilibrium model is obtained by writing the KKT conditions of the second- and first-stage problems, respectively. For the second-stage problem, we have

$$0 \leq x(k) - y(k,\ell,\omega) \perp \mu(k,\ell,\omega) \geq 0, \tag{13.18}$$

$$0 \leq \sum_{k \in K} y(k,\ell,\omega) + z(\ell,\omega) - d(\ell,\omega) \perp \pi(\ell,\omega) \geq 0, \tag{13.19}$$

$$0 \leq c(k,\omega) + \mu(k,\ell,\omega) - \pi(\ell,\omega) \perp y(k,\ell,\omega) \geq 0, \tag{13.20}$$

$$0 \leq PC(\omega) - \pi(\ell,\omega) \perp z(\ell,\omega) \geq 0. \tag{13.21}$$

The first-stage KKT conditions are stated as

$$0 \leq I(k) - E_p \sum_{\ell \in L} \tau(\ell)\, \mu(k,\ell,\omega) \perp x(k) \geq 0 \tag{13.22}$$

or after explicitly introducing the dependence of $\mu(k, \ell, \omega)$ on x, and rewriting $\mu(x, k, \omega) = \sum_{\ell \in L} \tau(\ell)\, \mu(x, k, \ell, \omega)$ and collecting all k investment criteria

$$0 \leq I(K) - E_p\, \mu(x, K, \omega) \perp x(K) \geq 0. \tag{13.23}$$

Again we sometimes also refer to $\mu(x, k) = E_p\, \mu(x, k, \omega)$ for the sake of notational convenience. Recall that $\mu(x, K)$ is the $|K|$-dimensional mapping formed by the $\mu(x, k)$. We refer to models of this type (relations (13.18), (13.19), (13.20), (13.21), and (13.22) or relation (13.23)) as the fixed demand model (FDM).

13.2.4 Focusing on the Investment Model

The complementarity condition (13.23) summarizes the whole equilibrium description of capacity expansion in the stochastic case. It takes the form of a stochastic complementarity problem and could easily be extended to a stochastic variational inequality if there were constraints on the x variables. While formulation (13.23) does not tell us anything that we could not have learned from the stochastic optimization model, it allows one to state questions that remain relevant when formulating an equilibrium model without being able to invoke an optimization problem to start with. The fundamental convexity property that underlies a significant part of the theory of stochastic programming carries through here by noting that $-\mu(x, K, \ell)$ and hence $-\mu(x, K)$ is monotone in x (see Appendix 3 for the definition of a monotone operator). The complementarity condition (13.23) resembles the stochastic equilibrium model introduced in Gürkan et al. (1999). Note, however, that the mapping $-\mu(x, K)$ is upper semi-continuous and not continuous as in Gürkan et al. and in the more usual complementarity theory (see Appendix 2.2). The next section discusses variations of this model.

13.3 Alternative Equilibrium Models of Capacity Expansion

13.3.1 Welfare Maximization Equilibrium Models

We noted that $-\mu(x, K)$ is monotone in x but not continuous. This creates technical difficulties if one wants to stick to the complementarity or variational inequality formulations, which are the more natural way to state equilibrium problems. The following introduces a variant of the model that leads to a continuous mapping $-\mu(x, K)$ while at the same time being more in line with standard economic thinking. The common assumption in economic theory is to suppose that demand varies with prices. Even then, the common wisdom in electricity modeling is to admit that the demand of electricity is price insensitive in the short run, as is effectively the case in the short-run operations model (13.18), (13.19), (13.20), and (13.21). This assumption is then commonly but unduly extended to the long run and embedded in the investment model. There are at least two reasons to question that extension. One is that demand, even if non-price sensitive today in the short run, is clearly price sensitive in the long run, which is the natural horizon for a capacity expansion model. Large industrial consumers will not keep their demand unchanged in the long run if they anticipate high electricity prices. Similarly, high electricity prices will induce conservation in household and tertiary and hence reduction of demand. The second reason is that the increasing interest for demand side management and the development of new technologies of the smart grid type will progressively introduce a true price response in the short run. Taking stock of these two reasons we modify the above model to encompass a demand function. We start with the optimization problem that we immediately state in a stochastic formulation.

Let $d(p(\ell), \ell, \omega)$ be the demand function in time segment ℓ and scenario ω. Both the demand and the price in time segment ℓ and scenario ω are one dimensional. Assuming, as in standard economics, that $d(\ell, \omega)$ is monotonically decreasing with $p(\ell, \omega)$, the function $d(p(\ell), \ell, \omega)$ can be inverted into $p(d(\ell), \ell, \omega)$ for each time segment ℓ and scenario ω. We do not know much today about this demand function with several studies giving widely diverging results. It thus makes sense to embed the uncertainty on the demand function in the model, which justifies our introducing a dependence on ω in $d(p(\ell), \ell, \omega)$ and $p(d(\ell), \ell, \omega)$.

The stochastic optimization model of the short-term welfare can then be stated as follows (in minimization form) where MSTW stands for minus short-term welfare:

$$\text{MSTW}(x, \omega) \equiv \min_{y} \sum_{\ell \in L} \tau(\ell) \left[\sum_{k \in K} c(k, \omega)\, y(k, \ell, \omega) - \int_0^{d(\ell,\omega)} p(\xi, \ell, \omega) d\xi \right], \tag{13.24}$$

s.t.

$$0 \leq x(k) - y(k, \ell, \omega), \qquad \tau(\ell)\mu(k, \ell), \tag{13.25}$$

$$0 \leq \sum_{k \in K} y(k, \ell, \omega) - d(\ell, \omega), \qquad \tau(\ell)\pi(\ell), \tag{13.26}$$

$$0 \leq y(k, \ell, \omega). \tag{13.27}$$

The long-term welfare optimization is similarly stated (in minimization form) as

$$\min_{x \geq 0} \sum_{k \in K} I(k)\, x(k) + E_p \text{MSTW}(x, \omega). \tag{13.28}$$

The equilibrium model is derived by writing the KKT conditions of that problem. We obtain for the short-run market

$$0 \leq x(k) - y(k, \ell, \omega) \perp \mu(k, \ell, \omega) \geq 0, \tag{13.29}$$

$$0 \leq \sum_{k \in K} y(k, \ell, \omega) - d(\ell, \omega) \perp \pi(\ell, \omega) \geq 0, \tag{13.30}$$

$$0 \leq c(k, \omega) + \mu(k, \ell, \omega) - \pi(\ell, \omega) \perp y(k, \ell, \omega) \geq 0, \tag{13.31}$$

$$0 \leq \pi(\ell, \omega) - p(d(\ell), \ell, \omega) \perp d(\ell, \omega) \geq 0, \tag{13.32}$$

while the investment criterion becomes

$$0 \leq I(k) - E_p \left[\sum_{\ell \in L} \tau(\ell)\, \mu(k, \ell, \omega) \right] \perp x(k) \geq 0 \tag{13.33}$$

or after introducing the dependence of $\mu(k, \ell, \omega)$ on x and defining $\mu(x, k, \omega) = \sum_{\ell} \tau(\ell)\, \mu(x, k, \ell, \omega)$ and assembling these relations for all k

$$0 \leq I(K) - E_p\, \mu(x, K, \omega) \perp x(K) \geq 0. \tag{13.34}$$

The main difference between (13.34) and (13.23) is in the properties of the mapping μ which is here a monotone continuous point-to-point mapping of the capacity x (see Appendix 2.3). We refer to this model (relations (13.29), (13.30), (13.31), (13.32), and (13.34)) as the variable demand model (VDM).

13.3.2 Optimization Problems That Do Not Extend to Equilibrium Models

The above short-run optimization problems are of the convex type and hence have a well-behaved dual. This allows one to write KKT conditions that can easily be interpreted in terms of perfect competition equilibrium. Not all capacity expansion optimization models have a convex second-stage optimization problem. This is in particular the case when the second stage involves unit commitment features (startup and shutdown of machines, minimum down- and up-time constraints). Considerable effort has been devoted to the analysis of these questions in short-term and hedging models (e.g., Eichhorn et al. 2010; Kuhn and Schultz 2009; Römisch and Vigerske 2010). We are not aware of any attempt to include them in capacity expansion models. Second-stage optimization models that include unit commitment features cannot be converted into complementarity problems and hence in complementarity models of equilibrium. Convexification of these effects such as elaborated in Gribik et al. (2007) could, however, lead to approximate equilibrium models.

13.3.3 Equilibrium Models That Do Not Derive from Optimization Problems

Leaving aside second-stage optimization problems that cannot be exactly converted into equilibrium models, we now consider equilibrium problems that cannot be obtained from KKT conditions of optimization problems. These abound in the electricity restructuring because of the diversity of organizations of the electricity market and its numerous submarkets (e.g., transmission and reserve of different quality). Different arrangements of other related markets such as the EU-ETS also easily lead to short-run equilibrium models that do not derive from optimization (e.g., Ehrenmann and Smeers 2010). In the interest of space we do not engage into that discussion here but simply illustrate our point by presenting an extension of the welfare maximization problem described by equations (13.29), (13.30), (13.31), (13.32), and (13.33) to a model where the demand model cannot be integrated into a willingness to pay function. This problem originates in the PIES model (Ahn and Hogan 1982) that was built by combining a large linear programming model and various heterogenous demand models that could not be integrated into a utility function (see Harker and Pang 1990, for the integrability property).

This variant of the equilibrium model can be stated as follows. Let $D(\omega) = (d(\ell,\omega), \ell \in L)$ and $P(\omega) = (p(\ell,\omega), \ell \in L)$ denote the vectors of demand and price in the different time segments. Smart grid technologies aim, among other goals, at introducing storage possibilities across the different time segments. In other words, demand $d(\ell,\omega)$ in time segment ℓ no longer depends on the sole price $p(\ell,\omega)$ in time segment ℓ, but on the whole price vector $P(\omega)$. The objective is to create at the demand side, storage possibilities that are so difficult to achieve at the generation side. Taking stock of that extension, we write the demand model as

$$D(P,\omega) \equiv \{d(P,\ell,\omega),\ \ell \in L\}.$$

Assuming that the vector function $D(P,\omega)$ can be inverted into a system $P(D,\omega)$

$$P(D,\omega) \equiv \{p(D,\ell,\omega), \ell \in L\}$$

but that $P(D,\omega)$ cannot be integrated into a willingness to pay function (equivalently $P(D,\omega)$ is not a gradient function) we replace the short-run welfare maximization problem by the following equilibrium conditions:

$$0 \le x(k) - y(k,\ell,\omega) \perp \mu(k,\ell,\omega) \ge 0, \tag{13.35}$$

$$0 \le \sum_{k\in K} y(k,\ell,\omega) - d(\ell,\omega) \perp \pi(\ell,\omega) \ge 0, \tag{13.36}$$

$$0 \le c(k,\omega) + \mu(k,\ell,\omega) - \pi(\ell,\omega) \perp y(k,\ell,\omega) \ge 0, \tag{13.37}$$

$$0 \le \pi(\ell,\omega) - p(D,\ell,\omega) \perp d(\ell,\omega) \ge 0. \tag{13.38}$$

These conditions cannot be obtained as KKT conditions of a welfare maximization problem. Still the investment criterion remains

$$0 \leq I(K) - E_p\left[\sum_{\ell \in L} \tau(\ell)\, \mu(K, \ell, \omega)\right] \perp x(K) \geq 0. \tag{13.39}$$

Applying a well-known integrability theorem (e.g., theorem 1.3.1 in Facchinei and Pang 2003), one can show that the model would have been a stochastic optimization problem if the inverted demand system $P(D, \omega)$ were integrable, that is, if it satisfied

$$\frac{\partial p(D, \ell, \omega)}{\partial d(\ell', \omega)} = \frac{\partial p(D, \ell', \omega)}{\partial d(\ell, \omega)} \quad \text{for all } \ell, \ell'.$$

This condition was not satisfied in the PIES model (Ahn and Hogan 1982); it will generally not be satisfied in multiperiod investment models where the demand models is "adapted" (demand in some period depends on past prices but not on future prices (e.g., Wu and Fuller 1996) and hence does not derive from a multiperiod willingness to pay function). Even in single period investment model the demand shifting properties created by smart grids will violate this integrability property. But even though we cannot rely on an optimization model, the investment problem retains the standard complementarity form (13.39) or with the usual short cut

$$0 \leq I(K) - \mu(x, K) \perp x(K) \geq 0. \tag{13.40}$$

Note that any demand system $P(D, \omega)$ that can be written as a gradient function satisfies the integrability property. This is in particular the case of multiperiod perfect foresight demand systems. Problem (13.40) can then be reformulated as an optimization model. The mapping $\mu(x, K)$ is continuous in x. Its monotonicity properties depend on the demand system. We do not discuss the question here (see Aghassi et al. 2006 for a discussion of convexity of asymmetric variational inequality problems).

13.4 Project Finance and Asset-Specific Discounting Rates

Investments in a risky environment require risk-adjusted discount factors. Specifically the two-stage investment model (13.1), (13.2), (13.3), (13.4), and (13.5) of Section 13.2.1 requires converting the total plant investment cost $\mathcal{I}(k)$ into the annual investment cost $I(k)$. This is commonly done, using the standard formula

$$I(k) \sum_{t=1}^{T} \frac{1}{(1 + r(k))^t} = \mathcal{I}(k) \;\; (\text{in €/MW}), \tag{13.41}$$

where T is the assumed life of the plant and $r(k)$ is a discount rate. This annual investment cost can be turned into an hourly capacity cost in €/MWh by dividing by 8760 (the number of hours in a year).

The tradition in the regulated industry is to apply a single discount rate r to all equipment in capacity expansion models of the type described by relations (13.1), (13.2), (13.3), (13.4), and (13.5). This discount rate reflects the risk exposure of the company and is meant to provide the adequate remuneration of investors and lenders. In order to simplify the discussion we assume a full equity financing (no debt) which means that r represents the cost of equity. Cost plus regulation limited the risk bearing on the monopoly company and the regulator accordingly decided the discount rate and the allowed return on capital of the company. The long life of the plants, the slowly evolving nature of the portfolio, and the relatively surprise free evolution of the economy in a still recent past further contributed to reduce risk.

The restructuring of the sector and the introduction of a project finance approach in investment changed this practice. A generation unit is valued on the basis of its own merit, which in particular means on the basis of its own risk exposure. The plant thus has its own risk-adjusted discount factor. A common practice (Graham and Harvey 2001) is to derive these risk factors through a CAPM-based approach, applied this time to each plant separately. The CAPM theory of financial assets is extensively discussed in any textbook of corporate finance; in contrast its application to physical assets is much less elaborated. A comprehensive treatment and appropriate references can be found in Armitage (2005), which points to several drawbacks of the use of the CAPM for project evaluation. Notwithstanding the reservations against the CAPM and its standard applications to physical assets found in the literature, we conduct the discussion in these terms because of the wide application of the method in practice.

The insertion of the project finance view in capacity expansion models raises different questions. Some are due to the multiperiod discounting of the cash flows. For the sake of brevity and in order to stick to our two-stage setup, we do not discuss them here and refer the reader to a companion paper (Ehrenmann and Smeers 2009). Other questions already arise in the two-stage context adopted in this chapter. Suppose that $I(k)$ is determined from the total investment cost $\mathcal{I}(k)$ of equipment k using formula (13.41). Both the fixed demand (FDM: relation: (13.23)) or variable demand models (VDM: relation (13.34)) can be formally implemented by assuming exogenously determined $r(k)$. This does not pose particular computational difficulty in a two-stage model. The question discussed below is whether it is a consistent treatment of risk.

We first note that the short-term equilibrium models (relations (13.18), (13.19), (13.20), and (13.21) for FDM and relations (13.29), (13.30), (13.31), and (13.32) for VDM) are unaffected by the choice of the discount rate. Each machine only appears in these short-term models through its capacity and variable cost; the market therefore finds the short-term equilibrium irrespectively of the investment cost of these units. Things are different in the investment model ((13.23) for FDM and (13.32) for VDM). The analogy (see Armitage (2005), chapter 6, for a treatment that goes beyond the analogy) between financial and physical assets that underlies the application of the CAPM to physical assets suggests the following reasoning. A unitary investment in plant k buys a capacity $1/I(k)$, a unit capacity in that plant generates a payoff

$$\sum_{\ell} \tau(\ell)[\pi(\ell,\omega) - c(k,\ell,\omega)]y(k,\ell,\omega). \tag{13.42}$$

Combining the two relations, the net return on investment in scenario ω of a unit investment in plant k can be written as

$$R(k,\omega) = \frac{\sum_{\ell} \tau(\ell)[\pi(\ell,\omega) - c(k,\ell,\omega)]\, y(k,\ell,\omega)}{I(k)}. \tag{13.43}$$

Note that $R(k,\omega)$ and hence $E_p[R(k,\omega)]$ are results of the model established with given $r(k)$. The question is whether the returns $R(k,\omega)$ are compatible with the discount rate $r(k)$ used for the equipment or in other words whether $r(k) = E_p[R(k,\omega)] - 1$. This requires explicitly invoking the underlying risk theory, which in this section is the usual CAPM.

In order to do so we expand the definition of the scenarios by explicitly introducing the "market" $M(\omega)$ and the return on the market $R(M,\omega)$ in scenario ω. These are additional data that need to be part of the scenarios. Limiting the discussion to the fixed demand model for the sake of brevity, scenario ω therefore encompasses assumptions on $c(k,\omega)$, $d(\ell,\omega)$, $PC(\omega)$, $M(\omega)$, and $R(M,\omega)$. In order to be consistent with the CAPM, $r(k)$ should be consistent with the return $R(k,\omega)$ accruing from an investment in plant k. This implies imposing the standard CAPM formula

$$r(k) = r^f + \frac{\operatorname{cov}(R(k,\omega), R(M,\omega))}{\sigma^2[R(M,\omega)]} E_p[R(M,\omega) - R^f], \tag{13.44}$$

where R^f is the gross risk-free rate (1 + the net risk-free rate r^f). The investment models (13.23) (for FDM) or (13.34) (for VDM) in a CAPM-based environment are then completed by adding the two relations

$$I(k)\sum_{t=1}^{T} \frac{1}{(1+r(k))^t} = \mathcal{I}(k) \text{ (in €/MW)}, \tag{13.45}$$

where $r(k)$ satisfies

$$r(k) = r^f + \frac{\operatorname{cov}(R(k,\omega)\, R(M,\omega))}{\sigma^2[R(M,\omega)]} E_p[R(M,\omega) - R^f]. \tag{13.46}$$

One can immediately see that the addition of these relations destroys the two-stage decomposition of the model by introducing a strong coupling between these stages. It also destroys any monotonicity property. The model can, however, still be posed as a fixed point problem of $r(K)$.

13.5 Linear Stochastic Discount Factors

The above approach requires solving a fixed point problem (a "circularity" problem in Armitage (2005) in order to find the $r(k)$) compatible with the endogenous risk exposure of the different plants. This is cumbersome. An alternative much lighter method introduced in Fama (1977) is to resort to CAPM-based deterministic equivalents of stochastic cash flows. We present this approach in the context of the more general theory of linear stochastic discount factors that embeds not only the CAPM but also the less used arbitrage pricing theory (APT) and the multitemporal consumer theory. As before we assume an expanded definition of scenarios that includes the "market" $M(\omega)$. We refer the reader to Cochrane (2005) or Armitage (2005) for discussions of linear stochastic discount rates and restrict ourselves to the elements of the theory necessary for constructing the stochastic equilibrium models.

13.5.1 A Primer on Linear Stochastic Discount Factors

Consider a two-stage setup where one wants to assess in time 0 a cash flow $X(\omega)$ accruing in time 1. A stochastic discount factor (or price kernel or state price vector) is a vector $m(\omega)$ such that the value of this cash flow in stage 0 is equal to $E_p[m(\omega) \times X(\omega)]$. Both economic and finance offer theories that allow one to construct the price kernel $m(\omega)$. Specifically (e.g., see Cochrane 2005) the CAPM leads to state a stochastic discount rate of the form

$$m(\omega) = a - b \times R(M, \omega),$$

where a and b are determined to satisfy

$$E_p[m(\omega) \,|\, \omega] = \frac{1}{R_f}$$

($m(\omega)$ prices the risk free asset $1(\omega)$ that redeems 1 in all scenarios ω), and

$$E_p[m(\omega) \times R(M, \omega)] = \frac{1}{R_f}$$

($m(\omega)$ prices the "market" $M(\omega)$). Because $R(M, \omega)$ and R_f are data of the problem, a and b and hence the stochastic discount factor are also data to the problem.

The pricing kernel $m(\omega)$ and the payoff $X(\omega)$ do not necessarily span the same space. Let $X = X_m + X_{\perp m}$ be a decomposition of the space of payoffs where X_m is the subspace spanned by $m(\omega)$ (which has a nonzero covariance with $m(\omega)$) and $X_{\perp m}$ the subspace orthogonal to X_m with respect to the scalar product $E_p(x \times y)$.

In standard financial language, $X_m(\omega)$ is the systematic risk embedded in the payoff $X(\omega)$ (the risk priced by $m(\omega)$ or the "market" in CAPM parlance) and $X_{\perp m}(\omega)$ is the idiosyncratic component of $X(\omega)$, that is, the component that is, priced to zero by $m(\omega)$.

We first explain that resorting to linear discount rates allows one to conduct all discounting at the risk-free rate and therefore eliminates the need for finding a risk-adjusted discount rate that reflects the risk exposure of the plants (in particular in order to find the annual value $I(k)$). We thus bypass the need to solve a fixed point problem. In order to conduct the discussion with sufficient generality, we depart from the two-stage paradigm and consider a three-stage setup where one invests in period 0 and collects random payoffs $X^1(\omega_1)$ and $X^2(\omega_1, \omega_2)$ in periods 1 and 2. The stochastic discounting approach extends as follows. Let $m^1(\omega_1)$ and $m^2(\omega_2|\omega_1)$ be the stochastic and conditional stochastic discount factors in stages 1 and 2, respectively. Let also $p^1(\omega_1)$ and $p^2(\omega_2|\omega_1)$ be the probabilities and conditional probabilities of the different states of the world in stages 1 and 2, respectively. Proceeding recursively from stage 2 to 0, the global value in 0 of the two cash flows is equal to

$$E_{p^1}\big[m^1(\omega_1) \times \big[X^1(\omega_1) + E_{p^2(|\omega_1)}\big[m^2(\omega_2|\omega_1) \times X^2(\omega_2|\omega_1)\big]\big]\big].$$

We now show how this expression can be restated in a form that only involves the risk-free discount factor.

The value in stage 1 of payoff $X^2(\omega_2|\omega_1)$ conditional on ω_1 can be written as

$$\begin{aligned}
&E_{p^2(|\omega_1)}[m_2(\omega_2|\omega_1)X^2(\omega_2|\omega_1)] \\
&\quad = \frac{1}{R^f}\left[E_{p^2(|\omega_1)}X^2(\omega_2|\omega_1) + R^f \operatorname{cov}[m_2(\omega_2|\omega_1), X^2(\omega_2|\omega_1)]\right] \\
&\quad = \frac{1}{R^f}\widetilde{X}^2(\omega_1),
\end{aligned}$$

where $\widetilde{X}^2(\omega_1)$ is a deterministic equivalent in period 1 of the conditional cash flow $X^2(\omega_2|\omega_1)$. The payoff accruing in stage 1 and to be assessed in stage 0 is thus

$$X^1(\omega_1) + \frac{1}{R^f}\widetilde{X}^2(\omega_1).$$

Conducting the same reasoning, the valuation in stage 0 of this random cash flow in stage 1 can be written as

$$\frac{1}{R^f}\widetilde{X}^{1,2}.$$

One can now restate these manipulations by expliciting the different contributions accruing in stages 1 and 2. Specifically

$$\widetilde{X}^2(\omega_1) = E_{p^2(|\omega_1)}X^2(\omega_2\,|\,\omega_1) + R^f \operatorname{cov}[m_2(\omega_2\,|\,\omega_1), X^2(\omega_2\,|\,\omega_1)]$$

is the deterministic equivalent in stage 2 of the conditional random cash flow $X_2(\omega_2 \,|\, \omega_1)$. Similarly

$$\widetilde{X}^1(\omega_1) = E_{p^1} X^1(\omega_1) + R^f \operatorname{cov}[m_1(\omega_1), X^1(\omega_1)]$$

is the deterministic equivalent in stage 1 (before discounting to 0 by $1/R^f$) of the random cash flow $X_1(\omega)$. One observes that all discounting can be conducted at the risk-free rate using CAPM-based deterministic equivalent cash flow (see Armitage 2005 for a detailed treatment). This justifies conducting all the valuations at the risk-free rate by working with risk-adjusted cash flows. The approach bypasses the "circularity" of the standard discounting practice, here presented through a fixed point model. This is the approach taken in the next section. It avoids resorting to the annoying fixed point problem and is always compatible with the CAPM theory. In contrast the more standard discounting approach suffers from other circularity aspects arising when projects are also partially financed by debt (Armitage 2005). Last, we shall see that the certainty equivalent cash flow offers a good chance (but no guarantee) of retaining the monotonicity properties of $\mu(x,k)$.

The models of Sections 13.5.1 and 13.5.2 are equivalent in this two-stage context, provided one makes the assumption that the annual investment cost can be obtained by a single $r(k)$ (unconditional CAPM). This equivalence no longer holds in more general multiperiod cases where the certainty equivalent approach (Fama 1977) can always be applied but the risk-adjusted discount rate requires some assumptions. We thus follow the certainly equivalent approach (Fama 1977) that even though less usual does not suffer from the criticism addressed to the standard risk-adjusted discounting.

13.5.2 Optimization and Equilibrium Models

Returning to the reference two-stage setup, we assume that the annuities $I(k)$ of plants k have been computed from the total investment costs $\mathcal{I}(k)$ at the risk-free rate R^f. We again begin with the optimization form of the capacity expansion problem where the second-stage objective function $Q(x,\omega)$ is given by the short-run model (13.1), (13.2), (13.3), and (13.4). Calling upon the notion of stochastic discount factor the value in stage 0 of the total random cost incurred in stage 1 can be stated as $R^f E_p[m(\omega)\, Q(x,\omega)]$ where $m(\omega)$ is the stochastic discount factor. The minimization of the total investment and operations cost can then be stated as

$$\min_{x \geq 0} \sum_{k \in K} I(k)\, x(k) + R^f E_p[m(\omega)\, Q(x,\omega)] \tag{13.47}$$

or

$$\min_{x \geq 0} \sum_{k \in K} I(k)\, x(k) + E_p[Q(x,\omega) + R^f \operatorname{cov}[m(\omega), Q(x,\omega)]. \tag{13.48}$$

Recall that the state price $m(\omega)$ is exogenous to the problem (as in a CAPM-based formulation) and that the expectation with the price kernel is a linear operator that therefore introduces minimal complications. The consequence is that the problem stated in the original variables x, y, and z retains the standard form of a convex stochastic optimization problem for which one can write KKT conditions. Because the definition of $Q(x, \omega)$ remains unchanged, its KKT conditions are also unchanged: they are restated below for the fixed demand model

$$0 \le x(k) - y(k, \ell, \omega) \perp \mu(k, \ell, \omega) \ge 0, \tag{13.49}$$

$$0 \le \sum_{k \in K} y(k, \ell, \omega) + z(\ell, \omega) - d(\ell, \omega) \perp \pi(\ell, \omega) \ge 0, \tag{13.50}$$

$$0 \le c(k, \omega) + \mu(k, \ell, \omega) - \pi(\ell, \omega) \perp y(k, \ell, \omega) \ge 0, \tag{13.51}$$

$$0 \le \mathrm{PC}(\omega) - \pi(\ell, \omega) \perp z(\ell, \omega) \ge 0. \tag{13.52}$$

The model is completed by the KKT conditions of the investment optimization problem. These become (using the abridged notation $\mu(x, k, \omega)$)

$$0 \le I(K) - R^f E_p[m(\omega)\, \mu(x, K, \omega)] \perp x(K) \ge 0. \tag{13.53}$$

Because $-\mu(x, K, \omega)$ is a monotone operator, the expression $-E_p[m(\omega)\, \mu(x, K, \ell, \omega)]$ would then also be a monotone operator (and problem (13.47) a convex problem in x) if the stochastic discount rate $m(\omega)$ were non-negative. This is not guaranteed, for instance, for the stochastic discount rate derived from the CAPM. It can then happen that the optimization problem (13.47) is unbounded because of those scenarios ω where $m(\omega)$ is negative. From an optimization point of view an unbounded problem does not have any dual solution. From an equilibrium point of view, $-E_p[m(\omega)\, \mu(x, K, \ell, \omega)]$ is no longer monotone and (in this particular case) (13.53) has no solution. The origin of the problem is that the commonly used CAPM method does not necessarily guarantee that all $m(\omega)$ are non-negative (it does not satisfy the stochastic dominance property). This is a matter of data on $R(M, \omega)$. The occurrence of negative $m(\omega)$ is thus entirely a CAPM matter and is independent of the power problem on hand; this can be checked ex ante.

Finally note that even though all annual investment costs $I(k)$ are computed from the total investment cost $\mathcal{I}(k)$ (including intermediary financial costs during construction) at a common risk-free rate, and the same state price $m(\omega)$ applies to all plants, the formulation effectively accounts for the risk exposure specific to each plant. The method is thus fully in line with a CAPM-based project finance approach that values each plant according to its risk exposure. This consistence is achieved through the deterministic equivalent cash flow:

$$E_p[m(\omega)\, \mu(x, k, \omega)] = E_p[\mu(x, k, \omega)] + R^f \operatorname{cov}[m(\omega), \mu(x, k, \omega)], \tag{13.54}$$

where $\operatorname{cov}[m(\omega), \mu(x, k, \omega)]$ is the plant-specific risk adjustment to the expected gross margin.

13.6 Non-linear Stochastic Discount Rates and Valuation Functions

Linear stochastic discount factors price systematic risk but not idiosyncratic risk. The capacity expansion model discussed in the preceding section is based on linear stochastic discount factors and hence only prices systematic risk. The equilibrium model therefore complies with economic theories that claim that idiosyncratic risk can only have zero price at equilibrium and hence can be discarded. It does not fit with other economic theories that explain that agency costs allow idiosyncratic risk to intervene in decisions. We assume in this section that agency costs may lead investors to account for idiosyncratic risk in their decisions. Linear discount factors that only span a subspace of the payoffs cannot, by construction, reflect that phenomenon. The modern theory of risk functions allows one to expand the above equilibrium model by introducing non-linear stochastic discount factors that we here apply to the subspace of the payoffs that are not priced by $m(\omega)$. The theory of coherent risk functions originates in Artzner et al. (1999); an extensive treatment of these functions in optimization is provided in Shapiro et al. (2009). We here insert coherent risk functions in equilibrium models. Because of the context of our model (focussing more on positive cash flows than positive losses) we conduct the discussion in terms of both risk and valuation functions; depending on the context we want to find the "value" of a cash flow or the "risk" of a cost.

13.6.1 A Primer on Coherent Risk/Valuation Functions

Consider a two-period problem where one wants to assess in stage 0 a cash flow $X(\omega)$ accruing in stage 1. A valuation function $\rho(X(\omega))$ gives a value to this cash flow. In order to remain as close as possible to the above discussion of stochastic discount factors we limit ourselves to coherent valuation functions introduced by Artzner et al. (1999) in a context of risk and now extensively developed in the literature. Coherent risk/valuation functions ρ satisfy four axioms that are recalled in Appendix 4. We are particularly interested in a representation theorem that states that every coherent valuation function can be represented as

$$\rho(X(\omega)) = \mathrm{Inf}_{m \in \mathcal{M}} E_p[m(\omega) \times X(\omega)], \tag{13.55}$$

where $\mathcal{M}$ is a convex set of probability measures.

The representation theorem implies that a valuation function generates a stochastic discount factor $m(\omega)$ which is the minimant of the above infimum problem. This stochastic discount factor differs from the one discussed in the preceding section in at least two respects. Because m is a probability measure, it is non-negative. This guarantees the monotonicity property (Axiom 2 in Appendix 4), that is, sometimes violated in stochastic discount factors derived from other theories such as the CAPM. Another major difference is that this stochastic discount rate is a function of the payoff $X(\omega)$ and hence the valuation process is a non-linear operator.

We accordingly write $\rho(X(\omega)) = E_p[m(X,\omega) \times X(\omega)]$ and note that $\rho(X(\omega))$ is a concave function of X in the case of the risk-adjusted valuation of a payoff (and a convex function for the risk-adjusted valuation of a cost). We also note that one of the axioms defining risk/valuation functions (Axiom 3 in Appendix 4) immediately implies that the value of a risk-free asset $1(\omega)$ that redeems 1 in all states of the world in stage 1 is equal to $1/R^f$, or in other words $\rho(1(\omega)) = 1/R^f$. The reader is referred to the original papers of Artzner et al. (1999) and to the extensive discussion and the many examples of risk functions (coherent or not) found in Shapiro et al. (2009). All this applies here whether to risk or value functions.

We conclude this brief summary by noting that, because risk/valuation functions give rise to stochastic discount factors, they also allow one to restate the value of a random cash flow as a deterministic equivalent where all discounting takes place at the risk-free rate. We briefly recall the reasoning developed in the previous Section 13.5.1 and adapt it to this non-linear case. We caution, however, that the multitemporal discounting with risk/valuation functions raises questions of time consistency of these functions that are not discussed here (see Artzner et al. 2007; Shapiro et al. 2009). In order to justify the discounting at the risk-free rate consider again a three-stage setup where one invests in period 0 and collects the random payoffs $X^1(\omega)$ and $X^2(\omega_1,\omega_2)$ in stages 1 and 2, respectively. Let $m^2(X^2|\omega_1,\omega_2|\omega_1)$ be the conditional stochastic discount factor derived from the valuation function in stage 2 for given ω_1. As before let $p^2(\omega_2|\omega_1)$ be the conditional probability of the different states of the world in stage 2. The value in stage 1 of payoff $X^2(\omega_2|\omega_1)$ conditional on ω_1 can be written as

$$\begin{aligned}
&E_{p^2|\omega_1}[m_2(X^2|\omega_1,\omega_2|\omega_1) \times X^2(\omega_2|\omega_1)] \\
&= \frac{1}{R^f}\left[E_{p^2|\omega_1}X^2(\omega_2|\omega_1) + R^f \operatorname{cov}[m_2(X^2|\omega_1,\omega_2|\omega_1), X^2(\omega_2|\omega_1)]\right] \\
&= \frac{1}{R^f}\widetilde{X}^2(\omega_1),
\end{aligned}$$

where $\widetilde{X}^2(\omega_1)$ is the deterministic equivalent in period 1 of the conditional cash flow $X^2(\omega_2|\omega_1)$.

Let now $p^1(\omega_1)$ be the probability measure of the different states of the world in stage 1. We need to assess in stage 0 a payoff

$$X^1(\omega_1) + \frac{1}{R^f}\widetilde{X}^2(\omega_1)$$

accruing in period 1. Conducting the same reasoning, the valuation in stage 0 of this random cash flow accruing in stage 1 can be written as

$$\frac{1}{R^f}\widetilde{X}^{1,2}.$$

As in Section 13.5.1, one sees that all deterministic equivalent cash flows are discounted at the risk-free rate provided these deterministic equivalents have been computed with the stochastic discount factors derived from the valuation function. The implication is that the annual investment cost $I(k)$ of a plant k can be computed from the total investment cost $\mathcal{I}(k)$ using the risk-free rate. This part of the reasoning would remain true if we were to account for the question of time consistency alluded to above. But the multitemporal discounting of the cash flows would be restricted to time-consistent valuation functions.

13.6.2 Coherent Risk/Valuation Functions and Stochastic Discount Factors

Consider again the decomposition of the payoff space $X = X_m + X_{\perp m}$ where $X_m(\omega)$ and $X_{\perp m}(\omega)$ are, respectively, the systematic and idiosyncratic components of the payoff. Different economic theories (e.g., CAPM, APT, multitemporal consumption model) can be used to construct a stochastic discount rate $m^\star(\omega)$ that prices the systematic risk while giving a zero value to the idiosyncratic risk. We introduce a price for the idiosyncratic risk embedded in $X(\omega)$ by writing the value of the payoff in stage zero as

$$\rho(X(\omega)) = E_p[m^\star(\omega) \times X_{m^\star}(\omega)] + \mathrm{Inf}_{m \in \mathcal{M}_{\perp m^\star}} E_p[m(\omega) \times X_{\perp m^\star}(\omega)], \quad (13.56)$$

where $m^\star$ prices the systematic risk (for instance, as specified in the CAPM) and $\mathcal{M}_{\perp m^\star}$ is a set of probability measures of $\mathcal{M}$ orthogonal to $m^\star$ that price the idiosyncratic risk.

This expression requires a decomposition of the payoff into its systematic (priced by $m^\star(\omega)$) and idiosyncratic parts (orthogonal to $m^\star(\omega)$). This is easily obtained by writing $X^\star_m(\omega)$ as the projection of $X(\omega)$ on $m^\star$ or

$$X(\omega) = \frac{\mathrm{cov}[X(\omega)\, m^\star(\omega)]}{\sigma^2(m^\star(\omega))} m^\star(\omega) + X_{\perp m^\star}(\omega).$$

The obtained expression $X_{\perp m^\star}(\omega)$ is clearly linear in $X(\omega)$.

13.6.3 Optimization Model

Consider now the risk-adjusted capacity expansion optimization model. This model is expressed in terms of costs, which justifies our using the risk interpretation of ρ. We thus apply the above risk function to the second-stage cost $Q(x, \omega)$ and define the following convex optimization problem:

$$\min_{x \geq 0} \sum_{k \in K} I(k)\, x(k) + R^f \rho[Q(x, \omega)]. \quad (13.57)$$

One obtains the KKT conditions of that optimization problem by concatenating the equilibrium conditions of the short-term problem (13.18), (13.19), (13.20), and (13.21) with the following risk-adjusted investment condition

$$0 \leq I(K) - R^f \frac{\partial \rho}{\partial \nu(\omega)} \frac{\partial Q(x, \omega)}{\partial x} \perp x(K) \geq 0. \tag{13.58}$$

Define the stochastic discount factor $\phi(x, \omega) = \partial \rho(Q(x, \omega))/\partial \nu(\omega) = m^\star(\omega) + m(Q(x, \omega))$ and recall that $\partial Q(x, \omega)/\partial x(k) = \mu(x, k, \omega)$. The investment criterion is stated as

$$0 \leq I(K) - R^f E_p(\phi(x, \omega) \times \mu(x, K, \omega)) \perp x(K) \geq 0. \tag{13.59}$$

Note that the representation theorem implies that $m(Q(x, \omega))$ is a probability measure. In contrast $m^\star$ need not be as the example of the CAPM discussed in the preceding section shows. $\mu(x, K) = E_p(\phi(x, \omega) \times \mu(x, K, \omega))$ is a monotone operator when $m^\star$ is non-negative since $\rho(X(\omega))$ is then a coherent risk function (see Appendix 4).

13.6.4 Equilibrium Models

The above formulation cannot be directly restated in competitive equilibrium terms. Specifically the dependence of the stochastic discount rate on the total system cost or welfare has a natural interpretation for a risk-averse monopoly company but it does not apply to agents of a market where the total cost of the system is the sum of the costs incurred by these agents. We first discuss tentative equilibrium interpretations of condition (13.58) by invoking different assumptions of market structures.

Suppose without loss of generality that the market only comprises a single electricity consumer (noted c) and the other agents (noted $o \in O$) own generation capacities $x^o(k), k \in K$ $(x(k)) = \sum_{o \in O} x^o(k)$ for all k. Take first the fixed demand model; the short-run cost optimization model (of Section 13.2.3) in state of world ω can be restated in dual form as

$$Q(x, \omega) \equiv \max \sum_{\ell} \tau(\ell) \left[\pi(\ell, \omega) d(\ell, \omega) - \sum_{k} \mu(k, \ell, \omega)\, x(k) \right], \tag{13.60}$$

s.t.

$$\pi(\ell, \omega) - \mu(k, \ell, \omega) \leq c(k, \omega), \tag{13.61}$$

$$\pi(\ell, \omega) \leq PC(\omega), \tag{13.62}$$

$$\pi(\ell, \omega) \geq 0, \;\; \mu(k, \ell, \omega) \geq 0. \tag{13.63}$$

The dual objective function is the sum of the total payment for electricity by the consumer $\Pi(\omega) = \sum_{\ell} \tau(\ell)\pi(\ell,\omega)d(\ell,\omega)$ minus the sum over all plants of the gross margins $\mu(x,k,\omega)x^o(k)$ made by the owners of the different capacities $x^o(k)$. The total profit of capacity owner o is then $\Pi^o(x,\omega) = \sum_k \mu(x,k,\omega)x^o(k)$.
We write

$$Q(x,\omega) = \Pi(\omega) - \sum_{o\in O} \Pi^o(x,\omega). \tag{13.64}$$

Recalling that the derivative of $Q(x,\omega)$ with respect to $x(k)$ is the marginal profit $\mu(x,k,\omega)$ accruing to every owner of capacity k in state of the world ω when increasing this capacity, the investment condition (13.59) could be associated with an owner o of capacity k if we could ascertain that $\partial\rho[Q(x,\omega)]/\partial\nu(\omega) = \phi(x,\omega)$ is the stochastic discount rate of every capacity owner o making a profit $\sum_k \mu(x,k,\omega)x^o(k)$. It remains to identify if and when this latter condition can hold.

Consider first the case where all capacity owners o have the same generation structure x^o, that is the same portfolio of generation plants (generators are identical up to a scaling factor). Their profits $\Pi^o(x,\omega)$ are then equal for all ω up to the scaling factor. Suppose also that they have the same risk function. Let $\phi^o(\Pi^o(x,\omega),\omega)$ be the stochastic discount factor of capacity owner o when it is making a profit $\Pi^o(x,\omega) = \sum_k \mu(x,k,\omega)x^o(k)$. Because these generators have the same share of different plants, the positive homogeneity of the coherent risk function ρ (Axiom 4 in Appendix 4) implies they also have the same stochastic discount factor. Unfortunately, this common value is not necessarily equal to $\partial\rho[Q(x,\omega)]/\partial\nu(\omega)$ because the total cost $Q(x,\omega)$ which is the argument of the risk function in (13.58) contains the term $\Pi(\omega)$ which is not proportional to $\Pi^o(x,\omega)$.

Consider now consumer c. The stochastic discount rate of this consumer is $\phi^c(\Pi(\omega),\omega)$, which is equal to $\partial\rho(\Pi(\omega))/\partial\nu(\omega)$. Suppose now that we can arrange for

$$\phi^o(\Pi^o(x,\omega),\omega) = \phi^c(\Pi(\omega),\omega) \quad \forall o, \tag{13.65}$$

then we also have

$$\phi^o(\Pi^o(x,\omega),\omega) = \phi^c(\Pi(\omega),\omega) = \frac{\partial\rho}{\partial\nu(\omega)}[Q(x,\omega)]. \tag{13.66}$$

and (13.58) would be a true equilibrium model with risk-averse consumers and generators having the same stochastic discount factor.

Financial products can in principle help achieve equality (13.65) (Ralph and Smeers 2010) even if generators are not identical up to a scaling factor. Suppose a market that trades bonds paying coupons indexed on electricity and fuel prices. The owner of a plant could hedge its revenue by a portfolio of these bonds. The consumer could also hedge its electricity payment $\Pi(\omega)$ with these bonds. The difficulty of this reasoning is that it does not apply to idiosyncratic risk that, by definition, cannot be traded. Stating model (13.57) as an equilibrium model may thus be an interesting counterfactual but it is only an approximation that holds if idiosyncratic risk is not too important.

The same reasoning applies to the variable demand model. Total welfare is the sum of consumer and producer surplus, a relation that we write as

$$\begin{aligned} -MSTW(x,\omega) = & \sum_{o \in O} \sum_{\ell \in L} \tau(\ell) \left[\sum_{k \in K} p(\ell,\omega) - c(k,\omega) \right] y^o(k,\ell,\omega) \\ & + \left[\int_0^{d(\ell,\omega)} p(\xi,\ell,\omega) d\xi - p(\ell,\omega)\, d(\ell,\omega) \right]. \end{aligned} \tag{13.67}$$

or $MTSW(x,\omega) = -\sum_o \Pi^o(x,\omega) - \Pi^c(\omega)$ where $\Pi^o(x,\omega)$ and $\Pi^c(\omega)$ are, respectively, generator o and consumer c surplus. We can therefore interpret the variable demand version of (13.58) as an equilibrium model if we can guarantee

$$\phi^o(\Pi^o(x,\omega),\omega) = \phi^c(\Pi^c(\omega),\omega) = \frac{\partial \rho}{\partial \nu(\omega)}(MSTW(x,\omega)). \tag{13.68}$$

Equality will hold for all generators if they have the same generation structure (again up to a scaling factor). But equality between $\phi^o(\Pi^o(x,\omega),\omega)$ and $\phi^c(\Pi^c(\omega),\omega)$ can only hold if consumers and generators can trade idiosyncratic risk. As before, this is not possible by definition.

In order to conclude the discussion consider an alternative market structure where each generator specializes in one technology. The set O of generators o then coincides with the set K of technologies k. By construction, we can no longer attempt to make the stochastic discount factors of the different generators equal. A financial market could still entail that equality if idiosyncratic risk could be traded between generators and the consumer. But this is again impossible by definition of the idiosyncratic risk. In conclusion we can retain model (13.59) completed by the short-run equilibrium conditions (13.6) to (13.9) or its variable demand model counterpart as an ideal counterfactual but need to consider alternative formulations that better account for the nature of idiosyncratic risk. The following presents different models that one can think of. They are all stated in the fixed demand model form but can easily be adapted to the variable demand model. The two first models have the same short-term market submodel (13.69), (13.70), (13.71), and (13.72); the third model has a slightly modified version that will be presented in due course:

$$0 \le x(k) - y(k,\ell,\omega) \perp \mu(k,\ell,\omega) \ge 0, \tag{13.69}$$

$$0 \le \sum_{k \in K} y(k,\ell,\omega) + z(\ell,\omega) - d(\ell,\omega) \perp \pi(\ell,\omega) \ge 0, \tag{13.70}$$

$$0 \le c(k,\omega) + \mu(k,\ell,\omega) - \pi(\ell,\omega) \perp y(k,\ell,\omega) \ge 0, \tag{13.71}$$

$$0 \le \mathrm{PC}(\omega) - \pi(\ell,\omega) \perp z(\ell,\omega) \ge 0. \tag{13.72}$$

13.6.4.1 Perfect Risk Trading or Perfectly Diversified Plant Portfolios

Whatever the shortcomings of the optimization model (13.57) for representing an equilibrium it remains a useful counterfactual because it represents an ideal situation of a complete market where idiosyncratic risk could be traded, for instance, through special insurance contracts. Also, because it is a convex optimization problem, it is easy to compute. Note that it may be unbounded (and hence not amenable to an equilibrium interpretation) when $m^*(\omega)$ has negative components. We simply restate the investment part (which is similar to the criterion of (13.53) but not with a risk adjustment on welfare instead of costs) of the model here without further discussion.

$$0 \leq I(K) - R^f \frac{\partial \rho(Q(x,\omega),\omega)}{\partial \nu(\omega)} \frac{\partial Q(x,\omega)}{\partial x} \perp x(K) \geq 0 \tag{13.73}$$

that is rewritten as

$$0 \leq I(K) - R^f E_p(\phi(x,\omega) \times \mu(x,K,\omega)) \perp x(K) \geq 0. \tag{13.74}$$

13.6.4.2 The Project Finance Approach

The project finance approach is a second counterfactual where each plant is valued on the basis of its own merit. Consider now a market where each agent invests in a particular technology. Let ρ^k be the valuation function of agent k. The investment criterion leads to a formulation where each investor k solves the following profit maximization problem:

$$\min_{x(k)\geq 0} [I(k) - R^f \rho^k[\mu(x,k,\omega)]]x(k), \tag{13.75}$$

where $\mu(x,k,\omega)$ retains its interpretation of marginal value of a unit capacity of plant k in the short-run equilibrium when the system portfolio is x. The corresponding investment condition for plant of type k is then

$$0 \leq I(k) - R^f E_p \phi^k(x,\omega)\mu(x,k,\omega) \perp x(k) \geq 0, \tag{13.76}$$

where $\phi^k(x,\omega) = m^\star(\omega) + m^k\left(\sum_{\ell\in L} \tau(\ell)\mu(x,k,\ell,\omega),\omega\right)$. This condition applies to all plants k each one being written with plant k specific stochastic discount factor ϕ^k (compare with Section 13.5 where a simple stochastic discount factor applies to all plants). This problem can easily be converted into a variational inequality problem if, as is often the case, there are constraints on investment possibilities. Each mapping $-E_p\phi^k(x,\omega)\mu(x,k,\omega)$ is monotone if $m^\star(\omega)$ is non-negative. Consider now the global investment criterion

$$0 \leq I(K) - F(x) \perp x(K) \geq 0, \tag{13.77}$$

where $F^k(x) = -R^f E_p \phi^k(x, \omega)\mu(x, k, \omega)$. Even though each $F^k(x)$ is monotone, the mapping $F(x)$ does not necessarily satisfy this property.

13.6.4.3 Diversified Portfolio Models

We now come back to the model with different generators o that each invest in and operate a mix of generating capacities. For the sake of realism we here complicate the problem and suppose that each operator has an existing generation fleet $\overline{x}^o(k)$ and is considering investing $x^o(k)$. We define

$$\overline{x}(k) = \sum_{o \in O} \overline{x}^o(k), \tag{13.78}$$

$$x(k) = \sum_{o \in O} x^o(k), \tag{13.79}$$

and modify the equilibrium conditions of the short-run market in order to account for existing capacities. These become

$$0 \leq \overline{x}(k) + x(k) - y(k, \ell, \omega) \perp \mu(k, \ell, \omega) \geq 0, \tag{13.80}$$

$$0 \leq \sum_{k \in K} y(k, \ell, \omega) + z(\ell, \omega) - d(\ell, \omega) \perp \pi(\ell, \omega) \geq 0, \tag{13.81}$$

$$0 \leq c(k, \omega) + \mu(k, \ell, \omega) - \pi(\ell, \omega) \perp y(k, \ell, \omega) \geq 0, \tag{13.82}$$

$$0 \leq \mathrm{PC}(\omega) - \pi(\ell, \omega) \perp z(\ell, \omega) \geq 0. \tag{13.83}$$

The profit of generator o in state of world ω is then

$$\Pi^o(\overline{x} + x, \omega) \equiv \sum_k [\overline{x}^o(k) + x^o(k)]\mu[\overline{x}(k) + x(k), \omega]. \tag{13.84}$$

The investment criterion needs now be rewritten for each generator o

$$0 \leq I(k) - R^f E_p[\phi^o[\Pi^o(\overline{x} + x, \omega); \omega]\mu[\overline{x}(k) + x(k), \omega] \perp x^o(k) \geq 0. \tag{13.85}$$

This criterion applies to the set of plants and can thus be rewritten as

$$0 \leq I(K) - F^0(x) \perp x^0(K) \geq 0, \tag{13.86}$$

where $F^0(x) = R^f E_p[\phi^0[\pi^0(\overline{x} + x, \omega)\mu[\overline{x}(K) + x(K), \omega]$. The criterion must be stated for each generator o. As for the project finance model, this mapping is not monotone in general. This problem can easily be converted into a variational inequality problem if, as is often the case, there are constraints on investment possibilities.

13.6.4.4 Comment

The loss of monotonicity in the "project finance approach" and the "diversified portfolio models" has an economic interpretation. It is related to the need to value idiosyncratic risks that by definition cannot be traded. This brings us into the domain of incomplete markets. We shall come back to this phenomenon in further papers.

13.7 Numerical Illustration

13.7.1 The Setup

In this section we illustrate the impact of the investment criteria discussed in the main text on a simple but realistic example that exhibits different relevant features. Very much like the rest of the chapter, the example illustrates a skeleton of real models. We first present the data for the deterministic model that we then extend to a stochastic version. Consider a two-stage problem: one invests in different types of capacities in stage 0 and operates them in stage 1. Following Joskow (2007), our benchmark example is a three-technology problem involving coal, combined cycle gas turbine (CCGT), and open cycle gas turbine (OCGT). Each equipment type is characterized by its annual investment and fixed operating costs and a CO_2 emission factor. Because risk premia are endogenous in the stochastic equilibrium model, investment costs are meant to be annualized from overnight construction costs at the risk-free rate (see Table 13.1).

All costs are assumed to be linear in capacity or operation levels implying that we neglect economies of scale in generation. This assumption is common in capacity expansion models. Emission factors for each plant are in tons of CO_2 per MWh. These figures are stylized views on costs and emission factors found in the European industry at the time of this writing. They do not correspond to particular projects but are realistic. The operating costs will be derived from fuel prices.

The deterministic model supposes one fuel price scenario and one scenario for CO_2 prices. We assume a simple price cap throughout the chapter. These are shown in Table 13.2.

Table 13.1 Fixed annual cost and emission in a three-technology world

	Coal	CCGT	OGGT
I: annual capacity and fixed operating cost (k Euro/MW)	140	80	60
e: emission t/MWh	1	0.35	0.6

Table 13.2 Variable cost, price cap and CO2 price

	Coal	CCGT	OGGT
$c(k)$: fuel and variable operating cost (Euro/MWh)	25	45	80
CO_2: price (Euro/ton)	20		
PC: price cap (Euro/MWh)	300		

Table 13.3 Reference load–duration curve and its decomposition in time segments

Power level and utilization						
d: MW	86,000	83,000	80,000	60,000	40,000	20,000
τ: duration (1000 h)	0.01	0.04	0.31	4.4	3	1

The load duration curve is segmented in six demand blocks in order to keep the model simple, while still guaranteeing sufficient detail for arriving at meaningful results. Table 13.3 gives the relevant figures and units.

13.7.2 Introducing Risk

The deterministic model only involves a single $\omega \in \Omega$ (Section 13.2.2). The stochastic model resorts to three probability spaces (load, fuel, and CO_2) and hence three probability measures. We generate three possible realizations for each risk factor multiplying the hourly load level, variable operating costs, and CO_2 prices with the factors listed in Table 13.4. Combining the three values we end up with 27 scenarios that are assumed equally probable.

13.7.3 Alternative Equilibrium Models: CO_2 Cap and Free Allocation

In order to illustrate the flexibility of the equilibrium framework we also consider slightly more complicated versions of the simple capacity expansion model used throughout the chapter where we include a simple representation of the EU-ETS (emission (of CO_2) trading scheme, Ehrenmann and Smeers 2010) (Section 13.3). Assuming then that the CO_2 price is determined in a cap and trade system we explicitly model the cap. Starting from the emissions obtained in the deterministic base case which are found equal to 219 mt, we create three scenarios around this figure by adding/subtracting 10 mt (Table 13.5).

As before, the three scenarios are equally likely.

Table 13.4 Scenario generation

	Low (%)	Central (%)	High (%)
Load	90	100	110
Fuel	70	100	130
CO_2 price	50	100	150

Table 13.5 Scenario generation

	Low	Central	High
CO_2 cap	209 mt	219 mt	229 mt

We complicate this EU-ETS model and depart from an optimization model (see Ehrenmann and Smeers 2010) for a discussion of that extension) by considering the case where free allowances are granted to investors. These are usually linked to the expected running hours and to a share of expected emissions covered by free allocation. We use expected running hours of 8000 h for coal, 6000 h CCGT, and of 1000 h for GT and assume a 20% coverage. The free allocation rule is then free allocation = expected running hours × emission rate × coverage.

13.7.4 Alternative Equilibrium Models: Elastic Demand

We consider two cases of elastic demand (Section 13.3). The first one is the standard situation where one assumes that there is no cross price elasticity: demand in one time segment only depends on the price in that time segment. The demand function is affine and calibrated as follows: we use a reference price equal to the long-run marginal cost of coal (LRMC) at 8000 h for the two lowest demand blocks, the LRMC of CCGT at 6000 h for the intermediate demand, and the LRMC of a GT at 1000 h. These are computed at the fuel and CO_2 prices of the deterministic case (Table 13.6).

For a given elasticity we calculate the electricity consumption in node i as a function of the price:

$$Q_i = \alpha_i - \beta_i(\text{price}_i).$$

We calibrate the coefficients α_i and β_i by writing

$$\beta_i = \frac{(\text{elasticity} * \text{demand}_i)}{\text{price}_i}$$

and

$$\alpha_i = \text{demand}_i + \text{price}_i * \beta_i.$$

The second case deals with the much more novel situation where one supposes that developments of the smart grid type allow for cross substitutions between time segments. We modify the demand function into

$$Q_i = \alpha_i - \sum_j \beta_{ij}(\text{price}_j),$$

Table 13.6 Reference load and price

Power level and utilization						
d: MW	86,000	83,000	80,000	60,000	40,000	20,000
Reference price	152	152	65.3	65.3	62.5	62.5
elasticity	0.5	0.5	0.5	0.5	0.5	0.5

Table 13.7 Elasticities

β_{61}	−0.1
β_{62}	−0.1
β_{51}	−0.1
β_{52}	−0.1

which makes demand in some time segment dependent on the electricity price in other time segments. We set the cross elasticities so that the off-peak demand increases if peak prices are very high, thereby reproducing the storage effect intended in smart grid developments (Table 13.7).

13.7.5 Systematic Risk and Linear Discount Factor

As explained in the text, the pricing of risk correlated to market risk requires inserting information on market returns in the scenarios (Section 13.5). In order to simplify matters we assume that the market return is highly correlated with electricity consumption, fuel prices, and, to a smaller extent with load. The results together with the impact on the stochastic discount factor are given in Table 13.8.

Table 13.8 Market return and linear stochastic discount factor

Fuel	Load	CO_2	Market return	$m(\omega)$
Low	Low	Low	0.55	1.421
Low	Low	Medium	0.6	1.380
Low	Low	High	0.65	1.338
Low	Medium	Low	0.95	1.087
Low	Medium	Medium	1.0	1.045
Low	Medium	High	1.05	1.003
Low	High	Low	1.35	0.753
Low	High	Medium	1.4	0.711
Low	High	High	1.45	0.669
Medium	Low	Low	0.65	1.338
Medium	Low	Medium	0.7	1.296
Medium	Low	High	0.75	1.254
Medium	Medium	Low	1.05	1.003
Medium	Medium	Medium	1.1	0.962
Medium	Medium	High	1.15	0.920
Medium	High	Low	1.45	0.669
Medium	High	Medium	1.5	0.627
Medium	High	High	1.55	0.585
High	Low	Low	0.75	1.254
High	Low	Medium	0.8	1.212
High	Low	High	0.85	1.171
High	Medium	Low	1.15	0.920
High	Medium	Medium	1.2	0.878
High	Medium	High	1.25	0.836
High	High	Low	1.55	0.585
High	High	Medium	1.6	0.543
High	High	High	1.65	0.502

13.7.6 Idiosyncratic Risk and Non-linear Discount Factor

The pricing of idiosyncratic requires a risk function (Section 13.6). We use the CVAR, which is now becoming standard and is quite amenable to computation. We account for risk aversion (in this case a prudent evaluation of the payoffs) by ignoring the highest 5% realizations. In order not to mix effects we run this case without pricing market risks (setting $m^{\star}(\omega)$ to zero). All the risk is thus assumed idiosyncratic.

13.7.7 Simulation Results

Table 13.9 summarizes the computational results; we only report the investments in the different types of plants, the total capacity, and the shortfall (see Joskow (2007) and the discussion of the "missing money" for an explanation of the shortfall) and the average baseload price:

The table provides a direct quantification of important effects. Moving from a deterministic to a stochastic environment reinforces the partial load capacities (CCGT). Introducing cross-time segment substitution (moving from elastic demand to elastic demand with DMS) implies a shift from peak (OCGT) to base (coal) units. The free allocation of permits (moving from CO_2 constraint to free allocation) effectively acts as a strong incentive to investments. Finally, the introduction of a CVaR shifts investments from high upfront capital expenditure (coal) to lower upfront capital expenses. All these phenomena are expected. The equilibrium model allows one to quantitatively explore them.

Table 13.9 Computational results

	Coal	CCGT	OCGT	Total	Max shortfall	Average baseload price
Deterministic (Section 13.2)	20,000	40,000	20,000	80,000	6000	60.98
Stochastic (Section 13.2.3)	20,000	46,000	6000	72,000	22.6	60.70
Elastic demand (Section 13.3.1)	20,340	40,780	0	61,120	0	60.89
Elastic demand DSM (Section 13.3.3)	24,120	37,690	0	61,810	0	60.91
CO2 constraint (Section 13.3.3)	27,200	38,800	6000	72,000	22,600	59.85
Free allocation (Section 13.5.2)	28,890	43,110	0	72,000	22,600	60.25
Linear discount factor (Section 13.5)	12,000	54,000	6000	72,000	22,600	61.71
CVAR (Section 13.6)	18,000	48,000	6000	72,000	22,600	61.16

13.8 Conclusion

The optimization capacity expansion models of the regulatory period have not yet found their counterpart in the restructured electricity markets. We present different adaptations of these former optimization models to the new competitive environment. Our objective is twofold: the model should be interpretable in terms of market equilibrium in order to fit with the new competition environment; it should also account for risk, which is becoming daunting in the industry. We begin with a simple standard two-stage version of a stochastic optimization capacity expansion model that we immediately convert into a parsimonious complementarity representation of the investment problem. The discussion then proceeds as follows. Even though we start with an optimization model, we show that the equilibrium formulation can encompass much more general models that do not derive from an optimization problem; this is important in a context where demand resources will play an increasing role. This is also important because of the flurry of new policies affecting the sector and distorting its environment from normal competition conditions. The rest of the chapter deals with different treatments of risk encountered in practice. While a pure risk-neutral version of the model is straightforward to construct, things become more difficult if one wants to account for risk aversion and how it is treated in corporate finance. CAPM-driven evaluations are common in practice and the philosophy of project finance requires to value each plant according to its own risk profile. This is usually done by discounting the cash flows of individual plants at a plant-specific risk-adjusted discount rate. Because the risk exposure of a plant depends on the development of the generation system, this approach effectively poses a problem of fixed point, a much more complex object than what is usually envisaged in practice. We bypass this difficulty by invoking stochastic discount factors that we directly embed in the equilibrium models. Discounting then takes place at the risk-free rate, therefore, eliminates difficulties that occur when different plants are affected by different discount rates. The counterpart is the need to compute risk-adjusted or deterministic equivalents of cash flows. We discuss two different approaches both based on stochastic discount factors. Linear discount factors come for standard economic theories like the CAPM, the APT, or the multitemporal consumer theory. They price what is commonly known as "systematic risk." They fully accommodate the requirements of project finance that each plant should be valued on the basis of its own risk exposure. But they limit this valuation to the risk priced by the market, something which is often, but not always accepted in practice. Difficulties arise when one insists on accounting for idiosyncratic risk. We do this by resorting to risk functions that we embed in the equilibrium context. The extension raises interesting questions: while the insertion of risk functions in a capacity expansion optimization model maintains convexity properties, this is not so in equilibrium models when we describe idiosyncratic risk by risk functions. The models can still be written without difficulties. But they lose their convexity properties. This chapter does not discuss the interpretation of this loss of convexity but it suggests that it is deep: because idiosyncratic risk cannot be traded, the market becomes incomplete, a phenomenon that is revealed by the loss the mathematical properties that guarantee existence and uniqueness of solution (here equilibrium). A numerical illustration illustrates the different features treated in the chapter.

Appendix 1: The Optimization Model as a Perfect Competition Equilibrium Model

Suppose a set of generators and a set of time segments. A perfect competition equilibrium is a vector of electricity prices (one price per time segment) and a vector of investments (one per type of capacity) and generation levels (one per type of capacity and time segment), such that investments and operations maximize generators' profits at prevailing prices and total generation matches demand in each time segment.

More explicitly in the single-stage view of the equilibrium, a perfect competition equilibrium consists of a set of electricity prices $\pi(\ell)$ and a set in investment and operation levels $x(k)$ and $y(k, \ell)$, such that $x(k)$ and $y(k, \ell)$ maximize the profit of generator of capacity k at prices $\pi(\ell)$ and the sum of the $y(k, \ell)$ over all k matches demand $d(\ell)$ in time segment ℓ.

Alternatively a two-stage view of the problem considers a set of investors in generation capacities that build/buy capacities and lease them (tolling agreement) to a set of operators of generation capacities that rent these capacities and sell electricity in a set of time segments. In this setup, a perfect competition equilibrium is a vector of electricity prices (one per time segment), a vector of capacity prices (one per time segment), a vector of investment levels (one per type of plant), and a vector of operation levels (one per type of plant and time segment), such that investment levels maximize investors' profits at prevailing capacity prices and operation levels maximize operators' profits at prevailing electricity and capacity prices.

More explicitly, in the operators' model, a perfect competition equilibrium in the second stage consists of a set of electricity prices $\pi(\ell)$, a set of rental capacity prices $\mu(k, \ell)$, a set of capacities $x(k)$ of plant k, and a set of operations level $y(k, \ell)$, such that the operator of technology maximizes its profit by renting $x(k)$ of capacity k at price $\mu(k, \ell)$ and selling $y(k, \ell)$ with that capacity at price $\pi(\ell)$ and the sum of the $y(k, \ell)$ over all k matches demand $d(\ell)$ in time segment ℓ. Similarly, in the investor's and overall models, a perfect competition equilibrium in the first stage consists of a set of capacity rental prices $\mu(k, \ell)$, a set of investments in capacity $x(k)$ such that $x(k)$ maximizes the profit of investors in technology k when rental price is $\mu(k, \ell)$. The two-stage perfect competition equilibrium is a set of electricity price $\pi(\ell)$ and rental price $\mu(k, \ell)$, investments $x(k)$, and operation levels $y(k, \ell)$ that form a perfect competition equilibrium in both the first and second stages.

Proposition (relation between the optimization and the perfect competition equilibrium problems) *The two definitions of the perfect competition equilibrium are equivalent in the sense that a perfect competition equilibrium of one type is also a perfect competition equilibrium of the other type. The solution of the optimization problem is a perfect competition equilibrium and conversely.*

Proof Verify that the complementarity conditions of the three problems are identical.

Appendix 2: Properties of $\mu(x, K)$

$\mu(x, K)$ *Is a Subgradient of* $Q(x)$

Rewrite problem (13.1) as

$$Q(x) = \sum_{\ell \in L} \tau(\ell)\, Q(x, \ell),$$

where

$$Q(x, \ell) = \min_{y(k,\ell), z(\ell)} \sum_{k \in K} [c(k)\, y(k, \ell) + \mathrm{PC}\, z(\ell)],$$

$$\begin{aligned} \text{s.t.} \quad & 0 \leq x(k) - y(k, \ell), && \mu(k, \ell), \\ & 0 \leq \sum_{k \in K} y(k, \ell) + z(\ell) - d(\ell), && \pi(\ell), \\ & 0 \leq y(k, \ell). \end{aligned}$$

One knows that

$$\mu(K, \ell) \in \partial_x Q(x, \ell)$$

and hence since

$$\begin{aligned} Q(x) &= \sum_{\ell \in L} \tau(\ell)\, Q(x, \ell), \\ \mu(K) &= \sum_{\ell} \tau(\ell)\, \mu(K, \ell) \in \partial Q(x). \end{aligned}$$

$\mu(x, K)$ *Is Multivalued*

Consider a time segment $\overline{\ell}$ and a set of capacities $x(k)$ such that

$$\begin{aligned} & 0 = x(k) - y(k, \overline{\ell}), && k = 1, \ldots, k^* \\ & 0 = y(k, \overline{\ell}), && k > k^* \\ & \textstyle\sum_k y(k, \overline{\ell}) = d(\overline{\ell}). \end{aligned}$$

One can check that

$$\begin{aligned}
&\pi(\overline{\ell}) = c(k^* - 1), \\
&\mu(k, \overline{\ell}) = \pi(\overline{\ell}) - c(k), \quad k \le k^* - 1, \\
&\mu(k, \overline{\ell}) = 0, \qquad\qquad\quad\;\; k \ge k^*
\end{aligned}$$

is a vector of dual variables satisfying the short-term model. Alternatively, one also verifies that

$$\begin{aligned}
&\pi(\overline{\ell}) = c(k^*), \\
&\mu(k, \overline{\ell}) = \pi(\overline{\ell}) - c(k), \quad k \le k^*, \\
&\mu(k, \overline{\ell}) = 0, \qquad\qquad\quad\;\; k > k^*
\end{aligned}$$

is also a vector of dual variables satisfying the short-term model.

$\mu(x, K)$ *Is a Gradient of* $MSTW(x, \omega)$

Rewrite problem (13.24), (13.25), (13.26), and (13.27)

$$\mathrm{MSTW}(x, \omega) = \sum_{\ell \in L} \tau(\ell)\, \mathrm{MSTW}(x, \omega, \ell),$$

where

$$\mathrm{MSTW}(x, \omega, \ell) = \min_{y(k,\ell)} \left[\sum_{k \in K} c(k, \omega)\, y(k, \ell, \omega) - \int_0^{d(\ell,\omega)} p(\xi, \ell, \omega) d\xi \right]$$

$$\begin{aligned}
\text{s.t.}\quad & 0 \le x(k) - y(k, \ell, \omega), && \mu(k, \ell), \\
& 0 \le \sum_{k \in K} y(k, \ell, \omega) - d(\ell, \omega), && \pi(\ell), \\
& 0 \le y(k, \ell, \omega).
\end{aligned}$$

The function is differentiable for any point $x(K)$ where the short-term complementarity conditions are strict, that is, such that $0 = x(k) - y(k, \ell)$ implies $y(k + 1, \ell) > 0$. Consider a point where this condition is not satisfied, that is,

$$\begin{aligned}
& y(k^*, \ell, \omega) = 0 < x(k^*), \\
& 0 < y(k^* - 1, \ell, \omega) = x(k^* - 1).
\end{aligned}$$

Because $d(\ell, \omega) > 0$ one has

$$\pi(\ell, \omega) = p(d(\ell, \omega),$$

which is uniquely determined and hence

$$\mu(k^* - 1, \ell, \omega) = \pi(\ell, \omega) - c(k^* - 1, \omega)$$

is also uniquely determined.

Last, $\mu(k^*, \ell, \omega) = 0$ since

$$x(k^*) - y(k^*, \ell, \omega) > 0.$$

The subdifferential boils down to a single point. Because the subdifferential is upper semicontinuous, this implies that it is continuous.

Appendix 3: Definition of Monotonicity

Let $\mu(x)$ be a mapping from R^n into R^n. $\mu(x)$ is monotone if

$$[\mu(x) - \mu(y)]^T (x - y) \geq 0 \text{ for all } x \text{ and } y.$$

The definition extends to point to set mapping as follows. Let $\mu(x)$ be a point to set mapping from R^n to R^n. $\mu(x)$ is monotone if for any x, y, there exists

$$\mu^x \in \mu(x); \quad \mu^y \in \mu(y),$$

such that

$$(\mu^x - \mu^y)^T (x - y) \geq 0.$$

Appendix 4: Definition of Coherent Risk Function

The following definitions are taken from Shapiro et al. (2009):

- **A.** Let $\mathcal{Z}$ be a linear space of random outcome $Z(\omega)$. $\rho(\mathcal{Z})$ satisfies the following axioms.
- **A1. Convexity:** $\rho(tZ + (1-t)Z') \leq t\rho(Z) + (1-t)\rho(Z')$ for all $t \in [0, 1]$ and all $Z, Z' \in \mathcal{Z}$.
- **A2. Monotonicity:** If $Z, Z' \in \mathcal{Z}$ and $Z \geq Z'$ the $\rho(Z) \geq \rho(Z')$.
- **A3. Translation equivariance** if $a \in R$ and $Z \in \mathcal{Z}$ then $\rho(Z + a) = \rho(Z) + a$.
- **A4. Positive homogeneity:** If $t > 0$ and $Z \in \mathcal{Z}$ then $\rho(tZ) = t\rho(Z)$.
  The move from a risk function to a valuation function obtains by replacing the convexity axiom A1 by a concavity axiom A′1.
- **A′1. Concavity:** $\rho(tZ + (1-t)Z') \geq t\rho(Z) + (1-t)\rho(Z')$ for all $t \in [0, 1]$ and all $Z, Z' \in \mathcal{Z}$.

References

Aghassi, M., Bertsimas, D. and G. Perakis (2006). Solving asymmetric variational inequalities via convex optimization. *Operations Research Letters*, 34, 481–490.

Ahn, B.-H. and W.W. Hogan (1982). On convergence of the PIES algorithm for computing equilibria. *Operations Research*, 30(2), 282–300.

Armitage, S. (2005). *The Cost of Capital, Intermediate Theory*. Cambridge University Press, Cambridge.

Artzner, P., Delbaen, F., Eber, J.-M. and D. Heath (1999). Coherent measures of risk. *Mathematical Finance*, 9(3), 203–228.

Artzner, P., Delbaen, F., Eber, J.-M. and H. Ku (2007). Coherent multiperiod risk adjusted values and Bellman's principle. *Annals of Operational Research*, 152, 5–22.

Cochrane, J.H. (2005). *Asset Pricing*. Princeton University Press, Princeton and Oxford.

Ehrenmann, A. and Y. Smeers (2009). Risk adjusted discounted cash flows in capacity expansion models. Presentation at the MPS 2009, Chicago. Slides available, paper in preparation.

Ehrenmann, A. and Y. Smeers (2010). Generation capacity expansion in a risky environment: a stochastic equilibrium analysis. Accepted for publication in *Operations Research.*

Eichhorn, A., Heitsch, H. and W. Römisch (2010). Stochastic optimization of electricity portfolios: scenario tree modeling and risk management. In P.M. Pardalos, M.V.F. Pereira, S. Rebennack and N.A. Iliadis (eds.), *Handbook of Power Systems*. Berlin, Springer, pp. 405–433.

Facchinei, F. and J.-S. Pang (2003). *Finite-Dimensional Variational Inequalities and Complementarity Problems*. New York, NY, Springer.

Fama, E. (1977). Risk-adjusted discount rates and capital budgeting under uncertainty. *Journal of Financial Economics*, 5(1), 3–24.

Grahan, J. and C. Harvey (2001). The theory and practice of corporate finance: Evidence form the field. *Journal of Financial Economics*, 60(2–3), 187–243.

Gribik, P.R., Hogan, W.W. and S.L. Pope (2007). Market-Clearing Electricity Prices and Energy Uplift. Draft Paper, Havard University, downloadable at: http://ksghome.harvard.edu

Gürkan, G., Özge, A.Y. and S.M. Robinson (1999). Sample-path solutions of stochastic variational inequalities. *Mathematical Programming*, 84, 313–333.

Gürkan, G., Özge, A.Y. and Y. Smeers (2009). Capacity investment in electricity markets: perfect competition. Discussion Paper, Department of Econometrics and Operations Research, Universiteit Tilburg, 2009.

Harker, P.T. and J.-S. Pang (1990). Finite-dimensional variational inequality and nonlinear complementarity problems: a survey of theory, algorithms and applications. *Mathematical Programming, Series A and B*, 48(2), 161–220.

Haurie, A., Legrand, J., Smeers, Y. and G. Zaccour (1988). A stochastic dynamic Nash Cournot model for the European Gas market. Les Cahiers du GERAD, C-87-24, October 1988.

Joskow, P.L. (2007). Competitive electricity markets and investments in new generation capacity. In D. Helm (ed.), *The New Energy Paradigm*. Oxford University Press, Oxford.

Kuhn, S. and R. Schultz (2009). Risk neutral and risk averse power optimization in electricity networks with dispersed generation. *Mathematical Methods of Operations Research*, 69(2), 353–367.

Lin, G.-H. and M. Fukushima (2009). Stochastic equilibrium problems and stochastic mathematical programs with equilibrium constraints. A survey. Technical Report, Department of Applied Mathematics and Physics, Kyoto University, March 2009.

Morlat, G. and F. Bessière (1971). *Vingt cinq ans d'Economie electrique*. Paris, Dunod.

Ralph, D. and Y. Smeers (2010). Risk averse structure equilibria: a perfectly competitive two-stage market in capacity, financial securities and production. Manuscript Judge Business School Cambridge.

Römisch, W. and S. Vigerske (2010). Recent progress in two-stage mixed-integer stochastic programming with applications in power production planning. In P.M. Pardalos, M.V.F. Pereira, S. Rebennack and N.A. Iliadis (eds.), *Handbook of Power Systems*. Berlin, Springer.

Shapiro, A., Dentcheva, D. and A. Ruszczynski (2009). *Lectures on Stochastic Programming: Modeling and Theory*, MPS/SIAM Series on Optimization 9. Philadelphia, PA, SIAM-Society for Industrial and Applied Mathematics.

Wu, Y.J. and J.D. Fuller (1996). An algorithm for the multiperiod market equilibrium model with geometric distributed lag demand. *Operations Research*, 44(6), 1002–1012.

Part III
Theory and Computation

Chapter 14
Scenario Tree Generation for Multi-stage Stochastic Programs

Holger Heitsch and Werner Römisch

Abstract We broaden the theoretical basis for generating scenario trees in multi-stage stochastic programming based on stability analysis. Numerical experience for constructing trees of demand and price scenarios in electricity portfolio management of a municipal power utility is also provided.

Keywords Stochastic programming · Scenario tree · Scenario reduction · Scenario generation · Stability · Multi-stage · Power portfolio · Electricity

14.1 Introduction

Many solution methods for stochastic programming models in finance and energy rely on approximating the underlying probability distribution P by a probability measure based on a finite number of scenarios (or atoms). In case of multi-stage models scenarios have to satisfy certain constraints due to the nonanticipativity of decisions. Such constraints lead to tree-structured scenarios. Hence, designing approximation schemes for multi-stage models requires the *generation* of *scenarios* as well as of *trees*.

There is a variety of methods for generating scenarios (see also the introductory overview (Römisch 2003) and Chapter 15, this volume), namely,

- Monte Carlo methods (applied to statistical models or data) (Shapiro 2003b),
- Quasi-Monte Carlo methods (see Lemieux (2009) for a recent exposition),
- optimal quantization methods for probability measures (see Graf and Luschgy (2000)),
- quadrature rules using sparse grids (e.g. Chen and Mehrotra (2008))

based on the underlying probability distribution P. In general, however, the generated scenarios will not exhibit tree structure.

Presently, there exist several approaches for generating such scenario trees. In a number of cases the tree structure is (partially) predetermined and scenarios

W. Römisch (✉)
Institute of Mathematics, Humboldt University Berlin, 10099 Berlin, Germany
e-mail: romisch@math.hu-berlin.de

M. Bertocchi et al. (eds.), *Stochastic Optimization Methods in Finance and Energy*, International Series in Operations Research & Management Science 163,
DOI 10.1007/978-1-4419-9586-5_14,

are generated sequentially by *(conditional) random sampling* (Dempster 2004; Kouwenberg 2001; Shapiro 2003a) or *quasi-Monte Carlo sampling* (Pennanen 2009). Alternatively, scenarios are generated by using bound-based constructions (Frauendorfer 1996; Kuhn 2005), using optimization methods such that certain moments are matched (Høyland and Wallace 2001; Gülpinar et al. 2004) or such that a best approximation of P is obtained in terms of a suitable probability metric (Pflug 2001; Hochreiter and Pflug 2007; Mirkov and Pflug 2007). We also refer to Dupačová et al. (2000) and the references therein for an overview on scenario tree generation.

The approach to scenario tree generation presented in the following does not require restrictive conditions on P. It starts with a number of scenarios generated by one of the methods mentioned above. The branching structure of the tree is not predetermined, but automatically detected by the algorithms such that a good approximation of P measured in terms of the closeness of optimal values is obtained. The whole approach is based on a quantitative stability analysis of (linear) multi-stage stochastic programs (see Heitsch et al. (2006) and Heitsch and Römisch (2010)). The algorithms rely on applying scenario reduction sequentially over time and are first analyzed in Heitsch and Römisch (2009a). In this chapter we review parts of our earlier work in Sections 14.2 and 14.4, develop new convergence results in Section 14.3, and report on an application to the electricity portfolio management of a municipal power company in Section 14.5. Readers which are mainly interested in algorithms and their application may skip Sections 14.2 and 14.3.

To state the mathematical optimization model, let the periods of the time horizon be denoted $t = 1, \ldots, T$ and $\xi = (\xi_t)_{t=1}^T$ be an $\mathbb{R}^d$-valued stochastic process on some probability space $(\Omega, \mathcal{F}, \mathbb{P})$. The filtration $(\mathcal{F}_t(\xi))_{t=1}^T$ associated with ξ is defined by $\mathcal{F}_t(\xi) := \sigma(\xi^t)$ with $\xi^t = (\xi_1, \ldots, \xi_t)$ for $t = 1, \ldots, T$. We assume that $\mathcal{F}_1 = \{\emptyset, \Omega\}$ and $\xi_t \in L_r(\Omega, \mathcal{F}, \mathbb{P})$, i.e., $\mathbb{E}(|\xi_t|^r) < +\infty$ for every $t = 1, \ldots, T$ and some $r \geq 1$. By $P = \mathbb{P}\xi^{-1}$ we denote the probability distribution of ξ and by P_t its tth marginal probability distribution for $t = 1, \ldots, T$, i.e.,

$$P_t(B_t) = \mathbb{P}\xi^{-1}(\Xi_1 \times \cdots \times \Xi_{t-1} \times B_t \times \Xi_{t+1} \times \cdots \times \Xi_T) \quad (B_t \in \mathcal{B}(\Xi_t)),$$

where $\Xi_t \subseteq \mathbb{R}^d$ denotes the support of ξ_t and $\mathcal{B}(\Xi_t)$ the σ-field of its Borel subsets. In particular, Ξ_1 denotes the singleton $\Xi_1 = \{\xi_1\}$.

A linear multi-stage stochastic program can be written as

$$\min\left\{\mathbb{E}\left(\sum_{t=1}^{T}\langle b_t(\xi_t), x_t\rangle\right)\middle|\begin{array}{l} x_t = x_t(\xi_1, \ldots, \xi_t),\ x_t \in X_t, \\ A_{t,0}x_t + A_{t,1}(\xi_t) = h_t(\xi_t)\end{array}\ (t = 1, \ldots, T)\right\}, \quad (14.1)$$

where x_t is an $\mathbb{R}^{m_t}$-valued decision vector for time period t. The latter is a Borel measurable function depending (only) on $(\xi_1, \ldots, \xi_t) \in \times_{\tau=1}^t \Xi_\tau = \Xi^t$, i.e., it depends on the data observed until time t (*nonanticipativity*). In particular, the components of x_1 are *here and now* decisions since x_1 may only depend on ξ_1 which was assumed to be deterministic. The decisions are subject to constraints: each x_t has to be chosen within a given polyhedral set X_t. Moreover, there are dynamic constraints involving matrices $A_{t,\tau}$, $\tau \in \{0, 1\}$, and right-hand sides h_t. The matrices $A_{t,1}(\cdot)$, the cost coefficients $b_t(\cdot)$, and right-hand sides $h_t(\cdot)$ may depend on ξ_t in an affinely linear way. $\mathbb{E}$ denotes expectation with respect to $\mathbb{P}$, i.e., the objective corresponds to the expected total costs of a decision vector $(x_1, \ldots, x_T)$.

14.2 Stability of Multi-stage Stochastic Programs

Studying stability of the multi-stage stochastic program (14.1) consists in regarding it as an optimization problem in the infinite dimensional linear space $\times_{t=1}^{T} L_{r'}(\Omega, \mathcal{F}, \mathbb{P}; \mathbb{R}^{m_t})$. This is a Banach space when endowed with the norm

$$\|x\|_{r'} := \left(\sum_{t=1}^{T} \mathbb{E}\left[|x_t|^{r'}\right]\right)^{1/r'} \quad \text{for } r' \in [1, +\infty),$$
$$\|x\|_{\infty} := \max_{t=1,\ldots,T} \operatorname{ess\,sup} |x_t|,$$

where $|.|$ denotes some norm on the relevant Euclidean spaces and $\operatorname{ess\,sup} |x_t|$ denotes the essential supremum of $|x_t|$, i.e., the smallest constant C such that $|x_t| \leq C$ holds $\mathbb{P}$-almost surely. Analogously, ξ can be understood as an element of the Banach space $\times_{t=1}^{T} L_r(\Omega, \mathcal{F}, \mathbb{P}; \mathbb{R}^d)$ with norm $\|\xi\|_r$. For the integrability numbers r and r' it will be imposed that

$$r := \begin{cases} \in [1, +\infty), & \text{if only costs or only right-hand sides are random} \\ 2, & \text{if only costs and right-hand sides are random} \\ T, & \text{if all technology matrices are random} \end{cases},$$
$$r' := \begin{cases} \frac{r}{r-1}, & \text{if only costs are random} \\ r, & \text{if only right-hand sides are random} \\ +\infty, & \text{if all technology matrices are random} \end{cases}, \tag{14.2}$$

with regard to problem (14.1). The choice of r and the definition of r' are motivated by the knowledge of existing moments of the input process ξ, by having the stochastic program well defined (in particular, such that $\langle b_t(\xi_t), x_t \rangle$ is integrable for every decision x_t and $t = 1, ..., T$), and by satisfying the conditions (A2) and (A3) (see below).

Since r' depends on r and our assumptions will depend on both r and r', we will add some comments on the choice of r and its interplay with the structure of the underlying stochastic programming model. To have the stochastic program well defined, the existence of certain moments of ξ has to be required. This fact is well known for the two-stage situation (see, e.g., Ruszczyński and Shapiro (2003), chapter 2). If either right-hand sides or costs in a multi-stage model (14.1) are random, it is sufficient to require $r \geq 1$. The flexibility in case that the stochastic process ξ has moments of order $r > 1$ may be used to choose r' as small as possible in order to weaken condition (A3) (see below) on the feasible set. If the linear stochastic program is fully random (i.e., costs, right-hand sides, and technology matrices are random), one needs $r \geq T$ to have the model well defined and no flexibility w.r.t. r' remains.

Next we introduce some notation. We set $s := Td$ and $m := \sum_{t=1}^{T} m_t$. Let

$$F(\xi, x) := \mathbb{E}\left[\sum_{t=1}^{T} \langle b_t(\xi_t), x_t\rangle\right]$$

denote the objective function defined on $L_r(\Omega, \mathcal{F}, \mathbb{P}; \mathbb{R}^s) \times L_{r'}(\Omega, \mathcal{F}, \mathbb{P}; \mathbb{R}^m)$ and let

$$\mathcal{X}(\xi) := \left\{x \in \times_{t=1}^{T} L_{r'}(\Omega, \sigma(\xi^t), \mathbb{P}; \mathbb{R}^{m_t}) \mid x_t \in \mathcal{X}_t(x_{t-1}; \xi_t) \text{ a.s. } \ (t = 1, \ldots, T)\right\}$$

denote the set of feasible elements of (14.1) with $x_0 \equiv 0$ and

$$\mathcal{X}_t(x_{t-1}; \xi_t) := \left\{x_t \in \mathbb{R}^{m_t} : x_t \in X_t,\, A_{t,0}x_t + A_{t,1}(\xi_t)x_{t-1} = h_t(\xi_t)\right\}$$

denoting the tth feasibility set for every $t = 1, \ldots, T$. This allows to rewrite the stochastic program (14.1) in the short form

$$\min\{F(\xi, x) : x \in \mathcal{X}(\xi)\}. \tag{14.3}$$

In the following, we need the optimal value

$$v(\xi) = \inf\{F(\xi, x) : x \in \mathcal{X}(\xi)\}$$

for every $\xi \in L_r(\Omega, \mathcal{F}, \mathbb{P}; \mathbb{R}^s)$ and, for any $\varepsilon \geq 0$, the *ε-approximate solution set* (level set)

$$S_\varepsilon(\xi) := \{x \in \mathcal{X}(\xi) : F(\xi, x) \leq v(\xi) + \varepsilon\}$$

of the stochastic program (14.3). Since, for $\varepsilon = 0$, the set $S_\varepsilon(\xi)$ coincides with the set solutions to (14.3), we will also use the notation $S(\xi) := S_0(\xi)$. The following conditions will be imposed on (14.3):

(A1) The numbers r, r' are chosen according to (14.2) and $\xi \in L_r(\Omega, \mathcal{F}, \mathbb{P}; \mathbb{R}^s)$.

(A2) There exists a $\delta > 0$ such that for any $\tilde{\xi} \in L_r(\Omega, \mathcal{F}, \mathbb{P}; \mathbb{R}^s)$ satisfying $\|\tilde{\xi} - \xi\|_r \leq \delta$, any $t = 2, \ldots, T$ and any $x_\tau \in L_{r'}(\Omega, \sigma(\tilde{\xi}^\tau), \mathbb{P}; \mathbb{R}^{m_\tau})$ ($\tau = 1, \ldots, t-1$) satisfying $x_\tau \in \mathcal{X}_\tau(x_{\tau-1}; \tilde{\xi}_\tau)$ a.s. (where $x_0 = 0$), there exists $x_t \in L_{r'}(\Omega, \sigma(\tilde{\xi}^t), \mathbb{P}; \mathbb{R}^{m_t})$ satisfying $x_t \in \mathcal{X}_t(x_{t-1}; \tilde{\xi}_t)$ a.s. (*relatively complete recourse locally around* ξ).

(A3) The optimal values $v(\tilde{\xi})$ of (14.3) with input $\tilde{\xi}$ are finite for all $\tilde{\xi}$ in a neighborhood of ξ and the objective function F is *level-bounded locally uniformly at* ξ, i.e., for some $\varepsilon_0 > 0$ there exists a $\delta > 0$ and a bounded subset B of $L_{r'}(\Omega, \mathcal{F}, \mathbb{P}; \mathbb{R}^m)$ such that $S_{\varepsilon_0}(\tilde{\xi})$ is contained in B for all $\tilde{\xi} \in L_r(\Omega, \mathcal{F}, \mathbb{P}; \mathbb{R}^s)$ with $\|\tilde{\xi} - \xi\|_r \leq \delta$.

For any $\tilde{\xi} \in L_r(\Omega, \mathcal{F}, \mathbb{P}; \mathbb{R}^s)$ sufficiently close to ξ in L_r, condition (A2) implies the existence of some feasible $\tilde{x}$ in $\mathcal{X}(\tilde{\xi})$ and (14.2) implies the finiteness of the objective $F(\tilde{\xi}, \cdot)$ at any feasible $\tilde{x}$. A sufficient condition for (A2) to hold is the *complete recourse condition* on every recourse matrix $A_{t,0}$, i.e., $A_{t,0}X_t = \mathbb{R}^{n_t}$, $t = 1, ..., T$. The locally uniform level-boundedness of the objective function F is quite standard in perturbation results for optimization problems (see, e.g., Rockafellar and Wets (1998), theorem 1.17). The finiteness condition on the optimal value $v(\xi)$ is not implied by the level boundedness of F for all relevant pairs (r, r'). In general, conditions (A2) and (A3) get weaker for increasing r and decreasing r', respectively.

The first stability result for multi-stage stochastic programs represents a quantitative continuity property of the optimal values. Its main observation is that multi-stage models behave stable at some stochastic input process ξ if both its probability distribution and its filtration are approximated with respect to the L_r-distance and the filtration distance

$$D_{\mathrm{f}}(\xi, \tilde{\xi}) := \sup_{\varepsilon > 0} \inf_{\substack{x \in S_\varepsilon(\xi) \\ \tilde{x} \in S_\varepsilon(\tilde{\xi})}} \sum_{t=2}^{T-1} \max\left\{ \left\| x_t - \mathbb{E}[x_t | \mathcal{F}_t(\tilde{\xi})] \right\|_{r'}, \left\| \tilde{x}_t - \mathbb{E}[\tilde{x}_t | \mathcal{F}_t(\xi)] \right\|_{r'} \right\}, \tag{14.4}$$

where $\mathbb{E}[\,\cdot\,|\mathcal{F}_t(\xi)]$ and $\mathbb{E}[\,\cdot\,|\mathcal{F}_t(\tilde{\xi})]$ $(t = 1, ..., T)$ are the corresponding conditional expectations, respectively. Note that for the supremum in (14.4) only small ε's are relevant and that the approximate solution sets are bounded for $\varepsilon \in (0, \varepsilon_0]$ according to (A3).

The following stability result for optimal values of program (14.3) is taken from Heitsch et al. (2006), theorem 2.1.

Theorem 1 *Let (A1), (A2), and (A3) be satisfied and the set X_1 be nonempty and bounded. Then there exist positive constants L and δ such that the estimate*

$$\left| v(\xi) - v(\tilde{\xi}) \right| \leq L \left(\|\xi - \tilde{\xi}\|_r + D_{\mathrm{f}}(\xi, \tilde{\xi}) \right) \tag{14.5}$$

holds for all random elements $\tilde{\xi} \in L_r(\Omega, \mathcal{F}, \mathbb{P}; \mathbb{R}^s)$ with $\|\tilde{\xi} - \xi\|_r \leq \delta$.

The result states that the changes of optimal values are at most proportional to the errors in terms of L_r- and filtration distance when approximating ξ. The corresponding constant L depends on $\|\xi\|_r$ (i.e. the rth moment of ξ), but is not known in general.

The filtration distance has a simpler representation if the approximation $\tilde{\xi}$ of ξ is ξ-*adapted*, i.e., if $\mathcal{F}_t(\tilde{\xi}) \subseteq \mathcal{F}_t(\xi)$ holds for every $t = 1, \ldots, T$. The latter is equivalent to the existence of measurable functions $f_t : \Xi^t \to \Xi^t$ such that

$$\tilde{\xi}^t = f_t(\xi^t) \quad (t = 1, \ldots, T).$$

For ξ-adapted approximations $\tilde{\xi}$ we have

$$D_{\mathrm{f}}(\xi, \tilde{\xi}) = \sup_{\varepsilon>0} \inf_{x \in S_\varepsilon(\xi)} \sum_{t=2}^{T-1} \left\| x_t - \mathbb{E}[x_t | \mathcal{F}_t(\tilde{\xi})] \right\|_{r'} \tag{14.6}$$

and if, in addition, the solution set $S(\xi)$ is nonempty, we obtain

$$D_{\mathrm{f}}(\xi, \tilde{\xi}) = \inf_{x \in S(\xi)} \sum_{t=2}^{T-1} \left\| x_t - \mathbb{E}[x_t | \mathcal{F}_t(\tilde{\xi})] \right\|_{r'}. \tag{14.7}$$

The latter representation allows the conclusion

$$D_{\mathrm{f}}(\xi, \tilde{\xi}) \leq \sum_{t=2}^{T-1} \left\| x_t - \mathbb{E}[x_t | \mathcal{F}_t(\tilde{\xi})] \right\|_{r'} \tag{14.8}$$

for any solution $x = (x_1, x_2, \ldots, x_T)$ of (14.3). For a given solution x of (14.3) there exist Borel measurable functions $g_t : \Xi^t \to \mathbb{R}^{m_t}$ such that

$$\mathbb{E}[x_t | \mathcal{F}_t(\tilde{\xi})] = \mathbb{E}[x_t | \tilde{\xi}^t] = g_t(\tilde{\xi}^t) \quad (t = 1, \ldots, T),$$

where $g_1(\tilde{\xi}_1) = g_1(\xi_1) = x_1$ and $g_t(\xi^t) = x_t$. In general, further properties of the functions

$$g_t(y_1, \ldots, y_t) = \mathbb{E}[x_t | \tilde{\xi}^t = (y_1, \ldots, y_t)] = \mathbb{E}[x_t | f_t(\xi^t) = (y_1, \ldots, y_t)],$$

i.e., of the *conditional expectations of* x_t *under the condition that* $\tilde{\xi}^t$ *equals* $(y_1, \ldots, y_t)$, are not known. Since x_t is a (measurable) function of ξ^t, the function value $g_t(y_1, \ldots, y_t)$ may be computed via the (multivariate) probability distribution P of ξ.

Unfortunately, in general, there is no solution $x \in S(\xi)$ such that the functions x_t depend continuously on ξ^t for every $t = 1, \ldots, T$ (cf. the discussion in Rockafellar and Wets (1974)). Sometimes, however, the functional dependence of x_t on ξ^t is of a specific form as in the following situation (Garstka and Wets 1974, theorem 4.3).

Proposition 1 *Assume that only right-hand sides in (14.1) are random and that* $S(\xi)$ *is nonempty. Then there exists* $x = (x_1, \ldots, x_T)$ *in* $S(\xi)$ *such that* $x_t = \varphi_t(h_1(\xi_1), \ldots, h_t(\xi_t))$ *and* φ_t *is a continuous, piecewise affine function for every* $t = 1, \ldots, T$*. In particular,* x_t *is Lipschitz continuous as a function of* ξ^t *for every* $t = 1, \ldots, T$.

This motivates the following condition on the conditional distributions of ξ and on the ξ-adapted approximation $\tilde{\xi}$ of ξ.

(A4) For each $t \in \{1, \dots, T\}$ and each pair (Φ_t, f_t) of Lipschitz continuous functions $\Phi_t : \Xi^t \to \mathbb{R}^{m_t}$ and $f_t : \Xi^t \to \Xi^t$, the function

$$g_t(y_1, \dots, y_t) = \mathbb{E}[\Phi_t(\xi^t) | f_t(\xi^t) = (y_1, \dots, y_t)] \tag{14.9}$$

is Lipschitz continuous on Ξ^t.

Then the following result is an immediate consequence of Theorem 1 and Proposition 1 if (A4) is imposed in addition.

Corollary 1 *Let (A1), (A2), (A3), and (A4) be satisfied and X_1 be nonempty and bounded. Assume that only right-hand sides in (14.1) are random and that $S(\xi) \neq \emptyset$. Then there exist positive constants $\hat{L}$ and δ such that*

$$\left| v(\xi) - v(\tilde{\xi}) \right| \leq \hat{L} \| \xi - \tilde{\xi} \|_r \tag{14.10}$$

whenever $\|\xi - \tilde{\xi}\|_r < \delta$ for any ξ-adapted $\tilde{\xi}$ such that $\tilde{\xi}^t = f_t(\xi^t)$, $t = 1, \dots, T$, with Lipschitz continuous functions $f_t : \Xi^t \to \mathbb{R}^{td}$.

Proof According to Proposition 1 there exists a solution $x = (x_1, \dots, x_T) \in S(\xi)$ such that $x_t = \Phi_t(\xi^t)$ is a Lipschitz continuous function of ξ^t for every $t = 1, \dots, T$. Let $\tilde{\xi}$ be ξ-adapted such that $\tilde{\xi}^t = f_t(\xi^t)$, $t = 1, \dots, T$, where the functions $f_t : \Xi^t \to \Xi^t$ are Lipschitz continuous on Ξ^t. According to (A4) the functions g_t from Ξ^t to $\mathbb{R}^{m_t}$, $t = 1, \dots, T$, given by (14.9) are Lipschitz continuous. Hence, there exist constants $K_t > 0$ such that

$$|g_t(y_1, \dots, y_t) - g_t(\tilde{y}_1, \dots, \tilde{y}_t)| \leq K_t \sum_{\tau=1}^{t} |y_\tau - \tilde{y}_\tau| \quad ((y_1, \dots, y_t), (\tilde{y}_1, \dots, \tilde{y}_t) \in \Xi^t)$$

and, hence,

$$D_{\mathrm{f}}(\xi, \tilde{\xi}) \leq \sum_{t=2}^{T-1} \left\| g_t(\xi^t) - g_t(\tilde{\xi}^t) \right\|_r \leq \sum_{t=2}^{T-1} K_t \sum_{\tau=1}^{t} \mathbb{E}(|\xi_\tau - \tilde{\xi}_\tau|^r)^{\frac{1}{r}} \leq K \| \xi - \tilde{\xi} \|_r$$

for some suitably large constant K. Together with Theorem 1 we obtain (14.10) with $\hat{L} = LK$. □

We note that our condition (A4) is similar to assumption 2.6 in Küchler (2008) and Corollary 1 reminds of Küchler (2008), theorem 3. We also note that in case of finite supports Ξ_t, $t = 1, \dots, T$, the functions g_t are necessarily Lipschitz continuous with possibly large Lipschitz constants K_t, $t = 1, \dots, T$, leading to a large constant $\hat{L}$ in Corollary 1.

Stability results of approximate solutions to (14.3) require a stronger version of the filtration distance D_{f}, namely,

$$D_f^*(\xi,\tilde{\xi}) := \sup_{x\in\mathcal{B}_\infty} \sum_{t=2}^{T} \left\| \mathbb{E}[x_t|\mathcal{F}_t(\xi)] - \mathbb{E}[x_t|\mathcal{F}_t(\tilde{\xi})] \right\|_{r'}, \tag{14.11}$$

where $\mathcal{B}_\infty := \{x : \Omega \to \mathbb{R}^m : x \text{ is } \mathcal{F}\text{-measurable}, |x(\omega)| \leq 1, \mathbb{P}\text{-almost surely}\}$. Notice that the sum is extended by the additional summand for $t = T$ and that the former infimum is replaced by a supremum with respect to a sufficiently large bounded set. If we require, in addition to assumption (A3), that for some $\varepsilon_0 > 0$ there exist constants $\delta > 0$ and $C > 0$ such that $|\tilde{x}(\omega)| \leq C$ for $\mathbb{P}$-almost every $\omega \in \Omega$ and all $\tilde{x} \in S_{\varepsilon_0}(\tilde{\xi})$ with $\tilde{\xi} \in L_r(\Omega, \mathcal{F}, \mathbb{P}; \mathbb{R}^s)$ and $\|\tilde{\xi} - \xi\|_r \leq \delta$, we have

$$D_f(\xi,\tilde{\xi}) \leq C\, D_f^*(\xi,\tilde{\xi}). \tag{14.12}$$

Sometimes it is sufficient to consider the unit ball in $L_{r'}$ rather than the corresponding ball $\mathcal{B}_\infty$ in L_∞ (cf. Heitsch and Römisch (2009a, 2010)). In contrast to D_f the distance D_f^* is a metric as it satisfies the triangle inequality.

Next we state a second stability result that represents a calmness property of approximate solution sets (Heitsch and Römisch (2010), theorem 2.4).

Theorem 2 *Let (A1), (A2), and (A3) be satisfied with $r' \in [1, +\infty)$ and the set X_1 be nonempty and bounded. Assume that $S(\xi) \neq \emptyset$. Then there exist $\bar{L} > 0$ and $\bar{\varepsilon} > 0$ such that*

$$d\!l_\infty\left(S_\varepsilon(\xi), S_\varepsilon(\tilde{\xi})\right) \leq \frac{\bar{L}}{\varepsilon}\left(\|\xi - \tilde{\xi}\|_r + D_f^*(\xi,\tilde{\xi})\right) \tag{14.13}$$

holds for every $\tilde{\xi} \in L_r(\Omega, \mathcal{F}, \mathbb{P}; \mathbb{R}^s)$ with $\|\xi - \tilde{\xi}\|_r \leq \delta$ (with $\delta > 0$ from (A3)) and $S(\tilde{\xi}) \neq \emptyset$, and for any $\varepsilon \in (0, \bar{\varepsilon})$. Here, $d\!l_\infty$ denotes the Pompeiu–Hausdorff distance of closed bounded subsets of $L_{r'} = L_{r'}(\Omega.\mathcal{F}, \mathbb{P}; \mathbb{R}^m)$ given by

$$d\!l_\infty(B, \tilde{B}) = \sup_{x\in L_{r'}} \left| d_B(x) - d_{\tilde{B}}(x) \right|$$

with $d_B(x)$ denoting the distance of x to B, i.e., $d_B(x) = \inf_{y\in B} \|x - y\|_{r'}$.

The most restrictive assumption in Theorem 2 is the existence of solutions to both problems. Notice that solutions always exist if the underlying random vector ξ has a finite number of scenarios or if $r' \in (1, +\infty)$. For a more thorough discussion we refer to Heitsch and Römisch (2010), section 2. Notice that the constant $\frac{\bar{L}}{\varepsilon}$ gets larger for decreasing ε and that, indeed, Theorem 2 does not remain true for the Pompeiu–Hausdorff distance of solution sets $S(\xi) = S_0(\xi)$ and $S(\tilde{\xi}) = S_0(\tilde{\xi})$, respectively.

14.3 Tree Approximation Framework and Convergence

We present a general framework for tree approximations of multi-stage stochastic programs in case that empirical estimates of the underlying probability distribution are available and prove convergence using the stability results of Section 14.2.

First, we show that sequences $(\xi^{(n)})$ of ξ-adapted random variables converging to ξ in L_r also converge to ξ in terms of the filtration distance D_{f}. Recall that $\xi^{(n)}$ is ξ-adapted if $\mathcal{F}_t(\xi^{(n)}) \subseteq \mathcal{F}_t(\xi)$ holds for every $t = 1, \ldots, T$.

Proposition 2 *Let (A1), (A2), and (A3) be satisfied,* $r' \in [1, +\infty)$ *and* $S(\xi)$ *be nonempty. Let* $(\xi^{(n)})$ *be a sequence of* ξ*-adapted random variables converging to* ξ *in* L_r *such that the* σ*-fields* $\mathcal{F}_t(\xi^{(n)})$ *are nondecreasing with respect to n for every* $t = 2, \ldots, T$. *Then it holds*

$$\lim_{n\to\infty} D_{\mathrm{f}}(\xi, \xi^{(n)}) = \lim_{n\to\infty} \inf_{x\in S(\xi)} \sum_{t=2}^{T-1} \left\| x_t - \mathbb{E}[x_t|\mathcal{F}_t(\xi^{(n)})]\right\|_{r'} = 0.$$

Proof By $\hat{\mathcal{F}}_t$ we denote the smallest σ-field containing $\mathcal{F}_t(\xi^{(n)})$ for every $n \in \mathbb{N}$, i.e.,

$$\hat{\mathcal{F}}_t := \sigma\left(\bigcup_{n\in\mathbb{N}} \mathcal{F}_t(\xi^{(n)})\right) \quad (t = 1, \ldots, T).$$

As $\mathcal{F}_t(\xi^{(n)}) \subseteq \mathcal{F}_t(\xi)$ holds for every $n \in \mathbb{N}$, we conclude

$$\hat{\mathcal{F}}_t \subseteq \mathcal{F}_t(\xi) \quad (\forall t = 1, \ldots, T)$$

due to the convergence of the sequence $(\xi^{(n)})$ to ξ in L_r. Furthermore, the filtration distance $D_{\mathrm{f}}(\xi, \xi^{(n)})$ allows the representation

$$D_{\mathrm{f}}(\xi, \xi^{(n)}) = \inf_{x\in S(\xi)} \sum_{t=2}^{T-1} \left\| x_t - \mathbb{E}[x_t|\mathcal{F}_t(\xi^{(n)})]\right\|_{r'}$$

due to (14.7). As the σ-fields $\mathcal{F}_t(\xi^{(n)})$ are nondecreasing with respect to n, we obtain for each $x \in S(\xi)$ and each $t = 2, \ldots, T$

$$\left\| x_t - \mathbb{E}[x_t|\mathcal{F}_t(\xi^{(n)})]\right\|_{r'} = \left\| \mathbb{E}[x_t|\hat{\mathcal{F}}_t] - \mathbb{E}[x_t|\mathcal{F}_t(\xi^{(n)})]\right\|_{r'} \longrightarrow 0$$

as $n \to \infty$ by classical convergence results of conditional expectations (e.g., Fetter (1977)). This completes the proof. □

The result remains true if the assumption that the σ-fields $\mathcal{F}_t(\xi^{(n)})$ are nondecreasing is replaced by the slightly weaker condition

$$\sigma\left(\bigcup_{k=1}^{\infty}\bigcap_{n=k}^{\infty}\mathcal{F}_t(\xi^{(n)})\right)=\sigma\left(\bigcap_{k=1}^{\infty}\bigcup_{n=k}^{\infty}\mathcal{F}_t(\xi^{(n)})\right)$$

for every $t = 2, \ldots, T$ (Fetter 1977). Next we show that ξ-adapted approximations can be obtained by certain discretization techniques.

14.3.1 Convergence of Discretizations

Let Ξ_t denote the closed subset supp (ξ_t) of $\mathbb{R}^d$ and $\Xi^t = \times_{\tau=1}^{t}\Xi_\tau$ for every $t = 1, \ldots, T$. Now, we consider ξ to be given on the probability space $(\Xi, \mathcal{B}(\Xi), P)$, where $\Xi = \Xi^T$ and $\mathcal{B}(\Xi)$ is the σ-field of Borel subsets of the sample space Ξ. Furthermore, the σ-fields $\mathcal{F}_t(\xi)$ are of the form

$$\mathcal{F}_t(\xi)=\left\{\xi^{-1}\left(B_t\times\Xi_{t+1}\times\cdots\times\Xi_T\right):B_t\in\mathcal{B}(\Xi^t)\right\}. \tag{14.14}$$

14.3.1.1 Decomposition of the Sample Space

We aim at approximating the stochastic process ξ by a certain sequence of discrete processes, i.e., by processes having a finite number of scenarios. The approach is based on finite decompositions of the sample space Ξ. Let us consider a sequence

$$\mathcal{D}_t^{(n)}\subset\mathcal{B}(\Xi_t),\quad n\in\mathbb{N}.$$

It will be called a *sequence of finite segmentations* of Ξ_t if the following conditions are satisfied:

(C1) The elements of $\mathcal{D}_t^{(n)}$ are pairwise disjoint for all $n \in \mathbb{N}$,
(C2) $\mathcal{D}_t^{(n)}$ is finite and it holds $P_t\left(\cup_{D_t\in\mathcal{D}_t^{(n)}}D_t\right)=1,\ n\in\mathbb{N}$.
(C3) For $\delta_{t,n} := \sup\left\{|\xi_t-\tilde{\xi}_t| : D_t\in\mathcal{D}_t^{(n)},\ \xi_t,\tilde{\xi}_t\in D_t,\ |\xi_t|,|\tilde{\xi}_t|\le n\right\}$ it holds $\lim_{n\to\infty}\delta_{t,n}=0$ for every $t = 1, \ldots, T$.

Conditions (C1) and (C2) ensure that the sets $\mathcal{D}_t^{(n)}$ define finite partitions of the sample space Ξ_t for every $n \in \mathbb{N}$ such that P-almost every element of Ξ_t can be associated with a unique set in $\mathcal{D}_t^{(n)}$. Condition (C3) says that the partition sets get arbitrarily small uniformly within increasing balls of radii n in Ξ_t.

Next we define a sequence of finite segmentations in Ξ by

$$\mathcal{D}^{(n)}:=\left\{D_1\times\cdots\times D_T : D_t\in\mathcal{D}_t^{(n)}, t=1,\ldots,T\right\},\quad n\in\mathbb{N}. \tag{14.15}$$

14.3.1.2 Discretization of the Stochastic Process

Using the sequence $\mathcal{D}^{(n)}$ we will define a sequence of approximate stochastic processes $\xi^{(n)} = (\xi_1^{(n)}, \ldots, \xi_T^{(n)})$. To this end, we select *nonanticipative* elements

$$\hat{\xi}_t^{D_1,\ldots,D_t,n} \in D_t \quad \text{with} \quad |\hat{\xi}_t^{D_1,\ldots,D_t,n}| \le C \max\{1, \inf\{|y_t| : y_t \in D_t\}\} \tag{14.16}$$

for every $n \in \mathbb{N}$, $t \in \{1, \ldots, T\}$ and every set $D_1 \times \cdots \times D_T \in \mathcal{D}^{(n)}$, where the boundedness condition in (14.16) has to be satisfied for some fixed constant $C \ge 1$. In this way we obtain a well-defined scenario

$$\hat{\xi}^{(n)}_{D_1 \times \cdots \times D_T} := \left(\hat{\xi}_1^{D_1,n}, \ldots, \hat{\xi}_T^{D_1,\ldots,D_T,n}\right)$$

for every $n \in \mathbb{N}$ and $D_1 \times \cdots \times D_T \in \mathcal{D}^{(n)}$ and define an approximate stochastic process by

$$\xi^{(n)}(\omega) := \hat{\xi}^{(n)}_{D_1 \times \cdots \times D_T} \text{ if } \omega \in \xi^{-1}(D_1 \times \cdots \times D_T) \tag{14.17}$$

and have $\xi^{(n)}$ well-defined on Ω $\mathbb{P}$-almost surely. The stochastic processes $\xi^{(n)}$ are approximations of ξ in the following sense.

Proposition 3 *Let $\xi \in L_r(\Omega, \mathcal{F}, \mathbb{P}; \mathbb{R}^s)$ and (C1), (C2), and (C3) be satisfied for each $t = 1, \ldots, T$. Then each stochastic process $\xi^{(n)}$ defined by (14.17) is ξ-adapted and it holds*

$$\lim_{n \to \infty} \|\xi - \xi^{(n)}\|_r = 0\,.$$

If, in addition, (A1), (A2), and (A3) are satisfied, $r' \in [1, +\infty)$, $S(\xi)$ is nonempty, and $\mathcal{D}_t^{(n+1)}$ is a refinement of $\mathcal{D}_t^{(n)}$ for each $n \in \mathbb{N}$, one has

$$\lim_{n \to \infty} D_{\mathrm{f}}(\xi, \xi^{(n)}) = 0\,.$$

Proof Due to the construction of $\xi^{(n)}$, the sets

$$\left\{\xi^{-1}\left(D_1 \times \cdots \times D_t \times \Xi_{t+1} \times \cdots \times \Xi_T\right) : D_\tau \in \mathcal{D}_\tau^{(n)}, \tau = 1, \ldots, t\right\}$$

generate the σ-fields $\mathcal{F}_t(\xi^{(n)})$. Thus, it holds $\mathcal{F}_t(\xi^{(n)}) \subseteq \mathcal{F}_t(\xi)$ according to (14.14) for every $n \in \mathbb{N}$ and $t = 1, \ldots, T$.

Next we show the L_r-convergence of $(\xi^{(n)})$. To this end, let

$$B_{\gamma_n}(0) := \left\{y \in \Xi : \max_{t=1,\ldots,T} |y_t| \le \gamma_n\right\}$$

the closed ball in Ξ around the origin with radius

$$\gamma_n := n - \max_{t=1\ldots,T} \delta_{t,n} .$$

Then, by using (C1) and (C2) we obtain

$$\begin{aligned}
\|\xi - \xi^{(n)}\|_r^r &= \int_\Omega \left|\xi(\omega) - \xi^{(n)}(\omega)\right|^r \mathbb{P}(d\omega) \\
&= \sum_{D_1\times\cdots\times D_T \in \mathcal{D}^{(n)}} \int_{D_1\times\cdots\times D_T} |\xi - \hat{\xi}^{(n)}_{D_1,\ldots,D_T}|^r P(d\xi) \\
&\leq c_r \sum_{D_1\times\cdots\times D_T \in \mathcal{D}^{(n)}} \int_{D_1\times\cdots\times D_T} \sum_{t=1}^T |\xi_t - \hat{\xi}_t^{D_1,\ldots,D_t,n}|^r P(d\xi),
\end{aligned}$$

where c_r is a suitable (norm equivalence) constant. Splitting the integration interval with respect to $B_{\gamma_n}(0)$ and its complement, using (C3) and the boundedness condition (14.16) allows to estimate

$$\begin{aligned}
\|\xi - \xi^{(n)}\|_r^r &\leq c_r \sum_{t=1}^T \delta_{t,n}^r + c_r \int_{\Xi\setminus B_{\gamma_n}(0)} \sum_{t=1}^T (|\xi_t|(1+C))^r P(d\xi) \\
&\leq \hat{C}\left(\max_{t=1,\ldots,T} \delta_{t,n}^r + \int_{\Xi\setminus B_{\gamma_n}(0)} |\xi|^r P(d\xi)\right),
\end{aligned}$$

where $\hat{C} > 0$ is some constant not depending on n. Because both summands of the last estimate tend to zero whenever n tends to infinity, the first part is proved. The second part is a consequence of Proposition 2. □

Proposition 3 and Theorem 1 provide a convergence result for discretizations of (14.1), namely,

$$\lim_{n\to\infty} v(\xi^{(n)}) = v(\xi) . \tag{14.18}$$

Of course, this is not surprising when recalling the convergence results for discretizations of multi-stage stochastic programs in Pennanen (2005, 2009).

To determine the probabilities of the scenarios of $\xi^{(n)}$ for some $n \in \mathbb{N}$, one has to compute

$$P(D_1 \times \cdots \times D_T) \quad \text{for every} \quad D_1 \times \cdots \times D_T \in \mathcal{D}^{(n)}.$$

This might be difficult in general. However, if additional structure on ξ is available, the discretization scheme may be adapted such that the probabilities are computationally accessible. For example, let the stochastic process ξ be driven by a finite number of mutually independent $\mathbb{R}^{d_t}$-valued random variables z_t with probability distributions P_t, $t = 2, \ldots, T$, i.e.,

$$\xi_t = g_t(\xi_1, \ldots, \xi_{t-1}, z_t),$$

where the g_t, $t = 2, \ldots, T$, denote certain measurable functions from $\mathbb{R}^{(t-1)d} \times \mathbb{R}^{d_t}$ to $\mathbb{R}^d$ (see, e.g., Kuhn (2005) and Pennanen (2009)). Then there exists a measurable function G such that $\xi = G(z_2, \ldots, z_T)$. If $\mathcal{D}_t^{(n)}$ is now a partition of the support of z_t in $\mathbb{R}^{d_t}$, $t = 2, \ldots, T$, then $\xi^{(n)}$ may be defined by

$$\xi_t^{(n)} = g_t(\xi_1^{(n)}, \ldots, \xi_{t-1}^{(n)}, z_t^{(n)}),$$

where $z_t^{(n)}$, $t = 2, \ldots, T$, has a finite number of given scenarios in every $D_t \in \mathcal{D}_t^{(n)}$. The probability distribution of $\xi^{(n)}$ is then known if $P_t(D_t)$ can be computed for all $D_t \in \mathcal{D}_t^{(n)}$, $t = 2, \ldots, T$. This covers situations, where ξ is a Gaussian process or is given by certain time series models.

14.3.2 Convergence of Discrete and Empirical Measures

Our next result deals with convergence properties of discrete probability distributions.

Proposition 4 *Let P be a probability distribution on $\mathbb{R}^{Td}$ supported by a finite set of scenarios $\Xi = \{\xi^1, \ldots, \xi^N\} \subseteq \mathbb{R}^{Td}$ with positive probabilities $p_i := P(\{\xi^i\})$. Moreover, let $(P^{(n)})_{n\in\mathbb{N}}$ be a sequence of probability distributions on $\mathbb{R}^{Td}$ supported by Ξ with probabilities $p_i^{(n)} := P^{(n)}(\{\xi^i\})$ such that*

$$\lim_{n\to\infty} p_i^{(n)} = p_i, \quad i = 1, \ldots, N.$$

Then there exist random variables ξ and $(\xi^{(n)})_{n\in\mathbb{N}}$ defined on some probability space $(\Omega, \mathcal{F}, \mathbb{P})$ having probability distributions P and $P^{(n)}$, $n \in \mathbb{N}$, respectively, such that

(i) $\|\xi - \xi^{(n)}\|_r \to 0 \quad (n \to \infty)$ *and*

(ii) $D_{\mathrm{f}}^*(\xi, \xi^{(n)}) = \sup_{x\in\mathcal{B}_\infty} \sum_{t=2}^{T} \left\|\mathbb{E}[x_t|\mathcal{F}_t(\xi)] - \mathbb{E}[x_t|\mathcal{F}_t(\xi^{(n)})]\right\|_{r'} \to 0 \quad (n \to \infty)$,

for all $1 \le r < \infty$ and $1 \le r' < \infty$, where $\mathcal{B}_\infty$ is given by

$$\mathcal{B}_\infty := \{x = (x_1, \ldots, x_T) \in L_{r'}(\Omega, \mathcal{F}, \mathbb{P}; \mathbb{R}^m) : |x_t(\omega)| \le 1,\ \mathbb{P}\text{-a.s.}\} \tag{14.19}$$

and denotes the set of all essentially bounded functions.

Proof The sequence $(P^{(n)})_{n\in\mathbb{N}}$ converges weakly to P on $\mathbb{R}^{Td}$. Hence, there exists a probability space $(\Omega, \mathcal{F}, \mathbb{P})$ and there exist $\mathbb{R}^{Td}$-valued random variables ξ and $\xi^{(n)}$, $n \in \mathbb{N}$, defined on it with probability distributions P and $P^{(n)}$, $n \in \mathbb{N}$, respectively, such that it holds

$$\xi^{(n)} \to \xi \quad \mathbb{P}\text{-a.s.}$$

(see, e.g., Dudley (1989), theorem 11.7.1). Since the random variables are supported by the finite set Ξ, almost sure convergence also implies convergence in the rth mean, i.e.,

$$\lim_{n\to\infty} \|\xi - \xi^{(n)}\|_r = 0\,.$$

It remains to show $\lim_{n\to\infty} D_{\mathrm{f}}^*(\xi, \xi^{(n)}) = 0$. To this end, we introduce partitions $\{E_{tk}\}_{k\in I_t}$ and $\{E_{tk}^{(n)}\}_{k\in I_t}$ in Ω, which generate the σ-fields $\mathcal{F}_t(\xi)$ and $\mathcal{F}_t(\xi^{(n)})$, respectively. Let

$$\begin{aligned} E_{tk} &:= \{\omega \in \Omega : (\xi_1, \ldots, \xi_t)(\omega) = (\xi_1^k, \ldots, \xi_t^k)\}, \quad k \in I_t, \\ E_{tk}^{(n)} &:= \{\omega \in \Omega : (\xi_1^{(n)}, \ldots, \xi_t^{(n)})(\omega) = (\xi_1^k, \ldots, \xi_t^k)\}, \quad k \in I_t, \end{aligned}$$

where $I_t \subseteq \{1, \ldots, N\}$ denotes the index set of distinguishable scenarios at time $t = 2, \ldots, T$. We set

$$\bar{E}_{tk}^{(n)} := E_{tk} \cap E_{tk}^{(n)} \quad \text{and} \quad \bar{\Omega}_t^{(n)} := \Omega \setminus \cup_{k\in I_t} \bar{E}_{tk}^{(n)},$$

and observe that

$$\begin{aligned} \int_\Omega |\xi - \xi^{(n)}|^r \mathbb{P}(d\omega) &\geq \int_{E_{tk}\setminus \bar{E}_{tk}^{(n)}} |\xi - \xi^{(n)}|^r \mathbb{P}(d\omega) + \int_{E_{tk}^{(n)}\setminus \bar{E}_{tk}^{(n)}} |\xi - \xi^{(n)}|^r \mathbb{P}(d\omega) \\ &\geq C_{\min} \left(\mathbb{P}(E_{tk} \setminus \bar{E}_{tk}^{(n)}) + \mathbb{P}(E_{tk}^{(n)} \setminus \bar{E}_{tk}^{(n)}) \right), \end{aligned}$$

where $C_{\min} := \min\left\{ |\xi^i - \xi^j|^r : i, j \in \{1, \ldots, N\}, \xi^i \neq \xi^j \right\}$ denotes the minimal distance between two varying scenarios. Hence, by the L_r-convergence of $\xi^{(n)}$ we conclude for all $k \in I_t$ that $\mathbb{P}(E_{tk} \setminus \bar{E}_{tk}^{(n)})$ and $\mathbb{P}(E_{tk}^{(n)} \setminus \bar{E}_{tk}^{(n)})$ tend to zero whenever n tends to infinity. Moreover, the latter implies that $\mathbb{P}(E_{tk}^{(n)})$ as well as $\mathbb{P}(\bar{E}_{tk}^{(n)})$ converge to $\mathbb{P}(E_{tk})$. Let $p_{tk} = \mathbb{P}(E_{tk})$ and $p_{tk}^{(n)} = \mathbb{P}(E_{tk}^{(n)})$. Then we have for any $x \in \mathcal{B}_\infty$

$$\begin{aligned} \left\|\mathbb{E}[x_t|\mathcal{F}_t(\xi)] - \mathbb{E}[x_t|\mathcal{F}_t(\xi^{(n)})]\right\|_{r'}^{r'} &= \int_\Omega \left|\mathbb{E}[x_t|\mathcal{F}_t(\xi)] - \mathbb{E}[x_t|\mathcal{F}_t(\xi^{(n)})]\right|^{r'} \mathbb{P}(d\omega) \\ &= \sum_{k\in I_t} \int_{\bar{E}_{tk}^{(n)}} \left|\mathbb{E}[x_t|\mathcal{F}_t(\xi)] - \mathbb{E}[x_t|\mathcal{F}_t(\xi^{(n)})]\right|^{r'} \mathbb{P}(d\omega) \\ &\quad + \int_{\bar{\Omega}_t^{(n)}} \left|\mathbb{E}[x_t|\mathcal{F}_t(\xi)] - \mathbb{E}[x_t|\mathcal{F}_t(\xi^{(n)})]\right|^{r'} \mathbb{P}(d\omega) \end{aligned}$$

$$\leq \sum_{k\in I_t} \mathbb{P}(\bar{E}_{tk}^{(n)}) \left| \frac{\int_{E_{tk}} x_t \mathbb{P}(d\omega)}{p_{tk}} - \frac{\int_{E_{tk}^{(n)}} x_t \mathbb{P}(d\omega)}{p_{tk}^{(n)}} \right|^{r'} + 2\mathbb{P}(\bar{\Omega}_t^{(n)})$$

$$\leq \sum_{k\in I_t} \mathbb{P}(\bar{E}_{tk}^{(n)}) \left(\left| \frac{\int_{\bar{E}_{tk}^{(n)}} x_t (p_{tk}^{(n)} - p_{tk}) \mathbb{P}(d\omega)}{p_{tk}^{(n)} p_{tk}} \right| \right.$$

$$\left. + \left| \frac{\int_{E_{tk}\setminus \bar{E}_{tk}^{(n)}} p_{tk}^{(n)} x_t \mathbb{P}(d\omega)}{p_{tk}^{(n)} p_{tk}} \right| + \left| \frac{\int_{E_{tk}^{(n)}\setminus \bar{E}_{tk}^{(n)}} p_{tk} x_t \mathbb{P}(d\omega)}{p_{tk}^{(n)} p_{tk}} \right| \right)^{r'}$$

$$+ 2\mathbb{P}(\bar{\Omega}_t^{(n)})$$

$$\leq \sum_{k\in I_t} \mathbb{P}(\bar{E}_{tk}^{(n)}) \left(\frac{\mathbb{P}(\bar{E}_{tk}^{(n)}) |p_{tk}^{(n)} - p_{tk}|}{p_{tk}^{(n)} p_{tk}} + \frac{\mathbb{P}(E_{tk} \setminus \bar{E}_{tk}^{(n)})}{p_{tk}} \right.$$

$$\left. + \frac{\mathbb{P}(E_{tk}^{(n)} \setminus \bar{E}_{tk}^{(n)})}{p_{tk}^{(n)}} \right)^{r'} + 2\mathbb{P}(\bar{\Omega}_t^{(n)}),$$

where we used the almost sure boundedness $|x_t| \leq 1$. The latter estimate does not depend on x and due to the fact that all summands tend to zero, $D_{\mathrm{f}}^*(\xi, \xi^{(n)})$ converges to 0. □

Proposition 4 will be used in the proof of Theorem 3 to compare two stochastic processes having identical scenarios, but different probabilities.

14.3.2.1 Empirical Distributions and Sampling

Let ξ be a Ξ-valued stochastic process defined on some probability space $(\Omega, \mathcal{F}, \mathbb{P})$ with induced distribution P. Furthermore, let $(\xi^k)_{k\in\mathbb{N}}$ be a sequence of independent and identically distributed Ξ-valued random variables on some probability space $(\Omega^*, \mathcal{F}^*, \mathbb{P}^*)$ such that $P = \mathbb{P}^* \xi_1^{-1}$. We consider the random empirical measures

$$P^{(k)}(\omega^*)(B) := \frac{1}{k} \sum_{j=1}^{k} \delta_{\xi^j(\omega^*)}(B), \qquad n \in \mathbb{N},\ \omega^* \in \Omega^*,\ B \in \mathcal{B}(\Xi), \tag{14.20}$$

where δ_z denotes the probability measure on Ξ placing unit mass at $z \in \Xi$. Then the sequence $(P^{(k)}(\omega^*))$ converges $\mathbb{P}^*$-almost surely to P in the sense of weak convergence (see, e.g., Dudley (1989), chapter 11.4). The portmanteau theorem (e.g., Dudley (1989), theorem 11.1.1) offers several equivalent characterizations of weak convergence, one of them is recalled in the following for later use.

Proposition 5 *Let $(P^{(k)})$ be a sequence of empirical measures according to (14.20). Then it holds*

$$\lim_{k\to\infty} P^{(k)}(\omega^*)(B) = P(B), \qquad \text{for all } B \in \mathcal{B}(\Xi) \text{ with } P(\partial B) = 0,$$

for $\mathbb{P}^*$-almost every $\omega^* \in \Omega^*$, where ∂B denotes the (topological) boundary of the Borel set B in the space $\mathbb{R}^s$.

Proposition 5 allows to estimate probabilities of Borel sets in $\mathbb{R}^s$ empirically, e.g., by sampling. In particular, the probability of any set belonging to a finite segmentation $\mathcal{D}^{(n)}$ of Ξ (see (14.15)) can be estimated by sampling if its boundary has Lebesgue measure zero and P is absolutely continuous.

14.3.3 Application to Scenario Tree Construction

The following conceptual algorithm represents a general approach to constructing scenario trees for multi-stage stochastic programs.

Algorithm 1 *Let ξ be the original Ξ-valued stochastic input process of the stochastic program (14.1) defined on some probability space $(\Omega, \mathcal{F}, \mathbb{P})$ and let P be the probability distribution of ξ.*

- ***Step [1]:*** *Determine a sequence of finite segmentations $\mathcal{D}^{(n)}$ in Ξ such that assumptions (C1), (C2), and (C3) are satisfied (cf. Sect. 14.3.1) and choose a reasonably large $n \in \mathbb{N}$.*
- ***Step [2]:*** *Determine the empirical measure $P^{(k)}$ based on k independent and identically P-distributed random variables.*
- ***Step [3]:*** *Compute the probabilities $p^{(n,k)} = P^{(k)}(D_1 \times \cdots \times D_T)$ for every $D_1 \times \cdots \times D_T \in \mathcal{D}^{(n)}$ according to formula (14.20).*
- ***Step [4]:*** *Choose nonanticipative scenarios whose tth components belong to D_t for any $D_1 \times \cdots \times D_T \in \mathcal{D}^{(n)}$ according to (14.16).*
- ***Step [5]:*** *Finally, define the stochastic scenario tree process $\xi_{\text{tr}} := \xi_{\text{tr}}^{(n,k)}$ with scenarios chosen in step 4 endowed with the empirical probabilities $p^{(n,k)}$ from step 3.*

Next we study the asymptotic behavior of the approximate scenario trees $(\xi_{\text{tr}}^{(n,k)})_{n,k\in\mathbb{N}}$ constructed by Algorithm 1. Note that the parameters n and k measure the quality of the discretization $\mathcal{D}^{(n)}$ of Ξ and of the empirical probabilities, respectively.

Theorem 3 *Let (A1), (A2), and (A3) be satisfied and X_1 be nonempty and bounded. Let $1 \le r' < \infty$ and assume that for some constant $C > 0$ the estimate*

$$D_{\mathrm{f}}(\hat{\xi}, \tilde{\xi}) \le C D_{\mathrm{f}}^*(\hat{\xi}, \tilde{\xi}) \tag{14.21}$$

holds for all $\hat{\xi}$ and $\tilde{\xi}$ in a neighborhood of ξ in L_r. Assume that the sequence $(\xi_{\text{tr}}^{(n,k)})_{n,k\in\mathbb{N}}$ is constructed by Algorithm 1. Furthermore, assume for the sequence

$(\mathcal{D}^{(n)})_{n\in\mathbb{N}}$ of partitions of Ξ that $\mathcal{D}^{(n+1)}$ is a refinement of $\mathcal{D}^{(n)}$ and that $P_t(\partial D_t) = 0$ for all $D_t \in \mathcal{D}_t^{(n)}$ $(t = 1, \ldots, T)$. Then it holds

$$\lim_{n\to\infty}\left(\lim_{k\to\infty} v(\xi_{\mathrm{tr}}^{(n,k)})\right) = v(\xi) \quad P^*\text{-almost surely,} \tag{14.22}$$

where $v(\xi_{tr}^{(n,k)})$ and $v(\xi)$ denote the optimal values of the stochastic program (14.1) with input $\xi_{\mathrm{tr}}^{(n,k)}$ and ξ, respectively, and $(\Omega^*, \mathcal{F}^*, \mathbb{P}^*)$ is the probability space on which the random empirical measures $P^{(k)}$, $k \in \mathbb{N}$, are defined.

Proof We use the constructions and notations of Section 14.3.1 and consider, for each $n \in \mathbb{N}$, the (discrete) stochastic processes $\xi^{(n)}$ possessing the scenarios $\hat{\xi}^{(n)}_{D_1\times\cdots\times D_T}$ with probabilities $P(D_1 \times \cdots \times D_T)$ for every $D_1 \times \cdots \times D_T \in \mathcal{D}^{(n)}$. Due to Proposition 3 the processes $\xi^{(n)}$ are ξ-adapted and it holds

$$\lim_{n\to\infty} \|\xi - \xi^{(n)}\|_r = 0 \quad \text{and} \lim_{n\to\infty} D_{\mathrm{f}}(\xi, \xi^{(n)}) = 0\,.$$

Hence, we obtain from (14.5) in Theorem 1

$$|v(\xi) - v(\xi^{(n)})| \leq \varepsilon_n$$

for some sequence (ε_n) tending to zero as $n \to \infty$.

Let $\hat{\Omega}^*$ be the subset of Ω^* such that $P^*(\hat{\Omega}^*) = 1$ and the sequence $(P^{(k)}(\omega^*))$ of random empirical measures converges weakly for every $\omega^* \in \hat{\Omega}^*$.

Now, let $\omega^* \in \hat{\Omega}^*$ and let $\xi_{\mathrm{tr}}^{(n,k)} = \xi_{\mathrm{tr}}^{(n,k)}(\omega^*)$ be the process determined by Algorithm 1. We will show that for any large $n \in \mathbb{N}$ there exists $k(n) \in \mathbb{N}$ (depending on ω^*) such that

$$|v(\xi^{(n)}) - v(\xi_{\mathrm{tr}}^{(n,k)}(\omega^*))| \leq \frac{1}{n}$$

for all $k \geq k(n)$. Then the triangle inequality would imply

$$|v(\xi) - v(\xi_{\mathrm{tr}}^{(n,k)}(\omega^*))| \leq \varepsilon_n + \frac{1}{n} \qquad (k \geq k(n))$$

and, hence, the proof was complete.

To this end, let $n \in \mathbb{N}$ be sufficiently large and fixed. We consider the processes $\xi^{(n)}$ and $\xi_{\mathrm{tr}}^{(n,k)}$ $(k \in \mathbb{N})$ and observe that both processes possess identical scenarios which do not depend on k. In general, only the probabilities

$$p^{(n)}_{D_1\times\cdots\times D_T} = P(D_1 \times \cdots \times D_T) \text{ and } p^{(n,k)}_{D_1\times\cdots\times D_T} = P^{(k)}(\omega^*)(D_1 \times \cdots \times D_T)$$

associated with scenario $\hat{\xi}^{(n)}_{D_1\times\cdots\times D_T}$ are different.

It holds $P(\partial(D_1 \times \cdots \times D_T)) \leq \sum_{t=1}^{T} P_t(\partial D_t) = 0$ since for every boundary point x of $D_1 \times \cdots \times D_T$ there exists $t \in \{1, \ldots, T\}$ such that $x_t \in \partial D_t$. Hence, Proposition 5 implies

$$p^{(n,k)}_{D_1 \times \cdots \times D_T} \to p^{(n)}_{D_1 \times \cdots \times D_T} \qquad (k \to \infty),$$

and Proposition 4 yields the existence of a common probability space such that

$$\|\xi^{(n)} - \xi_{\text{tr}}^{(n,k)}\|_r \to 0 \quad \text{and} \quad D_{\text{f}}^*(\xi^{(n)}, \xi_{\text{tr}}^{(n,k)}) \to 0 \quad (k \to \infty).$$

Then estimate (14.21) implies that the inequality

$$D_{\text{f}}(\xi^{(n)}, \xi_{\text{tr}}^{(n,k)}) \leq C D_{\text{f}}^*(\xi^{(n)}, \xi_{\text{tr}}^{(n,k)})$$

holds for large n and k. By making use of Theorem 1 (applied to $\xi^{(n)}$ instead of ξ), we obtain

$$|v(\xi^{(n)}) - v(\xi_{\text{tr}}^{(n,k)})| \leq L(\xi^{(n)})\big(\|\xi^{(n)} - \xi_{\text{tr}}^{(n,k)}\|_r + D_{\text{f}}^*(\xi^{(n)}, \xi_{\text{tr}}^{(n,k)})\big)$$

for some constant $L(\xi^{(n)})$ and all sufficiently large $k \in \mathbb{N}$. This implies

$$|v(\xi^{(n)}) - v(\xi_{\text{tr}}^{(n,k)})| \to 0 \qquad (k \to \infty)$$

and, in particular, the existence of $k(n)$ such that

$$|v(\xi^{(n)}) - v(\xi_{\text{tr}}^{(n,k)})| \leq \frac{1}{n}$$

for all $k \geq k(n)$. □

We note that the limits in (14.22) cannot be interchanged. This may be interpreted such that it makes no sense to choose n very large, i.e., to choose a very fine partition of Ξ if for some reason k is not sufficiently large. In Section 14.4 we will discuss a variant of the general scenario tree construction approach provided by Algorithm 1 that is based on successive scenario reduction.

14.4 Scenario Tree Construction Based on Scenario Reduction

Next we discuss a method for generating scenario trees which is developed in Heitsch and Römisch (2009a) and motivated by Algorithm 1. It is based on a procedure of successive scenario reduction and bundling steps for increasing time stages applied to a sufficiently large scenario set. The latter is typically obtained by sampling from the underlying distribution. The stage-wise reduction may be viewed as a simultaneous realization of steps 1, 3, and 4 of Algorithm 1. Before describing the details, we briefly recall the ideas of optimal scenario reduction.

14.4.1 Optimal Scenario Reduction

The basic idea of scenario reduction consists in determining a (nearly) best approximation in terms of a suitable probability metric of the underlying discrete probability distribution by a probability measure with smaller support. The metric is associated with the stochastic programming model in a natural way such that the model behaves stable with respect to changes of the probability distribution. Such *natural* metrics are provided in Römisch (2003) for several classes of stochastic programs.

Originally, the concept of scenario reduction was developed in Dupačová et al. (2003) and Heitsch and Römisch (2003). More recently, it has been improved for two-stage models in Heitsch and Römisch (2007) and extended to mixed-integer and chance-constrained models in Henrion et al. (2008, 2009) as well as to multi-stage models in Heitsch and Römisch (2009b). The concept does not impose special conditions on the underlying probability distribution except the existence of certain moments.

Scenario reduction aims at reducing the number of scenarios in an optimal way. If ξ is a given random vector on some probability space $(\Omega, \mathcal{F}, \mathbb{P})$ with finite support, i.e., represented by the scenarios ξ^i and probabilities p_i, $i = 1, \ldots, N$, then one may be interested in finding a suitable index subset $J \subset \{1, \ldots, N\}$ and a new random vector $\hat{\xi}$ supported only by the scenarios ξ^j, $j \notin J$, such that $\hat{\xi}$ is the best approximation to ξ. Here, we consider the norm $\|\cdot\|_r$ in L_r as the natural distance function.

If J is given, the best approximation to ξ can be given explicitly. To show this, let $A_i = \xi^{-1}(\xi^i) \in \mathcal{F}$, $i = 1, \ldots, N$. Then

$$\|\xi - \hat{\xi}\|_r^r = \sum_{i=1}^{N} \int_{A_i} |\xi^i - \xi^{j(i)}|^r \mathbb{P}(d\omega) = \sum_{i=1}^{N} p_i |\xi^i - \xi^{j(i)}|^r$$

for some mapping $j : \{1, \ldots, N\} \to \{1, \ldots, N\} \setminus J$ and the best approximation problem reads

$$\min \left\{ \sum_{i=1}^{N} p_i |\xi^i - \xi^{j(i)}|^r \,\middle|\, j : \{1, \ldots, N\} \to \{1, \ldots, N\} \setminus J \right\}. \tag{14.23}$$

Since the lower bound

$$\sum_{i=1}^{N} p_i |\xi^i - \xi^{j(i)}|^r \geq \sum_{i \in J} p_i \min_{j \notin J} |\xi^i - \xi^j|^r$$

is always valid, the minimum in (14.23) is attained for the mapping j : $\{1,\ldots,N\} \to \{1,\ldots,N\} \setminus J$ defined by

$$j(i) \in \arg\min_{j \notin J} |\xi^i - \xi^j|, \qquad i \in J.$$

Hence, the best approximation $\hat{\xi}$ is supported by the scenarios ξ^j with probabilities q_j, $j \notin J$, where

$$\|\xi - \hat{\xi}\|_r^r = \sum_{i \in J} p_i \min_{j \notin J} |\xi^i - \xi^j|^r, \tag{14.24}$$

$$q_j = p_j + \sum_{\substack{i \in J \\ j(i)=j}} p_i . \tag{14.25}$$

In other words, the *redistribution rule* (14.25) consists in assigning the new probability to a preserved scenario to be equal to the sum of its former probability and of all probabilities of deleted scenarios that are closest to it.

Finding the optimal index set J, say, with prescribed cardinality, such that it solves the combinatorial optimization problem

$$\min\left\{\sum_{i \in J} p_i \min_{j \notin J} |\xi^i - \xi^j|^r : J \subseteq \{1,\ldots,N\},\ |J| = N - n\right\} \quad (1 \leq n < N),$$

is much more complicated. The latter problem represents a metric k-median problem which is known to be NP-hard, hence (polynomial-time) approximation algorithms and heuristics become important. Simple heuristics may be derived from formula (14.24) for the approximation error. The results are two heuristic algorithms to compute nearly optimal index sets J with given cardinality $N - n$.

Algorithm 2 *(Forward selection)*

[*Initialization*]
Set $J := \{1,\ldots,N\}$.

[*Index Selection*]
Determine an index $l \in J$ such that

$$l \in \arg\min_{u \in J} \sum_{k \in J \setminus \{u\}} p_k \min_{j \notin J \setminus \{u\}} |\xi^k - \xi^j|^r$$

and set $J := J \setminus \{l\}$. If the cardinality of J equals $N - n$ go to the termination step. Otherwise continue with a further index selection step.

[*Termination*]
Determine scenarios ξ^j, $j \notin J$, and apply the redistribution rule (14.25) for the final index set J.

Algorithm 3 *(Backward reduction)*

[*Initialization*]
Set $J := \emptyset$.

[*Index Selection*]
Determine an index $u \notin J$ such that

$$u \in \arg\min_{l \notin J} \sum_{k \in J \cup \{l\}} p_k \min_{j \notin J \cup \{l\}} |\xi^k - \xi^j|^r$$

and set $J := J \cup \{u\}$. If the cardinality of J equals $N - n$ go to the termination step. Otherwise continue with a further index selection step.

[*Termination*]
Determine scenarios ξ^j, $j \notin J$, and apply the redistribution rule (14.25) for the final index set J.

Optimal scenario reduction allows two interesting interpretations. The first is that it may actually be considered as a problem of *optimal quantization of probability measures* (in the sense of Graf and Luschgy (2000)) when applied to a discrete probability measure (with N atoms) and using the rth order Wasserstein distance (see also Heitsch and Römisch (2009a), lemma 2.1). Second, scenario reduction leads to a canonical decomposition of the sample space $\mathbb{R}^s$. To illustrate this fact assume that $\{\xi^j : j \notin J\}$ has been computed to reduce the original scenario set $\{\xi^1, \ldots, \xi^N\}$ contained in $\mathbb{R}^s$. Then the so-called *Voronoi regions* defined by

$$V(\xi^j) := \left\{ \xi \in \mathbb{R}^s : |\xi - \xi^j| < \min_{k \notin J \cup \{j\}} |\xi - \xi^k| \right\}, \quad j \notin J$$

represent disjoint subsets of $\mathbb{R}^s$. It is known that for strictly convex norms $|\cdot|$ the union of the closures of $V(\xi^j)$ cover $\mathbb{R}^s$ and the boundaries $\partial V(\xi^j)$ have Lebesgue measure λ^s zero. The latter holds for the l_p-norms with $1 < p < \infty$ and for $p = 2$ Voronoi regions are even convex (see Graf and Luschgy (2000), section 1). Voronoi regions are a suitable choice for the sets D_t in Section 14.3.1.

Figure 14.1 shows the Voronoi decomposition of the space R^2 obtained by scenario reduction starting from $N = 1\,000$ samples from the two-dimensional standard normal distribution computed with Algorithm 2.

14.4.2 Scenario Tree Construction

The idea of the tree construction method is to apply the scenario reduction techniques to a set of scenarios successively for increasing and decreasing time stages, respectively. This leads to forward or backward variants of a tree generation method that aims at recovering the original information structure approximately. Next we present a detailed description of the forward variant, the backward approach may be found in Heitsch and Römisch (2009a).

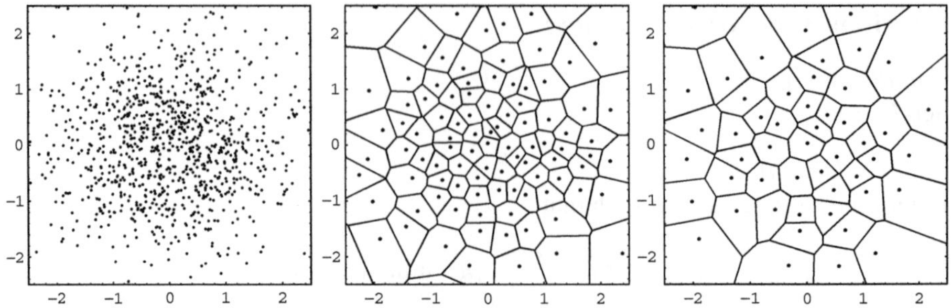

Fig. 14.1 Illustration of scenario reduction starting from 1000 sample scenarios of the two-dimensional standard normal distribution (*left*) reduced to 100 scenarios (*middle*) and further reduced to 50 scenarios (*right*). Displayed are the corresponding decompositions of $\mathbb{R}^2$ into Voronoi regions obtained by the forward selection algorithm with respect to the Euclidean norm

In the following, let $I := \{1, \ldots, N\}$ be the index set of the given set of scenarios ξ^i. Then the successive scenario reduction technique is applied to the time horizons $\{1, \ldots, t\}$ with increasing time $t \in \{1, \ldots, T\}$. It computes partitions of I of the form

$$\mathcal{C}_t := \{C_t^1, \ldots, C_t^{k_t}\}, \qquad k_t \in \mathbb{N},$$

successively such that

$$C_t^k \cap C_t^{k'} = \emptyset \quad \text{for} \quad k \neq k', \quad \text{and} \quad \bigcup_{k=1}^{k_t} C_t^k = I$$

holds for every t. The elements of a partition $\mathcal{C}_t$ are called (scenario) clusters. The following algorithm allows to generate different scenario tree processes depending on the parameter settings for the reductions in each step (Fig. 14.2).

Algorithm 4 (*Forward construction*)

[*Initialization*]
Define $\mathcal{C}_1 = \{I\}$ and set $t := 2$.

[*Cluster computation*]
Let $\mathcal{C}_{t-1} = \{C_{t-1}^1, \ldots, C_{t-1}^{k_{t-1}}\}$. For every $k \in \{1, \ldots, k_{t-1}\}$ apply scenario reduction to the scenario subsets $\{\xi_t^i\}_{i \in C_{t-1}^k}$ (at time t). This yields disjoint subsets of remaining and deleted scenarios I_t^k and J_t^k, respectively. Next, obtain the mappings $j_t^k : J_t^k \to I_t^k$ such that

$$j_t^k(i) \in \arg\min_{j \in I_t^k} |\xi_t^i - \xi_t^j|, \quad i \in J_t^k,$$

according to the reduction procedure (cf. Section 14.4.1). Finally, define an overall mapping $\alpha_t : I \to I$ by

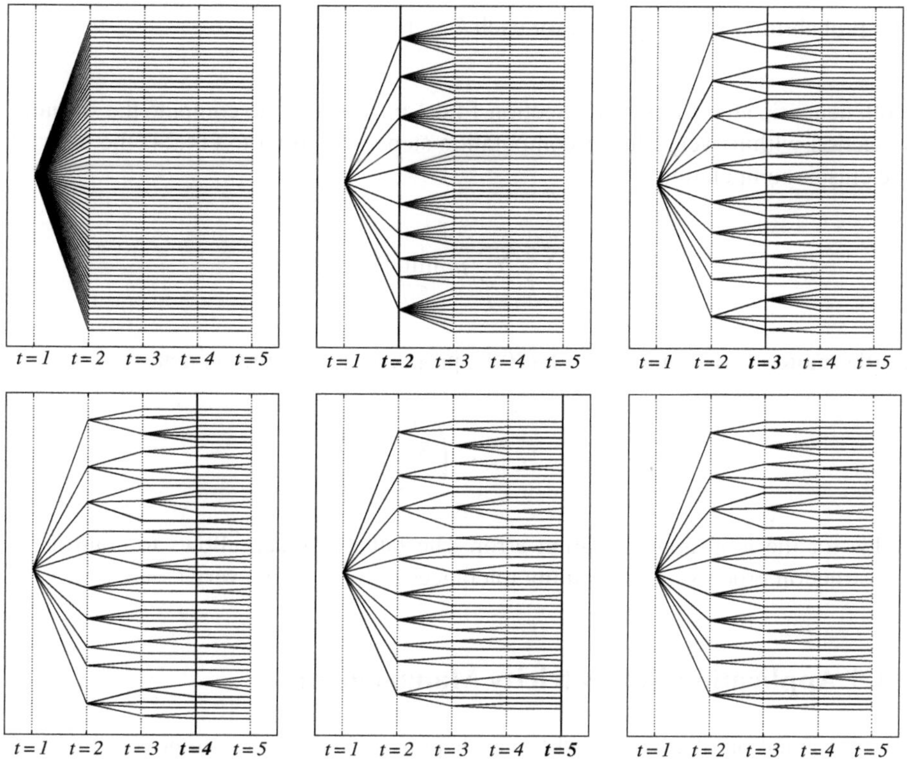

Fig. 14.2 Demonstration of the forward tree construction for an example containing $T = 5$ time periods. Displayed are the stepwise changes of the scenarios tree structure starting with a fan of individual scenarios

$$\alpha_t(i) = \begin{cases} j_t^k(i), & i \in J_t^k \text{ for some } k = 1, \dots, k_{t-1}, \\ i & \text{otherwise.} \end{cases} \tag{14.26}$$

A new partition at t is defined now by

$$C_t := \left\{ \alpha_t^{-1}(i) \,\middle|\, i \in I_t^k,\ k = 1, \dots, k_{t-1} \right\}$$

which is in fact a refinement of the partition C_{t-1}. If $t < T$ set $t := t + 1$ and continue with a further cluster computation step, otherwise go to the termination step.

[*Termination*]
According to the partition set C_T and mappings (14.26) define a scenario tree process ξ_{tr} supported by the scenarios

$$\xi_{\text{tr}}^k = \left(\xi_1^*, \xi_2^{\alpha_2(i)}, \dots, \xi_t^{\alpha_t(i)}, \dots, \xi_T^{\alpha_T(i)} \right) \quad \text{for any} \quad i \in C_T^k,$$

and probabilities $q_k := \sum_{i \in C_T^k} p_i$, for each $k = 1, \ldots, k_T$.

Both heuristic algorithms from Section 14.4.1 may be used to compute the scenario reduction within every cluster computation step. According to (14.24) the error of the cluster computation step t is

$$\text{err}_t := \sum_{k=1}^{k_{t-1}} \sum_{i \in J_t^k} p_i \min_{j \in I_t^k} |\xi_t^i - \xi_t^j|^r.$$

Furthermore, as shown in Heitsch (2007, proposition 6.6), the estimate

$$\|\xi - \xi_{\text{tr}}\|_r \leq \left(\sum_{t=2}^{T} \text{err}_t \right)^{\frac{1}{r}}$$

holds for the total approximation error. The latter estimate allows to control the construction process by prescribing tolerances ε_t for err_t for every $t = 2, \ldots, T$.

14.5 Application to Electricity Management

The deregulation of energy markets has lead to an increased awareness of the need for profit maximization with simultaneous consideration of financial risk, adapted to individual risk aversion policies of market participants. Mathematical modeling of such optimization problems with uncertain input data results in large-scale stochastic programming models with a risk functional in the objective. When considering a medium-term planning horizon, one is faced with consecutive decisions based on consecutive observations, thus, the stochastic programs need to be multi-stage.

Next we report on some experiences with constructing scenario trees for a multi-stage stochastic optimization model that is tailored to the requirements of a typical German municipal power utility, which has to serve an electricity demand and a heat demand of customers in a city and its vicinity. The power utility owns a combined heat and power (CHP) facility that can serve the heat demand completely and the electricity demand partly. Further electricity can be obtained by purchasing volumes for each hour at the (day ahead) spot market of the European Energy Exchange (EEX) or by signing a supply contract for a medium-term horizon with a larger power producer. The latter possibility is suspected to be expensive, but relying on the spot market only is known to be extremely risky. Spot price risk, however, may be reduced by obtaining electricity futures at EEX. The optimization aims to maximize the mean overall revenue and, simultaneously, to minimize a risk functional on a basis of a hourly discretized optimization horizon of 1 year. Details of the optimization model can be found in Eichhorn et al. (2005).

Electricity demand and heat demand as well as spot and future prices are not known in advance, but statistical information is available due to historical observations (Fig. 14.3). A very heterogeneous statistical model is employed. It is adapted

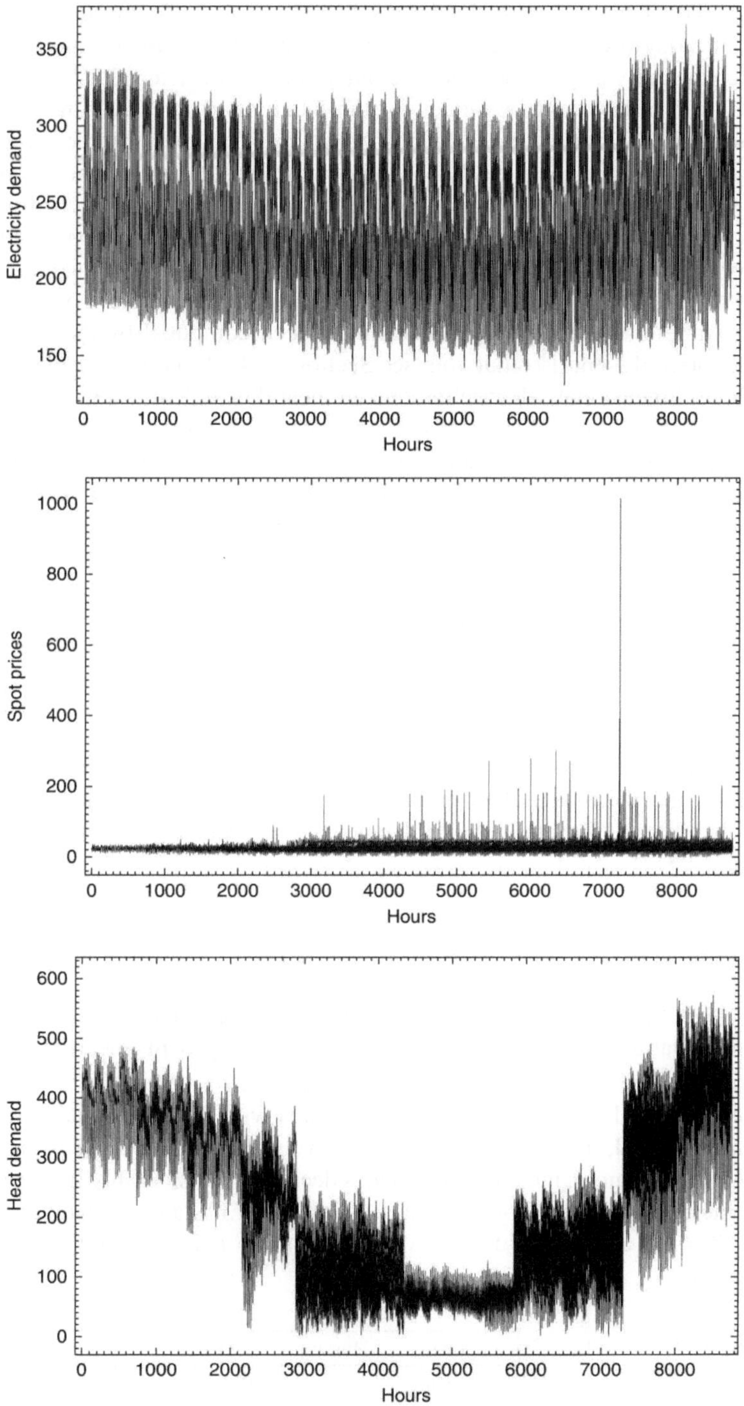

Fig. 14.3 Yearly scenario profiles of the trivariate stochastic process with components electricity demand (*top*), spot prices (*center*), and heat demand (*bottom*)

to historical data in a rather involved procedure. It consists of a cluster classification for the intra-day (demand and price) profiles and a three-dimensional time series model for the daily average values. The latter consists of deterministic trend functions and a trivariate ARMA model for the (stationary) residual time series; see Eichhorn et al. (2005) for further details. An arbitrary number of three-dimensional sample paths (scenarios) can easily be obtained by simulating white noise processes for the ARMA model and by adding on the trend functions and matched intra-day profiles from the clusters afterward. However, such a bunch of sample paths does not reflect the information structure in multi-stage stochastic optimization, i.e., it neglects the fact that information is revealed gradually over time. For this reason, finally, trivariate scenario tree processes have been computed by the approach of recursive forward scenario reduction (see Section 14.4). Table 14.1 displays the size of the three-dimensional (electricity demand, heat demand, and spot price) scenarios which serve as inputs for the tree construction (Algorithm 4). We performed a couple of test series for generating scenario trees. Due to the fact that electricity future products can only be traded monthly, branching was allowed only at the end of each month which leads to scenario trees of at most 12 stages. Because stochastic data enter both the objective and right-hand sides of the model Algorithm 4 is used with $r = r' = 2$ (cf. (14.2)). Moreover, different relative reduction levels $\varepsilon_{\mathrm{rel}}$ have been chosen. The relative levels are given by

$$\varepsilon_{\mathrm{rel}} := \frac{\varepsilon}{\varepsilon_{\max}} \qquad \text{and} \qquad \varepsilon_{\mathrm{rel},t} := \frac{\varepsilon_t}{\varepsilon_{\max}},$$

where $\varepsilon_{\max}$ is given as the best possible L_r-distance of the stochastic process represented by all scenarios and their probabilities and of one of its scenarios endowed with probability 1. The individual tolerances ε_t at branching points is computed such that

$$\varepsilon_t^r = \frac{2\varepsilon^r}{T-1}\left(q + (1-2q)\frac{t-2}{T-2}\right), \quad t = 2, \ldots, T, \tag{14.27}$$

where $q \in [0, 1]$ is a parameter that affects the branching structure of the constructed trees. Note that a value $q < \frac{1}{2}$ generates a sequence ε_t^r with linear growth while $q > \frac{1}{2}$ results in a decreasing sequence ε_t^r, $t = 1, \ldots, T$.

Table 14.2 displays the results of our test runs with different relative reduction levels. As expected, for very small reduction levels, the reduction affects only a few scenarios. Furthermore, the number of nodes decreases considerably if the reduction level is increased. The computing times of less than 30 s already include approximately 20 s for computing distances of all scenario pairs that are needed in all calculations.

Table 14.1 Dimension of simulation scenarios

Components	Horizon	Scenarios	Time steps	Nodes
3 (trivariate)	1 year	100	8760	875,901

Table 14.2 Results of Algorithm 4 for yearly demand-price scenario trees

	Scenarios		Nodes			
ε_{rel}	Initial	Tree	Initial	Tree	Stages	Time (s)
0.20	100	100	875,901	775,992	4	24.53
0.25	100	100	875,901	752,136	5	24.54
0.30	100	100	875,901	719,472	7	24.55
0.35	100	97	875,901	676,416	8	24.61
0.40	100	98	875,901	645,672	10	24.64
0.45	100	96	875,901	598,704	10	24.75
0.50	100	95	875,901	565,800	9	24.74
0.55	100	88	875,901	452,184	10	24.75
0.60	100	87	875,901	337,728	11	25.89

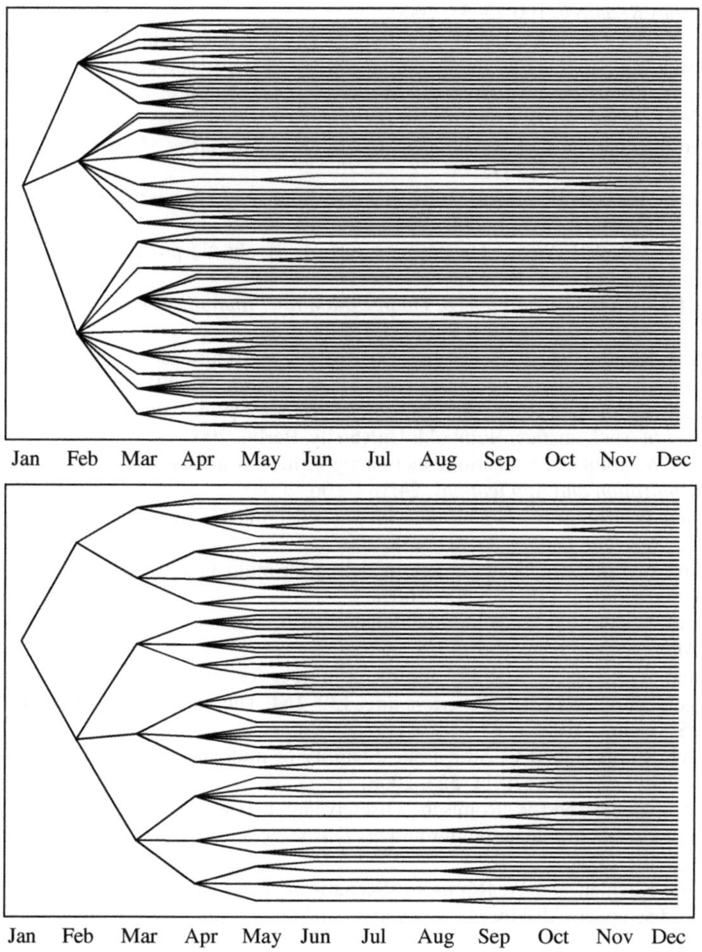

Fig. 14.4 Yearly demand–price scenario trees obtained by Algorithm 4

Figure 14.4 illustrates the scenario trees obtained for reduction levels of 40 and 55%, respectively. Observe that in all computations the maximal number of 12 stages is not reached even at higher reduction levels. This phenomenon could be caused by the low level of heat demand during the summer period (see Fig. 14.3) such that branching occurs less frequently.

References

M. Chen and S. Mehrotra. Epi-convergent scenario generation method for stochastic problems via sparse grid. *Stochastic Programming E-Print Series*, 7, 2008.

M. A. H. Dempster. Sequential importance sampling algorithms for dynamic stochastic programming. *Zapiski Nauchnykh Seminarov POMI*, 312:94–129, 2004.

R. M. Dudley. *Real Analysis and Probability*. Chapman & Hall, New York, NY, 1989.

J. Dupačová, G. Consigli, and S. W. Wallace. Scenarios for multistage stochastic programs. *Annals of Operations Research*, 100:25–53, 2000.

J. Dupačová, N. Gröwe-Kuska, and W. Römisch. Scenario reduction in stochastic programming: An approach using probability metrics. *Mathematical Programming*, 95:493–511, 2003.

A. Eichhorn, W. Römisch, and I. Wegner. Mean-risk optimization of electricity portfolios using multiperiod polyhedral risk measures. In *IEEE St. Petersburg PowerTech Proceedings*, 2005.

H. Fetter. On the continuity of conditional expectations. *Journal of Mathematical Analysis and Applications*, 61:227–231, 1977.

K. Frauendorfer. Barycentric scenario trees in convex multistage stochastic programming. *Mathematical Programming*, 75:277–293, 1996.

S. J. Garstka and R. J-B. Wets. On decision rules in stochastic programming. *Mathematical Programming*, 7:117–143, 1974.

S. Graf and H. Luschgy. *Foundations of Quantization for Probability Distributions*, volume 1730 of *Lecture Notes in Mathematics*. Springer, Berlin, 2000.

N. Gülpinar, B. Rustem, and R. Settergren. Simulation and optimization approaches to scenario tree generation. *Journal of Economic Dynamics & Control*, 28:1291–1315, 2004.

H. Heitsch. *Stabilität und Approximation stochastischer Optimierungsprobleme*. PhD thesis, Department of Mathematics, Humboldt University, Berlin, 2007.

H. Heitsch and W. Römisch. Scenario reduction algorithms in stochastic programming. *Computational Optimization and Applications*, 24:187–206, 2003.

H. Heitsch and W. Römisch. A note on scenario reduction for two-stage stochastic programs. *Operations Research Letters*, 35:731–738, 2007.

H. Heitsch and W. Römisch. Scenario tree modeling for multistage stochastic programs. *Mathematical Programming*, 118:371–406, 2009a.

H. Heitsch and W. Römisch. Scenario tree reduction for multistage stochastic programs. *Computational Management Science*, 6:117–133, 2009b.

H. Heitsch and W. Römisch. Stability and scenario trees for multistage stochastic programs. In G. Infanger, editor, *Stochastic Programming*, 139–164. The State of the Art, In Honor of G.B. Dantzig, Springer, 2010.

H. Heitsch, W. Römisch, and C. Strugarek. Stability of multistage stochastic programs. *SIAM Journal on Optimization*, 17:511–525, 2006.

R. Henrion, C. Küchler, and W. Römisch. Discrepancy distances and scenario reduction in two-stage stochastic mixed-integer programming. *Journal of Industrial and Management Optimization*, 4:363–384, 2008.

R. Henrion, C. Küchler, and W. Römisch. Scenario reduction in stochastic programming with respect to discrepancy distances. *Computational Optimization and Applications*, 43:67–93, 2009.

R. Hochreiter and G. Ch. Pflug. Financial scenario generation for stochastic multi-stage decision processes as facility location problems. *Annals of Operations Research*, 152:257–272, 2007.

K. Høyland and S. W. Wallace. Generating scenario trees for multi-stage decision problems. *Management Science*, 47:295–307, 2001.

R. Kouwenberg. Scenario generation and stochastic programming models for asset liability management. *European Journal of Operational Research*, 134:279–292, 2001.

C. Küchler. On stability of multistage stochastic programs. *SIAM Journal on Optimization*, 19:952–968, 2008.

D. Kuhn. *Generalized Bounds for Convex Multistage Stochastic Programs, Lecture Notes in Economics and Mathematical Systems*, volume 548. Springer-Verlag Berlin Heidelberg, 2005.

C. Lemieux. *Monte Carlo and Quasi-Monte Carlo Sampling*. Springer Series in Statistics. Springer, New York, NY, 2009.

R. Mirkov and G. Ch. Pflug. Tree approximations of dynamic stochastic programs. *SIAM Journal on Optimization*, 18:1082–1105, 2007.

T. Pennanen. Epi-convergent discretizations of multistage stochastic programs. *Mathematics of Operations Research*, 30:245–256, 2005.

T. Pennanen. Epi-convergent discretizations of multistage stochastic programs via integration quadratures. *Mathematical Programming*, 116:461–479, 2009.

G. Ch. Pflug. Scenario tree generation for multiperiod financial optimization by optimal discretization. *Mathematical Programming*, 89:251–271, 2001.

R. T. Rockafellar and R. J-B. Wets. Continuous versus measurable recourse in n-stage stochastic programming. *Journal of Mathematical Analysis and Applications*, 48:836–859, 1974.

R. T. Rockafellar and R. J-B. Wets. *Variational Analysis*, volume 317 of *Grundlehren der mathematischen Wissenschaften*. Springer, Berlin, 1st edition, 1998. (Corr. 2nd printing 2004).

W. Römisch. Stability of stochastic programming problems. In A. Ruszczyñski and A. Shapiro, editors, *Stochastic Programming*, Volume 10 of *Handbooks in Operations Research and Management Science*. Elsevier, Amsterdam, 1st edition, chapter 8, pages 483–55, 2003.

W. Römisch. Scenario generation. In J. J. Cochran et al., editors, *Encyclopedia of Operations Research and Management Science*. Wiley Encyclopedia of Operations Research and Management Science, Wiley, Chapter 1.5.4.2, 2010.

A. Ruszczyński and A. Shapiro, editors. *Stochastic Programming*, volume 10 of *Handbooks in Operations Research and Management Science*. Elsevier, Amsterdam, 1st edition, 2003.

A. Shapiro. Inference of statistical bounds for multistage stochastic programming problems. *Mathematical Methods of Operations Research*, 58:57–68, 2003a.

A. Shapiro. Monte Carlo sampling methods. In A. Ruszczyński and A. Shapiro, editors, *Stochastic Programming*, Volume 10 of *Handbooks in Operations Research and Management Science*. Elsevier, Amsterdam, 1st edition, chapter 6, pages 353–425, 2003b.

Chapter 15
Approximations for Probability Distributions and Stochastic Optimization Problems

Georg Ch. Pflug and Alois Pichler

Abstract In this chapter, an overview of the scenario generation problem is given. After an introduction, the basic problem of measuring the distance between two single-period probability models is described in Section 15.2. Section 15.3 deals with finding good single-period scenarios based on the results of the first section. The distance concepts are extended to the multi-period situation in Section 15.4. Finally, Section 15.5 deals with the construction and reduction of scenario trees.

Keywords Scenario generation · Probability distances · Optimal discretizations · Scenario trees · Nested distributions

15.1 Introduction

Decision making under uncertainty is based upon

(i) a probability model for the uncertain values,
(ii) a cost or profit function depending on the decision variables and the uncertainties, and
(iii) a probability functional, like expectation, median, etc., to summarize the random costs or profits in a real-valued objective.

We describe the decision model under uncertainty by

$$(Opt) \qquad \max\{F(x) = \mathcal{A}_P[H(x,\xi)] : x \in \mathbb{X}\}, \tag{15.1}$$

where $H(\cdot,\cdot)$ denotes the profit function, with x the decision and ξ the random variable or random vector modeling uncertainty. Both – the uncertain data ξ and the decision x – may be stochastic processes adapted to some filtration $\mathcal{F}$. $\mathcal{A}$ is the probability functional and $\mathbb{X}$ is the set of constraints. P denotes the underlying probability measure.

G. Ch. Pflug (✉)
Department of Statistics and Operations Research, University of Vienna A-1010, Wien – Vienna, Austria
e-mail: georg.pflug@univie.ac.at

M. Bertocchi et al. (eds.), *Stochastic Optimization Methods in Finance and Energy*, International Series in Operations Research & Management Science 163,
DOI 10.1007/978-1-4419-9586-5_15,

The most difficult part in establishing a formalized decision model of type (15.1) for a real decision situation is to find the probability model P. Typically, there is a sample of past data available, but not more. It needs two steps to come from the sample of observations to the scenario model:

(1) In the first step a *probability model* is identified, i.e., the description of the uncertainties as random variables or random processes by identifying the probability distribution. This step is based on statistical methods of model selection and parameter estimation. If several probability measures represent the data equally well, one speaks of *ambiguity*. In the non-ambiguous situation, one and only one probability model is selected and this model is the basis for the next step.
(2) In the following scenario generation step, a *scenario model* is found, which is an approximation of (15.1) by a *finite model* of lower complexity than the probability model.

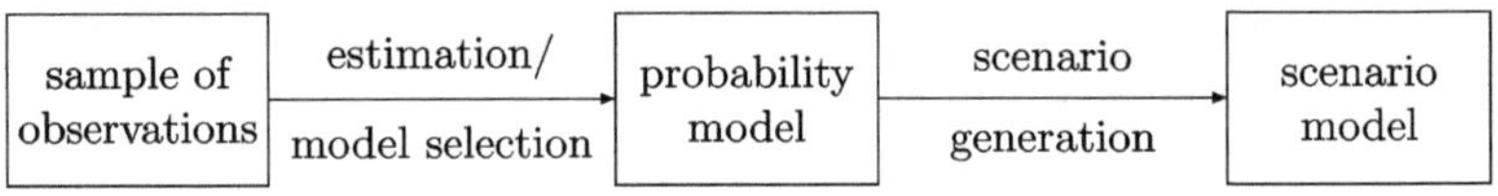

The scenario model differs from the original model (15.1) insofar as P is replaced by $\tilde{P}$, but – especially in multi-period problems – also the feasible set $\mathbb{X}$ is replaced by a smaller, feasible set $\tilde{\mathbb{X}}$:

$$\widetilde{(Opt)} \qquad \max\{\tilde{F}(\tilde{x}) = \mathcal{A}_{\tilde{P}}[H(\tilde{x}, \xi)] : \tilde{x} \in \tilde{\mathbb{X}}\}. \tag{15.2}$$

It is essential that the scenario model is finite (i.e., only finitely many values of the uncertainties are possible), but a good approximation to the original model as well. The finiteness is essential to keep the computational complexity low, and the approximative character is needed to allow the results to be carried over to the original model. In some cases, which do not need to be treated further here, finitely many scenarios are given right from the beginning (e.g., Switzerland will join the EU in 2015 – yes or no, or UK will adopt the EURO in 2020 and so on). The same is true, if the scenarios are just taken as the historic data without further transformation or information reduction. However, the justification of using just the empirical sample as the scenario model lies in the fact that these samples converge to the true underlying distribution if the sample size increases. For a quantification of this statement see Section 15.3.3 on Monte Carlo approximation below.

When stressing the approximative character of a scenario model we have to think about the consequences of this approximation and to estimate the error committed by replacing the probability model by the scenario model. Consider the diagram below: It is typically impossible to solve problem (15.1) directly due to its complexity. Therefore, one has to go around this problem and solve the simplified problem (15.2). Especially for multi-period problems, the solution of the approximate problem is not directly applicable to the original problem, but an extension function is needed, which extends the solution of the approximate problem to a solution of the original problem. Of course, the extended solution of the approximate problem is typically suboptimal for the original problem. The respective gap is called the approximation error. It is the important goal of scenario generation to make this gap acceptably small.

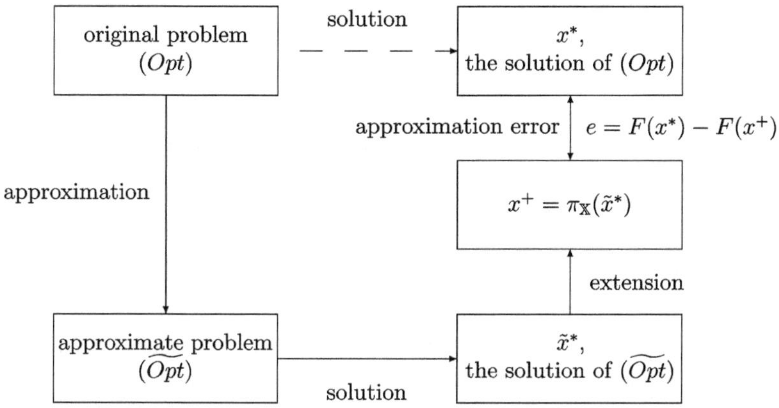

The quality of a scenario approximation depends on the distance between the original probability P and the scenario probability $\tilde{P}$. Well-known quantitative stability theorems (see, e.g., Dupacova (1990), Heitsch et al. (2006), and Rachev and Roemisch (2002)) establish the relation between distances of probability models on one side and distances between optimal values or solutions on the other side (see, e.g., cite stability). For this reason, distance concepts for probability distributions on $\mathbb{R}^M$ are crucial.

15.2 Distances Between Probability Distributions

Let $\mathcal{P}$ be a set of probability measures on $\mathbb{R}^M$.

Definition 1.1 A *semi-distance* on $\mathcal{P}$ is a function $\mathsf{d}(\cdot,\cdot)$ on $\mathcal{P} \times \mathcal{P}$, which satisfies (i) and (ii) as follows:

(i) **Nonnegativity:** For all $P_1, P_2 \in \mathcal{P}$

$$\mathsf{d}(P_1, P_2) \geq 0.$$

(ii) **Triangle inequality:** For all $P_1, P_2, P_3 \in \mathcal{P}$

$$\mathsf{d}(P_1, P_2) \leq \mathsf{d}(P_1, P_3) + \mathsf{d}(P_3, P_2).$$

If a semi-distance satisfies the strictness property

(iii) **Strictness:** If $\mathsf{d}(P_1, P_2) = 0$, then $P_1 = P_2$,
it is called a *distance*.

A general principle for defining semi-distances and distances consists in choosing a family of integrable functions $\mathcal{H}$ (i.e., a family of functions such that the integral $\int h(w)\, dP(w)$ exists for all $P \in \mathcal{P}$) and defining

$$\mathsf{d}_{\mathcal{H}}(P_1, P_2) = \sup\left\{ \left| \int h(w)\, dP_1(w) - \int h(w)\, dP_2(w) \right| : h \in \mathcal{H} \right\}.$$

We call $\mathsf{d}_{\mathcal{H}}$ the (semi-)distance *generated by* $\mathcal{H}$.

In general, $\mathsf{d}_{\mathcal{H}}$ is only a semi-distance. If $\mathcal{H}$ is *separating*, i.e., if for every pair $P_1, P_2 \in \mathcal{P}$ there is a function $h \in \mathcal{H}$ such that $\int h\, dP_1 \neq \int h\, dP_2$, then $\mathsf{d}_{\mathcal{H}}$ is a distance.

15.2.1 The Moment-Matching Semi-Distance

Let $\mathcal{P}_q$ be the set of all probability measures on $\mathbb{R}^1$ which possess the qth moment, i.e., for which $\int \max(1, |w|^q)\, dP(w) < \infty$. The moment-matching semi-distance on $\mathcal{P}_q$ is

$$\mathsf{d}_{M_q}(P_1, P_2) = \sup\left\{ \left| \int w^s\, dP_1(w) - \int w^s\, dP_2(w) \right| : s \in \{1, 2, \ldots, q\} \right\}. \tag{15.3}$$

The moment-matching semi-distance is not a distance, even if q is chosen to be large or even infinity. In fact, there are examples of two different probability measures on $\mathbb{R}^1$, which have the same moments of all orders. For instance, there are probability measures, which have all moments equal to those of the lognormal distribution, but are not lognormal (see, e.g., Heyde (1963)).

A widespread method is to match the first four moments (Hoyland and Wallace 2001), i.e., to work with d_{M_4}. The following example shows that two densities coinciding in their first four moments may exhibit very different properties.

Example Let P_1 and P_2 be the two probability measures with densities g_1 and g_2 where

$$\begin{aligned}
g_1(x) = 0.39876\, [\, & \exp(-|x + 0.2297|^3)\, \mathbb{1}_{\{x \le -1.2297\}} \\
& + \exp(-|x + 0.2297|)\, \mathbb{1}_{\{-1.2297 < x \le -0.2297\}} \\
& + \exp(-0.4024 \cdot (x + 0.2297))\, \mathbb{1}_{\{-0.2297 < x \le 2.2552\}} \\
& + 1.09848 \cdot (0.4024x + 0.29245)^{-6}\, \mathbb{1}_{\{2.2552 < x\}}], \\
g_2(x) = 0.5962\, & \mathbb{1}_{\{|x| \le 0.81628\}} + 0.005948\, \mathbb{1}_{\{0.81628 < |x| \le 3.05876\}}.
\end{aligned}$$

Both densities are unimodal and coincide in the first four moments, which are $m_1 = 0$, $m_2 = 0.3275$, $m_3 = 0$, $m_4 = 0.72299$ ($m_q(P) = \int w^q\, dP(w)$). The fifth moments, however, could not differ more: While P_1 has infinite fifth moment, the fifth moment of P_2 is zero. Density g_1 is asymmetric, has a sharp cusp at the point -0.2297 and unbounded support. In contrast, g_2 is symmetric around 0, has a flat density there, has finite support, and possesses all moments. The distribution functions and quantiles differ drastically: We have that $G_{P_1}(0.81) = 0.6257$ and $G_{P_1}(-0.81) = 0.1098$, while $G_{P_2}(0.81) = 0.9807$ and $G_{P_2}(-0.81) = 0.0133$. Thus the probability of the interval $[-0.81, 0.81]$ is only 51% under P_1, while it is 96% under P_2.

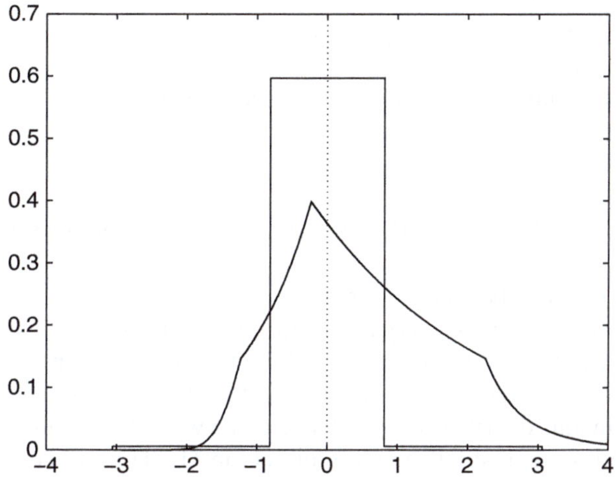

Two densities g_1 and g_2 with identical first four moments.

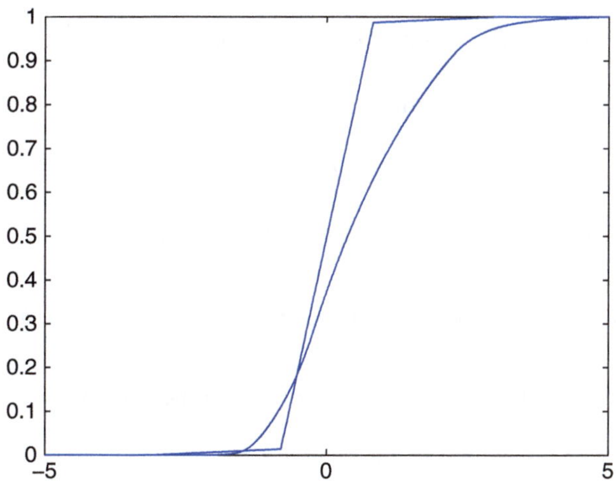

The two pertaining distribution functions G_1 and G_2.

Summarizing, the moment matching semi-distance is not well suited for scenario approximation, since it is not fine enough to capture the relevant quality of approximation.

The other extreme would be to choose as the generating class $\mathcal{H}$ all measurable functions h such that $|h| \le 1$. This class generates a distance, which is called the variational distance (more precisely twice the variational distance). It is easy to see that if P_1 resp. P_2 have densities g_1 resp. g_2 w.r.t. λ, then

$$\begin{aligned}
&\sup\left\{\int h\,dP_1 - \int h\,dP_2 : |h| \le 1, h \text{ measurable}\right\}\\
&= \int |g_1(x) - g_2(x)|\,d\lambda(x)\\
&= 2\sup\{|P_1(A) - P_2(A) : A \text{ measurable set}\}.
\end{aligned}$$

We call

$$\mathsf{d}_V(P_1, P_2) = \sup\{|P_1(A) - P_2(A)| : A \text{ measurable set}\} \tag{15.4}$$

the *variational distance* between P_1 and P_2.

The variational distance is a very fine distance, too fine for our applications: If P_1 has a density and P_2 sits on at most countably many points, then $\mathsf{d}_V(P_1, P_2) = 1$, independently on the number of mass points of P_2. Thus there is no hope to approximate any continuous distribution by a discrete one w.r.t. the variational distance.

One may, however, restrict the class of sets in (15.4) to a certain subclass. If one takes the class of half-unbounded rectangles in $\mathbb{R}^M$ of the form $(-\infty, t_1] \times (-\infty, t_2] \times \cdots \times (-\infty, t_M]$ one gets the *uniform distance*, also called *Kolmogorov distance*

$$\mathsf{d}_U(P_1, P_2) = \sup\{|G_{P_1}(x) - G_{P_2}(x)| : x \in \mathbb{R}^M\}, \tag{15.5}$$

where $G_P(t)$ is the distribution function of P,

$$G_P(t) = P\{(-\infty, t_1] \times \cdots \times (-\infty, t_M]\}.$$

The uniform distance appears in the Hlawka–Koksma inequality:

$$\int h(u)\,dP_1(u) - \int h(u)\,dP_2(u) \le \mathsf{d}_U(P_1, P_2) \cdot V(h), \tag{15.6}$$

where $V(h)$ is the Hardy–Krause variation of h. In dimension $M = 1$, V is the usual total variation

$$V(h) = \sup\left\{\sum_i |h(z_i) - h(z_{i-1})| : z_1 < z_2 < \cdots < z_n, n \text{ arbitrary}\right\}.$$

In higher dimensions, let $V^{(M)}(h) = \sup \sum_{J_1,\dots,J_n \text{ is a partition by rectangles } J_i} |\Delta_{J_i}(h)|$, where $\Delta_J(h)$ is the sum of values of h at the vertices of J, where adjacent vertices get opposing signs. The Hardy–Krause variation of h is defined as $\sum_{m=1}^{M} V^{(m)}(h)$. d_U is invariant w.r.t. monotone transformations, i.e., if T is a monotone transformation, then the image measures $P_i^T = P_i \circ T^{-1}$ satisfy

$$\mathsf{d}_U(P_1^T, P_2^T) = \mathsf{d}_U(P_1, P_2).$$

Notice that a unit mass at point 0 and at point $1/n$ are at a distance 1 both in the d_U distance, for any even large n. Thus it may be felt that also this distance is too fine for good approximation results. A minimal requirement of closedness is that integrals of bounded, continuous functions are close and this leads to the notion of weak convergence.

Definition 1.2 (Weak convergence) A sequence of probability measures P_n on $\mathbb{R}^M$ converges weakly to P (in symbol $P_n \Rightarrow P$), if

$$\int h\, dP_n \to \int h\, dP$$

as $n \to \infty$, for all bounded, continuous functions h. Weak convergence is the most important notion for scenario approximation: We typically aim at constructing scenario models which, for increasing complexity, finally converge weakly to the underlying continuous model.

If a distance d has the property that

$$\mathsf{d}(P_n, P) \to 0 \quad \text{if and only if } P_n \Rightarrow P$$

we say that *d metricizes weak convergence*.

The following three distances metricize the weak convergence on some subsets of all probability measures on $\mathbb{R}^M$:

(i) *The bounded Lipschitz metric*: If $\mathcal{H}$ is the class of bounded functions with Lipschitz constant 1, the following distance is obtained:

$$\mathsf{d}_{\mathrm{BL}}(P_1, P_2) = \sup\left\{\int h\, dP_1 - \int h dP_2 : |h(u)| \le 1, L_1(h) \le 1\right\}, \quad (15.7)$$

where

$$L_1(h) = \inf\{L : |h(u) - h(v)| \le L\|u - v\|\}.$$

This distance metricizes weak convergence for all probability measures on $\mathbb{R}^M$.

(ii) *The Kantorovich distance*: If one drops the requirement of boundedness of h, one gets the Kantorovich distance

$$\mathsf{d}_{\mathrm{KA}}(P_1, P_2) = \sup\left\{\int h\, dP_1 - \int h\, dP_2 : L_1(h) \le 1\right\}. \tag{15.8}$$

This distance metricizes weak convergence on sets of probability measures which possess uniformly a first moment. (A set $\mathcal{P}$ of probability measures has uniformly a first moment, if for every ϵ there is a constant C_ϵ such that $\int_{\|x\|>C_\epsilon} \|x\|\, dP(x) < \epsilon$ for all $P \in \mathcal{P}$.) See the review of Gibbs and Su (2002). On the real line, the Kantorovich metric may also be written as

$$d_{\mathrm{KA}}(P, \tilde{P}) = \int |G_P(u) - G_{\tilde{P}}(u)|\, du = \int |G_P^{-1}(u) - G_{\tilde{P}}^{-1}(u)|\, du,$$

where $G_P^{-1}(u) = \sup\{v : G_P(v) \le u\}$ (see Vallander (1973)).

(iii) *The Fortet–Mourier metric*: If $\mathcal{H}$ is the class of Lipschitz functions of order q, the Fortet–Mourier distance is obtained:

$$\mathsf{d}_{\mathrm{FM}_q}(P_1, P_2) = \sup\left\{\int h\, dP_1 - \int h dP_2 : L_q(h) \le 1\right\}, \tag{15.9}$$

where the Lipschitz constant of order q is defined as

$$L_q(h) = \inf\{L : |h(u) - h(v)| \le L||u - v|| \max(1, ||u||^{q-1}, ||v||^{q-1})\}. \tag{15.10}$$

Notice that $L_q(h) \le L_{q'}(h)$ for $q' \le q$. In particular, $L_q(h) \le L_1(h)$ for all $q \ge 1$ and therefore

$$\mathsf{d}_{\mathrm{FM}_q}(P_1, P_2) \ge \mathsf{d}_{\mathrm{FM}_{q'}}(P_1, P_2) \ge \mathsf{d}_{\mathrm{KA}}(P_1, P_2).$$

The Fortet–Mourier distance metricizes weak convergence on sets of probability measures possessing uniformly a qth moment. Notice that the function $u \mapsto \|u\|^q$ is q-Lipschitz with Lipschitz constant $L_q = q$. On $\mathbb{R}^1$, the Fortet–Mourier distance may be equivalently written as

$$\mathsf{d}_{\mathrm{FM}_q}(P_1, P_2) = \int \max(1, |u|^{q-1})|G_{P_1}(u) - G_{P_2}(u)|\, du$$

(see Rachev (1991), p. 93). If $q = 1$, the Fortet–Mourier distance coincides with the Kantorovich distance.

The Kantorovich distance is related to the mass transportation problem (Monge's problem – Monge (1781), see also Rachev (1991), p. 89) through the following theorem.

Theorem 1.3 (Kantorovich–Rubinstein)

$$\begin{aligned}\mathsf{d}_{\mathrm{KA}}(P_1, P_2) = \inf\{\mathbb{E}(|X_1 - X_2|) : &\ \textit{s.t. the joint distribution } (X_1, X_2)\\ &\ \textit{is arbitrary, but the marginal distributions are fixed}\\ &\ \textit{such that} X_1 \sim P_1, X_2 \sim P_2\}\end{aligned} \tag{15.11}$$

(*see* Rachev (1991), *theorems 5.3.2 and* 6.1.1).

The infimum in (15.11) is attained. The optimal joint distribution (X_1, X_2) describes how masses with distribution P should be transported with minimal effort to yield the new mass distribution $\tilde{P}$. The analogue of the Hlawka–Koksma inequality is here

$$\int |h(u)\, dP_1(u) - \int h(u)\, dP_2(u)| \leq \mathsf{d}_{\mathrm{KA}}(P_1, P_2) \cdot L_1(h). \tag{15.12}$$

The Kantorovich metric can be defined on arbitrary metric spaces R with metric d: If P_1 and P_2 are probabilities on this space, then $\mathsf{d}_{\mathrm{KA}}(P_1, P_2; d)$ is defined by

$$\mathsf{d}_{\mathrm{KA}}(P_1, P_2; d) = \sup\left\{\int h\, dP_1 - \int h\, dP_2 : |h(u) - h(v)| \leq d(u, v)\right\}. \tag{15.13}$$

Notice that the metric $\mathsf{d}(\cdot, \cdot; d)$ is compatible with the metric d in the following sense: If δ_u resp. δ_v are unit point masses at u resp. v, then

$$\mathsf{d}_{\mathrm{KA}}(\delta_u, \delta_v; d) = d(u, v).$$

The theorem of Kantorovich–Rubinstein extends to the general case:

$$\begin{aligned}\mathsf{d}_{\mathrm{KA}}(P_1, P_2; d) = \inf\{\mathbb{E}[d(X_1, X_2)] : &\ \text{where the joint distribution } (X_1, X_2)\\ &\ \text{is arbitrary, but the marginal distributions are fixed}\\ &\ \text{such that } X_1 \sim P_1, X_2 \sim P_2\}.\end{aligned} \tag{15.14}$$

A variety of Kantorovich metrics may be defined, if $\mathbb{R}^M$ is endowed with different metrics than the Euclidean one.

15.2.2 Alternative Metrics on $\mathbb{R}^M$

By (15.13), to every metric d on $\mathbb{R}^M$, there corresponds a distance of the Kantorovich type. For instance, let d_0 be the discrete metric

$$d_0(u, v) = \begin{cases} 1 & \text{if } u \neq v, \\ 0 & \text{if } u = v. \end{cases}$$

The set of all Lipschitz functions w.r.t. the discrete metric coincides with the set of all measurable functions h such that $0 \leq h \leq 1$ or its translates. Consequently the pertaining Kantorovich distance coincides with the variational distance

$$\mathsf{d}_{\mathrm{KA}}(P_1, P_2; d_0) = \mathsf{d}_V(P_1, P_2).$$

Alternative metrics on $\mathbb{R}^1$ can, for instance, be generated by nonlinear transform of the axis. Let χ be any bijective monotone transformation, which maps $\mathbb{R}^1$ into $\mathbb{R}^1$. Then $d(u, v) := |\chi(u) - \chi(v)|$ defines a new metric on $\mathbb{R}^1$. Notice that the family functions which are Lipschitz w.r.t the distance d and the Euclidean distance $|u - v|$ may be quite different.

In particular, let us look at the bijective transformations (for $q > 0$)

$$\chi_q(u) = \begin{cases} u & |u| \leq 1 \\ |u|^q \operatorname{sgn}(u) & |u| \geq 1 \end{cases}.$$

Notice that $\chi_q^{-1}(u) = \chi_{1/q}(u)$. Introduce the metric $d_{\chi_q}(u, v) = |\chi_q(u) - \chi_q(v)|$. We remark that $d_{\chi_1}(u, v) = |u - v|$. Denote by $\mathsf{d}_{\mathrm{KA}}(\cdot, \cdot; d_{\chi_q})$ the Kantorovich distance which is based on the distance d_{χ_q},

$$\mathsf{d}_{\mathrm{KA}}(P_1, P_2; d_{\chi_q}) = \sup \left\{ \int h \, dP_1 - \int h \, dP_2 : \|h(u) - h(v)\| \leq d_{\chi_q}(u, v) \right\}.$$

Let P^{χ_q} be the image measure of P under χ_q, that is P^{χ_q} assigns to a set A the value $P\{\chi_{1/q}(A)\}$. Notice that P^{χ_q} has distribution function

$$G_{P^{\chi_q}}(x) = G_P(\chi_{1/q}(x)).$$

This leads to the following identity:

$$\mathsf{d}_{\mathrm{KA}}(P_1, P_2; d_{\chi_q}) = \mathsf{d}_{\mathrm{KA}}(P_1^{\chi_q}, P_2^{\chi_q}; |\cdot|).$$

It is possible to relate the Fortet–Mourier distance d_{M_q} to the distance $\mathsf{d}_{\mathrm{KA}}(\cdot, \cdot; d_{\chi_q})$. To this end, we show first that

$$L_q(h \circ \chi_q) \leq q \cdot L_1(h) \tag{15.15}$$

and

$$L_1(h \circ \chi_{1/q}) \leq L_q(h). \tag{15.16}$$

If $L_1(h) < \infty$, then

$$\begin{aligned} |h(\chi_q(u)) - h(\chi_q(v))| &\leq L_1(h)|\chi_q(u) - \chi_q(v)| \\ &\leq L_1(h) \cdot q \cdot \max(1, |u|^{q-1}, |v|^{q-1})|u - v| \end{aligned}$$

which implies (15.15). On the other hand, if $L_q(h) < \infty$, then

$$
\begin{aligned}
&|h(\chi_{1/q}(u)) - h(\chi_{1/q}(v))| \\
&\le L_q(h) \max(1, |\chi_{1/q}(u)|^{q-1}, |\chi_{1/q}(v)|^{q-1}) |\chi_{1/q}(u) - \chi_{1/q}(v)| \\
&\le L_q(h) \max(1, \max(|u|, |v|)^{q-1}) \max(1, \max(|u|, |v|)^{(1-q)/q} |u - v| \\
&\le L_q(h) |u - v|
\end{aligned}
$$

and therefore (15.16) holds. As a consequence, we get the following relations:

$$
\begin{aligned}
\frac{1}{q} \mathsf{d}_{\mathrm{KA}}(P_1, P_2; d_{\chi_q}) &= \frac{1}{q} \mathsf{d}_{\mathrm{KA}}(G_{P_1} \circ \chi_{1/q}, G_{P_2} \circ \chi_{1/q}) \\
&\le \mathsf{d}_{\mathrm{FM}_q}(P_1, P_2) \\
&\le \mathsf{d}_{\mathrm{KA}}(G_{P_1} \circ \chi_{1/q}, G_{P_2} \circ \chi_{1/q}) = \mathsf{d}_{\mathrm{KA}}(P_1, P_2; d_{\chi_q}). \quad (15.17)
\end{aligned}
$$

Thus one sees that it is practically equivalent to measure the distance by the Fortet–Mourier distance of order q or by the Kantorovich distance (i.e., the Fortet–Mourier distance of order 1) but with the real line endowed with the metric d_{χ_q}.

15.2.3 Extending the Kantorovich Distance to the Wasserstein Distance

By the Kantorovich–Rubinstein Theorem, the Kantorovich metric is the optimal cost of a transportation problem. If the cost function is not the distance d itself, but a power d^r of it, we are led to the *Wasserstein metric* $\mathsf{d}_{\mathrm{W},r}$.

$$
\begin{aligned}
\mathsf{d}_{\mathrm{W},r}(P_1, P_2; d)^r = \inf \Big\{ \int d(x_1, x_2)^r \pi(dx_1, dx_2);\ &\text{s.t. the joint probability} \\
&\text{measure } \pi \text{ has marginals such that} \\
&\pi(A \times \mathbb{R}^M) = P_1(A) \text{ and } \pi(\mathbb{R}^M \times B) = P_2(B) \Big\}. \quad (15.18)
\end{aligned}
$$

This is a metric for $r \ge 1$. For $r < 1$, the Wasserstein distance is defined as

$$
\begin{aligned}
\mathsf{d}_{\mathrm{W},r}(P_1, P_2; d) = \inf \Big\{ \int d(x_1, x_2)^r \pi(dx_1, dx_2);\ &\text{where the joint probability} \\
&\text{measure } \pi \text{ has marginals such that} \\
&\pi(A \times \mathbb{R}^M) = P_1(A) \text{ and } \pi(\mathbb{R}^M \times B) = P_2(B) \Big\} \quad (15.19)
\end{aligned}
$$

(see Villani (2008)).

15.2.4 Notational Convention

When there is no doubt about the underlying distance d, we simply write

$$\mathsf{d}_r(P_1, P_2) \quad \text{instead of} \quad \mathsf{d}_{W,r}(P_1, P_2; d).$$

Recall that in this notation, the Kantorovich distance is $\mathsf{d}_{\mathrm{KA}}(P_1, P_2) = \mathsf{d}_1(P_1, P_2)$.

15.2.5 Wasserstein Distance and Kantorovich Distance as Linear Program

To compute the Wasserstein distance for discrete probability distributions amounts to solving a linear program, as we shall outline here:

Consider the discrete measures $P := \sum_s p_s \delta_{z_s}$ and $Q := \sum_t q_t \delta_{z'_t}$ and – within this setting – the problem

$$\begin{aligned}
\underset{(\text{in } \pi)}{\text{Minimize}} \quad & \sum_{s,t} \pi_{s,t} c_{s,t} \\
\text{subject to} \quad & \sum_t \pi_{s,t} = p_s, \\
& \sum_s \pi_{s,t} = q_t \text{ and} \\
& \pi_{s,t} \geq 0,
\end{aligned} \tag{15.20}$$

where $c_{s,t} := d\left(z_s, z'_t\right)^r$ is a matrix derived from the general distance function d. This matrix c represents the weighted costs related to the transportation of masses from z_s to z'_t.

Notice that (15.20) is a general linear program, which is already well defined by specifying the cost matrix $c_{s,t}$ and the probability masses p_s and q_t. So in particular the mass points z_s are not essential to formulate the problem; they may represent abstract objects.

Observe further that

$$\sum_{s,t} \pi_{s,t} = \sum_s p_s = \sum_t q_t = 1,$$

thus π is a probability measure, representing – due to the constraints – a transport plan from P to Q. The linear program (15.20) thus returns the minimal distance between those measures, and as a minimizing argument the optimal transport plan in particular for transporting masses from object z_s to z'_t.

The linear program (15.20) has a dual with vanishing duality gap. The theorem of Kantorovich–Rubinstein allows to further characterize (in the situation $r = 1$) the dual program; it amounts to just finding a Lipschitz-1 function h which maximizes the respective expectation:

$$\begin{array}{ll} \text{Maximize} \\ \text{(in } h) & \sum_s p_s h_s - \sum_t q_t h_t, \\ \text{subject to} & h_s - h_t \le d(z_s, z_t). \end{array}$$

Due to the vanishing duality gap the relation

$$\sum_s p_s h_s^* - \sum_t q_t h_t^* = \sum_{s,t} \pi_{s,t}^* c_{s,t}$$

holds true, where h^* is the optimal Lipschitz-1 function and π^* the optimal joint measure.

15.2.6 Bibliographical Remarks

The distance d_{KA} was introduced by Kantorovich in 1942 as a distance in general spaces. In 1948, he established the relation of this distance (in $\mathbb{R}^n$) to the mass transportation problem formulated by Gaspard Monge in 1781. In 1969, L. N. Wasserstein – unaware of the work of Kantorovich – reinvented this distance for using it for convergence results of Markov processes and 1 year later R. L. Dobrushin used and generalized this distance and initiated the name Wasserstein distance. S. S. Vallander studied the special case of measures in $\mathbb{R}^1$ in 1974 and this paper made the name Wasserstein metric popular. The distance d_U was introduced by Kolmogorov in 1933. It is often called Kolmogorov distance.

15.3 Single-Period Discretizations

In this section, we study the *discrete approximation problem:*

Let a probability measure P on $\mathbb{R}^M$ be given. We want to find a probability measure sitting on at most S points, such that some distance $\mathsf{d}(P, \tilde{P})$ is small.

We say "small" and not "minimal", since it may be computationally intractable to really find the minimum. Formally, define $\mathcal{P}_S$ as the family of all probability measures

$$\tilde{P} = \sum_{s=1}^{S} p_s \delta_{z_s}$$

on $\mathbb{R}^M$ sitting on at most S points. Here δ_z denotes the point mass at point z. We try to find a $\tilde{P} \in \mathcal{P}_S$ such that

$$\mathsf{d}(P, \tilde{P}) \text{ is close to } \min\{\mathsf{d}(P, Q) : Q \in \mathcal{P}_S\}.$$

There are several methods to attack the problem, which we order in decreasing order of computational effort and as well in decreasing order of approximation quality (see Fig. 15.1):

	Approximation quality	Computational ease
Optimal quantization	high	low
Quantization heuristics		
Quasi - Monte Carlo		
Monte Carlo	low	high

Fig. 15.1 Methods of scenario generation

While the probability measure P may have a density or may be discrete, the measure $\tilde{P}$ is always characterized by the list of probability masses p_s, $s = 1, \ldots, S$ and the list of mass points z_s, $s = 1, \ldots, S$. The discrete measure $\tilde{P}$ is represented by

$$\tilde{P} = \left[\frac{p_1 \; p_2 \; \cdots \; p_S}{z_1 \; z_2 \; \cdots \; z_S.} \right].$$

The approximation problem depends heavily on the chosen distance d, as will be seen. We treat the optimal quantization problem in Section 15.3.1, the quasi-Monte Carlo technique in Section 15.3.2 and the Monte Carlo technique in Section 15.3.3. Some quantization heuristics are contained in Chapter 3.

15.3.1 Optimal and Asymptotically Optimal Quantizers

The basic approximation problem, i.e., to approximate a distribution on $\mathbb{R}^M$ by a distribution sitting on at most S points leads to a non-convex optimization problem. To find its global solution is computationally hard. Moreover, no closed-form solution is known, even for the simplest distributions.

Recall the basic problem: Let d be some distance for probability measures and let $\mathcal{P}_S$ be the family of probability distributions sitting on at most S points. For a given probability $P \in \mathcal{P}_S$, the main question is to find the *quantization error*

$$q_{S,\mathsf{d}}(P) = \inf\{\mathsf{d}(P, Q) : Q \in \mathcal{P}_S\} \tag{15.21}$$

and the *optimal quantization* set

$$\mathcal{Q}_{S,\mathsf{d}}(P) = \text{argmin}\,\{\mathsf{d}(P, Q) : Q \in \mathcal{P}_S\} \tag{15.22}$$

(if it exists).

15.3.1.1 Optimal Quantizers for Wasserstein Distances d_r

Since all elements $\tilde{P}$ of $\mathcal{P}_S$ are of the form $\tilde{P} = \sum_{s=1}^{S} p_s \delta_{z_s}$, the quantization problem consists essentially of two steps:

(1) Find the optimal supporting points z_s.
(2) Given the supporting points z_s, find the probabilities p_s, which minimize

$$\mathsf{d}_r\left(P, \sum_{s=1}^{S} p_s \delta_{z_s}\right) \tag{15.23}$$

for the Wasserstein distance.

Recall the Wasserstein d_r distances contain the Kantorovich distance as the special case for $r = 1$.

We show that the second problem has an easy solution for the chosen distance: Let $z = (z_1, \ldots, z_S)$ be the vector of supporting points and suppose that they are all distinct. Introduce the family of *Voronoi diagrams* $\mathcal{V}(z)$ as the family of all partitions $(A_1, \ldots, A_S)$, where A_s are pairwise disjoint and $\mathbb{R}^M = \bigcup_{s=1}^{S} A_s$, such that

$$A_s \subseteq \{y : d(y, z_s) = \min_k d(y, z_k)\}.$$

Then the possible optimal probability weights p_s for minimizing $\mathsf{d}_r\left(P, \sum_{s=1}^{S} p_s \delta_{z_s}\right)$ can be found by

$$p = (p_1, \ldots, p_S), \text{ where } p_s = P(A_s); \quad (A_1, \ldots, A_S) \in \mathcal{V}(z).$$

The optimal probability weights are unique, iff

$$P\{y : d(y, z_s) = d(y, z_k), \text{ for some } s \neq k\} = 0.$$

One may also reformulate (15.23) in terms of random variables: Let $X \sim P$ and let $\tilde{X}$ be a random variable which takes only finitely many values $\tilde{X} \in \{z_1, \ldots, z_S\}$, then

$$\begin{aligned} D_{\mathsf{d}_r}(z) &:= \inf\{\mathbb{E}[d(X, \tilde{X})^r] : X \sim P, \tilde{X} \in \{z_1, \ldots, z_S\}\} \\ &= \inf\{\mathsf{d}_r(P, \tilde{P})^r : \text{supp}(\tilde{P}) \subseteq \{z_1, \ldots, z_S\}\} \\ &= \int \min_s d(x, z_s)^r \, dP(x). \end{aligned}$$

Unfortunately, the function $z \mapsto D_{\mathsf{d}_r}(z)$ is non-convex and typically has multiple local minima. Notice that the quantization error defined in (15.21) satisfies

$$q_{S,\mathsf{d}_r}(P) = \min\{[D_{\mathsf{d}_r}(z)]^{1/r} : z = (z_1, \ldots, z_S)\}.$$

Lemma 2.1 The mapping $P \mapsto q_{S,\mathsf{d}_r}(P)$ is concave.

Proof Let $P = \sum_{i=1}^{I} w_i P_i$ for some positive weights with $\sum_{i=1}^{I} w_i = 1$. Then

$$q_{S,\mathsf{d}_r}(P) = \sum_{i=1}^{I} w_i \int \min_s [d(x, z_s)]^r \, \mathrm{d}P_i(x) \geq \sum_{i=1}^{I} w_i q_{S,\mathsf{d}_r}(P_i).$$

□

Lemma 2.2 Let $\sum_{i=1}^{I} S_i \leq S$ and the weights w_i as above. Then

$$q_{S,\mathsf{d}_r}\left(\sum_{i=1}^{I} w_i P_i\right) \leq \sum_{i=1}^{I} w_i q_{S_i,\mathsf{d}_r}(P_i).$$

Proof Let $\tilde{P}_i$ such that $\mathsf{d}_r(P_i, \tilde{P}_i) = q_{S_i,\mathsf{d}_r}(P)$. Then

$$q_{S,\mathsf{d}_r}\left(\sum_{i=1}^{I} w_i P_i\right) \leq \mathsf{d}_r\left(\sum_{i=1}^{I} w_i P_i, \sum_{i=1}^{I} w_i \tilde{P}_i\right) \leq \sum_{i=1}^{I} w_i q_{S_i,\mathsf{d}_r}(P_i).$$

□

For the next lemma, it is necessary to specify the distance d to

$$d_r(x, y) = \left[\sum_{m=1}^{M} |x_m^r - y_m^r|\right]^{1/r}.$$

Lemma (Product quantizers) Let $\prod_{i=1}^{I} S_i \leq S$. Then

$$q_{S,\mathsf{d}_r}(\otimes_{i=1}^{I} P_i) \leq \sum_{i=1}^{I} q_{S,\mathsf{d}_r}(P_i).$$

Proof Let $\tilde{P}_i \in \mathcal{P}_S$ such that $\mathsf{d}_r(P_i, \tilde{P}_i) = q_{S_i,\mathsf{d}_r}(P)$. Then

$$q_{S,\mathsf{d}_r}(\otimes_{i=1}^{I} P_i) \leq \mathsf{d}_r(\otimes_{i=1}^{I} P_i, \otimes_{i=1}^{I} \tilde{P}_i) = \sum_{i=1}^{I} q_{S_i,\mathsf{d}_r}(P_i).$$

□

15.3.1.2 Interpretation as Facility Location Problems on $\mathbb{R}$

On $\mathbb{R}^1$ one may w.l.o.g. assume that the points z_s are in ascending order $z_1 \leq z_2 \leq \cdots \leq z_S$. The pertaining Voronoi tessellation then is

$$V_s = (b_{s-1}, b_s], \qquad s = 1, \ldots S,$$

with $b_0 = -\infty$, $b_S = +\infty$ and for $1 \le s \le S$; the points b_s are chosen such that $d(b_s, z_s) = d(b_s, z_{s+1})$, and the optimal masses are then $p_s = G(b_s) - G(b_{s-1})$. For the Euclidean distance on $\mathbb{R}$, the b_s's are just the midpoints

$$z_s = \frac{1}{2}(b_{s-1} + b_s), \qquad s = 1, \dots, S.$$

In this case, the Wasserstein distance is

$$\mathsf{d}_r(P, \tilde{P})^r = \sum_{s=1}^{S} \int_{\frac{z_{s-1}+z_s}{2}}^{\frac{z_s+z_{s+1}}{2}} |u - z_s|^r \, dG(u), \tag{15.24}$$

with $z_0 = -\infty$ and $z_{S+1} = +\infty$.

While step (2) of (15.23) is easy, step (1) is more involved: Let, for $z_1 \le z_2 \le \dots \le z_S$,

$$D(z_1, \dots, z_S) = \int \min_s d(x, z_s)^r \, dG(x) \tag{15.25}$$

The global minimum of this function is sought for. Notice that D is non-convex, i.e., the global minimization of D is a hard problem. The function D can be viewed as the travel mean costs of customers distributed along the real line with d.f. G to travel to the nearest location of one of the facilities, placed at $z_1, \dots, z_S$, if the travel costs from u to v are $d(u, v)^r$.

Facility Location The problem of minimizing the function D in (15.25) is an instance of the classical Weber location problem, which was introduced by A. Weber in 1909 (English translation published 1929 (Weber 1929)). A recent state-of-the-art overview of the theory and applications was published by Drezner and Hamacher (2002).

Explicit Optimal Solutions Only in exceptional cases explicit solutions of the optimal quantization problems (15.21) resp. (15.22) are known. Here are some examples; for more details, see Graf and Luschgy (2000).

Example (The Laplace distribution in $\mathbb{R}^1$) For the Euclidean distance on $\mathbb{R}$ and the Laplace distribution with density $f(x) = \frac{1}{2}\exp(-|x|)$, the optimal supporting points for even $S = 2k$ are

$$z_s = \begin{cases} 2\log\left(\frac{s}{\sqrt{k^2+k}}\right), & 1 \le s \le k, \\ 2\log\left(\frac{\sqrt{k^2+k}}{S+1-s}\right), & k+1 \le s \le S. \end{cases}$$

The quantization error is

$$q_{S,\mathsf{d}_1}(P) = \begin{cases} \log\left(1+\frac{2}{S}\right), & S \text{ even },\\ \frac{2}{S+1}, & S \text{ odd.} \end{cases}$$

Example (The Exponential distribution in $\mathbb{R}^1$) For the Euclidean distance on $\mathbb{R}$ and the Exponential distribution with density $f(x) = \exp(-x)\mathbb{1}_{x\geq 0}$, the optimal supporting points are

$$z_s = 2\log\left(\frac{\sqrt{S^2+S}}{S+1-s}\right); \quad s = 1,\dots,S.$$

The quantization error is

$$q_{S,\mathsf{d}_1}(P) = \log\left(1+\frac{1}{S}\right).$$

Example (The uniform distribution in $\mathbb{R}^M$) To discuss the uniform distribution in higher dimension, it is necessary to further specify the distance on the sample space: As distances on $\mathbb{R}^M$ we consider the p-norms $d_p(u,v) = \left[\sum_{m=1}^M |u_m - v_m|^p\right]^{1/p}$ for $1 \leq p < \infty$ resp. $d_\infty(u,v) = \max_{1\leq m\leq M} |u_m - v_m|$.

Based on d_p, we consider the Wasserstein distances $\mathsf{d}_{p,r}$

$$\mathsf{d}_{p,r}(P_1,P_2)^r = \inf\Big\{\int d_p(x_1,x_2)^r \pi(dx_1,dx_2); \text{ where the joint probability measure } \pi \text{ has marginals such that } \pi(A\times\mathbb{R}^M) = P_1(A) \text{ and } \pi(\mathbb{R}^M\times B) = P_2(B)\Big\}. \tag{15.26}$$

Let $q^{(M)}_{S,\mathsf{d}_{p,r}} := q_{S,\mathsf{d}_{p,r}}(\mathcal{U}[0,1]^M)$ be the minimal error of approximation to the uniform distribution on the M-dimensional unit cube $[0,1]^M$. The exact values are only known for

$$q^{(M)}_{1,\mathsf{d}_{p,r}} = \begin{cases} \frac{M}{(1+r)2^r}, & 1 \leq p < \infty,\\ \frac{M}{(M+r)2^r}, & p = \infty \end{cases}$$

and

$$q^{(M)}_{1,\mathsf{d}_{\infty,r}} = \frac{M}{(M+r)2^r}.$$

15.3.1.3 Asymptotically Optimal Quantizers

In most cases, the solutions of (15.21) and (15.22) are unknown and cannot be expressed in an analytic way. Therefore one may ask the simpler "asymptotic" questions:

- What is the rate in which $q_{S,\mathsf{d}_r}(P)$ converges to zero as S tends to infinity?
- Is there a constructive way to find a sequence being asymptotically optimal, i.e., a sequence (P_S^+) such that

$$\frac{\mathsf{d}(P, P_S^+)}{q_{S,\mathsf{d}_r}(P)} \to 1$$

as $S \to \infty$?

To investigate these questions introduce the quantities

$$\bar{q}_{\mathsf{d}_{p,r}}^{(M)} := \inf_S S^{r/M} q_{S,\mathsf{d}_{p,r}}(\mathcal{U}[0,1]^M), \tag{15.27}$$
$$\bar{q}_{\mathsf{d}_{p,r}}(P) := \inf_S S^{r/M} q_{S,\mathsf{d}_{p,r}}(P),$$

where $\mathcal{U}[0,1]^M$ is the uniform distribution on the M-dimensional unit cube $[0,1]^M$.

It deduces from the concavity of $P \mapsto q_{S,\mathsf{d}_{r,p}}(P)$ that the mappings $P \mapsto \bar{q}_{\mathsf{d}_{p,r}}(P)$ are concave as well.

The quantities $\bar{q}_{\mathsf{d}_{p,r}}^{(M)}$ are found to be universal constants; however, they are known in special cases only:

- $\bar{q}_{\mathsf{d}_{p,r}}^{(1)} = \frac{1}{(1+r)2^r}$,
- $\bar{q}_{\mathsf{d}_{\infty,r}}^{(M)} = \frac{M}{(M+r)2^r}$.

The next theorem links a general distribution P with the uniform distribution, providing a relation between $\bar{q}_{\mathsf{d}_{p,r}}(P)$ and $\bar{q}_{\mathsf{d}_{p,r}}^{(M)}$. As for a proof we refer the reader to Graf and Luschgy (2000), theorem 6.2. or Na and Neuhoff (1995).

Theorem (Zador–Gersho formula) *Suppose that P has a density g with $\int |u|^{r+\delta} g(u)\,du < \infty$ for some $\delta > 0$. Then*

$$\bar{q}_{\mathsf{d}_{p,r}}(P) = \inf_S S^{r/M} q_{S,\mathsf{d}_{p,r}}(P) = \bar{q}_{\mathsf{d}_{p,r}}^{(M)} \cdot \left[\int_{\mathbb{R}^M} g(x)^{\frac{M}{M+r}}\, dx \right]^{\frac{M+r}{M}}, \tag{15.28}$$

where $\bar{q}_{\mathsf{d}_{p,r}}^{(M)}$ is given by (15.27).

For the real line, this specializes to

$$\bar{q}_{\mathsf{d}_{p,r}}(P) = \frac{1}{(1+r)2^r} \left(\int_{\mathbb{R}} g(u)^{\frac{1}{1+r}}\, du \right)^{1+r}$$

for all p, since all distances d_p are equal for $M = 1$. Specializing further to $r = 1$, one gets

$$\bar{q}_{d_{p,1}}(P) = \frac{1}{4}\left(\int_{\mathbb{R}} \sqrt{g(u)}\, du\right)^2.$$

Heuristics The Zador–Gersho formula (15.28), together with identity (15.24) give rise that the optimal, asymptotic point density for z_s is proportional to $g^{\frac{M}{M+r}}$. In $\mathbb{R}^1$ again this translates to solving the quantile equations

$$\int_{-\infty}^{z_s} g^{\frac{1}{1+r}}(x)\, dx = \frac{2s-1}{2S} \cdot \int_{-\infty}^{\infty} g^{\frac{1}{1+r}}(x)\, dx$$

for $s = 1, 2, \ldots S$ and then to put

$$p_s := \int_{\frac{z_{s-1}+z_s}{2}}^{\frac{z_s+z_{s+1}}{2}} g(x)\, dx$$

as above. It can be proved that the points z_s specified accordingly are indeed asymptotically optimal and in practice this has proven to be a very powerful setting.

Example If P is the univariate normal distribution $N(\mu, \sigma^2)$, then $\bar{q}_{\mathsf{d}_{p,1}}(N(\mu, \sigma) = \sigma\sqrt{\frac{\pi}{2}}$. For the M-dimensional multivariate normal distribution $N(\mu, \Sigma)$ one finds that

$$\bar{q}_{\mathsf{d}_{p,r}}(N(\mu, \Sigma) = \bar{q}_{\mathsf{d}_{p,r}}^{(M)}(2\pi)^{r/2}\left(\frac{M+r}{M}\right)^{(M+r)/2}(\det\Sigma)^{r/(2M)}.$$

15.3.1.4 Optimal Quantizers for the Uniform (Kolmogorov) Distance

In dimension $M = 1$, the solution of the uniform discrete approximation problem is easy: Notice that the uniform distance is invariant w.r.t. monotone transformations, which means that the problem can be mapped to the uniform distribution via the quantile transform and then mapped back.

The uniform distribution $\mathcal{U}[0, 1]$ is optimally approximated by

$$\tilde{P} = \left[\begin{array}{ccccc} 1/S & 1/S & 1/S & \cdots & 1/S \\ \hline \frac{1}{2S} & \frac{3}{2S} & \frac{5}{2S} & \cdots & \frac{2S-1}{2S} \end{array}\right]$$

and the distance is

$$\mathsf{d}_U(\mathcal{U}[0, 1], \tilde{P}) = 1/S.$$

For an arbitrary P with distribution function G, the optimal approximation is given (by virtue of the quantile transform) by

$$z_s = G^{-1}\left(\frac{2s-1}{2S}\right), \quad s = 1, \ldots, S,$$
$$p_s = 1/S, \quad s = 1, \ldots, S.$$

The distance is bounded by $1/S$. If G is continuous, then the distance is exactly equal to $1/S$. In this case, the uniform distance approximation problem leads always to approximating distributions, which put the same mass $1/S$ at all mass points. This implies that the tails of P are not well represented by $\tilde{P}$. This drawback is not present when minimizing the Kantorovich distance.

15.3.1.5 Heavy Tail Control

Minimizing the Kantorovich distance minimizes the difference in expectation for all Lipschitz functions. However, it may not lead to a good approximation for higher moments or even the variance. If a good approximation of higher moments is also required, the Fortet–Mourier distance may be used.

Another possibility, which seems to be even more adapted to control tails, is to choose the Wasserstein r-distance instead of the Kantorovich distance for some appropriately chosen $r > 1$.

To approximate tails and higher moments are necessary for many real-world applications especially in the area of financial management.

Example Suppose we want to approximate the t-Student distribution P with 2 degrees of freedom, i.e., density $(2 + x^2)^{-3/2}$ by a discrete distribution sitting on 5 points. The optimal approximation with the uniform distance is

$$\tilde{P}_1 = \left[\frac{0.2 \quad\quad 0.2 \quad\quad 0.2 \quad 0.2 \quad\quad 0.2}{-1.8856 \; -0.6172 \;\; 0 \;\; 0.6172 \; 1.8856}\right],$$

while minimizing the Kantorovich distance, the approximated distribution is

$$\tilde{P}_2 = \left[\frac{0.0446 \; 0.2601 \; 0.3906 \; 0.2601 \; 0.0446}{-4.58 \;\; -1.56 \quad\; 0 \quad\;\; 1.56 \quad 4.58}\right].$$

These results can be compared visually in Fig. 15.2 where one can see clearly that the minimization of the Kantorovich distance leads to a much better approximation of the tails.

But even this approximation can be improved by heavy tail control: Applying the stretching function χ_2 to the t-distribution, one gets the distribution with density (Fig. 15.3)

$$\begin{cases} (2 + x^2)^{-3/2} & \text{if } |x| \leq 1, \\ \frac{1}{2\sqrt{x}}(2 + |x|)^{-3/2} & \text{if } |x| > 1. \end{cases}$$

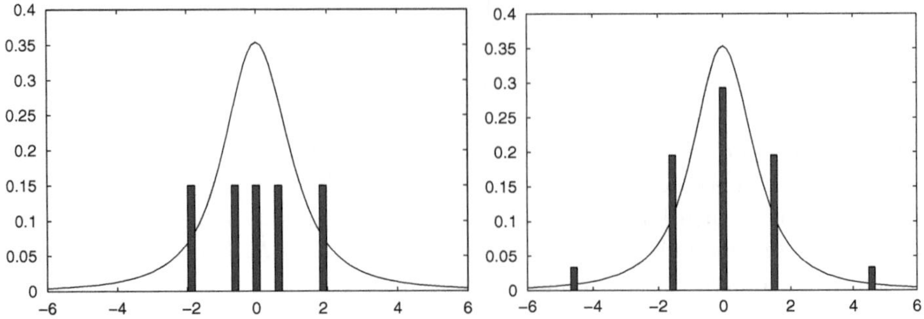

Fig. 15.2 Approximation of the $t(2)$-distribution: uniform (*left*) and Kantorovich (*right*)

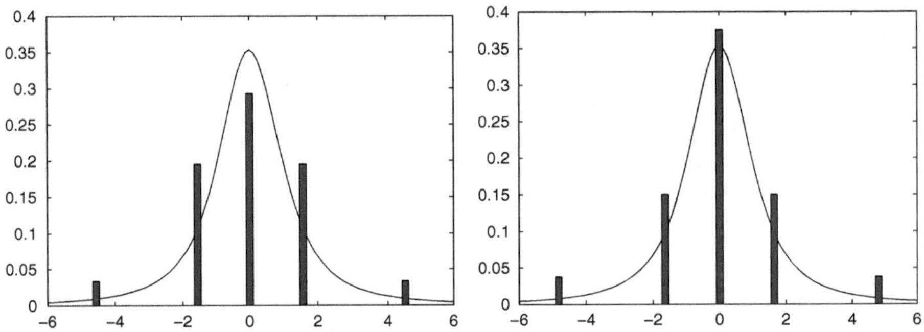

Fig. 15.3 Approximation of the $t(2)$-distribution: Kantorovich with Euclidean distance (*left*) and Kantorovich with distorted distance d_{χ_2} (*right*)

This distribution has heavier tails than the original t-distribution. We approximate this distribution w.r.t the Kantorovich distance by a 5-point distribution and transform back the obtained mass points using the transformation $\chi_{1/2}$ to get

$$\tilde{P} = \left[\frac{0.05 \quad\;\; 0.2 \quad\;\; 0.5 \;\; 0.2 \quad\; 0.05}{-4.8216\; -1.6416\;\; 0\;\; 1.6416\; 4.8216} \right].$$

Compared to the previous ones, this approximation has a better coverage of the tails. The second moment is better approximated as well.

15.3.2 *Quasi-Monte Carlo*

As above, let $\mathcal{U}[0,1]^M$ be the uniform distribution on $[0,1]^M$ and let $\tilde{P}_S = \frac{1}{S}\sum_{s=1}^{S}\delta_{z_s}$. Then the uniform distance $d_U(\mathcal{U}[0,1]^M, \tilde{P}_S)$ is called the *star discrepancy* $D_S^*(z_1,\ldots,z_S)$ of $(z_1,\ldots,z_S)$

$$D_S^*(z_1,\ldots,z_S) = \sup\left\{\frac{1}{S}\sum_{s=1}^{S} \mathbb{1}_{\{[0,a_1]\times\cdots\times[0,a_M]\}}(z_s) - \prod_{m=1}^{M} a_m : 0 < a_m \le 1\right\}. \tag{15.29}$$

A sequence $z = (z_1, z_2, \ldots)$ in $[0,1]^M$ is called a *low-discrepancy sequence*, if

$$\limsup_{S\to\infty} \frac{S}{\log(S)^M} D_S^*(z_1,\ldots,z_S) < \infty,$$

i.e., if there is a constant C such that for all S,

$$D_S^*(z) = D_S^*(z_1,\ldots,z_S) \le C\frac{\log(S)^M}{S}.$$

A low-discrepancy sequence allows the approximation of the uniform distribution in such a way that

$$d_U(\mathcal{U}[0,1]^M, \tilde{P}_S) \le C\frac{\log(S)^M}{S}.$$

Low-discrepancy sequences exist. The most prominent ones are as follows:

- The van der Corput sequence (for $M = 1$):
  Fix a prime number p and let, for every s

  $$s = \sum_{k=0}^{L} d_k(s)p^k$$

  be the representation of n in the p-adic system. The sequence z_n is defined as

  $$z_n^{(p)} = \sum_{k=0}^{L} d_k(s)p^{-k-1}.$$

- The Halton sequence (for $M > 1$):
  Let $p_1, \ldots, p_M$ be different prime numbers. The sequence is

  $$(z_s^{(1)}, \ldots, z_s^{(M)}),$$

  where $z_s^{(p_m)}$ are van der Corput sequences.
- The Hammersley sequence:
  For a given Halton sequence $(z_s^{(1)}, \ldots, z_s^{(M)})$, the Hammersley sequence enlarges it to $\mathbb{R}^{M+1}$ by

  $$(z_s^{(1)}, \ldots, z_s^{(M)}, s/S).$$

- The Faure sequence.
- The Sobol sequence.
- The Niederreiter sequence.

(For a thorough treatment of QMC sequences, see Niederreiter (1992).) The quasi-Monte Carlo method was used in stochastic programming by Pennanen (2009). It is especially good if the objective is the expectation, but less advisable if tail or rare event properties of the distribution enter the objective or the constraints.

15.3.2.1 Comparing Different Quantizations

We[1] have implemented the following portfolio optimization problem: Maximize the expected average value at risk ($\mathbb{AV@R}$, CVaR) of the portfolio return under the constraint that the expected return exceeds some threshold and a budget constraint. We used historic data of 13 assets. The "true distribution" is their weekly performance for 10 consecutive years. The original data matrix of size 520×13 was reduced by scenario generation (scenario approximation) to a data matrix of size 50×13. We calculated the optimal asset allocation for the original problem (1) and for the scenario methods: (2) minimal Kantorovich distance, (3) moment matching, and (4) quasi-Monte Carlo with a Sobol sequence. The Kantorovich distance was superior to all others, see Fig. 15.4.

15.3.3 Monte Carlo

The Monte Carlo approximation is a random approximation, i.e., $\tilde{P}$ is a random measure and the distance $\mathsf{d}(P, \tilde{P})$ is a random variable. This distinguishes the Monte Carlo method from the other methods.

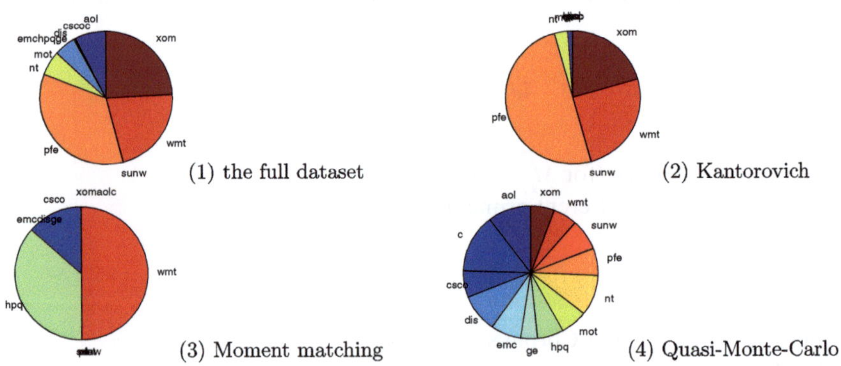

Fig. 15.4 Asset allocation using different scenario approximation methods. The pie charts show the optimal asset mix

[1] The programming of these examples was done by R. Hochreiter.

Let $X_1, X_2, \ldots, X_S$ be an i.i.d. sequence distributed according to P. Then the Monte Carlo approximation is

$$\tilde{P}_S = \frac{1}{S}\sum_{s=1}^{S}\delta_{X_s}.$$

15.3.3.1 The Uniform Distance of the MC Approximation

Let $X_1, X_2, \ldots$ be an i.i.d. sequence from a uniform [0,1] distribution. Then by the famous Kolmogorov–Smirnov theorem

$$\lim_{S\to\infty} P\{\sqrt{S}D_S^*(X_1,\ldots,X_S) > t\} = 2\sum_{k=1}^{\infty}(-1)^{k-1}\exp(-2k^2t^2).$$

15.3.3.2 The Kantorovich Distance of the MC Approximation

Let $X_1, X_2, \ldots$ be an i.i.d. sequence from a distribution with density g. Then

$$\lim_{S\to\infty} P\{S^{1/M}d_{\mathrm{KA}}(P,\tilde{P}_S) \le t\} = \int [1-\exp(-t^M b_M g(x))]g(x)\,dx,$$

where $b_M = \frac{2\pi^{M/2}}{M\Gamma(M/2)}$ is the volume of the M-dimensional Euclidean ball (see Graf and Luschgy (2000), theorem 9.1).

15.4 Distances Between Multi-period Probability Measures

Let P be a distribution on

$$\mathbb{R}^{MT} = \underbrace{\mathbb{R}^M \times \mathbb{R}^M \times \cdots \times \mathbb{R}^M}_{T \text{ times}}.$$

We interpret the parameter $t = 1, \ldots T$ as time and the parameter $m = 1, \ldots M$ as the multivariate dimension of the data.

While the multivariate dimension M does not add to much complexity, the multi-period dimension T is of fundamentally different nature. Time is the most important source of uncertainty and the time evolution gives increasing information. We consider the conditional distributions given the past and formulate scenario approximations in terms of the conditional distributions.

15.4.1 Distances Between Transition Probabilities

Any probability measure P on $\mathbb{R}^{MT}$ can be dissected into the chain of transition probabilities

$$P = P_1 \circ P_2 \circ \cdots \circ P_T,$$

where P_1 is the unconditional probability on $\mathbb{R}^M$ for time 1 and P_t is the conditional probability for time t, given the past. We do not assume that P is Markovian, but bear in mind that for Markovian processes, the past reduces to the value one period before.

We will use the following notation: If $u = (u_1, \ldots, u_T) \in \mathbb{R}^{MT}$, with $u_t \in \mathbb{R}^M$, then $u^t = (u_1, \ldots, u_t)$ denotes the history of u up to time t.

The chain of transition probabilities are defined by the following relation: $P = P_1 \circ P_2 \circ \cdots \circ P_T$, if and only if for all sequence of Borel sets $A_1, A_2, \cdots, A_T$,

$$\begin{aligned} & P(A_1 \times \cdots \times A_T) \\ & = \int_{A_1} \cdots \int_{A_T} P_T(du_T|u^{T-1}) \cdots P_3(du_3|u^2) P_2(du_2|u_1) P_1(du_1). \end{aligned}$$

We assume that d makes $\mathbb{R}^M$ a complete separable metric space, which implies that the existence of regular conditional probabilities is ensured (see, e.g., Durrett (1996), ch. 4, theorem 1.6): Recall that if R_1 and R_2 are metric spaces and $\mathcal{F}_2$ is a σ-algebra in R_2, then a family of regular conditional probabilities is a mapping $(u, A) \mapsto P(A|u)$, where $u \in R_1$, $A \in \mathcal{F}_2$, which satisfies

- $u \mapsto P(A|u)$ is (Borel) measurable for all $A \in \mathcal{F}_2$, and
- $A \mapsto P(A|u)$ is a probability measure on $\mathcal{F}_2$ for each $u \in R_1$.

We call regular conditional probabilities $P(A|u)$ also *transition probabilities*.

The notion of probability distances may be extended to transition probabilities: Let P and Q be two transition probabilities on $R_1 \times \mathcal{F}_2$ and let d be some probability distance. Define

$$\mathsf{d}(P, Q) := \sup_u \mathsf{d}(P(\cdot|u), Q(\cdot|u)). \tag{15.30}$$

In this way, all probability distances like d_U, d_{KA}, d_{FM} may be extended to transition probabilities in a natural way.

The next result compares the Wasserstein distance $\mathsf{d}_r(P, Q)$ with the distances of the components $\mathsf{d}_r(P_t, Q_t)$. To this end, introduce the following notation. Let d be some distance on $\mathbb{R}^M$. For a vector $w = (w_1, \ldots, w_T)$ of nonnegative weights we introduce the distance $d^t(u^t, v^t)^r := \sum_{s=1}^t w_s \cdot d(u_s, v_s)^r$ on $\mathbb{R}^{tM}$ for all t.

Then the following theorem holds true:

Theorem 3.1 *Assume that P and Q are probability measures on $\mathbb{R}^{TM}$, such that for all conditional measures P_t resp. Q_t we have that*

$$\mathsf{d}_r(P_{t+1}[.|u^t], \mathbb{Q}_{t+1}[.|v^t]) \leq \tau_{t+1} + K_{t+1} \cdot d^t(u^t, v^t) \tag{15.31}$$

for $t = 1, \ldots, T-1$. Then

$$\mathsf{d}_r(P^{t+1}, Q^{t+1})^r \leq w_{t+1}\tau_{t+1} + (1 + w_{t+1}K_{t+1})\mathsf{d}_r(P^t, Q^t)^r \tag{15.32}$$

and therefore

$$\mathsf{d}_r(P, Q)^r \leq \mathsf{d}_r(P_1, Q_1)^r \prod_{s=1}^{T}(1 + w_s K_s) + \sum_{s=1}^{T} \tau_s w_s \prod_{j=1}^{s}(1 + w_j K_j). \tag{15.33}$$

Proof Let π^t denote the optimal joint measure (optimal transportation plan) for $\mathsf{d}_r(P^t, Q^t)$ and let $\pi_{t+1}[.|u, v]$ denote the optimal joint measure for $\mathsf{d}_r(P_{t+1}[.|u], Q_{t+1}[.|v])$ and define the combined measure

$$\pi^{t+1}\left[A_t \times B^t, C_t \times D^t\right] := \int_{A_t \times C_t} \pi_{t+1}\left[B^t \times D^t | u^t, v^t\right] \pi^t\left[du^t, dv^t\right].$$

Then

$$\begin{aligned}
\mathsf{d}_r(P^{t+1}, Q^{t+1})^r &\leq \int d^{t+1}(u^{t+1}, v^{t+1})^r \,\pi^{t+1}\left[\mathrm{d}u^{t+1}, \mathrm{d}v^{t+1}\right] \\
&= \int \left[d^t(u^t, v^t)^r + w_{t+1}d_{t+1}(u_{t+1}, v_{t+1})^r\right] \\
&\qquad \pi_t\left[\mathrm{d}u_{t+1}, \mathrm{d}v_{t+1}|u^t, v^t\right]\pi^t\left[du^t, dv^t\right] \\
&= \int d^t\left(u^t, v^t\right)^r \pi^t\left[du^t, dv^t\right] + \\
&\qquad + w_{t+1}\int d_{t+1}\left(u_{t+1}, v_{t+1}\right)^r \pi_t\left[du_{t+1}, dv_{t+1}|u^t, v^t\right]\pi^t\left[du^t, dv^t\right] \\
&= \int d^t\left(u^t, v^t\right)^r \,\pi^t\left[du^t, dv^t\right] + \\
&\qquad + w_{t+1}\int \mathsf{d}_r\left(P_{t+1}\left[.|u^t\right], Q_{t+1}\left[.|v^t\right]\right)^r \,\pi^t\left[du^t, dv^t\right] \\
&\leq \int d^t\left(u^t, v^t\right)^r \,\pi^t\left[du^t, dv^t\right] + \\
&\qquad + w_{t+1}\int \tau_{t+1} + K_{t+1}d^t\left(u^t, v^t\right)^r \,\pi^t\left[du^t, dv^t\right] \\
&= \mathsf{d}_r\left(P^t, Q^t\right)^r + w_{t+1}\left(\tau_{t+1} + K_{t+1}\mathsf{d}_r\left(P^t, Q^t\right)^r\right) \\
&= w_{t+1}\tau_{t+1} + (1 + w_{t+1}K_{t+1})\,\mathsf{d}_r\left(P^t, Q^t\right)^r.
\end{aligned}$$

This demonstrates (15.32). Applying this recursion for all t leads to (15.33). □

Definition 3.2 Let P be a probability on $\mathbb{R}^{MT}$, dissected into transition probabilities $P_1, \ldots, P_T$. We say that P has the $(K_2, \ldots, K_T)$-Lipschitz property, if for $t = 2, \ldots, T$,

$$\mathsf{d}_r(P_t(\cdot|u), P_t(\cdot|v)) \leq K_t d(u, v).$$

Example (The normal distribution has the Lipschitz property) Consider a normal distribution P on $\mathbb{R}^{m_1+m_2}$, i.e.,

$$P = \mathcal{N}\left(\begin{pmatrix}\mu_1\\ \mu_2\end{pmatrix}, \begin{pmatrix}\Sigma_{11} & \Sigma_{12}\\ \Sigma_{12} & \Sigma_{22}\end{pmatrix}\right).$$

The conditional distribution, given $u \in \mathbb{R}^{m_1}$, is

$$P_2(\cdot|u) = \mathcal{N}\left(\mu_1 - \Sigma_{12}\Sigma_{11}^{-1}(u - \mu_1), \Sigma_{22} - \Sigma_{12}\Sigma_{11}\Sigma_{12}^{-1}\right).$$

We claim that

$$\mathsf{d}_{\mathrm{KA}}(P_2(\cdot|u), P_2(\cdot|v)) \le \|\Sigma_{12}\Sigma_{11}^{-1}(u - v)\|. \tag{15.34}$$

Let $Y \sim \mathcal{N}(0, \Sigma)$ and let $Y_1 = a_1 + Y$ and $Y_2 = a_2 + Y$. Then $Y_1 \sim \mathcal{N}(a_1, \Sigma)$, $Y_2 \sim \mathcal{N}(a_2, \Sigma)$ and therefore

$$\mathsf{d}_{\mathrm{KA}}(\mathcal{N}(a_1, \Sigma), \mathcal{N}(a_2, \Sigma)) \le \mathbb{E}(\|Y_1 - Y_2\|) = \|a_1 - a_2\|.$$

Setting $a_1 = \mu_1 - \Sigma_{12}\Sigma_{11}^{-1}(u - \mu_1)$, $a_2 = \mu_1 - \Sigma_{12}\Sigma_{11}^{-1}(v - \mu_1)$ and $\Sigma = \Sigma_{22} - \Sigma_{12}\Sigma_{11}\Sigma_{12}^{-1}$, the result (15.34) follows.

Corollary 3.3 Condition (15.31) is fulfilled, if the following two conditions are met:

(i) P has the $(K_2, \ldots, K_T)$-Lipschitz property in the sense of Definition 3.2
(ii) Closeness of all P_t and Q_t in the following sense:

$$\mathsf{d}_r(P_t, Q_t) = \sup_u \mathsf{d}_r(P_t(\cdot|u), Q_t(\cdot|u) \le \tau_t$$

The Lipschitz assumption for P is crucial. Without this condition, no relation between $\mathsf{d}_r(P, Q)$ and the $\mathsf{d}_r(P_t, Q_t)$'s holds. In particular, the following example shows that $\mathsf{d}_1(P_t, Q_t)$ may be arbitrarily small and yet $\mathsf{d}_1(Q, P)$ is far away from zero.

Example Let $P = P_1 \circ P_2$ be the measure on $\mathbb{R}^2$, such that

$$P_1 = \mathcal{U}[0, 1],$$
$$P_2(\cdot|u) = \begin{cases}\mathcal{U}[0, 1] & \text{if } u \in \mathbb{Q},\\ \mathcal{U}[1, 2] & \text{if } u \in \mathbb{R} \setminus \mathbb{Q},\end{cases}$$

where $\mathcal{U}[a, b]$ denotes the uniform distribution on $[a, b]$. Obviously, P is the uniform distribution on $[0, 1] \times [1, 2]$. (P_1, P_2) has no Lipschitz property. Now, let $Q^{(n)} = Q_1^{(n)} \circ Q_2^{(n)}$, where

$$Q_1^{(n)} = \sum_{k=1}^{n} \frac{1}{n} \delta_{k/n},$$

$$Q_2^{(n)}(\cdot|u) = \begin{cases} \sum_{k=1}^{n} \frac{1}{n}\delta_{k/n} & \text{if } u \in \mathbb{Q}, \\ \sum_{k=1}^{n} \frac{1}{n}\delta_{1+k/n} & \text{if } u \in \mathbb{R} \setminus \mathbb{Q}. \end{cases}$$

Note that $Q^{(n)} = Q_1^{(n)} \circ Q_2^{(n)}$ converges weakly to the uniform distribution on $[0, 1] \times [0, 1]$. In particular, it does not converge to P and the distance is $\mathsf{d}_1(P, Q^{(n)}) = 1$. However,

$$\mathsf{d}_1(P_1, Q_1^{(n)}) = 1/n; \quad \mathsf{d}_1(P_2, Q_2^{(n)}) = 1/n.$$

15.4.2 Filtrations and Tree Processes

In many stochastic decision problems, a distinction has to be made between the scenario process (which contains decision-relevant values, such as prices, demands, and supplies) and the information which is available. Here is a simple example due to Philippe Artzner.

Example A fair coin is tossed three times. Situation A: The payoff process $\xi^{(A)}$ is

$$\xi_1^{(A)} = 0; \quad \xi_2^{(A)} = 0;$$

$$\xi_3^{(A)} = \begin{cases} 1 & \text{if heads is shown at least two times,} \\ 0 & \text{otherwise.} \end{cases}$$

We compare this process to another payoff process (situation B)

$$\xi_1^{(B)} = 0, \quad \xi_2^{(B)} = 0,$$

$$\xi_3^{(B)} = \begin{cases} 1 & \text{if heads is shown at least the last throw,} \\ 0 & \text{otherwise.} \end{cases}$$

The available information is relevant: It makes a difference, whether we may observe the coin tossing process or not. Suppose that the process is observable. Then, after the second throw, we know the final payoff, if two times the same side appeared. Thus our information is complete although the experiment has not ended yet. Only in case that two different sides appeared, we have to wait until the last throw to get the final information.

Mathematically, the information process is described by a filtration.

Definition 3.4 Let $(\Omega, \mathcal{F}, P)$ be probability space. A *filtration* $\mathcal{F}$ on this probability space is an increasing sequence of σ-fields $\mathcal{F} = (\mathcal{F}_1, \ldots, \mathcal{F}_T)$ where for all t,

$\mathcal{F}_t \subseteq \mathcal{F}_{t+1} \subseteq \mathcal{F}$. We may always add the trivial σ-field $\mathcal{F}_0 = \{\emptyset, \Omega)$ as the zeroth element of a filtration.

A scenario process $\xi = (\xi_1, \ldots, \xi_T)$ is *adapted* to the filtration $\mathcal{F}$, if ξ_t is $\mathcal{F}_t$ measurable for $t = 1, \ldots, T$. If ξ_t is measurable w.r.t. $\mathcal{F}_t$, we use the notation

$$\xi_t \lhd \mathcal{F}_t;$$

if the process ξ is adapted to $\mathcal{F}$, we use the similar notation

$$\xi \lhd \mathcal{F}.$$

The general situation is as follows: We separate the information process from the value process. Information is modeled by a filtration $\mathcal{F}$, while the scenario values are modeled by a (real or vector values process) ξ, which is adapted to $\mathcal{F}$. We call a pair $(\xi, \mathcal{F})$ a *process-and-information pair*. In general, the filtration $\mathcal{F}$ is larger than the one generated by the process ξ.

In finite probability spaces, there is a one-to-one correspondence between filtrations and probability trees. A process is adapted to such a filtration, iff it is a function of the nodes of the tree. Figures 15.5 and 15.6 show the tree processes and the adapted scenario process sitting on the tree for Artzner's example and situations (A) and (B).

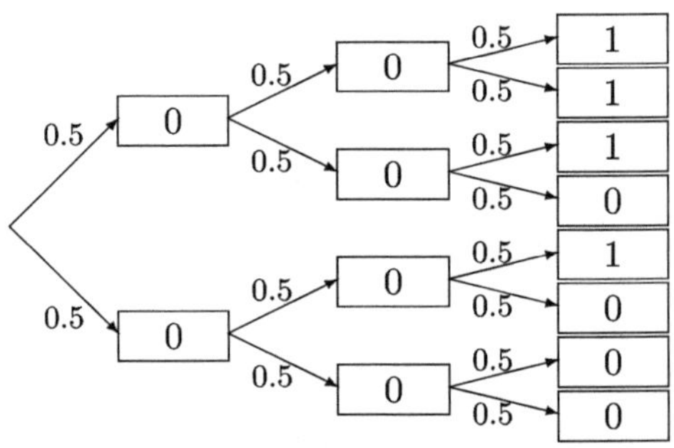

Fig. 15.5 The payoff process $\xi^{(A)}$ sitting on the tree (the filtration) generated by the coin process

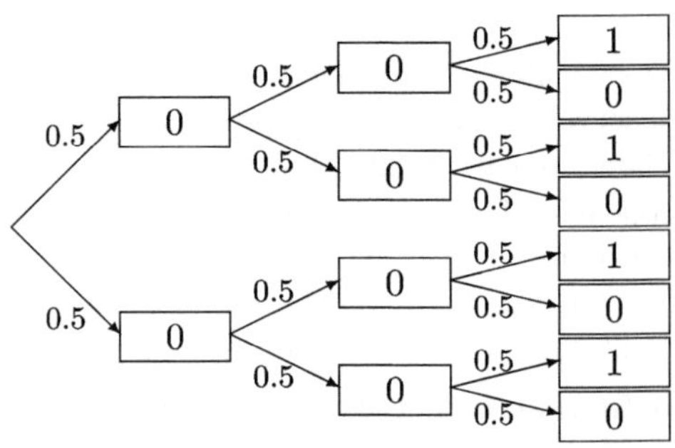

Fig. 15.6 The payoff process $\xi^{(B)}$ sitting on the tree (the filtration) generated by the coin process

15.4.2.1 Tree Processes

Definition 3.5 A stochastic process $\nu = (\nu_1, \ldots, \nu_T)$ is called a *tree process*, if $\sigma(\nu_1), \sigma(\nu_2), \ldots, \sigma(\nu_T)$ is a filtration. A tree process is characterized by the property that the conditional distribution of $\nu_1, \ldots, \nu_t$ given ν_{t+1} is degenerated, i.e., concentrated on a point. Notice that any stochastic process $\eta = (\eta_1, \ldots, \eta_T)$ can be transformed into a tree process by considering its history process

$$\begin{aligned} \nu_1 &= \eta_1, \\ \nu_2 &= (\eta_1, \eta_2), \\ &\vdots \\ \nu_T &= (\eta_1, \ldots, \eta_T). \end{aligned}$$

Let (ν_t) be a tree process and let $\mathcal{F} = \sigma(\nu)$. It is well known that any process ξ adapted to $\mathcal{F}$ can be written as $\xi_t = f_t(\nu_t)$ for some measurable function f_t. Thus a stochastic program may be formulated using the tree process ν and the functions f_t.

Let ν be a tree process according to Definition 3.5. We assume throughout that the state spaces of all tree processes are Polish. In this case, the joint distribution P of $\nu_1, \ldots, \nu_T$ can be factorized into the chain of conditional distributions, which are given as transition kernels, that is

$$\begin{aligned} P_1(A) &= P\{\nu_1 \in A\}, \\ P_2(A|u) &= P\{\nu_2 \in A|\nu_1 = u\}, \\ &\vdots \\ P_t(A|u_1, \ldots, u_{t-1}) &= P\{\nu_t \in A|\nu_1 = u_1, \ldots, \nu_{t-1} = u_{t-1}\}. \end{aligned}$$

Definition 3.6 (Equivalence of tree processes) Let ν resp. $\bar{\nu}$ two tree processes, which are defined on possibly different probability spaces $(\Omega, \mathcal{F}, P)$ resp. $(\bar{\Omega}, \bar{\mathcal{F}}, \bar{P})$. Let $P_1, P_2, \ldots, P_T$ resp. $\bar{P}_1, \bar{P}_2, \ldots, \bar{P}_T$ be the pertaining chain of conditional distributions. The two tree processes are *equivalent*, if the following properties hold: There are bijective functions $k_1, \ldots, k_T$ mapping the state spaces of ν to the state spaces of $\bar{\nu}$, such that the two processes $\nu_1, \ldots, \nu_T$ and $k_1^{-1}(\bar{\nu}_1), \ldots, k_T^{-1}(\bar{\nu}_T)$ coincide in distribution.

Notice that for tree processes defined on the same probability space, one could define equivalence simply as $\nu_t = k_t(\bar{\nu}_t)$ a.s. The value of Definition 3.6 is that it extends to tree processes on different probability spaces.

Definition 3.7 (Equivalence of scenario process-and-information pairs) Let (ξ, ν) be a scenario process-and-information pair, consisting of a scenario process ξ and a tree process, such that $\xi \lhd \sigma(\nu)$. Since $\xi \lhd \sigma(\nu)$, there are measurable functions f_t such that

$$\xi_t = f_t(\nu_t), \qquad t = 1, \ldots T.$$

Two pairs (ξ, ν) and $(\bar{\xi}, \bar{\nu})$ are called *equivalent*, if ν and $\bar{\nu}$ are equivalent in the sense of Definition 3.6 with equivalence mappings k_t such that

$$\bar{\xi}_t = f_t(k_t^{-1}(\xi_t)) \qquad \text{a.s.}$$

We agree that equivalent process-and-information pairs will be treated as equal. Notice that by drawing a finite tree with some values sitting on its nodes we automatically mean an equivalence class of process-and-information pairs.

If two stochastic optimization problems are defined in the same way, but using equivalent process-and-information pairs, then they have equivalent solutions. It is easy to see that an optimal solution of the first problem can be transformed into an optimal solution of the second problem via the equivalence functions k_t.

Figures 15.7 and 15.8 illustrate the concept of equivalence. The reader is invited to find the equivalence functions k_t in this example.

15.4.2.2 Distances Between Filtrations

In order to measure the distances between two discretizations on the same probability space one could first measure the distance of the respective filtrations. This

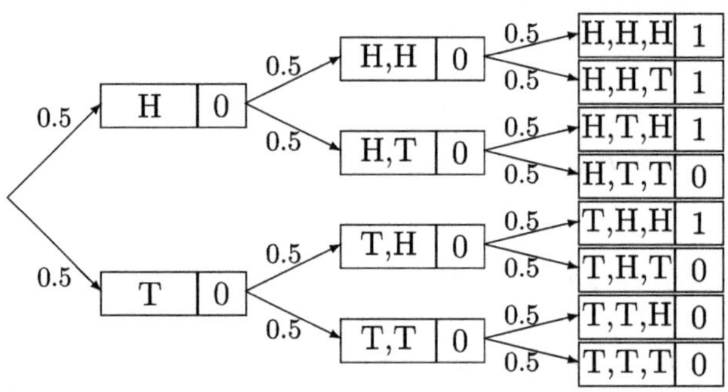

Fig. 15.7 Artzner's example: The tree process is coded by the history of the coin throws (H = head,T = tail), the payoffs are functions of the tree process

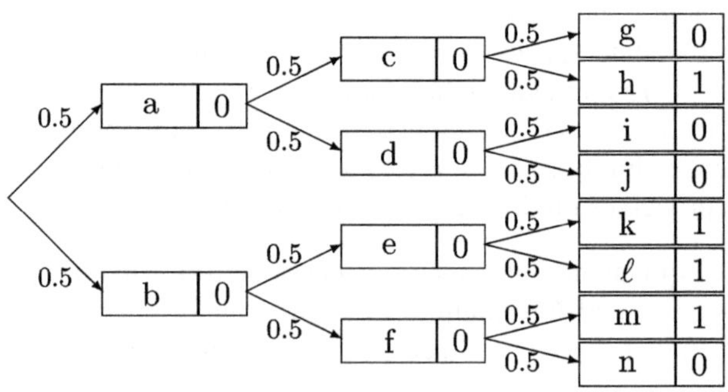

Fig. 15.8 A process-and-information pair which is equivalent to Artzner's example in Fig. 15.4. The coding of the tree process is just by subsequent letters

concept of filtration distance for scenario generation has been developed by Heitsch and Roemisch (2003) and Heitsch et al. (2006).

Let $(\Omega, \bar{\mathcal{F}}, P)$ be a probability space and let $\mathcal{F}$ and $\tilde{\mathcal{F}}$ be two sub-sigma algebras of $\bar{\mathcal{F}}$. Then the *Kudo* distance between $\mathcal{F}$ and $\tilde{\mathcal{F}}$ is defined as

$$\mathfrak{D}(\mathcal{F}, \tilde{\mathcal{F}}) = \max(\sup\{\|\mathbb{1}_A - \mathbb{E}(\mathbb{1}_A|\tilde{\mathcal{F}})\|_p : A \in \mathcal{F}\}, \sup\{\|\mathbb{1}_B - \mathbb{E}(\mathbb{1}_B|\mathcal{F})\|_p : A \in \tilde{\mathcal{F}}\}). \tag{15.35}$$

The most important case is the distance between a sigma algebra and a sub-sigma algebra of it: If $\tilde{\mathcal{F}} \subseteq \mathcal{F}$, then (15.35) reduces to

$$\mathfrak{D}(\mathcal{F}, \tilde{\mathcal{F}}) = \sup\{\|\mathbb{1}_A - \mathbb{E}(\mathbb{1}_A|\tilde{\mathcal{F}})\|_p : A \in \mathcal{F}\}. \tag{15.36}$$

If $\mathcal{F} = (\mathcal{F}_1, \mathcal{F}_2, \ldots)$ is an infinite filtration, one may ask, whether this filtration converges to a limit, i.e., whether there is a $\mathcal{F}_\infty$, such that

$$\mathfrak{D}(\mathcal{F}_t, \mathcal{F}_\infty) \to 0 \quad \text{for } t \to \infty.$$

If $\mathcal{F}_\infty$ is the smallest σ-field, which contains all $\mathcal{F}_t$, then $\mathfrak{D}(\mathcal{F}_t, \mathcal{F}_\infty) \to 0$. However, the following example shows that not every discrete increasing filtration converges to the Borel σ-field.

Example Let $\mathcal{F}$ be the Borel-sigma algebra on [0,1) and let $\tilde{\mathcal{F}}_n$ be the sigma algebra generated by the sets $\left[\frac{k}{2^n}, \frac{k+1}{2^n}\right)$, $k = 0, \ldots, 2^n - 1$. Moreover, let $A_n = \bigcup_{k=0}^{2^n-1} \left[\frac{k}{2^n}, \frac{2k+1}{2^{n+1}}\right)$. Then $\mathbb{E}(\mathbb{1}_{A_n}|\tilde{\mathcal{F}}_n) = \frac{1}{2}$ and $\|\mathbb{1}_{A_n} - \mathbb{E}(\mathbb{1}_{A_n}|\tilde{\mathcal{F}}_n)\|_p = 1/2$ for all n. While one has the intuitive feeling that $\tilde{\mathcal{F}}_n$ approaches $\mathcal{F}$, the distance is always $1/2$.

15.4.3 Nested Distributions

We define a nested distribution as the distribution of a process-and-information pair, which is invariant w.r.t. to equivalence. In the finite case, the nested distribution is the distribution of the scenario tree or better of the class of equivalent scenario trees.

Let d be a metric on $\mathbb{R}^M$, which makes it a complete separable metric space and let $\mathcal{P}_1(\mathbb{R}^M, d)$ be the family of all Borel probability measures P on $(\mathbb{R}^M, d)$ such that

$$\int d(u, u_0)\, dP(u) < \infty \tag{15.37}$$

for some $u_0 \in \mathbb{R}^M$.

For defining the nested distribution of a scenario process $(\xi_1, \ldots, \xi_T)$ with values in $\mathbb{R}^M$, we define the following spaces in a recursive way:

$$\Xi_1 := \mathbb{R}^M,$$

with distance $\mathbb{d}_1(u, v) = d(u, v)$:

$$\Xi_2 := \mathbb{R}^M \times \mathcal{P}_1(\Xi_1)$$

with distance $\mathbb{d}_2((u, P), (u, Q)) := d(u, v) + \mathsf{d}_{\mathrm{KA}}(P, Q; \mathbb{d}_1)$:

$$\Xi_3 := \mathbb{R}^M \times \mathcal{P}_1(\Xi_2) = (\mathbb{R}^M \times \mathcal{P}_1(\mathbb{R}^M \times \mathcal{P}_1(\mathbb{R}^M)))$$

with distance $\mathbb{d}_3((u, \mathbb{P}), (v, \mathbb{Q})) := d(u, v) + \mathsf{d}_{\mathrm{KA}}(\mathbb{P}, \mathbb{Q}; \mathbb{d}_2)$. This construction is iterated until

$$\Xi_T = \mathbb{R}^M \times \mathcal{P}_1(\Xi_{T-1})$$

with distance $\mathbb{d}_T((u, \mathbb{P}), (v, \mathbb{Q})) := d(u, v) + \mathsf{d}_{\mathrm{KA}}(\mathbb{P}, \mathbb{Q}; \mathbb{d}_{T-1})$.

Definition 3.8 (Nested distribution) A Borel probability distribution $\mathbb{P}$ with finite first moments on Ξ_T is called a *nested distribution of depth T*; the distance $\mathbb{d}_T$ is called the *nested distance* (Pflug 2009).

In discrete cases, a nested distribution may be represented as a recursive structure: Recall that we represent discrete probabilities by a list of probabilities (in the first row) and the values (in the subsequent rows).

Notice that *one* value or vector of values may only carry *one* probability, the following structure on the left does not represent a valid distribution, the structure right is correct:

$$\left[\begin{array}{cccc} 0.1 & 0.2 & 0.4 & 0.3 \\ \hline 3.0 & 3.0 & 1.0 & 5.0 \end{array}\right] \qquad \left[\begin{array}{ccc} 0.3 & 0.4 & 0.3 \\ \hline 3.0 & 1.0 & 5.0 \end{array}\right].$$

In the same manner, but in a recursive way, we may now represent a nested distribution as a structure, where some values are distributions themselves:

$$\left[\begin{array}{ccc} 0.2 & 0.3 & 0.5 \\ \hline 3.0 & 3.0 & 2.4 \\ \left[\begin{array}{ccc} 0.4 & 0.2 & 0.4 \\ \hline 6.0 & 4.7 & 3.3 \end{array}\right] & \left[\begin{array}{c} 1.0 \\ \hline 2.8 \end{array}\right] & \left[\begin{array}{cc} 0.6 & 0.4 \\ \hline 1.0 & 5.1 \end{array}\right] \end{array}\right].$$

This recursive structure encodes the tree shown in Fig. 15.9.

For any nested distribution $\mathbb{P}$, there is an embedded multivariate distribution P. We explain the projection from the nested distribution to the embedded multivariate distribution just for the depth 3, the higher depths being analogous:

Let $\mathbb{P}$ be a nested distribution on Ξ_3, which has components η_1 (a real random variable) and μ_2 (a random distribution on Ξ_2, which means that it has in turn components η_2 (a random variable) and μ_3, a random distribution on $\mathbb{R}^m$).

Fig. 15.9 Example tree

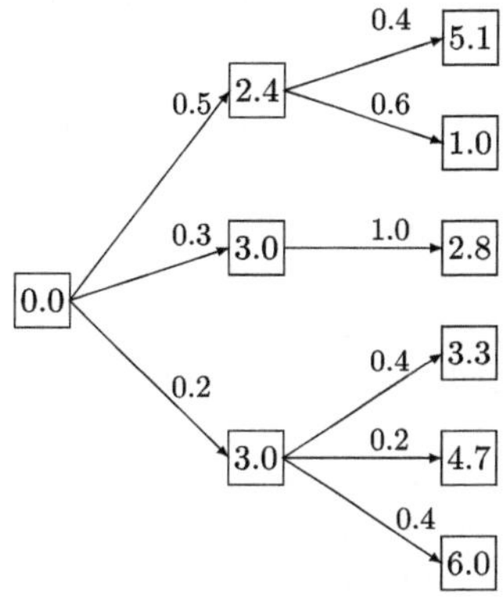

Then the pertaining multivariate distribution on $\mathbb{R}^m \times \mathbb{R}^m \times \mathbb{R}^m$ is given by

$$P(A_1 \times A_2 \times A_3) = \mathbb{E}_{\mathbb{P}}[\mathbb{1}_{\{\eta_1 \in A_1\}} \mathbb{E}_{\mu_2}(\mathbb{1}_{\{\eta_2 \in A_2\}} \mu_2(A_3))]$$

for A_1, A_2, A_2 Borel sets in $\mathbb{R}^m$.

The mapping from the nested distribution to the embedded distribution is not injective: There may be many different nested distributions which all have the same embedded distribution, see Artzner's example: $\xi^{(A)}$ and $\xi^{(B)}$ in Figs. 15.5 and 15.6 have different nested distributions, but the same embedded distributions. Here is another example:

Example Consider again the process-and-information pair (the tree) of Fig. 15.9. The embedded multivariate but non-nested distribution of the scenario process is

$$\begin{bmatrix} 0.08 & 0.04 & 0.08 & 0.3 & 0.3 & 0.2 \\ \hline 3.0 & 3.0 & 3.0 & 3.0 & 2.4 & 2.4 \\ 6.0 & 4.7 & 3.3 & 2.8 & 1.0 & 5.1 \end{bmatrix}.$$

Evidently, this multivariate distribution has lost the information about the nested structure. If one considers the filtration generated by the scenario process itself and forms the pertaining nested distribution, one gets

$$\left[\begin{array}{cc} 0.5 & 0.5 \\ \hline 3.0 & 2.4 \\ \left[\begin{array}{cccc} 0.16 & 0.08 & 0.16 & 0.6 \\ \hline 6.0 & 4.7 & 3.3 & 2.8 \end{array}\right] & \left[\begin{array}{cc} 0.6 & 0.4 \\ \hline 1.0 & 5.1 \end{array}\right] \end{array}\right]$$

since in this case the two nodes with value 3.0 have to be identified.

Lemma 3.9 Let (ξ, ν) be equivalent to $(\bar{\xi}, \bar{\nu})$ in the sense of Definition 3.7. Then both pairs generate the same nested distribution.

For a proof see Pflug (2009).

15.4.3.1 Computing the Nested Distance

Recall that a nested distribution is the distribution of a process-and-information pair, which is invariant w.r.t. equivalence. In the finite case, the nested distribution is the distribution of the scenario tree (or better of the class of equivalent scenario trees), which was defined in a recursive way.

To compute the nested distance of two discrete nested distributions, P and $\tilde{P}$ say, corresponds to computing the distance of two tree processes ν and $\tilde{\nu}$. We further elaborate the algorithmic approach here, which is inherent is the recursive definition of nested distributions and the nested distance.

To this end recall first that the linear program (15.20) was defined so general, allowing to compute the distance of general objects whenever probabilities p_s, p_t, and the cost matrix $c_{s,t}$ were specified – no further specifications of the samples z are needed. We shall exploit this fact; in the situation we describe here the samples z_s and z_t represent nested distributions with finitely many states, or equivalently entire trees.

Consider the nodes in the corresponding tree: Any node is either a terminal node, or a node with children, or the root node of the respective tree. The following recursive treatment is in line with this structure:

- **Recursion start** (discrete nested distribution of depth 1) In this situation the nested distributions themselves are elements of the underlying space, $P \in \Xi_T$ ($\tilde{P} \in \Xi_T$ resp.), for which we have that

  $$\mathrm{dl}_T(P, \tilde{P}) = d(P, \tilde{P})$$

  according to the definition.
  In the notion of trees this situation corresponds to trees which consist of a single node only, which are terminal nodes (leaf nodes).
- **Recursive step** (discrete nested distribution of depth $t > 1$) In this situation the nested distribution may be dissected in

  $$P_t = \left(n_t, \sum_s p_s^t \delta_{z_s^{t+1}}\right) \text{ and } \tilde{P}_t = \left(\tilde{n}_t, \sum_{\tilde{s}} \tilde{p}_{\tilde{s}}^t \delta_{\tilde{z}_{\tilde{s}}^{t+1}}\right),$$

  where $z_s^{t+1} \in \Xi_{t+1}$ and $\tilde{z}_{\tilde{s}}^{t+1} \in \Xi_{t+1}$.
  We shall assume that dl_{t+1} is recursively available, thus define the matrix

  $$c_{s,\tilde{s}} := d(n_t, \tilde{n}_t) + \mathrm{dl}_{t+1}\left(z_s^{t+1}, \tilde{z}_{\tilde{s}}^{t+1}\right)$$

  and apply the linear program (15.20) together with the probabilities $p = (p_s)_s$ and $\tilde{p} = (\tilde{p}_{\tilde{s}})_{\tilde{s}}$, resulting with the distance $\mathrm{dl}_t(P_t, \tilde{P}_t)$.

  In the notion of trees, again, n_t and $\tilde{n}_t$ represent nodes with children. The children are given by the first components of z_s^{t+1} ($\tilde{z}_{\tilde{s}}^{t+1}$, resp.); further notice that

$(z_s^{t+1})_s$ are just all subtrees of n_t, all this information is encoded in the nested distribution P_t.

- **Recursive end** (nested distribution of depth $t = T$) The computation of the distance has already been established in the recursion step.

Note, that the corresponding trees in this situation represent the entire tree process here.

15.4.3.2 The Canonical Construction

Suppose that $\mathbb{P}$ and $\tilde{\mathbb{P}}$ are nested distributions of depth T. One may construct the pertaining tree processes in a canonical way: The canonical name of each node is the nested distribution pertaining of the subtree, for which this node is the root. Two valuated trees are equivalent, iff the respective canonical constructions are identical.

Example 3.10 Consider the following nested distribution:

$$\mathbb{P}_\epsilon = \left[\begin{array}{cc} 0.5 & 0.5 \\ \hline 2 & 2+\mathbb{E} \\ \left[\begin{array}{c} 1.0 \\ \hline 3 \end{array}\right] & \left[\begin{array}{c} 1.0 \\ \hline 1 \end{array}\right] \end{array}\right].$$

We construct a tree process such that the names of the nodes are the corresponding nested distributions of the subtrees:

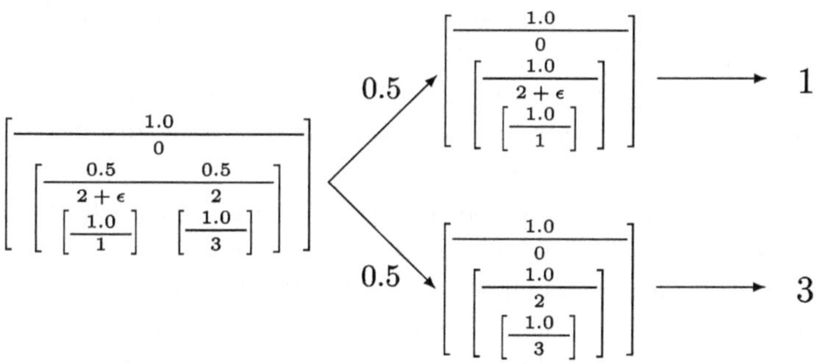

The nested distance generates a much finer topology than the distance of the multivariate distributions.

Example (see Heitsch et al. (2006)) Consider the nested distributions of Example 3.10 and

$$\mathbb{P}_0 = \left[\begin{array}{c} 1.0 \\ \hline 2 \\ \left[\begin{array}{cc} 0.5 & 0.5 \\ \hline 3 & 1 \end{array}\right] \end{array}\right].$$

Note that the pertaining multivariate distribution of $\mathbb{P}_\epsilon$ on $\mathbb{R}^2$ converges weakly to the one of $\mathbb{P}_0$, if $\epsilon \to 0$. However, the nested distributions do not converge: The nested distance is $\mathsf{dl}(\mathbb{P}_\epsilon, \mathbb{P}_0) = 1 + \epsilon$ for all ϵ.

Lemma 3.11 Let $\mathbb{P}$ be a nested distribution, P its multivariate distribution, which is dissected into the chain $P = P_1 \circ P_2 \circ \cdots \circ P_T$ of conditional distributions. If P has the $(K_2, \ldots, K_T)$-Lipschitz property (see Definition 3.2), then

$$\mathsf{dl}(\mathbb{P}, \tilde{\mathbb{P}}) \le \sum_{t=1}^{T} \mathsf{d}_1(P_t, \tilde{P}_t) \prod_{s=t+1}^{T+1} (K_s + 1).$$

For a proof see Pflug (2009).

Theorem 3.12 *Let $\mathbb{P}$ resp. $\tilde{\mathbb{P}}$ be two nested distributions and let P resp. $\tilde{P}$ be the pertaining multi-period distributions. Then*

$$\mathsf{d}_1(P, \tilde{P}) \le \mathsf{dl}(\mathbb{P}, \tilde{\mathbb{P}}).$$

Thus the mapping, which maps the nested distribution to the embedded multi-period distribution, is Lipschitz(1). The topologies generated by the two distances are not the same: The two trees in Figs. 15.5 resp. 15.6 are at multivariate Kantorovich distance 0, but their nested distance is $\mathsf{dl} = 0.25$.

Proof To find the Kantorovich distance between P and $\tilde{P}$ one has to find a joint distribution for two random vectors ξ and $\tilde{\xi}$ such that the marginals are P resp. $\tilde{P}$. Among all these joint distributions, the minimum of $d(\xi_1, \tilde{\xi}_t) + \cdots + d(\xi_T, \tilde{\xi}_T)$ equals $d(P, \tilde{P})$. The nested distance appears in a similar manner; the only difference is that for the joint distribution also the conditional marginals should be equal to those of $\mathbb{P}$ resp. $\tilde{\mathbb{P}}$. Therefore, the admissible joint distributions for the nested distance are subsets of the admissible joint distributions for the distance between P and $\tilde{P}$ and this implies the assertion. □

Remark The advantage of the nested distance is that topologically different scenario trees with completely different values can be compared and their distance can be calculated. Consider the example of Fig. 15.10. The conditional probabilities $P(\cdot|u)$ and $Q(\cdot|v)$ are incomparable, since no values of u coincide with values of v. Moreover, the Lipschitz condition of Definition 3.2 does not make sense for trees and therefore Theorem 3.1 is not applicable. However, the nested distance is well defined for two trees.

By calculation, we find the nested distance $\mathsf{dl} = 1.217$.

Example Here is an example of the nested distance between a continuous process and a scenario tree process. Let $\mathbb{P}$ be the following continuous nested distribution: $\xi_1 \sim N(0, 1)$ and $\xi_2|\xi_1 \sim N(\xi_1, 1)$. Let $\tilde{\mathbb{P}}$ be the discrete nested distribution

$$\left[\begin{array}{ccc} 0.30345 & 0.3931 & 0.30345 \\ \hline -1.029 & 0.0 & 1.029 \\ \left[\begin{array}{ccc} 0.30345 & 0.3931 & 0.30345 \\ \hline -2.058 & -1.029 & 0.0 \end{array}\right] & \left[\begin{array}{ccc} 0.30345 & 0.3931 & 0.30345 \\ \hline -1.029 & 0.0 & 1.029 \end{array}\right] & \left[\begin{array}{ccc} 0.30345 & 0.3931 & 0.30345 \\ \hline 0.0 & 1.029 & 2.058 \end{array}\right] \end{array}\right]$$

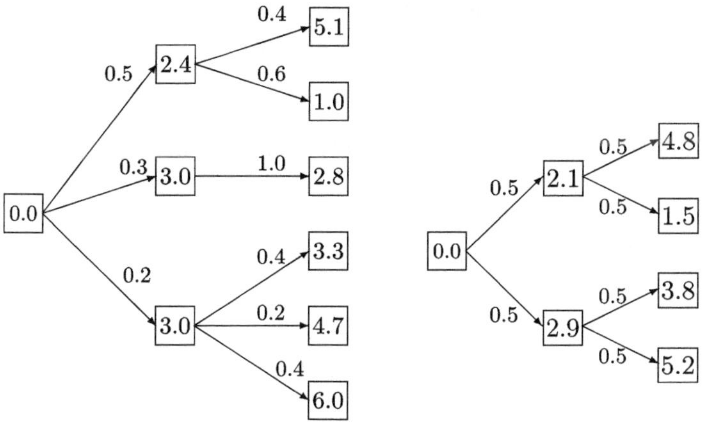

Fig. 15.10 Two valuated trees

Then the nested distance is $d(\mathbb{P}, \tilde{\mathbb{P}}) = 0.8215$, which we found by numerical distance calculation.

15.5 Scenario Tree Construction and Reduction

Suppose that a scenario process (ξ_t) has been estimated. In this section, we describe how to construct a scenario tree out of it. We begin with considering the case, where the information is the one which is generated by the scenario process and the method is based on the conditional probabilities. Later, we present the case where a tree process representing the information and a scenario process sitting on the tree process has been estimated.

15.5.1 Methods Based on the Conditional Distance

The most widespread method to construct scenario trees is to construct them recursively from root to leaves. The discretization for the first stage is done as described in Section 15.2. Suppose a chosen value for the first stage is x. Then the reference process ξ_t is conditioned to the set $\{\omega : |\xi_1 - x| \leq \epsilon\}$, where ϵ is chosen in such a way that enough trajectories are contained in this set. The successor nodes of x are then chosen as if x were the root and the conditional process were the total process.

The original distributions may either be continuous (e.g., estimated distributions) or more commonly discrete (e.g., simulated with some random generation mechanism which provides some advantages for practical usage). Different solution techniques for continuous as well as discrete approximation have been outlined in Pflug (2001) and Hochreiter and Pflug (2007).

Example The typical algorithm for finding scenario trees based on the Kantorovich distance can be illustrated by the following pictures.

Step 1. Estimate the probability model and simulate future trajectories

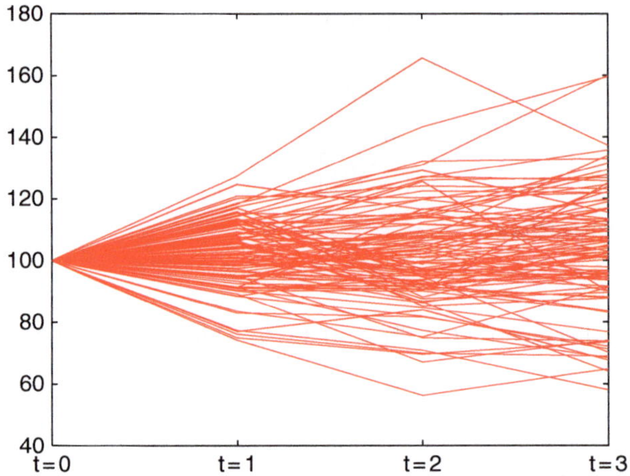

Step 2. Approximate the data in the first stage by minimal Kantorovich distance

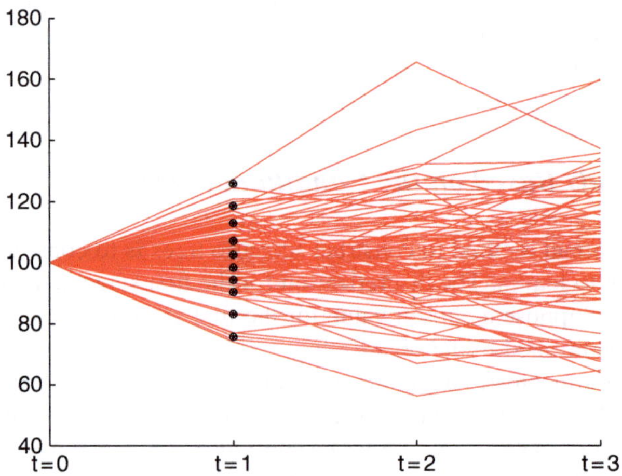

Step 3. For the subsequent stages, use the paths which are close to its predecessor ...

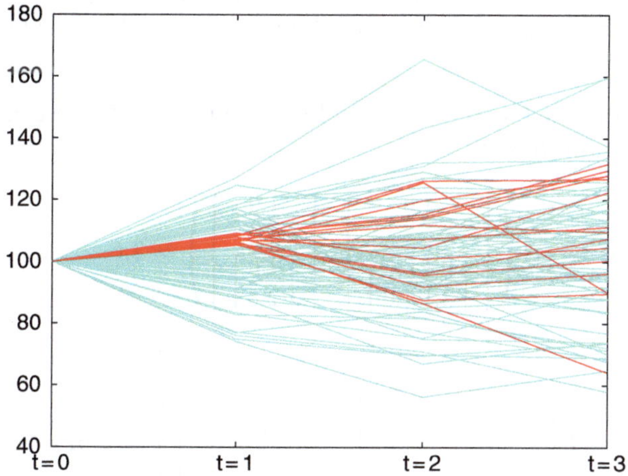

Step 4. .. and iterate this step through the whole tree.

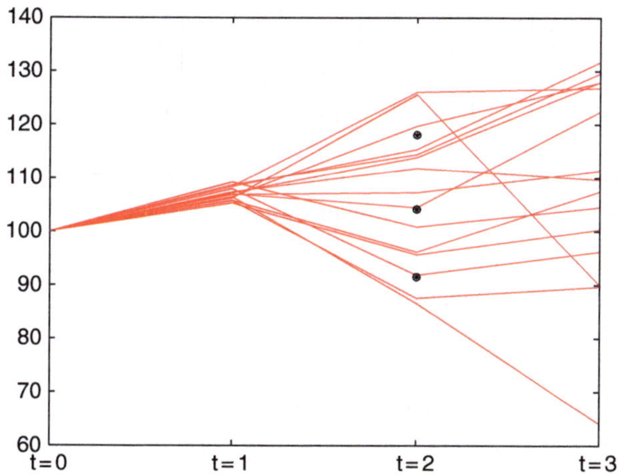

15.5.2 Methods Based on the Nested Distance

The general scenario tree problem is as follows:

15.5.2.1 The Scenario Tree Problem

Given a nested distribution $\mathbb{P}$ of depth T, find a valuated tree (with nested distance $\tilde{\mathbb{P}}$) with at most S nodes such that the distance $\mathrm{dl}(\mathbb{P}, \tilde{\mathbb{P}})$ is minimal.

Unfortunately, this problem is such complex, that is, it is impossible to be solved for medium tree sizes. Therefore, some heuristic approximations are needed. A possible method is a stepwise and greedy reduction of a given tree. Two subtrees of the tree may be collapsed, if they are close in terms of the nested distance. Here is an example:

Example The three subtrees of level 2 of the tree in Fig. 15.9. We calculated their nested distances as follows:

$$\mathrm{d\!l}\left(\begin{bmatrix} \dfrac{1.0}{2.4} \\ \begin{bmatrix} \dfrac{0.4\ 0.6}{5.1\ 1.0} \end{bmatrix} \end{bmatrix}, \begin{bmatrix} \dfrac{1.0}{3.0} \\ \begin{bmatrix} \dfrac{1.0}{2.8} \end{bmatrix} \end{bmatrix}\right) = 2.6,$$

$$\mathrm{d\!l}\left(\begin{bmatrix} \dfrac{1.0}{2.4} \\ \begin{bmatrix} \dfrac{0.4\ 0.6}{5.1\ 1.0} \end{bmatrix} \end{bmatrix}, \begin{bmatrix} \dfrac{1.0}{3.0} \\ \begin{bmatrix} \dfrac{0.4\ 0.2\ 0.4}{3.3\ 4.7\ 6.0} \end{bmatrix} \end{bmatrix}\right) = 2.62,$$

$$\mathrm{d\!l}\left(\begin{bmatrix} \dfrac{1.0}{2.4} \\ \begin{bmatrix} \dfrac{0.4\ 0.6}{5.1\ 1.0} \end{bmatrix} \end{bmatrix}, \begin{bmatrix} \dfrac{1.0}{3.0} \\ \begin{bmatrix} \dfrac{0.4\ 0.2\ 0.4}{3.3\ 4.7\ 6.0} \end{bmatrix} \end{bmatrix}\right) = 1.86.$$

To calculate a nested distance, we have to solve $\sum_{t=1}^{T} \#(N_t) \cdot \#(\tilde{N}_t)$ linear optimization problems, where $\#(N_t)$ is the number of nodes at stage t.

Example For the valued trees of Fig. 15.11
we get the following distances:

$$\mathrm{d\!l}(\mathbb{P}^{(1)}, \mathbb{P}^{(2)}) = 3.90; d(P^{(1)}, P^{(2)}) = 3.48,$$
$$\mathrm{d\!l}(\mathbb{P}^{(1)}, \mathbb{P}^{(3)}) = 2.52; d(P^{(1)}, P^{(3)}) = 1.77,$$
$$\mathrm{d\!l}(\mathbb{P}^{(2)}, \mathbb{P}^{(3)}) = 3.79; d(P^{(2)}, P^{(3)}) = 3.44.$$

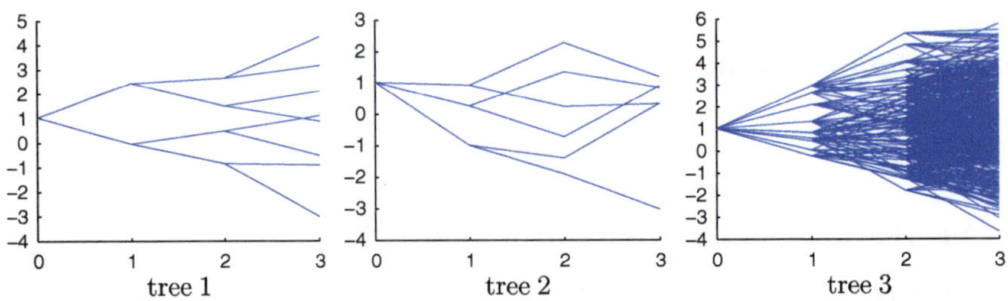

Fig. 15.11 Distances between nested distributions

Example Let $\mathbb{P}$ be the (nested) probability distribution of the pair $\xi = (\xi_1, \xi_2)$ be distributed according to a bivariate normal distribution

$$\begin{pmatrix} \xi_1 \\ \xi_2 \end{pmatrix} = N\left(\begin{pmatrix} 0 \\ 0 \end{pmatrix}, \begin{pmatrix} 1 & 1 \\ 1 & 2 \end{pmatrix} \right).$$

Note that the distribution of (ξ_1, ξ_2) is the same as the distribution of $(\zeta_1, \zeta_1 + \zeta_2)$, where ζ_i are independently standard normally distributed.

We know that the optimal approximation of a $N(\mu, \sigma^2)$ distribution in the Kantorovich sense by a three-point distribution is

$$\left[\frac{0.30345 \quad 0.3931 \quad 0.30345}{\mu - 1.029\sigma \quad \mu \quad \mu + 1.029\sigma} \right].$$

The distance is $0.3397\,\sigma$. Thus the optimal approximation to ξ_1 is

$$\left[\frac{0.30345 \; 0.3931 \; 0.30345}{-1.029 \quad 0.0 \quad 1.029} \right]$$

and the conditional distributions given these values are $N(-1.029, 1)$, $N(0, 1)$, and $N(1.029, 1)$. Thus the optimal two-stage approximation is

$$\tilde{\mathcal{P}}^{(1)} = \left[\frac{0.30345 \qquad\qquad 0.3931 \qquad\qquad 0.30345}{\begin{matrix} -1.029 & 0.0 & 1.029 \\ \left[\frac{0.30345 \; 0.3931 \; 0.30345}{-2.058 \; -1.029 \quad 0.0}\right] & \left[\frac{0.30345 \; 0.3931 \; 0.30345}{-1.029 \quad 0.0 \quad 1.029}\right] & \left[\frac{0.30345 \; 0.3931 \; 0.30345}{0.0 \quad 1.029 \quad 2.058}\right] \end{matrix}} \right].$$

The nested distance is $\mathsf{dl}(\mathbb{P}, \tilde{\mathbb{P}}^{(1)}) = 0.76$.

Figure 15.12 shows on the left the density of the considered normal model $\mathbb{P}$ and on the right the discrete model $\tilde{\mathbb{P}}^{(1)}$.

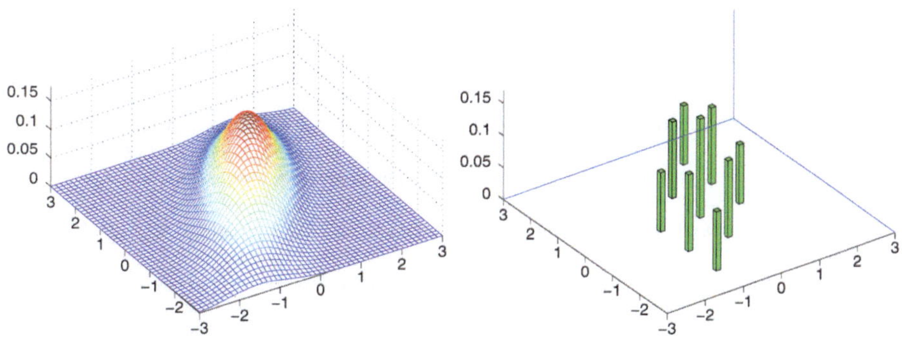

Fig. 15.12 Optimal two stage approximation

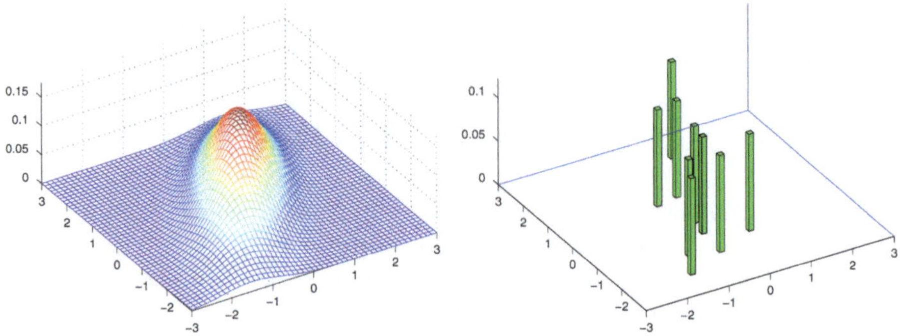

Fig. 15.13 Randomly sampled approximation

For comparison, we have also sampled nine random points from $\mathbb{P}$. Let $\tilde{\mathbb{P}}^{(2)}$ be the empirical distribution of these nine points. Figure 15.13 shows on the right side the randomly sampled discrete model $\tilde{\mathbb{P}}^{(2)}$. The nested distance is $\mathbb{d}(\mathbb{P}, \tilde{\mathbb{P}}^{(2)}) = 1.12$.

15.6 Summary

In this contribution we have outlined some funding components to get stochastic optimization available for computational treatment. At first stage the quantization of probability measures is studied, which is essential to capture the initial problem from a computational perspective. Various distance concepts are available in general, however, only a few of them insure that the quality of the solution of the approximated problem is in line with the initial problem.

This concept then is extended to processes and trees. To quantify the quality of an approximating process, the concept of a nested distance is introduced. Some examples are given to illustrate this promising concept.

The collected ingredients give a directive, how to handle a stochastic optimization problem in general. Moreover – and this is at least as important – they allow to qualify solution statements of any optimization problem, which have stochastic elements incorporated.

References

Z. Drezner and H.W. Hamacher. *Facility Location: Applications and Theory*. Springer, New York, NY, 2002.

J. Dupacova. Stability and sensitivity analysis for stochastic programming. *Annals of Operations Research*, 27:115–142, 1990.

Richard Durrett. *Probability: Theory and Examples*. Duxbury Press, Belmont, CA, second edition, 1996.

A. Gibbs and F.E. Su. On choosing and bounding probability metrics. *International Statistical Review*, 70(3):419–435, 2002.

S. Graf and H. Luschgy. *Foundations of Quantization for Probability Distributions*. Lecture Notes in Math. 1730. Springer, Berlin, 2000.

H. Heitsch and W. Roemisch. Scenario reduction algorithms in stochastic programming. *Computational Optimization and Applications*, 24:187–206, 2003.

H. Heitsch, W. Roemisch, and C. Strugarek. Stability of multistage stochastic programs. *SIAM Journal on Optimization*, 17:511–525, 2006.

C.C. Heyde. On a property of the lognormal distribution. *Journal of Royal Statistical Society Series B*, 25(2):392–393, 1963.

R. Hochreiter and G. Ch. Pflug. Financial scenario generation for stochastic multi-stage decision processes as facility location problems. *Annals of Operation Research*, 152(1):257–272, 2007.

K. Hoyland and S.W. Wallace. Generating scenario trees for multistage decision problems. *Management Science*, 47:295–307, 2001.

S. Na and D. Neuhoff. Bennett's integral for vector quantizers. *IEEE Transactions on Information Theory*, 41(4):886–900, 1995.

H. Niederreiter. *Random Number Generation and Quasi-Monte Carlo Methods*, volume 23. CBMS-NSF Regional Conference Series in Applied Math., Soc. Industr. Applied Math., Philadelphia, PA, 1992.

T. Pennanen. Epi-convergent discretizations of multistage stochastic programs via integration quadratures. *Mathematical Programming*, 116:461–479, 2009.

G. Ch. Pflug. Scenario tree generation for multiperiod financial optimization by optimal discretization. *Mathematical Programming, Series B*, 89:251–257, 2001.

G. Ch. Pflug. Version-independence and nested distributions in multistage stochastic optimization. *SIAM Journal on Optimization*, 20(3):1406–1420, 2009.

S.T. Rachev and W. Roemisch. Quantitative stability in stochastic programming: The method of probability metrics. *Mathematics of Operations Research*, 27(4):792–818, 2002.

S.T. Rachev. *Probability Metrics and the Stability of Stochastic Models*. Wiley, New York, NY, 1991.

S.S. Vallander. Calculation of the Wasserstein distance between probability distributions on the line. *Theor. Prob. Appl.*, 18:784–786, 1973.

C. Villani. *Optimal Transport*. Springer, Berlin, Heidelberg, 2008.

A. Weber. *Theory of the Location of Industries*. The University of Chicago Press, Chicago, IL 1929.

Chapter 16
Comparison of Sampling Methods for Dynamic Stochastic Programming

Michael A.H. Dempster, Elena A. Medova, and Yee Sook Yong

Abstract In solving a scenario-based dynamic (multistage) stochastic programme scenario generation plays a critical role, as it forms the input specification to the optimization process. Computational bottlenecks in this process place a limit on the number of scenarios employable in approximating the probability distribution of the paths of the underlying uncertainty. Traditional scenario generation approaches have been to find a sampling method that best approximates the path distribution in terms of some probability metrics such as minimization of moment deviations or Wasserstein distance. Here, we present a Wasserstein-based heuristic for discretization of a continuous state path probability distribution. The chapter compares this heuristic to the existing methods in the literature (Monte Carlo sampling, moment matching, Latin hypercube sampling, scenario reduction, and sequential clustering) in terms of their effectiveness in suppressing sampling error when used to generate the scenario tree of a dynamic stochastic programme. We perform an extensive computational investigation into the impact of scenario generation techniques on the in-sample and out-of-sample stability of a simplified version of a four-period asset–liability management problem employed in practice (Chapter 2, this volume). A series of out-of-sample tests are carried out to evaluate the effect of possible discretization biases. We also attempt to provide a motivation for the popular utilization of left-heavy scenario trees based on the Wasserstein distance criterion. Empirical results show that all methods outperform normal MC sampling. However, when evaluated against each other these methods essentially perform equally well, with second-order moment matching showing only marginal improvements in terms of in-sample decision stability and out-of-sample performance. The out-of-sample results highlight the problem of under-estimation of portfolio risk which results from insufficient samples. This discretization bias induces overly aggressive portfolio balance recommendations which can impair the performance of the model in real-world applications. Thus this issue needs to be carefully addressed in future research, see e.g. Dempster et al. (2010).

The views expressed are solely of the authors and not of Credit Suisse.

M.A.H. Dempster (✉)
Centre for Financial Research, Statistical Laboratory, Department of Pure Mathematics and Mathematical Statistics, University of Cambridge, Cambridge, UK; Cambridge Systems Associates Ltd., Cambridge, UK
e-mail: mahd2@cam.ac.uk

M. Bertocchi et al. (eds.), *Stochastic Optimization Methods in Finance and Energy*, International Series in Operations Research & Management Science 163,
DOI 10.1007/978-1-4419-9586-5_16,

Keywords Scenario generation · Sampling methods · Discretization error · Scenario-based approximation · Stochastic programming · In-sample and out-of-sample tests

16.1 Introduction

A *dynamic* (multistage) *stochastic programme* (DSP) can be formally written as

$$\min_{x \in \mathcal{X}} \int_{\Omega} f(x, \omega) P(d\omega), \tag{16.1}$$

where f is the value function defined in terms of both the uncertainty and decision spaces. Here P represents the probability measure on the *path space* Ω of the underlying multivariate stochastic process and $\mathcal{X}$ denotes the set of first-stage *implementable decisions* that satisfy feasibility in the remaining stages. Modelling of the complex features of a practical problem quickly removes analytical tractability and thus requires the numerical computation of optimal solutions. This in turn requires discrete approximation of the continuous state probability space. A vast literature has been devoted towards finding the best approximation of the continuous distribution; efforts concentrating, for example, on having state-space distribution moments matched (Hoyland et al. 2003; Hoyland and Wallace 2001) or minimizing Wasserstein probability metrics (Heitsch and Romisch 2005, Hochreiter and Pflug 2007; Romisch 2003; Chapters 14 and 17, this volume). All these efforts aim to find an approximation that best matches specific statistical properties of the discretized version to the theoretical one. These methods will be reviewed in Section 16.2 where we also present a heuristic for minimizing the Wasserstein distance. Section 16.3 examines these sampling methods by analysing the statistical properties of the resulting scenario trees. However, a practically more important criterion is to evaluate empirically the impact of these methods on the stability of the objective function and implementable decision values and to test against possible discretization biases (Dempster et al. 2010; Geyer et al. 2010; Kaut and Wallace 2003). The aim of this chapter is to compare these methods in terms of their effectiveness in suppressing the sampling error present in a small-sample scenario-based dynamic stochastic programme. The Pioneer guaranteed return fund problem presented in Chapter 2 (this volume) serves as a good vehicle to evaluate in Section 16.5 the comparative performance of these methods using a series of in-sample and out-of-sample tests. Section 16.6 concludes.

16.2 A Short Review of Commonly Used Scenario Sampling Methods

A commonly used technique in generating discrete samples from a continuous distribution is to use *Monte Carlo* (MC) sampling where uniformly distributed pseudo-random numbers are appropriately transformed to the target distribution

(Ross 2002). However, in asset–liability management (ALM) applications, we can computationally only afford a small sample size and inevitably face the sampling error problem. *Moment matching* (Hoyland et al. 2003; Hoyland and Wallace 2001; Villaverde 2003) has been extensively employed in the literature to address this problem. *Latin hypercube sampling* (LHS) (McKay et al. 1979), a form of stratified sampling for multi-dimensional distributions which uses the univariate *inverse cumulative density function* (ICDF), may also be employed. In this chapter, the term LHS and ICDF are used interchangeably. Wasserstein-based discretization has been suggested in (Heitsch and Romisch 2005; Hochreiter and Pflug 2007; Romisch 2003; Chapters 14 and 17, this volume). The sampling methods that we compare are MC sampling, first- and second-order moment matching, LHS/ICDF, and Wasserstein-based sampling methods as described in the following section.

16.2.1 The Wasserstein Distance

Given two discrete (sample) probability measures $\mathbb{P} := \{(p_i, x^{(i)}) : i = 1, 2, \ldots, M\}$ with discrete probabilities p_i at $x^{(i)}$ and $\mathbb{Q} := \{(q_j, y^{(j)}) : j = 1, 2, \ldots, N\}$ with discrete probabilities q_j at $y^{(j)}$, the *Wasserstein* (also known as the *Kantorovich*) metric computation for $\mathbb{P}$ and $\mathbb{Q}$ can be written as a *mass transportation problem* for the discrete probability mass given by the linear programming problem

$$\begin{aligned}
\zeta(\mathbf{X}, \mathbf{Y}) &:= \min_{\eta_{ij}} \sum_{i=1}^{M} \sum_{j=1}^{N} c\left(x^{(i)}, y^{(j)}\right) \eta_{ij}, \\
\text{s.t.} \quad & \sum_{j=1}^{N} \eta_{ij} = p_i \quad \text{for} \quad i = 1, \ldots, M, \\
& \sum_{i=1}^{M} \eta_{ij} = q_j \quad \text{for} \quad j = 1, \ldots, N, \\
& \left(\sum_{i=1}^{M} p_i = 1 \right), \\
& \left(\sum_{j=1}^{N} q_j = 1 \right), \\
& \eta_{ij} \geq 0 \quad \text{for} \quad i = 1, \ldots, M, \quad j = 1, \ldots, N.
\end{aligned} \tag{16.2}$$

Constraints in brackets in the sequel denote constraints that are to be satisfied by the input data. Here probability mass one from M sources (indexed by i) is to be transported to N destinations (indexed by j) with $c\left(x^{(i)}, y^{(j)}\right)$ denoting the corresponding cost of transporting a unit mass.

16.2.2 Minimizing the Wasserstein Distance

The motivation for minimizing the Wasserstein distance between the sampled and the continuous underlying path space in generating scenario trees for a DSP can be found in (Heitsch and Romisch 2005; Hochreiter and Pflug 2007; Romisch 2003; Chapters 14 and 17). The problem of interest is to find a small sample approximation that best approximates the path space in terms of the Wasserstein distance. For simplicity we will concentrate in our exposition on a random variable (sample point) formulation. However, these sample points can also be thought of as referring to a decision time point along the sample paths when approximating the path space of the underlying multivariate stochastic process.

When using the discrete mass transport formulation of Wasserstein distance (i.e. the mass transportation problem), we first need to approximate the underlying (absolutely) continuous distribution of interest by a distribution with large (but finite) support of cardinality M as the source. We then aim to find a much coarser set of (destination) locations and the corresponding probabilities such that the Wasserstein distance between the small sample probability measure and the large sample proxy of the true underlying probability measure are minimized.

This *approximation* task can be written as

$$
\begin{aligned}
\inf_{\mathbb{Q}} \zeta(\mathbb{P}, \mathbb{Q}) = & \inf_{\eta_{ij}, y^{(j)}, q_j} \sum_{i=1}^{M} \sum_{j=1}^{N} c\left(x^{(i)}, y^{(j)}\right) \eta_{ij}, \\
\text{s.t.} \quad & \sum_{j=1}^{N} \eta_{ij} = p_i \quad \text{for} \quad i = 1, \ldots, M, \\
& \sum_{i=1}^{M} \eta_{ij} = q_j \quad \text{for} \quad j = 1, \ldots, N, \\
& \left(\sum_{i=1}^{M} p_i = 1\right) \\
& \sum_{j=1}^{N} q_j = 1 \\
& q_j \geq 0 \quad \text{for} \quad j = 1, \ldots, N, \\
& \eta_{ij} \geq 0 \quad \text{for} \quad i = 1, \ldots, M, \quad j = 1, \ldots, N.
\end{aligned}
\tag{16.3}
$$

Assume that the optimum solution to the above problem is given by $y^{(j)*}$ and η_{ij}^*, then the optimal approximation of the initial discretization of the continuous space is given by placing sample points at $y^{(j)*}$ with probabilities $q_j^* = \sum_{i=1}^{M} \eta_{ij}^*$.

Problem 16.3 is a non-convex nonlinear programming problem. Finding the optimum solution is nontrivial as it involves two inter-related processes of deciding on (1) the destination locations (samples) and (2) the corresponding transported probability mass (probabilities) to these locations. The placement of the optimal destination locations requires an examination of how the probability mass at the source is to be distributed to these locations. However, the optimal distribution scheme cannot be determined before the knowledge of these destination locations. In addition, the transportation cost is undetermined before the destination locations are fixed.

In fact solving for $y^{(j)*}$ and q_j^* simultaneously can be formulated as a mixed 0–1 programming problem when a fixed number N $y^{(j)}$'s are chosen from the M $x^{(i)}$'s with $M \gg N$ as suggested by Hochreiter and Pflug (2007). This is the *uncapacitated facility location problem* (Nemhauser and Wolsey 1988).

The facility location problem optimizes a subset of location proposals (N in total from a set of M proposals) such that the service costs incurred by the clients (e.g. travel costs) are minimized. Using the same notation as before, let $\{x^{(i)}\}$ denote the locations of the clients utilizing the facilities, each requiring p_i units of service, and let $\{y^{(j)}\}$ denote the facility location proposals. Then $c\left(x^{(i)}, y^{(j)}\right)$ corresponds to the cost incurred by client i obtaining service from j and u_j is a binary variable set to 1 if the facility is placed at $y^{(j)}$ and 0 otherwise.

If we let the facility location proposals be the set of sample points which are to proxy the underlying continuous state space (i.e. $y^{(j)} = x^{(j)}$) for $j = 1, \ldots, M$), then the optimal solution to the following formulation of the (appropriate) uncapacitated facility location problem gives the desired (optimal) small sample approximation:

$$\begin{aligned}
\min_{\eta_{ij}, u_j \in \{0,1\}, q_j, y^{(j)}} \quad & \sum_{i=1}^{M} \sum_{j=1}^{M} c\left(x^{(i)}, y^{(j)}\right) \eta_{ij}, \\
\text{s.t.} \quad & \sum_{j=1}^{M} \eta_{ij} = p_i \quad \text{for} \quad i = 1, \ldots, M, \\
& \sum_{i=1}^{M} \eta_{ij} = u_j q_j \quad \text{for} \quad j = 1, \ldots, M, \\
& \left(\sum_{i=1}^{M} p_i = 1 \right), \\
& \sum_{j=1}^{M} u_j q_j = 1, \\
& \sum_{j=1}^{M} u_j = N \qquad (M \gg N) \\
& (y^{(j)} = x^{(j)}) \quad \text{for} \quad j = 1, \ldots, M,
\end{aligned} \tag{16.4}$$

$$q_j \geq 0 \quad \text{for} \quad j = 1, \ldots, M,$$
$$\eta_{ij} \geq 0 \quad \text{for} \quad i, j = 1, \ldots, M.$$

Problem 16.4 is a 0–1 mixed integer (nonlinear) programme which is $\mathcal{NP}$-hard. Hence in the next section we introduce a novel heuristic which finds a local minimum to the problem, compromising accuracy (global optimum) for speed, as fast construction of the underlying path space is desired for efficient application to DSP models.

16.2.3 Approximate Wasserstein Distance Minimization Algorithm

As we have previously seen, minimizing the Wasserstein distance requires the simultaneous determination of q_j and $y^{(j)}$ for $j = 1, \ldots, N$ which is $\mathcal{NP}$-hard. In this section we present the sequential Wasserstein distance minimization (`SMWD`) algorithm which alternates between two steps in which $\{q_j\}$ and $\{y^{(j)}\}$ are alternately kept fixed as parameters using the values from the previous iteration. The proposed optimization procedure iterates between two optimization problems:

- `Step 1`: Fixed location problem: We fix $\{y^{(j)}\}$ as parameters and let $\{q_j\}$ be variables. Given the destination points $\{y^{(j)}\}$ we optimize the destination probabilities $\{q_j\}$ such that the Wasserstein distance is minimized.
- `Step 2`: Fixed flow problem: We fix $\{q_j\}$ as parameters and let $\{y^{(j)}\}$ be variables. Given the destination probabilities $\{q_j^*\}$ and the optimum flow $\{\eta_{ij}^*\}$ from `step 1`, compute the destination locations $\{y^{(j)}\}$ such that the Wasserstein distance is further minimized.

The `SMWD` algorithm alternates between `step 1` and `step 2` until it converges. The Wasserstein distance is guaranteed to decrease at each step until a *local* minimum is reached and this occurs when the transportation flow stabilizes.

Step 1: Fixed location problem: Referring to the mass transportation formulation of the Wasserstein distance in Section 16.2.1, we now treat the $\{q_j\}$ as variables and the optimization problem becomes

$$\min_{\eta_{ij}, q_j} \quad \sum_{i=1}^{M} \sum_{j=1}^{N} c\left(x^{(i)}, y^{(j)}\right) \eta_{ij}, \tag{16.5}$$

$$\text{s.t.} \quad \sum_{j=1}^{N} \eta_{ij} = p_i \quad \text{for} \quad i = 1, \ldots, M, \tag{16.6}$$

$$\sum_{i=1}^{M} \eta_{ij} = q_j \quad \text{for} \quad j = 1, \ldots, N, \tag{16.7}$$

$$\left(\sum_{i=1}^{M} p_i = 1\right), \tag{16.8}$$

$$\sum_{j=1}^{N} q_j = 1, \tag{16.9}$$

$$q_j \geq 0 \quad \text{for} \quad j = 1, \ldots, N, \tag{16.10}$$

$$\eta_{ij} \geq 0 \quad \text{for} \quad i = 1, \ldots, M \text{ and } j = 1, \ldots, N. \tag{16.11}$$

Constraints (16.6) and (16.11) imply that the probability mass at the source needs to be transported to the destination without reservation. This makes constraints (16.7) redundant. Observe that (16.10) is also redundant in light of (16.7) and (16.11). Since the input p_i's are probability masses summing to one, constraint (16.8) is satisfied by the data and from (16.7) constraint (16.9) is automatically satisfied. Hence, the equivalent optimization problem becomes

$$\begin{aligned} \min_{\eta_{ij}} \quad & \sum_{i=1}^{M}\sum_{j=1}^{N} c\left(x^{(i)}, y^{(j)}\right)\eta_{ij}, \\ \text{s.t.} \quad & \sum_{j=1}^{N} \eta_{ij} = p_i \quad \text{for} \quad i = 1, \ldots, M, \\ & \eta_{ij} \geq 0 \quad \text{for} \quad i = 1, \ldots, M \text{ and } j = 1, \ldots, N. \end{aligned} \tag{16.12}$$

Let η_{ij}^* be the (unique) optimal transportation flow for problem (16.12). The optimal probabilities for the destination samples will then be

$$q_j^* = \sum_{i=1}^{M} \eta_{ij}^*, \tag{16.13}$$

and the corresponding Wasserstein distance is

$$\zeta_{\texttt{step 1}}(\{q_j^*\}) = \sum_{i=1}^{M}\sum_{j=1}^{N} c\left(x^{(i)}, y^{(j)}\right)\eta_{ij}^*. \tag{16.14}$$

We have $\zeta_{\texttt{step 1}}(\{q_j^*\}) \leq \zeta_{\texttt{step 1}}(\{q_j\})\, \forall q_j$. Hence the Wasserstein distance of the given points with the newly assigned probabilities is not greater than that of the old probabilities (or any other probabilities) simply from the principle of optimization.

Proposition 2.1 *The optimal solution of* `step 1` *is*

$$\eta_{ij}^* = \begin{cases} p_i & j = \arg\min_{l\in\{1,\ldots,N\}} c\left(x^{(i)}, y^{(l)}\right), \\ 0 & \text{otherwise.} \end{cases} \tag{16.15}$$

In other words, the probability mass from each of the source locations is transported in whole *to the* nearest *destination location in terms of the* $c\left(x^{(i)}, y^{(j)}\right)$ *distance measure.*

Proof The optimization problem (16.12) of `step 1` can be decomposed into M independent sub-problems (for $i = 1, \ldots, M$):

$$\begin{aligned}
&\min_{\eta_{ij}} \sum_{j=1}^{N} c\left(x^{(i)}, y^{(j)}\right) \eta_{ij}, \\
\text{s.t.} \quad &\eta_{ij} \geq 0 \quad \text{for} \quad j = 1, \ldots, N, \\
\text{s.t.} \quad &\sum_{j=1}^{N} \eta_{ij} = p_i,
\end{aligned} \tag{16.16}$$

where $\sum_{i=1}^{M} p_i = 1$. The Lagrangian of each of the problems (16.16), for $i = 1, \ldots, M$, is given by

$$\mathcal{L}(\{\eta_{ij}\}, \lambda_0, \{\lambda_j\}) = \sum_{j=1}^{N} c\left(x^{(i)}, y^{(j)}\right) \eta_{ij} + \lambda_0 \left(p_i - \sum_{j=1}^{N} \eta_{ij} \right) + \sum_{j=1}^{N} \lambda_j \eta_{ij}. \tag{16.17}$$

The necessary conditions for a minimum are given by

$$\frac{\partial \mathcal{L}}{\partial \eta_{ij}} = c\left(x^{(i)}, y^{(j)}\right) - \lambda_0 + \lambda_j = 0 \quad \text{for} \quad j = 1, \ldots, N, \tag{16.18}$$

$$\frac{\partial \mathcal{L}}{\partial \lambda_0} = p_i - \sum_{j=1}^{N} \eta_{ij} = 0, \tag{16.19}$$

$$\eta_{ij} \geq 0, \lambda_j \geq 0 \quad \text{and} \quad \lambda_j \eta_{ij} = 0 \quad \text{for} \quad j = 1, \ldots, N. \tag{16.20}$$

Assuming that $\mathcal{K}$ is a set of indices such that $\eta_{ik} \neq 0$ for $k \in \mathcal{K}$, from (16.20) we have $\lambda_k = 0$ for $k \in \mathcal{K}$. Substituting this into (16.18) gives $\lambda_0 = c\left(x^{(i)}, y^{(k)}\right)$ for $k \in \mathcal{K}$. However, the value of λ_0 is unique. Therefore, the set $\mathcal{K}$ contains only a single element. Let this be denoted by k^* and we have $\lambda_0 = c\left(x^{(i)}, y^{(k^*)}\right)$. Substituting this into (16.18) gives

$$\lambda_j = \begin{cases} 0, & j = k^* \\ c\left(x^{(i)}, y^{(j)}\right) - c\left(x^{(i)}, y^{(k^*)}\right) \geq 0, & j \neq k^* \end{cases}. \tag{16.21}$$

Hence, to satisfy all the conditions in (16.21)

$$k^* = \arg \min_{l \in \{1, \ldots, N\}} c\left(x^{(i)}, y^{(l)}\right). \tag{16.22}$$

Substituting (16.22) and (16.21) into (16.19), we have

$$\eta_{ij}^* = \begin{cases} p_i & j = \arg\min_{l\in\{1,\ldots,N\}} c\left(x^{(i)}, y^{(l)}\right) \\ 0 & \text{otherwise} \end{cases} .$$

Intuitively, this result is expected – if we were to transport the probability mass to the destinations without any constraint on how they should be allocated across these locations, the distribution scheme that yields the minimum cost is to, respectively, transport each probability mass to their nearest neighbour. □

Step 2: Fixed-flow problem: Referring to the mass transportation formulation of the Wasserstein distance in (16.2), we now let the destination locations $\{y^{(j)}\}$ be variables and set the transportation flow and the destination probabilities to be the optimum values from `step 1`. With the transportation flow η_{ij}^* already determined, the remaining optimization problem collapses to minimizing the following unconstrained problem

$$\min_{y^{(j)}} \sum_{i=1}^{M}\sum_{j=1}^{N} c\left(x^{(i)}, y^{(j)}\right)\eta_{ij}^*. \tag{16.23}$$

This is a simple unconstrained nonlinear convex problem and the minimum can be found by differentiating (16.23) with respect to the decision variables and equating them to zero.

l_1-Norm as the cost function: If the cost function is defined as the l_1-norm, i.e.,

$$c\left(x^{(i)}, y^{(j)}\right) := \sum_{d'=1}^{d}\left|x_{d'}^{(i)} - y_{d'}^{(j)}\right|, \tag{16.24}$$

then the objective function at `step 2` of the `SMWD` algorithm can be written as

$$\min \sum_{i=1}^{M}\sum_{j=1}^{N}\sum_{d'=1}^{d}\left|x^{(i)}{}_{d'} - y_{d'}^{(j)}\right|\eta_{ij}^*. \tag{16.25}$$

We can partition the above sum into two regions, namely

$$\min \sum_{j=1}^{N}\sum_{d'=1}^{d}\left[\sum_{i\in\{x^{(i)}{}_{d'}\geq y_{d'}^{(j)}\}}\left(x^{(i)}{}_{d'} - y_{d'}^{(j)}\right)\eta_{ij}^* + \sum_{i\in\{x^{(i)}{}_{d'}< y_{d'}^{(j)}\}}\left(y_{d'}^{(j)} - x^{(i)}{}_{d'}\right)\eta_{ij}^*\right]. \tag{16.26}$$

Differentiating (16.26) with respect to $y_{d'}^{(j)}$, equating to zero and letting the optimal location be denoted by $y_{d'}^{(j)*}$, we have

$$-\sum_{i\in\{x^{(i)}{}_{d'}\geq y^{(j)*}_{d'}\}} \eta^*_{ij} + \sum_{i\in\{x^{(i)}{}_{d'}<y^{(j)*}_{d'}\}} \eta^*_{ij} = 0. \tag{16.27}$$

Recall that the optimal flow η^*_{ij} from `step 1` is to transport the probability mass from the source to the nearest neighbour at the destination. Combining (16.15) and (16.27) we have

$$\sum_{\{i=1,\ldots,M:\eta_{ij}>0:x^{(i)}{}_{d'}<y^{(j)*}_{d'}\}} p_i = \sum_{\{i=1,\ldots,M:\eta_{ij}>0:x^{(i)}{}_{d'}>y^{(j)*}_{d'}\}} p_i. \tag{16.28}$$

It follows that the optimal destination location $y^{(j)*}_{d'}$ is set to a *median* of the source points transporting probability mass to the destination under consideration.

16.2.3.1 Squared Normalized Euclidean Distance as the Cost Function

Now, let $c\left(x^{(i)}, y^{(j)}\right)$ be defined as the *squared normalized Euclidean distance*, i.e.

$$c\left(x^{(i)}, y^{(j)}\right) := \sum_{d'=1}^{d} \frac{\left(x^{(i)}{}_{d'} - y^{(j)}_{d'}\right)^2}{\sigma_{d'}{}^2}. \tag{16.29}$$

The objective function of `step 2` becomes

$$\min \sum_{i=1}^{M}\sum_{j=1}^{N}\sum_{d'=1}^{d} \frac{\left(x^{(i)}{}_{d'} - y^{(j)}_{d'}\right)^2}{\sigma_{d'}{}^2}\eta^*_{ij}. \tag{16.30}$$

Differentiating (16.30) with respect to $y^{(j)}_{d'}$ and equating to zero gives optimal locations

$$y^{(j)*}_{d'} = \frac{\sum_{i=1}^{M}\eta^*_{ij}x^{(i)}{}_{d'}}{\sum_{i=1}^{M}\eta^*_{ij}} = \frac{\sum_{i=1}^{M}\eta^*_{ij}x^{(i)}{}_{d'}}{q^*_j} \tag{16.31}$$
$$\text{for} \quad d' = 1,\ldots,d \quad \text{and} \quad j = 1,\ldots,N.$$

The optimum destination location in (16.31) thus corresponds to the *centre of mass* of the source points transporting probability mass to the location under consideration.

We have previously assumed a sample from a known continuous distribution to start with, but this algorithm can also be applied to obtaining a small sample approximation of a large sample empirical distribution by letting the empirical samples be the source points.

Note that the SMWD algorithm presented here is related to the deterministic iteration (DI) algorithm in Chapter 17 (this volume). The DI algorithm is formulated in the continuous space setting and requires the predefinition of partitions of the uncertainty space at each iteration of the algorithm. However, this partitioning is difficult to implement in practice. In addition, the computation of sample probabilities requires multi-dimensional integration of the underlying probability distribution for each of these partitions and thus poses a challenge to employing this algorithm in practical use.

Algorithmi 1 The SMWD algorithm.

1: **Initialize** a large (finite) approximation of the multivariate distribution of interest at the source.
2: **Initialize** N destination locations.
3: **While** $\Delta\zeta$ < tolerance
4: `step 1`: Fixed location problem – update the destination probabilities (16.15).
5: `step 2`: Fixed flow problem – update the destination locations (16.31).
6: **end while**
7: Output destination locations and probabilities.

In the following, we investigate further some of the properties of the resulting sample outputs of the SMWD algorithm. For simplicity of exposition we assume sampling of a univariate distribution but the results have a straightforward extension to the multivariate case.

Proposition 2.2 *The mean of the destination sample is equal to that of the source, i.e.*

$$\sum_{j=1}^{N} y^{(j)*} q_j^* = \sum_{i=1}^{M} x^{(i)} p_i. \tag{16.32}$$

Proof

$$\sum_{j=1}^{N} y^{(j)*} q_j^* = \sum_{j=1}^{N} \left(\frac{\sum_{i=1}^{M} \eta_{ij}^* x^{(i)}}{\sum_{i=1}^{M} \eta_{ij}^*} \right) \left(\sum_{i=1}^{M} \eta_{ij}^* \right)$$
$$= \sum_{i=1}^{M} x^{(i)} \left(\sum_{j=1}^{N} \eta_{ij}^* \right) = \sum_{i=1}^{M} x^{(i)} p_i.$$

□

Proposition 2.3 *The variance of the destination sample is not greater than the variance of the source sample i.e.*

$$\sum_{j=1}^{N} q_j^* y^{(j)*2} - \left(\sum_{j=1}^{N} q_j^* y^{(j)*} \right)^2 \leq \sum_{i=1}^{M} p_i x^{(i)2} - \left(\sum_{i=1}^{M} p_i x^{(i)} \right)^2. \tag{16.33}$$

Proof Since the destination sample mean is equal to the source mean from (16.32), (16.33) can be simplified to

$$\sum_{j=1}^{N} q_j^* y^{(j)*2} \leq \sum_{i=1}^{M} p_i x^{(i)2}. \tag{16.34}$$

We know that the optimal solution satisfies the following:

$$q_j^* = \sum_{i=1}^{M} \eta_{ij}^*, \tag{16.35}$$

$$p_i = \sum_{j=1}^{N} \eta_{ij}^*, \tag{16.36}$$

$$y^{(j)*} = \frac{\sum_{i=1}^{M} \eta_{ij}^* x^{(i)*}}{\sum_{i=1}^{M} \eta_{ij}^*}. \tag{16.37}$$

Consider the term $\sum_{i=1}^{M} p_i x^{(i)*2} - \sum_{j=1}^{N} q_j^* y^{(j)*2}$. Using (16.15), we can write

$$\sum_{i=1}^{M} p_i x^{(i)2} - \sum_{j=1}^{N} q_j^* y^{(j)*2} = \sum_{i=1}^{M} \sum_{j=1}^{N} \eta_{ij}^* x^{(i)2} - \sum_{j=1}^{N} q_j^* \left(y^{(j)*}\right)^2.$$

Using (16.35) and (16.37) we have

$$\begin{aligned}
&\sum_{i=1}^{M} p_i x^{(i)2} - \sum_{j=1}^{N} q_j^* y^{(j)*2} \\
&\quad = \sum_{j=1}^{N} \left[\sum_{i=1}^{M} \eta_{ij}^* x^{(i)2} - 2y^{(j)*} \left(\sum_{i=1}^{M} \eta_{ij}^* x^{(i)} \right) + \sum_{i=1}^{M} \eta_{ij}^* y^{(j)*2} \right] \\
&\quad = \sum_{i=1}^{M} \sum_{j=1}^{N} \eta_{ij}^* \left(x^{(i)} - y^{(j)*} \right)^2.
\end{aligned} \tag{16.38}$$

Since $\eta_{ij}^* \geq 0$ and $\left(x^{(i)} - y^{(j)*} \right)^2 \geq 0$ therefore

$$\sum_{i=1}^{M} p_i x^{(i)2} - \sum_{j=1}^{N} q_j^* y^{(j)*2} \geq 0. \tag{16.39}$$

□

Hence, the variance of the destination samples are less than or equal to that of the source. Although the SMWD algorithm guarantees to improve the Wasserstein metric, it has the major drawback of under-estimating the underlying variance. This problem may be circumvented by applying second-order moment matching on top of the SMWD algorithm which is an algebraic $\mathcal{O}(M^2)$ operation (see e.g. Yong, 2007, pp. 60–63). The issue of this under-estimation when applying the SMWD algorithm to generating scenario trees for dynamic stochastic programmes will be investigated in the empirical studies presented in the sequel.

The SMWD algorithm is similar to the K-mean cluster algorithm used in cluster analysis (MacQueen 1967). Instead of updating the centroids after all the points have been clustered, we may update the centroids as each additional point is being presented. The aim is to make use of the information of the points that have been previously seen to update the centroids so that future points may be better clustered. This potentially increases the rate of convergence compared to the SMWD approach and has been adopted as a scenario generation method in (Dupacova et al. 2000; Chapter 17, this volume). Here, we refer to this variant as *sequential clustering* (SC). We also adopt the simultaneous backward reduction algorithm presented in (Chapter 14, this volume). However, since we generate the scenario tree conditionally forward, the reduction algorithm must be employed at each node and we refer to this technique as *nodal reduction* (NR).

16.3 Statistical Properties of Discretized Processes

In this section we investigate the statistical properties using these scenario sampling techniques when simulating multi-period trees. We simulate a seven-dimensional Gaussian diffusion process which drives the price returns of the equities and bond instruments used in the Pioneer guaranteed return fund problem presented in Section 16.5 (see also Chapter 2, this volume).

We investigate how the probability metrics vary with different tree structures. Even though our description of the Wasserstein distance deals only with random variables, in principle this can be easily extended to a path space by augmenting the random variables to include the time instances of interest. We will now see how this is effected considering of the sample paths of a stochastic process in the form of a scenario tree.

$\mathbf{X}$ and $\mathbf{Y}$ now represent d-dimensional vector stochastic processes rather than random vectors. Let $y_{d',t}^{(j)}$ denote the d'th co-ordinate sample value of the stochastic process at time t of the jth scenario for $j = 1, \ldots, N$. Also, let $y^{(j)} := \left[y_1^{(j)}, y_2^{(j)}, \ldots, y_T^{(j)} \right]'$ denote the jth scenario path, where $y_t^{(j)} := \left[y_{1,t}^{(j)}, y_{2,t}^{(j)}, y_{3,t}^{(j)}, \ldots, y_{d,t}^{(j)} \right]'$ is the vector whose elements are the time t realizations of all co-ordinates. Let $\{x^{(i)}, i = 1, \ldots, M\}$ be the large set of sample paths which is assumed to serve as the proxy for the continuous space process paths. The small sample approximation of the stochastic processes is given by $\{y^{(j)}, j = 1, \ldots, N\}$. We define the *distance between scenarios* $x^{(i)}$ *and* $y^{(j)}$ to be

$$c\left(x^{(i)}, y^{(j)}\right) := \sum_{t=1}^{T} \phi_t c_t \left(x^{(i)}{}_t, y_t^{(j)}\right), \tag{16.40}$$

where the $c_t\left(x^{(i)}{}_t, y_t^{(j)}\right)$'s are appropriate *cost functions* at a particular time instance and the ϕ_t's are appropriate *time weightings* that sum to unity. In our investigation we set $\phi_t := \frac{1}{T}$ but these weights can be set to other values, for instance by setting higher values for the initial periods, if the probability distances at these times are to be emphasized.

The Wasserstein distance, denoted by $\zeta(\mathbf{X}, \mathbf{Y})$, between a small sample scenario tree and a large sample approximation of the underlying path space follows the same definition as (16.2) with the distance measure augmented to include all the sampling times along a scenario path (cf. (16.40)):

$$\zeta(\mathbf{X}, \mathbf{Y}) := \min_{\eta_{ij}} \sum_{t=1}^{T} \sum_{i=1}^{M} \sum_{j=1}^{N} c_t\left(x^{(i)}, y^{(j)}\right) \eta_{ij} \tag{16.41}$$

subject to the same constraints as in (16.2).

Proposition 3.1 *The Wasserstein distance between the scenario trees $\zeta := \zeta\left(\eta_{ij}^*\right)$ defined by (16.41) is bounded below by the sum of the Wasserstein distances for each time period $\zeta_t := \zeta_t\left(\eta_{ij}^{(t)^*}\right)$, i.e.*

$$\zeta \geq \sum_t \zeta_t, \tag{16.42}$$

where $\zeta_t(\cdot)$ and $\zeta(\cdot)$ denote the transportation costs at time t and of the whole scenario tree, respectively, η_{ij}^* corresponds to the flow that minimizes the Wasserstein distance between the whole scenario tree and $\eta_{ij}^{(t)^*}$ denotes the optimal flow obtained from decoupling the scenarios into individual time points and solving for the Wasserstein distance for each particular period t.

Proof As the scenario path distance in (16.40) may be decomposed into a summation of the distances at different time periods, the Wasserstein distance between distributions for the whole scenario tree in terms of paths can be decomposed into the different time periods with the addition of the constraint:

$$\min_{\eta_{ij}^{(1)}, \eta_{ij}^{(2)}, \ldots, \eta_{ij}^{(T)}} \frac{1}{T} \sum_{t=1}^{T} c_t\left(x_t^{(i)}, y_t^{(j)}\right) \eta_{ij}^{(t)} \quad \text{s.t.} \quad \eta_{ij}^{(t)} = \eta_{ij} \tag{16.43}$$
$$\forall t = 1, \ldots, T \quad \text{and } i = 1, \ldots, M, \ j = 1, \ldots, N.$$

This constraint ensures that the same flow is used for the realizations along the entire set of scenario pairs. We know that $\eta_{ij}^{(t)^*}$ is the optimal flow in the Wasserstein distance computation at time t, giving a minimum value ζ_t. Thus since $\zeta_t\left(\eta_{ij}\right) \geq \zeta_t \; \forall \eta_{ij}$

$$\zeta_t(\eta_{ij}^*) \geq \zeta_t, \tag{16.44}$$

where $\zeta_t\left(\eta_{ij}\right)$ denotes the transportation costs at time t with η_{ij} units to be transported from source i to destination j. Summing over all time periods, we have

$$\zeta := \sum_{t=1}^{T} \zeta_t\left(\eta_{ij}^*\right) \geq \sum_{t=1}^{T} \zeta_t. \tag{16.45}$$

The equality is obtained when the flow for the whole scenario tree case is the same for each sub-period, i.e. $\eta_{ij}^{(t)^*} = \eta_{ij}^* \; \forall t \in \{1, \ldots, T\}$. □

Effectively, if we are able to minimize the summation of the Wasserstein distances at each time period, we are able to improve a lower bound of the Wasserstein distance of the tree. If the bound in (16.42) is tight, this would in turn help us to improve the Wasserstein distance between the small sample approximation and the continuous path space of the underlying stochastic process. We will investigate this empirically below.

From Proposition 3.1 we see that the summation of the Wasserstein distances at each of the time period gives a *lower* bound on the Wasserstein distance between the approximate sample path space and its large sample underlying counterpart. Empirically we found that the Wasserstein distance is a monotonic decreasing function of the sample size when it is obtained using an *unbiased* sampling method. It follows that the Wasserstein distance at each decision stage may be improved by increasing the number of realizations at each time period. For a tree with a fixed total number of scenarios, this can be achieved by employing a left-heavy branching structure. The lower bound will tend to a minimum when we have a fan scenario tree; but does this also lead to a minimum of the Wasserstein distance for the whole scenario tree? Also, how tight is the bound in Proposition 3.1?

To begin to address these questions, we have performed an empirical analysis on how these two metrics vary with different tree structures. We generate three period trees with different structures, each having approximately 500 scenarios. The Wasserstein distance is evaluated with respect to trees of 5000 scenarios with various tree structures over 3 periods generated using the `LHS/ICDF` method (see Section 16.2). The results, which are averages over 30 replications, are shown in Tables 16.1 and 16.2.

Table 16.1 shows that left-heavy scenario trees yield lower whole tree Wasserstein distances. For instance, the scenario tree with *branching structure* with equal 8-8-8 branching throughout all stages yields the worst Wasserstein distance. As the first stage branching increases, the Wasserstein distance generally decreases. The Wasserstein distance of the small sample scenario tree is influenced by the total number of nodes at each stage with the earlier stages dominating. This distinction

Table 16.1 Variation of the Wasserstein metric with different tree structures

Tree structure	# Scenarios	#Nodes at depth				Wasserstein distance			Lower bound $\sum_t W_t$		
		1	2	3	Total	MC	mean + cov	SMWD	MC	mean + cov	SMWD
8-8-8	512	8	64	512	584	13.79	11.71	10.45	11.05	9.26	8.81
23-11-2	506	23	253	506	782	11.45	9.81	9.21	7.86	6.79	7.12
28-6-3	504	28	168	504	708	11.45	9.94	9.32	7.78	6.78	7.16
42-4-3	504	42	168	504	714	11.11	9.59	9.15	7.19	6.29	6.86
63-4-2	504	63	252	504	819	10.58	9.19	8.86	6.37	5.75	6.46
125-2-2	500	125	250	500	875	10.21	8.76	8.81	5.62	5.28	6.24
500-1-1	500	500	500	500	1500	9.68	8.93	8.96	4.41	5.82	6.17

Table 16.2 Variation of the first four moments and Kolmogorov metric with different tree structures

Tree		SMWD						mean + cov					
		Moment metric						Moment metric					
structure	Stage	1	2	3	4	Corr	Kolmogorov	1	2	3	4	Corr	Kolmogorov
	1	0.01	0.33	0.51	0.40	0.36	0.33	0.00	0.02	0.50	0.39	0.37	0.30
8-8-8	2	0.04	0.33	0.36	0.26	0.31	0.25	0.05	0.01	0.41	0.27	0.32	0.19
	3	0.09	0.34	0.36	0.20	0.41	0.21	0.11	0.01	0.48	0.21	0.40	0.14
	1	0.00	0.18	0.25	0.28	0.31	0.24	0.00	0.01	0.37	0.28	0.32	0.21
63-4-2	2	0.04	0.31	0.31	0.20	0.27	0.23	0.05	0.03	0.38	0.26	0.28	0.17
	3	0.09	0.41	0.35	0.17	0.37	0.23	0.10	0.14	0.45	0.21	0.38	0.16

is clearly illustrated by comparing the 23-11-2 and 42-4-3 trees. Even though the number of nodes for 23-11-2 are greater at the later stages, its Wasserstein distance is worse due to its initial branching factor which is only half that of the 42-4-3 tree. As another example, consider the 28-6-3 and 23-11-2 tree structures which have similar initial branching. However, the marginal disadvantage of a slightly lower initial branching for the 23-11-2 tree is quickly compensated by the much higher number of nodes at the later stages. This leads to the 23-11-2 tree having a slightly lower Wasserstein distance (when sampled using the SMWD algorithm and MC with second-order moment matching) than the 28-6-3 tree. Therefore, as a rule of thumb, scenario trees with a greater number of nodes at the earlier stages tend to have lower Wasserstein distances.

From the above observations, we would expect the *fan* scenario tree (500-1-1) to yield the optimum lower bound. This is the case when MC sampling is used. However, when we sample a flat scenario tree conditionally and perform local mean matching, the sampling of the stochastic processes after the second period is deterministic. The randomness of the processes in subsequent stages is not captured as the stochastic variations are not propagated down the tree. This explains the degradation in the Wasserstein distances for the SMWD algorithm and MC sampling with moment matching as shown in the last row of Table 16.1.

Table 16.2 shows the absolute moment deviation of the moments and the *Kolmogorov* distance (which here measures the average over the individual seven factors of the maximum deviation between the discrete densities) between the large and small sample state distributions at different stages of the scenario trees.[1] Since the SMWD algorithm matches the sample mean to the large sample proxy for the continuous distribution at the source, the deviations in the first moment are small for the first period. However, the sample mean of the stochastic processes in the subsequent periods are not matched as we perform mean matching *locally*. The second-order moment deviation also increases with each period following propagation of sampling bias. Note that there is a minimum number of branches required to perform second-order matching, namely, the dimension of the underlying stochastic process, which by virtue of the Farkas lemma is also a necessary requirement to prevent spurious arbitrage in the optimization process (Geyer et al. 2010). Therefore, the 8-8-8 tree is the only arbitrage-free tree structure with second-order moment matching at *all* stages. For the 63-4-2 tree second-order moment matching is therefore only performed at the initial stage.

Table 16.1 shows in general that lower Wasserstein distances are obtained when scenario trees are generated using the SMWD algorithm as compared with mean and covariance matching. However, this difference is not significant, especially for left-heavy tree structures. The large deviations in the second-order moments observed in scenario trees sampled with the SMWD algorithm coupled with the slightly worse Kolmogorov distances compared to moment-matched trees lead us to conclude that

[1]The correlation column is the summation of the absolute deviations of all unique correlation matrix entries of the seven-dimensional factor process (see Yong (2007) for more details).

performing *covariance matching on top of MC sampling* is an easy and efficient, yet effective, sampling method for reducing sampling error (measured in terms of both moment deviations and Wasserstein distances) in tractable scenario trees for dynamic stochastic programming problems. Ideally, second-order moments should be locally matched at all nodes in a scenario tree, which also guarantees the removal of spurious arbitrage. However, in practical ALM problems this is not always feasible and other methods, based on mean matching across *all* scenarios between decision stages, can be used for this purpose.

16.4 Impact of Different Scenario Generation Methods on SP Solutions

We have seen how different scenario generation methods affect the statistical properties of the scenario tree relative to the underlying probability distribution of the stochastic data process. Among the more commonly used criteria is the stability of the objective function value (Kaut and Wallace 2003). However, in financial planning applications we are mainly interested in a measure of the stability of the implementable decisions which include an initial portfolio balance. Such a criterion is *not* generally used in the literature, implicitly based on the argument that two different decisions that yield similar objective function values should not be penalized as being unstable. This may occur if the surface of the objective function is a flat plateau or has multiple optima of similar values.[2] However, this argument obviously relies heavily on the assumption that the surface of the objective function in a scenario-based dynamic stochastic programme is accurately represented. Often, sampling error causes this surface to vary across different Monte Carlo replications of scenario trees. So, similarity in sample objective function values does not truly measure the stability of the stochastic programme. Two distinct implementable decisions with similar objective function values for the sample problem might evaluate to very different values on the "true" objective function surface.[3] Therefore, in-sample stability of a DSP should be measured with respect to criteria based on both objective function values and implementable and forward decision values.

There are different ways of defining stability of a decision vector across different sample trees. For example, we could take the standard deviation of the Euclidean norm of decision component differences or compute the standard deviation of each decision component difference and find the maximum. For an asset–liability management problem the relevant implementable decisions are the initial optimal asset mix which in turn can be characterized by the initial expected portfolio return

[2] This latter case will not be the case for a multistage *linear* stochastic programme.

[3] Even the theoretical literature on the asymptotic epi-convergence of sampled problems as unlimited tree branching structures tend to the underlying continuous path space in a suitable sense concentrate on asymptotic objective function values, see, for example, Shapiro (2003) and the references therein.

and volatility. This provides a convenient way of measuring stability of the implementable decisions in an ALM problem. For example, when we have multiple asset classes whose returns are highly correlated or all have nearly the same risk and return characteristics, different asset mixes may yield similar portfolio characteristics as a whole and therefore should be viewed as similar optimal solutions in terms of decision stability.

An *in-sample* stability criterion tests the variability of the computed optimal solutions due to sampling variability in the scenario trees. However, it can say nothing about the proximity of the *optimal solutions* to the true one. We know that the objective function value obtained is an upward biased estimate of the true value (in a maximization problem). Therefore, scenario generation methods that give lower in-sample optimal values lead to DSP problems which have optimal values closer to the *true* one for a continuous data process. However, this does not guarantee closeness of the optimal decisions. The in-sample optimal decision vectors also need to be tested against possible bias. A scenario generation technique that does not introduce bias would tend to allow for a replication of the in-sample behaviour *out-of-sample*. The comparison of in-sample and out-of-sample estimates thus allows us to investigate the degree to which our optimized implementable decisions suffer from sampling error in a small-size scenario tree and how severe (if any) is the biasedness resulting from small sample discretization and spurious arbitrage. Therefore, out-of-sample performance measurement allows us to investigate the closeness of the optimal solutions obtained to the true values. This performance measurement can be effected via a series of unbiased out-of-sample tests in a series of pseudo-historical backtests. For a *pseudo-historical backtest* we simulate a number of scenarios out-of-sample and use each as a proxy history for which an *historical backtest* is performed.

16.5 Case Study of the Pioneer Guaranteed Return Fund Problem

In this section, we apply the scenario generation techniques discussed above to the Pioneer guaranteed fund problem (Chapter 2, this volume) documented in the Appendix using US data.[4] We also investigate the effect of tree structure on the root-node implementable decision to develop a feel for how the tree structure may be chosen. First, we will explain some salient characteristics of the model formulation. The model considers a portfolio optimization of a *closed-end* fund with a *nominal* return guarantee of G per annum with annual decisions, return generation, and risk measurement over a 4 year horizon.

[4] We use monthly data from June 1995 to December 2003 on the S&P 500, Russell 2000, MSCI REIT and MSCI EAFE indices and the prices of the US Treasury bonds of maturity 1, 2, 3, 4, 5, 10, and 30 years to calibrate GBM models for the indices and a 3-factor yield curve model for the bonds using Kalman filtering. This yields seven stochastic factors for forward simulation (see Yong (2007) and Chapter 2, this volume for more details).

The model aims to maximize the performance of the fund taking into account the risk associated with falling short of the guarantee level. The parameter β represents a trade-off between risk control and maximizing portfolio wealth. A low value of β corresponds to very tight risk control and when $\beta = 0$ no incentive is given to having high return. The optimizer will choose decisions that yield the lowest in-sample expected maximum shortfall (across time and scenarios). As we vary β upwards, the optimal policy becomes increasingly risky. Higher return performance assets are chosen at the expense of greater risk. The trade-off between having higher expected maximum shortfall and terminal wealth is shown in Fig. 16.1. This mimics the efficient frontier of mean–variance analysis. In this setting, however, the risk measure is given by the value of expected maximum shortfall instead of variance.

Volatility of the root node portfolio allocation increases with β as more incentive is given to accumulation of wealth than the penalty incurred in dipping below the barrier. When β increases to unity only portfolio wealth is considered in the objective. The solution to the limiting problem is to invest in the asset class that offers the greatest expected return. The range of β in which the most variation in the initial portfolio allocation is observed is very much problem dependent. It is mostly affected by (1) the characteristics of the investable assets, such as their returns and volatilities, and (2) the tightness of the guaranteed return problem, whether it is difficult to achieve the target given the selection of investable assets. For our problem, the range of β of interest for implementable decision stability is from 0 to 0.15.

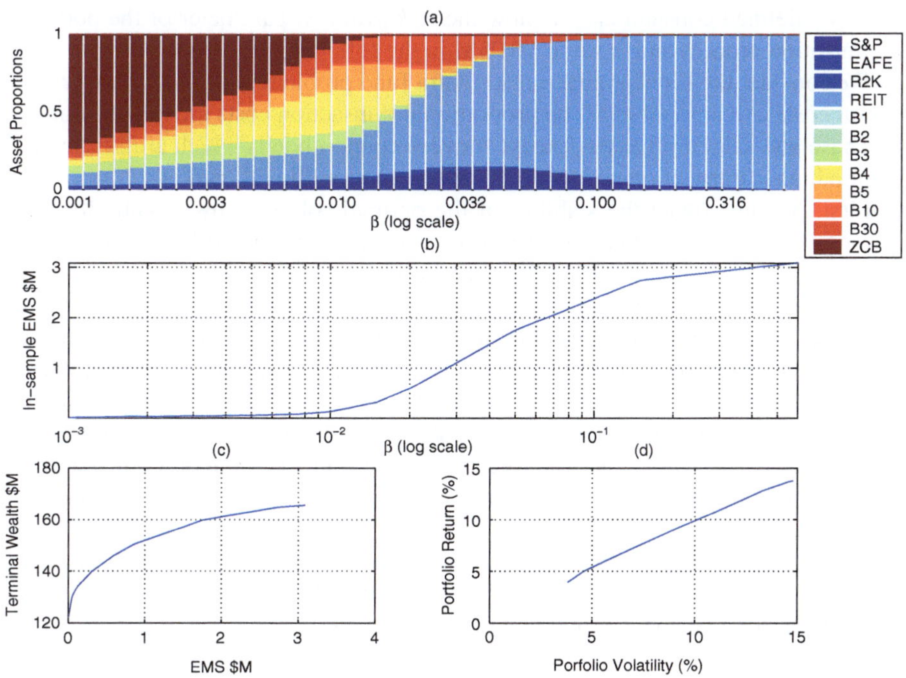

Fig. 16.1 (**a**) Variation of root node portfolio allocation with β. (**b**) In-sample efficient frontier. (**c**) Expected one-period root node portfolio return and volatility

16.5.1 Tree Structure Effect on the Multi-stage Problem

In this section we address the question of how to decide on the tree structure of a DSP by varying the branching factors at each stage of the scenario tree and we investigate the effect of this choice on (1) the stability of optimal solutions and (2) possible bias in the implementable root node recommendations.

For a fixed value of β the in-sample expected maximum shortfall is tree dependent and hence the initial portfolio risk characteristics vary with the choice of tree structure. To obtain a consistent comparison, we constrain the maximum shortfall to be the same for the different tree structures evaluated. This can be done by adding the following maximum shortfall constraint

$$\sum_{\omega \in \Omega} p(\omega) H(\omega) \leq c_{\max} \tag{16.46}$$

to the model formulation and replacing the objective function with one that only aims to maximize wealth at each period, i.e.

$$\sum_{\omega \in \Omega} \sum_{t=1}^{T} p(\omega) \hat{W}_t(\omega). \tag{16.47}$$

The user-defined constant $c_{\max}$ is now the *risk aversion* parameter of the portfolio optimization process and is set to \$0.05M in the following investigation.

We consider the Pioneer guaranteed return fund problem with and without *transaction costs*. An amount of 50 bp proportional transaction costs are imposed on the purchase and sale of all assets except for zero coupon bonds which are assumed to incur a higher transaction cost of 75 bp due to a more illiquid market.

First, we investigate the stability of the optimal solutions by solving 30 replications of the Pioneer guaranteed return fund problem with different tree structures kept to approximately 10,000 scenarios. The in-sample results are tabulated in Table 16.3. The result shows that left-heavy scenario trees produce more stable in-sample estimates of terminal wealth, hitting probability, and allocation recommendations. Based on these stability criteria, the 1250-2-2-2 scenario tree is preferred for problems without transaction costs and the 100-25-2-2 tree for problems with transaction costs. The introduction of transaction costs exerts a downward pressure on the wealth level, resulting in a more conservative allocation in order to guarantee staying above the barrier and two moments are matched for two stages (instead of only the first).

However, choosing the tree structure based on the stability criterion alone might not be sufficient. As we have seen earlier in Section 16.3, the choice of tree structure affects the statistical properties of the simulated process sample and consequently might influence the optimal root node (initial implementable) decisions. For our portfolio optimization problem, for example, we are mainly interested in the root node portfolio recommendations as they are implemented at each portfolio

Table 16.3 Variation of the optimal solutions with different tree structures

	Expected terminal wealth ($)		Barrier hitting probability[a]		Initial portfolio returns (%)		Initial portfolio volatility (%)	
Tree structure	Avg	St dev	Avg	St dev	Avg	St dev	Avg	St dev
No transaction costs								
10-10-10-10	142.41	(3.75)	0.099	(0.034)	6.0	(1.6)	5.9	(2.2)
30-7-7-7	143.47	(1.45)	0.062	(0.024)	5.2	(0.7)	4.7	(0.8)
80-5-5-5	146.13	(0.70)	0.057	(0.014)	4.9	(0.3)	4.2	(0.4)
100-25-2-2	143.16	(0.57)	0.057	(0.014)	4.9	(0.2)	4.3	(0.4)
1250-2-2-2	150.91	(0.11)	0.060	(0.004)	4.8	(0.1)	4.1	(0.1)
50 bp transaction costs for all assets except ZCB at 75 bp								
10-10-10-10	128.38	(3.39)	0.048	(0.009)	4.4	(1.1)	4.1	(1.3)
30-7-7-7	130.32	(1.43)	0.059	(0.013)	4.0	(0.3)	3.6	(0.3)
80-5-5-5	134.95	(0.77)	0.061	(0.012)	4.0	(0.2)	3.6	(0.2)
100-25-2-2	135.76	(0.38)	0.068	(0.010)	4.1	(0.1)	3.6	(0.2)
1250-2-2-2	145.29	(0.10)	0.069	(0.004)	4.3	(0.0)	3.8	(0.0)

[a] This is defined as $P(H(\omega) > 0)$ where $H(\omega)$ is the maximum shortfall over the four periods, as defined in the Appendix, Table 16.7

rebalance in practice. An equivalent stochastic programme then will be solved at the next period to obtain the corresponding investment strategy and this process is repeated until the fund/product horizon is reached. It is therefore essential to ensure that the root node recommendations obtained from small-sample scenario trees are as close as possible to the true (though normally unobtainable) solutions. Care is needed to select a tree structure that does not introduce sampling bias to the root node recommendations.

For problems with low coupling[5] between stages, the choice of branching factor at the later stages does not affect the root node decision significantly. In this specific case, a left-heavy tree is preferred. The Pioneer problem without transaction costs fits into this setting. Only the wealth and shortfall values are propagated from one stage to another. At each time period, the optimal policy is to invest in the assets which give the best immediate risk-adjusted returns. To see this more clearly, the top row of Fig. 16.2 compares the expected portfolio evolution of scenario trees 100-20-10-10 and 100-25-2-2. We intentionally keep the branching factors of the two tree structures to be approximately the same and vary the branching of the final two stages to see how branching at the later stages impacts the root node decisions. The root node decisions of the two trees are relatively similar, despite the considerable difference in the allocations at the final two stages (see the top row of Fig. 16.2). This implies that the root node decisions are fairly independent of the subsequent decisions, i.e. the problem is weakly coupled.

[5] *Low* (*high*) *coupling* indicates that the dependency of the decisions between stages is weak (strong).

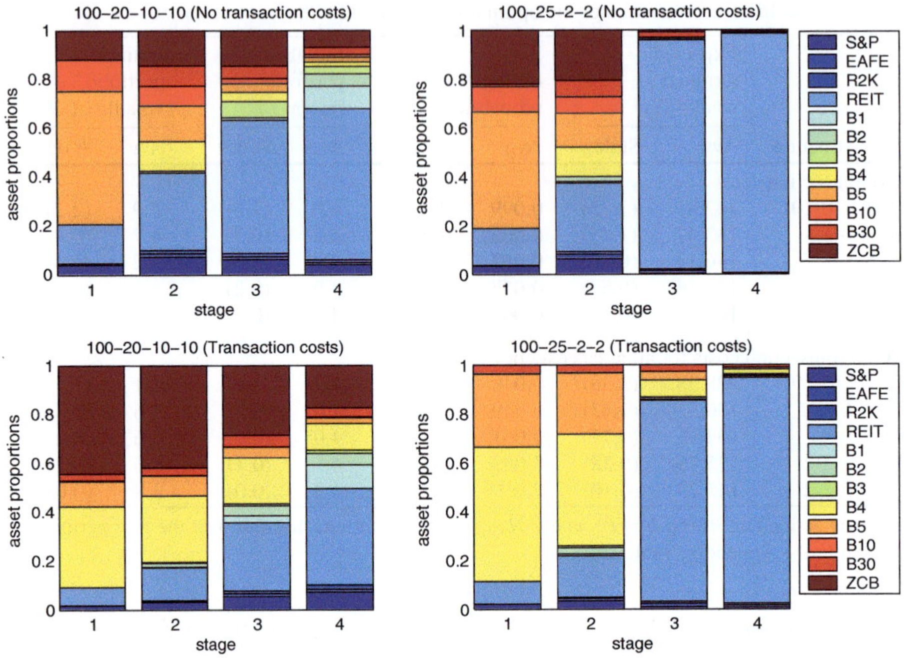

Fig. 16.2 Root node allocations for different tree structures (*Top row*: no transaction costs; *bottom row*: with transaction costs)

In the contrary case, when the coupling between stages is influential, it is not advisable to give up branching at the later stages for a heavy initial branching. The Pioneer problem with transaction costs is a good illustration. When transaction costs are present, there is a need to balance the objective of maximizing returns with minimizing the excessive transaction costs from large changes in portfolio holdings. The root node decisions become coupled to the decisions of the second stage, which in turn influence the decisions of the later stages. In this case, we have to ensure sufficient branching at each of the stages to minimize sampling bias. Otherwise, the propagation of the bias at the later stages to the root node (due to the strong coupling effect) might impair the performance of the optimization model.

Let us look at the following example. The bottom row of Fig. 16.2 plots the initial investment strategies derived from the 100-20-10-10 and 100-25-2-2 scenario trees with transaction costs. The recommended allocation for the former is to invest heavily in the 4-year zero coupon bond (ZCB) whereas no investment in the ZCB is made for the 100-25-2-2 tree. When there are only two branches emanating from a node at the final two stages of the 100-25-2-2 tree, the optimal decisions of the corresponding nodes are to invest heavily in the REIT index (the asset with the highest return) as the risks are not sufficiently represented by the limited samples. This sudden change to highly risky assets from an initially more conservative portfolio makes the ZCB less attractive due to its higher transaction cost. As a result of the bias in the decisions at the last two stages, it follows that the recommended investment strategy of the 100-25-2-2 tree is riskier than that of the 100-20-10-10 tree[6]. Hence, even though a left-heavy tree is usually preferred, the degree of left-heaviness needs to be suitably adjusted, particularly for problems with high coupling between stages, so as to avoid introducing bias to the root node recommendations to be implemented.

16.5.2 Effectiveness of Scenario Generation Techniques

The aim of this section is to ascertain which of the probability metrics introduced above plays an important role in influencing the solution quality of dynamic stochastic programmes. The various sampling techniques are applied to generate scenario trees for the four-period Pioneer guaranteed fund problem (with transaction costs). The performance of the various methods is evaluated based on both the in- and out-of-sample criteria outlined in Section 16.4. We revert to β as the risk-adjusting parameter, as the maximum shortfall constraint applied previously may yield an infeasible problem in the out-of-sample tests. This might occur when a bad decision is made, causing the out-of-sample portfolio wealth to dip way below the barrier level for which no feasible investment strategy satisfying the maximum shortfall constraint exists. To clearly illustrate the effect of sampling error, we have *deliberately* chosen scenario trees with relatively small branching factors in which second-order local moment matching is just possible and no spurious arbitrages exist.

16.5.2.1 In-Sample Results

To investigate the effectiveness of various sampling techniques in improving the in-sample stability of a stochastic programme, we generate scenario trees of size 7-7-7-7 using these methods and compare the standard deviation of the corresponding solutions obtained from 100 repetitions (scenario trees generated using alternative simulation seeds). We consider problems of different risk attitudes by varying β from 0.001 to 0.60. The in-sample expected terminal wealth is plotted against the expected maximum shortfall for this range of β, forming the efficient frontiers shown in Fig. 16.3. The two large crosses, respectively, correspond to the true solution for the two limiting cases, $\beta := 0$ and $\beta := 1$. The optimal strategy for $\beta := 1$ is to invest entirely in the highest yielding asset (REIT) at each stage, leading to the expected maximum shortfall and terminal wealth level given by the cross at the top right corner of Fig. 16.3. In contrast, $\beta := 0$ corresponds to minimizing the expected maximum shortfall without any reward for achieving greater returns. The optimal investment strategy is then to hold a sufficient amount of ZCBs to pay the guaranteed amount upon maturity. The ZCB holding required for a guaranteed return of 2.8% per annum for four periods after taking account of transaction costs is approximately \$100M (i.e. the (approximate) optimal investment strategy when $\beta := 0$ is 100% ZCB). This yields the cross at the lower left corner of Fig. 16.3.

[6] Note that the 4-year (sovereign) zero coupon bond is effectively a risk-free asset for a 4-year horizon problem if it is held to maturity.

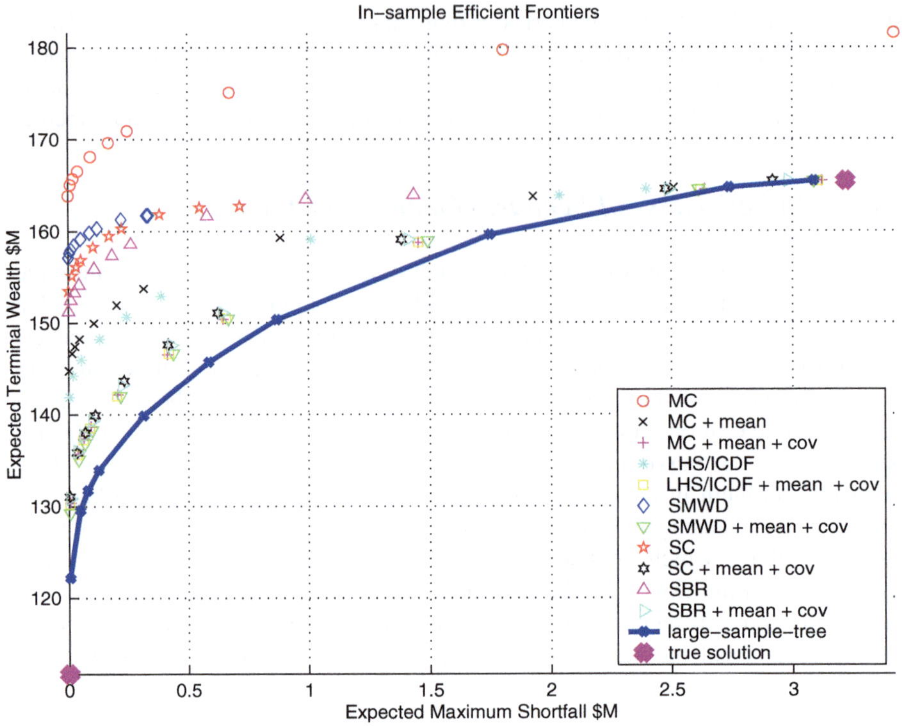

Fig. 16.3 In-sample efficient frontiers for the four-stage Pioneer guaranteed return fund problem

The blue line corresponds to the frontier generated by solving a 100-20-10-10 tree of 200,000 scenarios generated using LHS/ICDF with second-order moment matching and no spurious arbitrages and is henceforth referred to as the *large-sample tree reference* in the sequel.

Due to the upward bias in the in-sample terminal wealth estimates and the downward bias in maximum shortfall estimates, the various sampling methods produce efficient frontiers lying above the large-sample tree reference. Figure 16.3 shows that the variance reduction techniques (i.e. moment matching and LHS/ICDF methods) and the Wasserstein-based heuristics help reduce this sampling bias over MC sampling. In particular, methods with second-order moment matching produce frontiers that are close to the large-sample tree reference. As the 7-7-7-7 tree involves only 2401 scenarios compared with 200,000 scenarios of the large-sample tree, substantial time and memory savings are achieved using these methods.

Tables 16.4 and 16.5, respectively, show the in-sample terminal wealth, maximum shortfall, and hitting probability and the root node allocations (initial portfolio returns and volatilities) for $\beta := 0.075$ (risk averse) and $\beta := 0.60$ (risk loving). In general, methods such as the SMWD algorithm, LHS/ICDF, and moment matching show improvement in the stability of the terminal wealth and maximum shortfall estimates over MC sampling. Significant improvement is observed for $\beta := 0.6$ when mean and covariance matching are applied. This is because at high β, the

Table 16.4 In-sample stability results for the four-stage Pioneer guaranteed return fund problem

		Expected terminal wealth ($M)		Expected maximum shortfall ($M)		Barrier hitting probability	
β	Method	Avg	St dev	Avg	St dev	Avg	St dev
0.0075	`MC`	165.68	(16.08)	0.02	(0.04)	0.02	(0.03)
	`MC + mean`	147.43	(8.46)	0.03	(0.03)	0.03	(0.03)
	`MC + mean + cov`	137.33	(5.80)	0.06	(0.03)	0.05	(0.02)
	`SMWD`	157.92	(4.27)	**0.01**	(0.03)	0.01	(0.03)
	`SMWD + mean + cov`	136.89	(6.09)	0.07	(0.03)	0.06	(0.02)
	`SC`	156.05	(7.17)	0.03	(0.06)	0.01	(0.03)
	`SC + mean + cov`	138.03	(7.66)	0.07	(0.06)	0.04	(0.03)
	`NR`	153.29	(8.74)	0.03	(0.05)	0.02	(0.03)
	`NR + mean + cov`	137.72	(7.05)	0.06	(0.04)	0.05	(0.03)
	`LHS/ICDF`	145.12	(8.14)	0.03	(0.03)	0.03	(0.03)
	`LHS/ICDF + mean + cov`	137.24	(5.93)	0.06	(0.03)	0.05	(0.02)
	Large-sample tree reference	**131.72**	**(5.01)**	**0.08**	**(0.03)**	**0.06**	**(0.02)**
0.60	`MC`	181.57	(10.35)	3.42	(1.99)	0.29	(0.13)
	`MC + mean`	164.69	(0.77)	2.51	(1.20)	0.28	(0.09)
	`MC + mean + cov`	165.44	(0.04)	3.12	(0.53)	0.33	(0.07)
	`SMWD`	161.71	(0.19)	**0.33**	(0.33)	0.11	(0.08)
	`SMWD + mean + cov`	165.44	(0.04)	3.09	(0.55)	0.32	(0.07)
	`SC`	162.70	(2.80)	**0.72**	(0.60)	0.13	(0.10)
	`SC + mean + cov`	165.46	(0.09)	2.92	(0.66)	0.28	(0.09)
	`NR`	163.92	(3.15)	1.43	(0.98)	0.20	(0.10)
	`NR + mean + cov`	165.45	(0.06)	2.98	(0.58)	0.31	(0.08)
	`LHS/ICDF`	164.53	(0.48)	2.40	(0.82)	0.29	(0.08)
	`LHS/ICDF + mean + cov`	154.43	(0.04)	3.11	(0.53)	0.32	(0.06)
	Large-sample tree reference	**165.44**	**(0.03)**	**3.09**	**(0.43)**	**0.32**	**(0.05)**

optimum policy is to invest in the highest return asset. As long as we ensure that the in-sample expected returns of the assets match their theoretical counterparts, we successfully suppress sampling error and improve stability. Note that the true conditional expected returns depend predominantly on the means and variances of the underlying Brownian motions. Therefore, substantial improvement is achieved by performing a simple mean-matching procedure.

The operating range of the in-sample expected maximum shortfall is vastly different for these techniques as shown in Fig. 16.3 and Table 16.4. In particular, the Wasserstein-based heuristics (`SMWD`, `SC`, and `NR`) yield much lower expected maximum shortfalls. This is due to the under-estimation of the underlying variance in the small sample approximations sampled using the `SMWD` algorithm (see (16.33)) which further leads on to an under-estimation of the in-sample portfolio risk.

The resulting under-estimation is evident in that the recommended allocations are relatively aggressive for low values of β. This under-estimation, as we have seen, may be reduced by performing second-order moment matching on top of these techniques to yield more conservative initial portfolio allocations.

Table 16.5 In-sample portfolio allocation stability results for the four-stage Pioneer guaranteed return fund problem

		Total equities		Total bonds		Portfolio return		Portfolio volatility	
β	Method	Avg	St dev	Avg	St dev	Avg	St dev	Avg	St dev
0.0075	MC	59.4	(39.0)	40.6	(39.0)	8.53	(3.90)	10.43	(5.50)
	MC + mean	59.0	(35.0)	41.0	(35.0)	8.78	(3.57)	9.37	(4.29)
	MC + mean + cov	34.8	(22.5)	65.2	(22.5)	6.37	(2.26)	6.27	(2.74)
	SMWD	89.1	(21.1)	10.9	(21.1)	12.26	(2.28)	13.15	(2.62)
	SMWD + mean + cov	34.5	(23.5)	65.5	(23.5)	6.35	(2.38)	6.19	(2.84)
	NR	70.1	(33.7)	29.9	(33.7)	10.00	(3.48)	10.96	(4.35)
	NR + mean-cov	38.6	(27.3)	61.4	(27.3)	6.77	(2.78)	6.75	(3.27)
	SC	80.3	(29.4)	19.7	(29.4)	11.13	(3.09)	12.26	(3.81)
	SC + mean	80.6	(28.3)	19.4	(28.3)	11.24	(2.99)	12.16	(3.54)
	SC + mean + cov	40.4	(29.5)	59.6	(29.5)	6.93	(3.03)	7.03	(3.57)
	LHS/ICDF	53.9	(33.5)	46.1	(33.5)	8.2	(3.4)	8.6	(4.1)
	LHS/ICDF + mean + cov	34.9	(23.0)	65.1	(23.0)	6.36	(2.34)	6.25	(2.85)
	Large-sample tree reference	**23.6**	**(16.9)**	**60.0**	**(17.2)**	**5.25**	**(1.73)**	**4.90**	**(2.05)**
0.6	MC	96.7	(17.0)	3.3	(17.0)	12.12	(2.40)	17.43	(3.72)
	MC + mean	100.0	(0.0)	0.0	(0.0)	13.69	(0.11)	14.83	(0.45)
	MC + mean + cov	100.0	(0.0)	0.0	(0.0)	13.70	(0.00)	14.80	(0.00)
	SMWD	100.0	(0.0)	0.0	(0.0)	13.70	(0.00)	14.80	(0.00)
	SMWD + mean + cov	100.0	(0.0)	0.0	(0.0)	13.70	(0.00)	14.80	(0.00)
	SC	100.0	(0.0)	0.0	(0.0)	13.50	(0.53)	15.36	(1.81)
	SC + mean	100.0	(0.0)	0.0	(0.0)	13.70	(0.00)	14.80	(0.00)
	SC + mean + cov	100.0	(0.0)	0.0	(0.0)	13.70	(0.00)	14.80	(0.00)
	NR	100.0	(0.0)	0.0	(0.0)	13.37	(0.66)	15.80	(2.41)
	NR + mean + cov	100.0	(0.0)	0.0	(0.0)	13.70	(0.00)	14.80	(0.00)
	LHS/ICDF	100.0	(0.0)	0.0	(0.0)	13.70	(0.00)	14.80	(0.00)
	LHS/ICDF + mean + cov	0.0	(0.0)	0.0	(0.0)	13.70	(0.00)	14.80	(0.00)
	Large-sample tree reference	**100.0**	**(0.0)**	**0.0**	**(0.0)**	**13.70**	**(0.00)**	**14.80**	**(0.00)**

16.5.2.2 Out-of-Sample Results

We now investigate how sampling error affects the out-of-sample performance of a DSP by performing *telescoping-horizon pseudo-history* backtests on 1000 out-of-sample scenarios generated using the LHS/ICDF method. For an unbiased comparison, the stochastic processes simulated are calibrated to the same parameters used for generating scenario trees for the in-sample experiments. These parameters are assumed to be non-changing over the time horizon we are interested in and the usual recalibration needed in practice is therefore not needed here.

The tree sizes at each roll-forward period are 7-7-7-7, 7-7-7, 7-7, and 7 where the branching is intentionally kept low to investigate how inadequate sampling affects model performance. We compute the *realized terminal wealth* (sample average of the terminal wealth values) and *realized maximum shortfall* (sample average of the maximum deficits from the barrier levels) and plot the realized efficient frontier in Fig. 16.4. The heavy frontier line corresponds to running the pseudo-history backtests with scenario trees of 10,000 scenarios. The tree structures are 10-10-10-10, 25-20-20, 100-100, 10,000 at each of the roll-forward periods. Since the number of scenarios in these trees are substantial, the realized frontier is therefore referred to as the *large-sample tree reference frontier*.

When a limited number of scenarios are used to represent the continuous path space of the stochastic processes in scenario-based stochastic programmes, the optimization process exploits every opportunity (spurious profits) presented in the small-sample trees, even when spurious arbitrage is ruled out.

MC sampling in particular is very susceptible to this. The over-optimistic exploitation results in a higher-than-average in-sample terminal wealth as previously seen in Fig. 16.3 but a lower-than-average out-of-sample terminal wealth and excessive shortfalls shown in Fig. 16.4. The other sampling techniques show significant improvements over MC sampling and their frontiers lie somewhat closer to the large-sample tree reference.

Several interesting observations can be made by comparing the in-sample and out-of-sample efficient frontiers in Figs. 16.3 and 16.4, respectively. An unbiased scenario generation technique will successfully replicate the in-sample performance in the out-of-sample tests. At high β, all methods (except MC sampling) yield

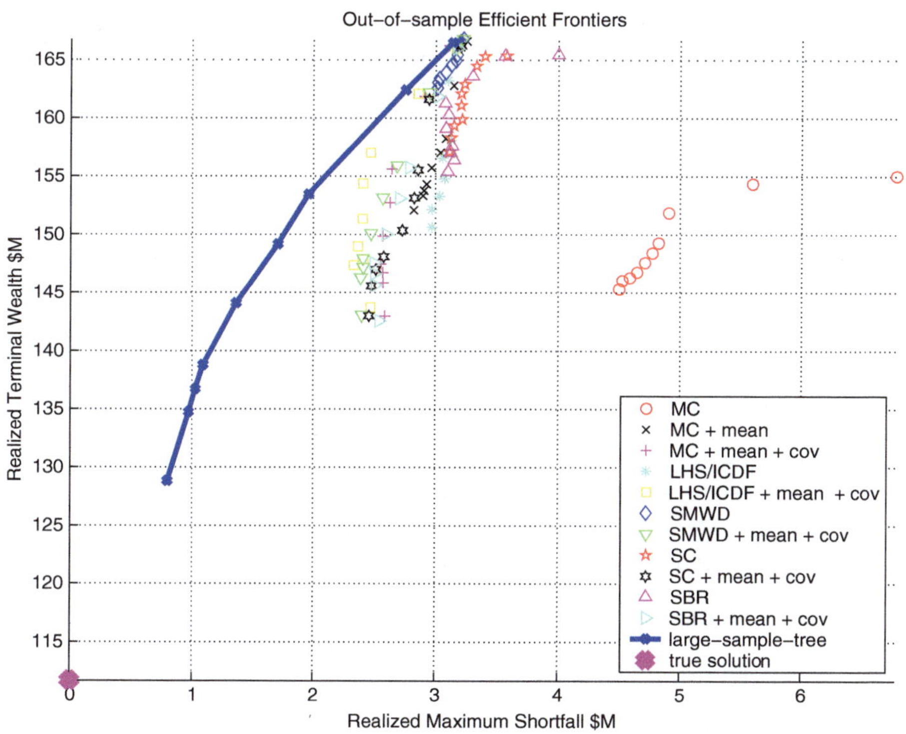

Fig. 16.4 Out-of-sample efficient frontiers for the four-stage Pioneer guaranteed return fund problem

terminal wealth and shortfall levels that are close to that predicted by the in-sample scenario trees. This is because the optimal policy is to invest in assets of the highest returns and a simple second-order moment matching, as previously explained in Section 16.5.2.1, is able to eliminate the sampling error to yield consistent portfolio allocation recommendations at each roll-forward period.

However, for tight risk control regions (low β values), small sample bias becomes apparent as none of the techniques are successful in replicating the in-sample performance even for trees with 10,000 scenarios. The in-sample expected maximum shortfall is greatly underestimated in a small sample scenario tree, especially for the Wasserstein-based heuristics (SMWD, SC, and NR). The miscalculated risk results in an overly aggressive implementable root node portfolio, causing the realized maximum shortfall to deviate greatly from that predicted in-sample.

Table 16.6 shows the realized shortfall values. The out-of-sample maximum shortfall for $\beta := 0.0075$ ranges from \$1.03M to \$4.59M and these differ greatly

Table 16.6 Out-of-sample results for the telescoping horizon pseudo-backtests. The scenario trees are 7-7-7-7, 7-7-7, 7-7, and 7, respectively, at each roll-forward period

		Realized terminal wealth (\$M)		Realized maximum shortfall (\$M)		Barrier hitting probability
β	Method	Avg	St dev	Avg	St dev	Avg
0.0075	MC	146.25	(43.94)	4.59	(7.62)	0.47
	MC + mean	153.77	(42.33)	2.91	(5.79)	0.36
	MC + mean + cov	146.69	(36.75)	2.57	(4.93)	0.39
	SMWD + mean	163.37	(44.22)	3.03	(6.32)	0.32
	SMWD + mean + cov	147.16	(35.85)	2.41	(4.59)	0.39
	NR	157.01	(42.93)	3.12	(6.40)	0.35
	NR + mean + cov	146.60	(37.04)	2.50	(4.68)	0.39
	SC	159.32	(43.49)	3.15	(6.44)	0.34
	SC + mean + cov	147.00	(37.45)	2.51	(4.99)	0.38
	LHS/ICDF	152.68	(40.85)	2.99	(5.88)	0.36
	LHS/ICDF + mean + cov	147.83	(36.34)	2.41	(4.62)	0.37
	Large-sample tree reference	**136.69**	**(28.61)**	**1.03**	**(2.21)**	**0.32**
0.60	MC	155.01	(51.03)	6.78	(11.23)	0.44
	MC + mean	166.61	(46.06)	3.25	(6.88)	0.32
	MC + mean + cov	166.75	(45.95)	3.22	(6.85)	0.31
	SMWD	166.75	(45.95)	3.22	(6.85)	0.31
	SMWD + mean + cov	166.74	(45.95)	3.22	(6.85)	0.31
	NR	165.41	(48.20)	4.00	(8.10)	0.33
	NR + mean + cov	166.75	(45.95)	3.22	(6.85)	0.31
	SC	165.31	(45.05)	3.58	(7.44)	0.32
	SC + mean + cov	166.75	(45.95)	3.22	(6.85)	0.31
	LHS/ICDF	166.74	(45.95)	3.22	(6.85)	0.31
	LHS/ICDF + mean + cov	166.75	(45.95)	3.22	(6.85)	0.31
	Large-sample tree reference	**166.75**	**(45.95)**	**3.22**	**(6.85)**	**0.31**

from the in-sample estimates which are of the order of \$0.1M. The in-sample estimation is definitely a poor estimation of the actual portfolio risk! Even though second-order moment matching is able to reduce this bias, the slight improvement is barely significant considering that the risk is still greatly under-estimated. This bias also prevails in the barrier hitting probability estimation. The out-of-sample hitting probability for a low β is about 32% for the 10,000 scenario trees. Even when using a scenario tree with 200,000 scenarios (the large-sample reference tree for the in-sample experiments, cf. Table 16.4) the prediction from the in-sample scenario trees is only 6%![7]

Figure 16.5 shows the out-of-sample wealth distributions of different roll forward to the horizon instances. The wealth distribution remains symmetrical at all stages for high β. However, for low β, it becomes increasingly skewed to the right as the stiffer penalty for dipping below the zero coupon barrier causes more conservative portfolio holdings.

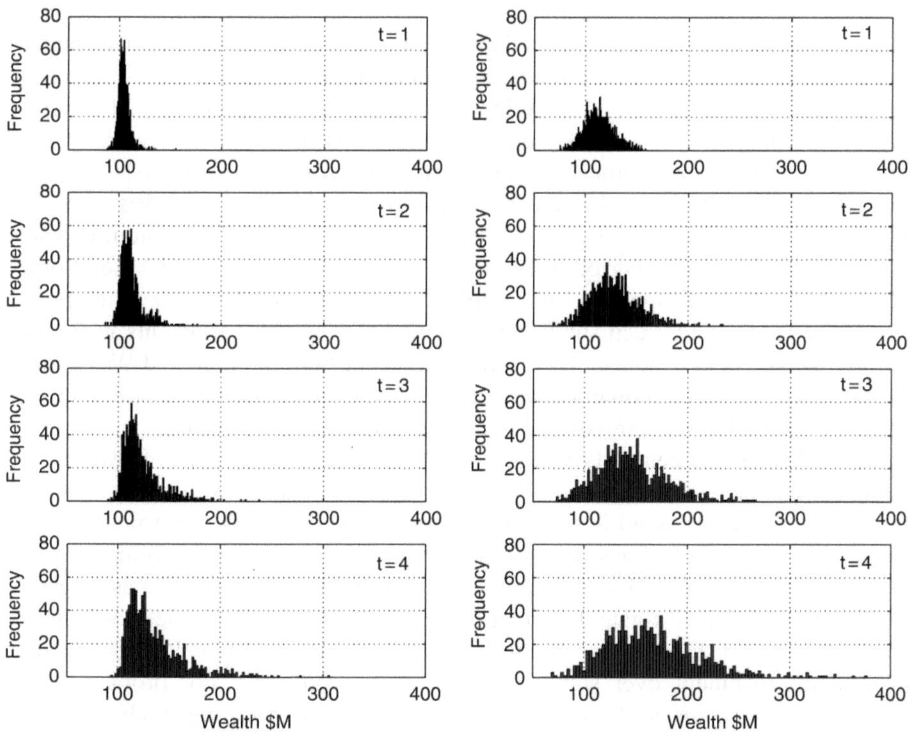

Fig. 16.5 Out-of-sample wealth distributions at different time instances (*left* $\beta := 0.0075$; *right* $\beta := 0.60$)

[7] We know from Chapter 2 (this volume) that these effects are at least partially mitigated by monthly return simulation and risk management with downward skewed jumps.

In summary, a limited sample size causes major under-estimation of the in-sample portfolio risk in tightly risk-controlled problems.

This results in practice in unwarranted risk taking and consequently leads to a degradation of fund performance. The sampling techniques studied here improve on the solution stability of the stochastic programme over MC sampling. Second-order moment matching turns out to be the most suitable technique to be applied in the case of the Pioneer guaranteed return fund problem to reduce sampling error. However, these techniques fail to fully address the issue of bias in the root node recommendations when using a small set of scenarios to approximate the continuous path space. We look at how this problem might be tackled from another perspective in Dempster et al. (2010).

16.6 Conclusion

Among the scenario generation methods, first- and second-order moment matching and stratified sampling are relatively computationally inexpensive. Among the Wasserstein-metric heuristics, SMWD requires the least computation. On the other hand, the complexity of nodal reduction algorithms increases substantially with the number of sample points. Investigation of the statistical properties of simulated processes illustrates that the Wasserstein distance is mainly driven by the number of samples used. Though SMWD yields the lowest Wasserstein distance for small-sample single-period trees, its comparative advantage over second-order moment matching disappears when applied to generating left-heavy multi-period trees. The latter yield much smaller deviation in second-order moments with comparable Wasserstein distance. Therefore, in terms of both moment and Wasserstein metrics, second-order moment matching proves to be an easy algebraic and yet effective method in reducing discretization error present in sampling.

A left-heavy tree generally yields a smaller Wasserstein distance. However, the degree of left-heaviness ought to be suitably adjusted to ensure sufficient branching at the later stages to avoid introducing discretization bias to the root-node recommendations, especially for stochastic problems with significant coupling between stages. The four-period Pioneer guaranteed return fund problem serves as an excellent case study to investigate the effect of sampling error in stochastic programmes. Empirical results show that all the techniques perform equally well, showing substantial improvement over standard Monte Carlo sampling. For problems with wealth maximization as the objective, simple moment matching is able to yield results close to the true solution as long as the expected returns of each asset are captured correctly in the scenario tree. However, for tight risk control, these techniques fail to alleviate a more prominent issue – underestimation of portfolio risk – associated with insufficient branching when a small sample scenario tree is used. The dearth of sample realizations causes an under-estimation of in-sample portfolio risk which results in unjustified risk taking in investment recommendations and impairs the performance of the fund using the model. Even though second-order moment matching helps alleviate this problem slightly, the remaining bias warrants attentive treatment of this issue in a companion paper (Dempster et al. 2010).

Acknowledgements This chapter is based in part on the third author's PhD thesis, written under the supervision of the first two. An earlier version was presented to the Cambridge Finance Workshop in the University of Cambridge and we are grateful to the participants and to a referee for helpful comments. We also wish to acknowledge the use of Cambridge System Associates Limited STOCHASTIC SUITE™ DSP generation and optimization software and of their computing facilities for the extensive computations involved in the preparation of this chapter.

Appendix: Pioneer Guaranteed Return Fund Model Formulation

The Pioneer guaranteed return fund model (Chapter 2, this volume) is a portfolio optimization of a *closed-end* fund with a *nominal* return guarantee of G per annum. At each period the model aims to maximize the performance of the fund taking into account the risk associated with falling short of the guaranteed value level. The formulation presented here follows closely (Chapter 2, this volume) except that we only impose transaction costs on the change in portfolio holdings. Selling the *off-the-run* bonds and replacing them with *on-the-run* bonds are assumed here not to incur any transaction cost. The model parameter and variable definitions are given in Table 16.7.

Objective

$$\max \quad \sum_{\omega\in\Omega} p(\omega)\left(\beta\sum_{t\in T^d}\hat{W}_t(\omega)-(1-\beta)H(\omega)\right). \tag{16.48}$$

Cash Balance Constraints

- Running cash balance constraints

$$\sum_{a\in A} P_{t,a}^{(\text{buy})}(\omega)q_{t,a}(\omega)+\sum_{a\in A}\left(\tau_a P_{t,a}^{(\text{buy})}(\omega)s_{t,a}^{+}(\omega)+\tau_a P_{t,a}^{(\text{sell})}(\omega)s_{t,a}^{-}(\omega)\right)$$

$$=\sum_{a\in A}\left(P_{t,a}^{(\text{sell})}(\omega)q_{t-1,a}(\omega)+D_{t,a}(\omega)q_{t-1,a}(\omega)\right) \tag{16.49}$$

$$\forall t\in T^d\backslash\{1\}\quad \forall\omega\in\Omega.$$

Table 16.7 Model parameters and variables

Sets definition	
$T^d = \{1, \ldots, T+1\}$	Set of decision/simulation times
A	Set of all assets
Ω	Set of scenarios
Parameter definitions	
β	Risk aversion attitude ($\beta = 1$ corresponding to risk loving)
Q_a	Initial asset holdings (in units) of asset $a \in A$
c_1	Initial cash amount
Stochastic parameter definitions	
$P^{(\mathrm{buy})}_{t,a}(\omega)/P^{(\mathrm{sell})}_{t,a}$	Buy/sell price of asset $a \in A$ at time t in scenario ω
$D_{t,a}(\omega)$	Annual coupon of bond $a \in A$ paid (in arrears) at time t in scenario ω
$Z_t(\omega)$	Zero coupon bond price at time t in scenario ω
$L_t(\omega) = W_1(1+G)^T Z_t(\omega)$	Barrier level at time t of guarantee G per annum in scenario ω
$p(\omega)$	Probability of scenario ω
Decision variable definitions	
$q_{t,a}(\omega)$	Quantity held in asset $a \in A$ over period $[t, t+1)$ in scenario ω
$q^+_{t,a}(\omega)/q^-_{t,a}(\omega)$	Quantity bought/sold in asset $a \in A$ at time t in scenario ω
$s^+_{t,a}(\omega)/s^-_{t,a}(\omega)$	Increment/decrement (in units) of asset $a \in A$ at time t in scenario ω
$W_t(\omega)$	Financial wealth *before* portfolio rebalancing at time t in scenario ω
$\hat{W}_t(\omega)$	Financial wealth *after* portfolio rebalancing at time t in scenario ω
$h_t(\omega) = \max(0, \hat{W}_t(\omega) - L_t(\omega))$	Shortfall at time t in scenario ω
$H(\omega)$	Maximum shortfall in scenario ω

- Initial cash balance constraints

$$\sum_{a\in A} P^{(\mathrm{buy})}_{1,a}(\omega) q_{1,a}(\omega) + \sum_{A} \left(\tau_a P^{(\mathrm{buy})}_{1,a}(\omega) s^+_{1,a}(\omega) + \tau_a P^{(\mathrm{sell})}_{1,a}(\omega) s^-_{1,a}(\omega) \right)$$

$$= c_1 + \sum_{A} \left(P^{(\mathrm{sell})}_{1,a}(\omega) Q_a + D_{1,a}(\omega) Q_a \right). \qquad (16.50)$$

Quantity Balance Constraints

- Running quantity balance constraints

$$q_{t,a}(\omega) = q_{t-1,a}(\omega) + q^+_{t,a}(\omega) - q^-_{t,a}(\omega) \qquad (16.51)$$

$$\forall t \in T^d \backslash \{1\} \quad \forall a \in A, \quad \forall \omega \in \Omega.$$

- Initial quantity balance constraint

$$q_{1,a}(\omega) = Q_a + q^+_{1,a}(\omega) - q^-_{1,a}(\omega) \quad \forall a \in A, \quad \forall \omega \in \Omega. \tag{16.52}$$

Annual Bond Roll-Over Constraints

The *off-the-run* bonds are sold and the new *on-the-run* bonds are bought. Note that we do not incur transaction costs on buying and selling resulting from annual rolling. Transaction costs are only incurred on changes in asset holdings.

$$\begin{aligned} q^-_{t,a}(\omega) &= q_{t-1,a}(\omega) \quad \forall t \in T^d \backslash \{1\}, \quad \forall a \in A, \quad \forall \omega \in \Omega, \\ q^-_{1,a}(\omega) &= Q_a \quad \forall a \in A, \quad \forall \omega \in \Omega. \end{aligned} \tag{16.53}$$

This constraint implies that

$$q_{t,a}(\omega) = q^+_{t,a}(\omega) \qquad \forall t \in T^d, \qquad \forall a \in A, \qquad \forall \omega \in \Omega. \tag{16.54}$$

Liquidation Constraints

The financial portfolio is liquidated in cash at the final horizon for at least the guarantees to be paid to the clients:

$$q_{T,a}(\omega) = 0 \qquad \forall a \in A, \quad \forall \omega \in \Omega. \tag{16.55}$$

This equation implies that

$$\begin{aligned} s^+_{T,a}(\omega) &= 0 \qquad \forall a \in A, \quad \forall \omega \in \Omega, \\ s^-_{T,a}(\omega) &= q_{T-1,a}(\omega) \quad \forall a \in A, \quad \forall \omega \in \Omega. \end{aligned} \tag{16.56}$$

Wealth Accounting Constraints

- Wealth before rebalancing

$$W_t(\omega) = \sum_A \left(P^{(\text{sell})}_{t,a}(\omega) q_{t-1,a}(\omega) + D_{t,a}(\omega) q_{t-1,a}(\omega) \right) \tag{16.57}$$

$$\forall t \in T^d \backslash \{1\}, \quad \forall \omega \in \Omega,$$

$$W_1(\omega) = \sum_A \left(P^{(\text{sell})}_{1,a}(\omega) Q_a + D_{1,a}(\omega) Q_a \right) + c_1. \tag{16.58}$$

- Wealth after rebalancing

$$\hat{W}_t(\omega) = \sum_A P_{t,a}^{\text{buy}}(\omega) q_{t,a}(\omega) \quad \forall t \in T^d \backslash \{T+1\}, \quad \forall \omega \in \Omega. \tag{16.59}$$

$$\hat{W}_T(\omega) = \sum_A \left(P_{T,a}^{(\text{sell})}(\omega)(q_{T-1,a}(\omega) - \tau_a s_{T,a}^-(\omega)) + D_{T,a}(\omega) q_{T-1,a}(\omega) \right)$$

$$\forall \omega \in \Omega. \tag{16.60}$$

Portfolio Change Constraints

We calculate the portfolio change (in units) through the following constraints:

- Decrement in asset position

$$q_{t,a}^+(\omega) - q_{t,a}^-(\omega) + s_{t,a}^-(\omega) \geq 0 \quad \forall t \in T^d, \quad \forall a \in A, \quad \forall \omega \in \Omega. \tag{16.61}$$

- Increment in asset position

$$q_{t,a}^-(\omega) - q_{t,a}^+(\omega) + s_{t,a}^+(\omega) \geq 0 \quad \forall t \in T^d, \quad \forall a \in A, \quad \forall \omega \in \Omega. \tag{16.62}$$

Barrier Constraints

We use the wealth after rebalance to evaluate whether it is above the barrier. The wealth after rebalancing is used because in the real world where the product is sold to the client, the fund manager will need to liquidate the financial portfolio in cash to pay the clients at least the amount they are guaranteed. Taking transaction costs into consideration, this will drive the portfolio strategies to be more conservative.

- Shortfall constraint

$$h_t(\omega) + \hat{W}_t(\omega) \geq L_t(\omega) \qquad \forall t \in T^d, \qquad \forall \omega \in \Omega. \tag{16.63}$$

- Maximum shortfall constraint

$$H(\omega) \geq h_t(\omega) \qquad \forall t \in T^d, \qquad \forall \omega \in \Omega. \tag{16.64}$$

Non-anticipativity Constraints

The non-anticipativity of the decision variables are implicit once we represent the stochastic processes using the scenario tree format. Therefore, no additional constraints are required.

References

M A H Dempster, E A Medova, and Y S Yong. Stabilization of implementable decisions in dynamic stochastic programming with applications to asset liability management. *Submitted*, 2010.

J Dupacova, C Consigli, and S W Wallace. Scenarios for multistage stochastic programs. *Annals of Operations Research*, 100:25–53, 2000.

A Geyer, M Hanke, and A Weissensteiner. No-arbitrage conditions, scenario trees, and multi-asset financial optimization. *European Journal of Operational Research*, 206:609–613, 2010.

H Heitsch and W Romisch. Generation of multivariate scenario trees to model stochasticity in power management. *IEEE St. Petersburg Power Tech*, 2005.

R Hochreiter and G Pflug. Financial scenario generation for stochastic multi-stage decision processes as facility location problems. *Annals of Operations Research*, 152(1):257–272, 2007.

K Hoyland and S W Wallace. Generating scenario trees for multi-stage decision problems. *Management Science*, 47:295–307, 2001.

K Hoyland, M Kaut, and S W Wallace. A heuristic for moment-matching scenario generation. *Computational Optimization and Applications*, 47:295–307, 2003.

M Kaut and S Wallace. Evaluation of scenario-generation methods for stochastic programming. *Stochastic Programming E-Print Series (http://www.speps.org)*, 14, 2003.

C L MacQueen. Some methods for classification and analysis of multivariate observations. In *Proceedings of 5-th Berkeley Symposium on Mathematical Statistics and Probability*, pages 281–297. Berkeley, University of California Press, 1967.

M D McKay, R J Beckman, and W J Conover. A comparison of three methods for selecting values of input variables in the analysis of output from a computer code. *Technometrics*, 21(2):239–245, 1979.

G L Nemhauser and L A Wolsey. *Integer and Combinatorial Optimization*. Wiley-Interscience Series in Discrete Mathematics and Optimization, New York, NY, 1988.

W Romisch. *Stability of Stochastic Programming Problems*. Chapter 8 in *Stochastic Programming*, Volume 10 of Handbooks in Operations Research and Management Science, A Ruszczynski and A Shapiro, eds. Elsevier, Amsterdam, 2003, pp. 483–554.

S M Ross. *Simulation*. Academic, London, 3rd edition, 2002.

A Shapiro. *Monte Carlo Sampling Methods*. Chapter 6 in *Stochastic Programming*, Volume 10 of Handbooks in Operations Research and Management Science, A Ruszczynski and A Shapiro, eds. Elsevier, Amsterdam, 2003.

M Villaverde. *Stochastic Optimization Approaches to Pricing, Hedging and Investment in Incomplete Markets*. PhD thesis, Judge Institute of Management, University of Cambridge, 2003.

Y S Yong. *Scenario Generation for Dynamic Fund Management*. PhD thesis, Centre for Financial Research, Judge Business School, University of Cambridge, 2007, pp. 353–425.

Chapter 17
Convexity of Chance Constraints with Dependent Random Variables: The Use of Copulae

René Henrion and Cyrille Strugarek

Abstract We consider the convexity of chance constraints with random right-hand side. While this issue is well understood (thanks to Prékopa's Theorem) if the mapping operating on the decision vector is componentwise concave, things become more delicate when relaxing the concavity property. In an earlier paper, the significantly weaker r-concavity concept could be exploited, in order to derive *eventual convexity* (starting from a certain probability level) for feasible sets defined by chance constraints. This result heavily relied on the assumption of the random vector having independent components. A generalization to arbitrary multivariate distributions is all but straightforward. The aim of this chapter is to derive the same convexity result for distributions modeled via copulae. In this way, correlated components are admitted, but a certain correlation structure is imposed through the choice of the copula. We identify a class of copulae admitting eventually convex chance constraints.

Keywords Copula · Chance constraints · Log-exp concavity · Probabilistic constraints · Convexity

17.1 Introduction

Chance constraints or probabilistic constraints represent an important tool for modeling restrictions on decision making in the presence of uncertainty. They take the typical form

$$\mathbb{P}(h(x,\xi) \geq 0) \geq p, \tag{17.1}$$

where $x \in \mathbb{R}^n$ is a decision vector, $\xi : \Omega \to \mathbb{R}^m$ is an m-dimensional random vector defined on some probability space $(\Omega, \mathcal{A}, \mathbb{P})$, $h : \mathbb{R}^n \times \mathbb{R}^m \to \mathbb{R}^s$ is a vector-valued mapping, and $p \in [0, 1]$ is some probability level. Accordingly, a decision vector x (which is subject to optimization with respect to some given objective function)

R. Henrion (✉)
Weierstrass Institute for Applied Analysis and Stochastics, 10117 Berlin, Germany
e-mail: henrion@wias-berlin.de

M. Bertocchi et al. (eds.), *Stochastic Optimization Methods in Finance and Energy*, International Series in Operations Research & Management Science 163,
DOI 10.1007/978-1-4419-9586-5_17,

is declared to be feasible, if the random inequality system $h(x, \xi) \geq 0$ is satisfied at least with probability p. A compilation of practical applications in which constraints of the type (17.1) play a crucial role may be found in the standard references (Prékopa 1995, 2003). Not surprisingly, one of the most important theoretical questions related to such constraints is that of convexity of the set of decisions x satisfying (17.1).

In this chapter, we shall be interested in chance constraints with separated random vectors, which come as a special case of (17.1) by putting $h(x, \xi) = g(x) - \xi$. More precisely, we want to study convexity of a set of feasible decisions defined by

$$M(p) = \{x \in \mathbb{R}^n | \mathbb{P}(\xi \leq g(x)) \geq p\}, \tag{17.2}$$

where $g : \mathbb{R}^n \to \mathbb{R}^m$ is some vector-valued mapping. With $F : \mathbb{R}^m \to \mathbb{R}$ denoting the distribution function of ξ, the same set can be rewritten as

$$M(p) = \{x \in \mathbb{R}^n | F(g(x)) \geq p\}. \tag{17.3}$$

It is well known from Prékopa's classical work (see, e.g., Prékopa (1995), Th. 10.2.1) that this set is convex for *all* levels p provided that the law $\mathbb{P} \circ \xi^{-1}$ of ξ is a log-concave probability measure on $\mathbb{R}^m$ and that the components g_i of g are concave. The latter condition is satisfied, of course, in the important case of g being a linear mapping. However, the concavity assumption may be too strong in many engineering applications, so the question arises, whether it can be relaxed to some weaker type of (quasi-) concavity. To give an example, the function e^x is not concave, while its log is. It has been shown in an earlier paper that such relaxation is possible indeed, but it comes at the price of a rather strong assumption of ξ having independent components and of a slightly weaker result, in which convexity of $M(p)$ cannot be guaranteed for any level p but only for sufficiently large levels. We shall refer to this property as *eventual convexity*. Note, however, that this is not a serious restriction in chance-constrained programming, because probability levels are chosen there close to 1 anyway. To provide an example (see Example 4.1 in Henrion and Strugarek (2008)), let the two-dimensional random vector ξ have a bivariate standard normal distribution with independent components and assume that

$$g_i(x_1, x_2) = 1/(x_1^2 + x_2^2 + 0.1) \quad (i = 1, 2).$$

Then, for $p^* \approx 0.7$ one has that $M(p)$ is nonconvex for $p < p^*$ whereas it is convex for $p \geq p^*$. Note that though the multivariate normal distribution is log concave (see Prékopa (1995)), the result by Prékopa mentioned above does not apply because the g_i are not concave. As a consequence, $M(p)$ is not convex for all p as it would be guaranteed by that result. Nonetheless, eventually convexity can be verified by the tools developed in Henrion and Strugarek (2008).

The aim of this chapter is to go beyond the restrictive independence assumption made in Henrion and Strugarek (2008). While to do this directly for arbitrary multivariate distributions of ξ seems to be very difficult, we shall see that positive results can be obtained in case that this distribution is modeled by means of a copula. Copulae allow to represent dependencies in multivariate distributions in a more efficient way than correlation does. They may provide a good approximation to the true underlying distribution just on the basis of its one-dimensional margins. This offers a new perspective also to modeling chance constraints not considered extensively so far to the best of our knowledge. The chapter is organized as follows: in a first section, some basics on copulae are presented and the concepts of r-concave and r-decreasing functions introduced. The following section contains our main result on eventual convexity of chance constraints defined by copulae. In a further section, log exp concavity of copulae, a decisive property in the mentioned convexity result, is discussed. Finally, a small numerical example illustrates the application of the convexity result in the context of chance-constrained programming.

17.2 Preliminaries

17.2.1 Basics on Copulae

In this section, we compile some basic facts about copulae. We refer to the introduction (Nelsen 2006) as a standard reference.

Definition 1 A copula is a distribution function $C : [0,1]^m \to [0,1]$ of some random vector whose marginals are uniformly distributed on $[0,1]$.

A theorem by Sklar states that for any distribution function $F : \mathbb{R}^m \to [0,1]$ with marginals $(F_i)_{1\le i\le m}$ there exists a copula C for F satisfying

$$\forall x \in \mathbb{R}^m, \ F(x) = C\left(F_1(x_1), \ldots, F_m(x_m)\right). \tag{17.4}$$

Moreover, if the marginals F_i are continuous, then the copula C is uniquely given by

$$C(u) = F\left(F_1^{-1}(u_1), \ldots, F_m^{-1}(u_m)\right); \quad F_i^{-1}(t) := \inf_{F_i(r)\ge t} r. \tag{17.5}$$

Note that Sklar's theorem may be used in two directions: either in order to identify the copula of a given distribution function or to define a distribution function via a given copula (for instance, a copula with desirable properties could be used to fit a given distribution function). The following are first examples for copulae:

- *Independent* or *product copula*: $C(u) = \Pi_{i=1}^m u_i$
- *Maximum* or *comonotone copula*: $C(u) = \min_{1\le i\le m} u_i$
- *Gaussian* or *normal copula*: $C^\Sigma(u) = \Phi^\Sigma\left(\Phi^{-1}(u_1), \ldots, \Phi^{-1}(u_m)\right)$

Here, Φ^{Σ} refers to a multivariate standard normal distribution function with mean zero, unit variances, and covariance (=correlation) matrix Σ. Φ is just the one-dimensional standard normal distribution function.

With regard to (17.4), the independent copula simply creates a joint distribution with independent components from given marginals. The theoretical interest of the maximum copula (also explaining its name) comes from the fact that any copula C is dominated by the maximum copula:

$$C(u) \leq \min_{1 \leq i \leq m} u_i. \tag{17.6}$$

The Gaussian copula has much practical importance (similar to the normal distribution function). It allows to approximate a multivariate distribution function F having marginals F_i which are not necessarily normally distributed, by a composition of a multivariate normal distribution function with a mapping having components $\Phi^{-1} F_i$ defined via one-dimensional distribution functions (Klaassen and Wellner 1997).

A lot of other practically important copulae is collected in a whole family:

Definition 2 A copula C is called Archimedean if there exists a continuous strictly decreasing function $\psi : [0, 1] \to \mathbb{R}^+$ such that $\psi(1) = 0$ and

$$C(u) = \psi^{-1}\left(\sum_{i=1}^{m} \psi(u_i)\right),$$

ψ is called the generator of the Archimedean copula C. If $\lim_{u \to 0} \psi(u) = +\infty$, then C is called a strict Archimedean copula.

Conversely, we can start from a generator ψ to obtain a copula. The following proposition provides sufficient conditions to get a copula from a generator:

Proposition 1 *Let $\psi : [0, 1] \to \mathbb{R}^+$ such that*

1. *ψ is strictly decreasing, $\psi(1) = 0$, $\lim_{u \to 0} \psi(u) = +\infty$*
2. *ψ is convex*
3. *$(-1)^k \frac{d^k}{dt^k} \psi^{-1}(t) \geq 0 \quad \forall k = 0, 1, \ldots, m, \ \forall t \in \mathbb{R}^+$.*

Then, $C(u) = \psi^{-1}\left(\sum_{i=1}^{m} \psi(u_i)\right)$ is a copula.

Example 1 We have the following special instances of Archimedean copulae:

1. The independent copula is a strict Archimedean copula whose generator is $\psi(t) = -\log(t)$.
2. Clayton copulas are the Archimedean copulas defined by the strict generator $\psi(t) = \theta^{-1}(t^{-\theta} - 1)$, with $\theta > 0$.

3. Gumbel copulas are the Archimedean copulas defined by the strict generator $\psi(t) = (-\log(t))^{\theta}$, with $\theta \geq 1$.
4. Frank copulas are the Archimedean copulas defined by the strict generator $\psi(t) = -\log\left(\frac{e^{-\theta t}-1}{e^{-\theta}-1}\right)$, with $\theta > 0$.

17.2.2 r-Concavity and r-Decreasing Density Functions

We recall the definition of an r-concave function (see, e.g., Prékopa (1995)):

Definition 3 A function $f : \mathbb{R}^s \rightarrow (0, \infty)$ is called r-concave for some $r \in [-\infty, \infty]$, if

$$f(\lambda x+(1-\lambda)y) \geq [\lambda f^r(x)+(1-\lambda)f^r(y)]^{1/r} \quad \forall x, y \in \mathbb{R}^s,\ \forall \lambda \in [0, 1]. \quad (17.7)$$

In this definition, the cases $r \in \{-\infty, 0, \infty\}$ are to be interpreted by continuity. In particular, 1-concavity amounts to classical concavity, 0-concavity equals log concavity (i.e., concavity of $\log f$), and $-\infty$-concavity identifies quasi-concavity (this means that the right-hand side of the inequality in the definition becomes $\min\{f(x), f(y)\}$). We recall, that an equivalent way to express log concavity is the inequality

$$f(\lambda x + (1-\lambda)y) \geq f^{\lambda}(x) f^{1-\lambda}(y) \quad \forall x, y \in \mathbb{R}^s,\ \forall \lambda \in [0, 1]. \qquad (17.8)$$

For $r < 0$, one may raise (17.7) to the negative power r and recognize, upon reversing the inequality sign, that this reduces to convexity of f^r. If f is r^*-concave, then f is r-concave for all $r \leq r^*$. In particular, concavity implies log concavity. We shall be mainly interested in the case $r \leq 1$.

The following property will be crucial in the context of this chapter:

Definition 4 We call a function $f : \mathbb{R} \rightarrow \mathbb{R}$ **r-decreasing** for some $r \in \mathbb{R}$, if it is continuous on $(0, \infty)$ and if there exists some $t^* > 0$ such that the function $t^r f(t)$ is strictly decreasing for all $t > t^*$.

Evidently, 0-decreasing means strictly decreasing in the classical sense. If f is a nonnegative function like the density of some random variable, then r-decreasing implies r'-decreasing whenever $r' \leq r$. Therefore, one gets narrower families of r-decreasing density functions with $r \rightarrow \infty$. If f is not just continuous on $(0, \infty)$ but happens even to be differentiable there, then the property of being r-decreasing amounts to the condition

$$tf'(t) + rf(t) < 0 \quad \text{for all } t > t^*.$$

It turns out that most one-dimensional distribution functions of practical use share the property of being r-decreasing for any $r > 0$ (an exception being the density of the Cauchy distribution which is r-decreasing only for $r < 2$). Table 17.1,

Table 17.1 t_r^*- values in the definition of r-decreasing densities for a set of common distributions

Law	Density	t_r^*
Normal	$\frac{1}{\sqrt{2\pi}\sigma}\exp\left(-\frac{(t-\mu)^2}{2\sigma^2}\right)$	$\frac{\mu+\sqrt{\mu^2+4r\sigma^2}}{2}$
Exponential	$\lambda\exp(-\lambda t) \quad (t>0)$	$\frac{r}{\lambda}$
Weibull	$abt^{b-1}\exp\left(-at^b\right) \quad (t>0)$	$\left(\frac{b+r-1}{ab}\right)^{1/b}$
Gamma	$\frac{b^a}{\Gamma(a)}\exp(-bt)\,t^{a-1} \quad (t>0)$	$\frac{a+r-1}{b}$
χ	$\frac{1}{2^{n/2-1}\Gamma(n/2)}t^{n-1}\exp\left(-\frac{t^2}{2}\right) \quad (t>0)$	$\sqrt{n+r-1}$
χ^2	$\frac{1}{2^{n/2}\Gamma(n/2)}t^{n/2-1}\exp\left(-\frac{t}{2}\right) \quad (t>0)$	$n+2r-2$
Log normal	$\frac{1}{\sqrt{2\pi}\sigma t}\exp\left(-\frac{(\log t-\mu)^2}{2\sigma^2}\right) \quad (t>0)$	$e^{\mu+(r-1)\sigma^2}$
Maxwell	$\frac{2t^2}{\sqrt{2\pi}\sigma^3}\exp\left(-\frac{t^2}{2\sigma^2}\right) \quad (t>0)$	$\sigma\sqrt{r+2}$
Rayleigh	$\frac{2t}{\lambda}\exp\left(-\frac{t^2}{\lambda}\right) \quad (t>0)$	$\sqrt{\frac{r+1}{2}\lambda}$

borrowed from Henrion and Strugarek (2008), for the readers convenience, compiles a collection of one-dimensional densities all of which are r-decreasing for any $r>0$ along with the respective t^*-values from Definition 4 labeled as t_r^* to emphasize their dependence on r:

We shall need as an auxiliary result the following lemma whose proof is easy and can be found in Henrion and Strugarek (2008, lemma 3.1):

Lemma 1 *Let $F:\mathbb{R}\to[0,1]$ be a distribution function with $(r+1)$-decreasing density f for some $r>0$. Then, the function $z\mapsto F(z^{-1/r})$ is concave on $\left(0,(t^*)^{-r}\right)$, where t^* refers to Definition 4. Moreover, $F(t)<1$ for all $t\in\mathbb{R}$.*

17.3 Main Result

The purpose of our analysis is to investigate convexity of the set

$$M(p)=\left\{x\in\mathbb{R}^n\,|C\left((F_1(g_1(x)),\ldots,F_m(g_m(x))\right)\geq p\right\}. \tag{17.9}$$

Here, g is as in (17.3), the F_i are the one-dimensional marginals of some m-dimensional distribution function F, and C is a copula. If C is the exact copula associated with F via (17.4) or (17.5), respectively, then the feasible set (17.9) coincides with the chance constraint (17.3). In general, we rather have in mind the idea that F is unknown and C serves its approximation. Then, (17.9) can be understood as a chance constraint similar to (17.3) or (17.2) but with the distribution function F occurring there replaced by the distribution function $C\circ H$, where $H_i(x):=F_i(x_i)$ for $i=1,\ldots,m$. We introduce the following property of a copula which will be crucial for our convexity result:

Definition 5 Let $q \in (0,1)^m$. A copula $C : [0,1]^m \to [0,1]$ is called log exp-concave on $\Pi_{i=1}^m [q_i, 1)$ if the mapping $\tilde{C} : \mathbb{R}^m \to \mathbb{R}$ defined by

$$\tilde{C}(u) = \log C\left(e^{u_1}, \ldots, e^{u_m}\right), \tag{17.10}$$

is concave on $\Pi_{i=1}^m [\log q_i, 0)$.

The following statement of our main result makes reference to Definitions 3, 4, and 5.

Theorem 1 *In (17.9), we make the following assumptions for $i = 1, \ldots, m$:*

1. *There exist $r_i > 0$ such that the components g_i are $(-r_i)$-concave.*
2. *The marginal distribution functions F_i admit $(r_i + 1)$-decreasing densities f_i.*
3. *The copula C is* log exp *concave on $\Pi_{i=1}^m [F_i(t_i^*), 1)$, where the t_i^* refer to Definition 4 in the context of f_i being $(r_i + 1)$-decreasing.*

Then, $M(p)$ in (17.9) is convex for all $p > p^* := \max\{F_i(t_i^*) | 1 \leq i \leq m\}$.

Proof Let $p > p^*$, $\lambda \in [0,1]$, and $x, y \in M(p)$ be arbitrary. We have to show that $\lambda x + (1-\lambda)y \in M(p)$. We put

$$q_i^x := F_i(g_i(x)) < 1, \quad q_i^y := F_i(g_i(y)) < 1 \quad (i = 1, \ldots, m), \tag{17.11}$$

where the strict inequalities rely on the second statement of Lemma 1. In particular, by (17.11), the inclusions $x, y \in M(p)$ mean that

$$C(q_1^x, \ldots, q_m^x) \geq p, \quad C(q_1^y, \ldots, q_m^y) \geq p. \tag{17.12}$$

Now, (17.11), (17.6), (17.12) and the definition of p^* entail that

$$1 > q_i^x \geq p > F_i(t_i^*) \geq 0, \quad 1 > q_i^y \geq p > F_i(t_i^*) \geq 0 \quad (i = 1, \ldots, m). \tag{17.13}$$

For $\tau \in [0,1]$, we denote the τ-quantile of F_i by

$$\tilde{F}_i(\tau) := \inf\{z \in \mathbb{R} | F_i(z) \geq \tau\}.$$

Note that, for $\tau \in (0,1)$, $\tilde{F}_i(\tau)$ is a real number. Having a density, by assumption 2, the F_i are continuous distribution functions. As a consequence, the quantile functions $\tilde{F}_i(\tau)$ satisfy the implication

$$q > F_i(z) \Longrightarrow \tilde{F}_i(q) > z \quad \forall q \in (0,1) \, \forall z \in \mathbb{R}.$$

Now, (17.11) and (17.13) provide the relations

$$g_i(x) \geq \tilde{F}_i(q_i^x) > t_i^* > 0, \quad g_i(y) \geq \tilde{F}_i(q_i^y) > t_i^* > 0 \quad (i = 1, \ldots, m) \tag{17.14}$$

(note that $t_i^* > 0$ by Definition 4). In particular, for all $i = 1, \ldots, m$, it holds that

$$\left[\min\{\tilde{F}_i^{-r_i}(q_i^x), \tilde{F}_i^{-r_i}(q_i^y)\}, \max\{\tilde{F}_i^{-r_i}(q_i^x), \tilde{F}_i^{-r_i}(q_i^y)\}\right] \subseteq \left(0, (t_i^*)^{-r_i}\right). \quad (17.15)$$

Along with assumption 1, (17.14) yields for $i = 1, \ldots, m$:

$$\begin{aligned} g_i(\lambda x + (1-\lambda)y) &\geq \left(\lambda g_i^{-r_i}(x) + (1-\lambda) g_i^{-r_i}(y)\right)^{-1/r_i} \\ &\geq \left(\lambda \tilde{F}_i^{-r_i}(q_i^x) + (1-\lambda)\tilde{F}_i^{-r_i}(q_i^y)\right)^{-1/r_i}. \end{aligned} \quad (17.16)$$

The monotonicity of distribution functions allows to continue by

$$F_i(g_i(\lambda x + (1-\lambda)y)) \geq F_i\left(\left(\lambda \tilde{F}_i^{-r_i}(q_i^x) + (1-\lambda)\tilde{F}_i^{-r_i}(q_i^y)\right)^{-1/r_i}\right) \quad (17.17)$$
$$(i = 1, \ldots, m).$$

Owing to assumption 2, Lemma 1 guarantees that the functions $z \mapsto F_i(z^{-1/r_i})$ are concave on $\left(0, (t_i^*)^{-r_i}\right)$. In particular, these functions are log concave on the indicated interval, as this is a weaker property than concavity (see Section 17.2.2). By virtue of (17.15) and (17.8), this allows to continue (17.17) as

$$F_i(g_i(\lambda x + (1-\lambda)y)) \geq \left[F_i\left(\tilde{F}_i(q_i^x)\right)\right]^{\lambda} \left[F_i\left(\tilde{F}_i(q_i^y)\right)\right]^{1-\lambda} \quad (i = 1, \ldots, m).$$

Exploiting the fact that the F_i as continuous distribution functions satisfy the relation $F_i(\tilde{F}_i(q)) = q$ for all $q \in (0,1)$, and recalling that $q_i^x, q_i^y \in (0,1)$ by (17.13), we may deduce that

$$F_i(g_i(\lambda x + (1-\lambda)y)) \geq \left[q_i^x\right]^{\lambda} \left[q_i^y\right]^{1-\lambda} \quad (i = 1, \ldots, m).$$

Since the copula C as a distribution function is increasing w.r.t. the partial order in $\mathbb{R}^m$, we obtain that

$$\begin{aligned} &C\left(F_1(g_1(\lambda x + (1-\lambda)y)), \ldots, F_m(g_m(\lambda x + (1-\lambda)y))\right) \geq \\ &\qquad C\left(\left[q_1^x\right]^{\lambda}\left[q_1^y\right]^{1-\lambda}, \ldots, \left[q_m^x\right]^{\lambda}\left[q_m^y\right]^{1-\lambda}\right) \end{aligned} \quad (17.18)$$

According to assumption 3 and (17.13), we have

$$\log C\left(\left[q_1^x\right]^\lambda\left[q_1^y\right]^{1-\lambda},\dots,\left[q_m^x\right]^\lambda\left[q_m^y\right]^{1-\lambda}\right) =$$

$$\log C\left(\exp\left(\log\left(\left[q_1^x\right]^\lambda\left[q_1^y\right]^{1-\lambda}\right)\right),\dots,\exp\left(\log\left(\left[q_m^x\right]^\lambda\left[q_m^y\right]^{1-\lambda}\right)\right)\right) =$$

$$\log C\left(\exp\left(\lambda\log\left[q_1^x\right]+(1-\lambda)\log\left[q_1^y\right]\right),\dots,\exp\left(\lambda\log\left[q_m^x\right]+(1-\lambda)\log\left[q_m^y\right]\right)\right) \geq$$

$$\lambda\log C(q_1^x,\dots,q_m^x)+(1-\lambda)\log C(q_1^y,\dots,q_m^y) \geq$$

$$\lambda\log p+(1-\lambda)\log p =$$

$$\log p,$$

where the last inequality follows from (17.12). Combining this with (17.18) provides

$$C\left(F_1(g_1(\lambda x+(1-\lambda)y)),\dots,F_m(g_m(\lambda x+(1-\lambda)y))\right) \geq p.$$

Referring to (17.9), this shows that $\lambda x+(1-\lambda)y \in M(p)$.

Remark 1 The critical probability level p^* beyond which convexity can be guaranteed in Theorem 1 is completely independent of the mapping g; it just depends on the distribution functions F_i. In other words, for given distribution functions F_i, the convexity of $M(p)$ in (17.2) for $p > p^*$ can be guaranteed for a whole class of mappings g satisfying the first assumption of Theorem 1. Therefore, it should come at no surprise that, for specific mappings g even smaller critical values p^* may apply.

17.4 Log Exp-Concavity of Copulae

Having a look to the assumptions of Theorem 1, the first one is easily checked from a given explicit formula for the mapping g, while the second one can be verified, for instance, via Table 17.1. Thus, the third assumption, namely, log exp-concavity of the copula, remains the key for applying the theorem. The next proposition provides some examples for log exp-concave copulae:

Proposition 2 *The independent, the maximum, and the Gumbel copulae are log exp-concave on* $\left[q,1\right)^m$ *for any* $q>0$.

Proof For the independent copula we have that

$$\log C\left(e^{u_1},\dots,e^{u_m}\right)=\log\prod_{i=1}^m e^{u_i}=\sum_{i=1}^m u_i$$

and for the maximum copula it holds that

$$\log C\left(e^{u_1},\ldots,e^{u_m}\right)=\log \min_{1\le i\le m} e^{u_i} = \min_{1\le i\le m} \log e^{u_i} = \min_{1\le i\le m} u_i .$$

Both functions are concave on $(-\infty,0)^m$, hence the assertion follows. For the Gumbel copula (see Example 1), the strict generator $\psi(t):=(-\log t)^\theta$ (for some $\theta\ge 1$) implies that $\psi^{-1}(s)=\exp\left(-s^{1/\theta}\right)$, whence, for $u_i\in(-\infty,0)$,

$$\begin{aligned}\log C\left(e^{u_1},\ldots,e^{u_m}\right)&=\log\psi^{-1}\left(\sum_{i=1}^m \psi\left(e^{u_i}\right)\right)=\log\psi^{-1}\left(\sum_{i=1}^m (-u_i)^\theta\right)\\&=\log\exp\left(-\left(\sum_{i=1}^m (-u_i)^\theta\right)^{1/\theta}\right)=-\left(\sum_{i=1}^m |u_i|^\theta\right)^{1/\theta}\\&=-\|u\|_\theta .\end{aligned}$$

Since $\|\cdot\|_\theta$ is a norm for $\theta\ge 1$ and since a norm is convex, it follows that $-\|\cdot\|_\theta$ is concave on $(-\infty,0)^m$. Thus, the Gumbel copula too is log exp-concave on $[q,1)^m$ for any $q>0$.

Contrary to the previous positive examples, we have the following negative case:

Example 2 The Clayton copula is not log exp-concave on any domain $\prod_{i=1}^m [q_i,1)$ where $q\in(0,1)^m$. Indeed, taking the generator $\psi(t):=\theta^{-1}\left(t^{-\theta}-1\right)$, we see that $\psi^{-1}(s)=(\theta s+1)^{-1/\theta}$ and calculate

$$\begin{aligned}\log C\left(e^{u_1},\ldots,e^{u_m}\right)&=\log\psi^{-1}\left(\sum_{i=1}^m \psi\left(e^{u_i}\right)\right)=\log\psi^{-1}\left(\sum_{i=1}^m \theta^{-1}\left(e^{-\theta u_i}-1\right)\right)\\&=-\theta^{-1}\log\left(1-m+\sum_{i=1}^m\left(e^{-\theta u_i}\right)\right).\end{aligned}$$

Now, for any $t<0$, define

$$\varphi(t):=\log C\left(e^t,\ldots,e^t\right)=-\theta^{-1}\log\left(1-m+me^{-\theta t}\right).$$

Its second derivative calculates as

$$\varphi''(t)=\frac{\theta m(m-1)e^{\theta t}}{\left(m-(m-1)e^{\theta t}\right)^2}.$$

Now, if C was log exp-concave on any $\prod_{i=1}^m [q_i,1)$ where $q\in(0,1)^m$, then, in particular, φ would be concave on the interval $[\tau,0)$, where $\tau:=\max_{i=1,\ldots,m}\log q_i<0$. This, however, contradicts the fact that $\varphi''(t)>0$ for any t.

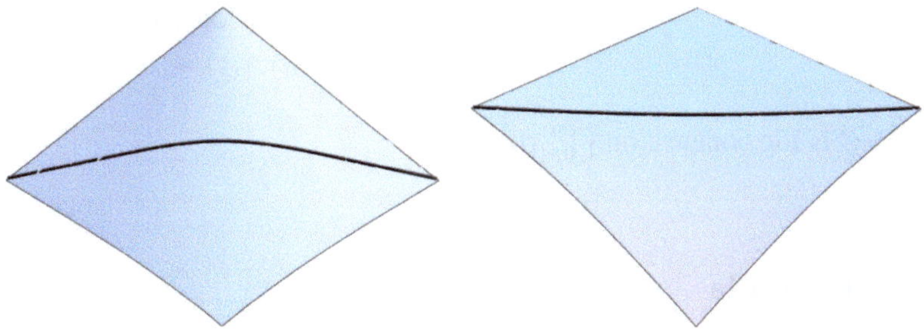

Fig. 17.1 Plot of the function $\log C(e^x, e^y)$ for the bivariate Gaussian copula C in case of positive (*left*) and negative (*right*) correlations

Concerning the Gaussian copula it seems to be difficult to check log exp-concavity even in the bivariate case. At least, numerical evidence shows that the bivariate Gaussian copula is log exp-concave for non-negative correlation coefficient but fails to be so for negative correlation coefficient. This fact is illustrated in Fig. 17.1, where the thick curve in the right diagram represents a convex (rather than concave) piece of the function $\log C(e^x, e^y)$.

The next proposition provides a necessary condition for a copula to be log exp-concave. It relates those copulae with the well-known family of log-concave distribution functions:

Proposition 3 *If a copula C is log exp-concave on $\prod_{i=1}^m [q_i, 1)$, then it is log concave on the same domain..*

Proof By definition of log exp-concavity, we have that

$$\log C(\exp(\lambda y_1 + (1-\lambda) z_1), \ldots, \exp(\lambda y_m + (1-\lambda) z_m)) \geq$$
$$\lambda \log C(\exp y_1, \ldots, \exp y_m) + (1-\lambda) \log C(\exp z_1, \ldots, \exp z_m) \quad (17.19)$$

for all $\lambda \in [0, 1]$ and all $y, z \in \prod_{i=1}^m [\log q_i, 0)$. Now, let $\lambda \in [0, 1]$ and $u, v \in \prod_{i=1}^m [q_i, 1)$ be arbitrary. By concavity of log and monotonicity of exp, we have that

$$\exp(\log(\lambda u_i + (1-\lambda) v_i)) \geq \exp(\lambda \log u_i + (1-\lambda) \log v_i) \quad (i = 1, \ldots, m). \quad (17.20)$$

Since C as a distribution function is nondecreasing with respect to the partial order of $\mathbb{R}^m$, the same holds true for $\log C$ by monotonicity of log. Consequently, first (17.20) and then (17.19) yield that

$$\log C(\lambda u + (1-\lambda) v) =$$
$$\log C(\exp(\log(\lambda u_1 + (1-\lambda) v_1)), \ldots, \exp(\log(\lambda u_m + (1-\lambda) v_m))) \geq$$
$$\log C(\exp(\lambda \log u_1 + (1-\lambda) \log v_1), \ldots, \exp(\lambda \log u_m + (1-\lambda) \log v_m)) \geq$$

$$\lambda \log C(u_1, \ldots, u_m) + (1-\lambda) \log C(v_1, \ldots, v_m) = \lambda \log C(u) + (1-\lambda) \log C(v).$$

Hence, C is log concave on $\prod_{i=1}^{m}[q_i, 1)$.

17.5 An Example

A small numerical example shall illustrate the application of Theorem 1. Consider a chance constraint of type (17.2):

$$\mathbb{P}\left(\xi_1 \leq x_1^{-3/4}, \xi_2 \leq x_2^{-1/4}\right) \geq p.$$

Assume that ξ_1 has an exponential distribution with parameter $\lambda = 1$ and ξ_2 has a Maxwell distribution with parameter $\sigma = 1$ (see Table 17.1). Assume that the joint distribution of (ξ_1, ξ_2) is not known and is approximated by means of a Gumbel copula C with parameter $\theta = 1$. Then, the feasible set defined by the chance constraint above is replaced by set (17.9) with F_1, F_2 being the cumulative distribution functions of the exponential and Maxwell distribution, respectively, and with $g_1(x_1, x_2) = x_1^{-3/4}$, $g_2(x_1, x_2) = x_2^{-1/4}$. When checking the assumptions of Theorem 1, we observe first, that the third one is satisfied due to Proposition 2. Concerning the first assumption, note that the components g_1 and g_2 fail to be concave (they are actually convex). According to Section 17.2.2, the g_i would be at least r-concave, if for some $r < 0$ the function g_i^r was convex. In our example, we may choose $r = -4/3$ for g_1 and $r = -4$ for g_2. Thus, we have checked the first assumption of the Theorem with $r_1 = 4/3$ and $r_2 = 4$. It remains to verify the second assumption. This amounts to require the density of the exponential distribution to be $7/3$-decreasing and the density of the Maxwell distribution to be 5-decreasing. According to Table 17.1 and with the given parameters of these distributions, these properties hold true in the sense of Definition 4 with a t^*-value of $t_1^* = (r_1+1)/\lambda = 7/3$ in case of the exponential distribution and with a t^*-value of $t_2^* = \sigma\sqrt{(r_2+1)+2} = \sqrt{7}$ in case of the Maxwell distribution. Now, Theorem 1 guarantees convexity of the feasible set (17.9), whenever $p > p^*$, where

$$p^* = \max\left\{F_1(7/3), F_2\left(\sqrt{7}\right)\right\}.$$

Using easily accessible evaluation functions for the cumulative distribution functions of the exponential and Maxwell distribution, we calculate $F_1(7/3) \approx 0.903$ and $F_2\left(\sqrt{7}\right) \approx 0.928$. Hence, $M(p)$ in (17.9) is definitely convex for probability levels larger than 0.928.

Figure 17.2 illustrates the resulting feasible set for different probability levels (feasible points lie below the corresponding level lines). By visual analysis,

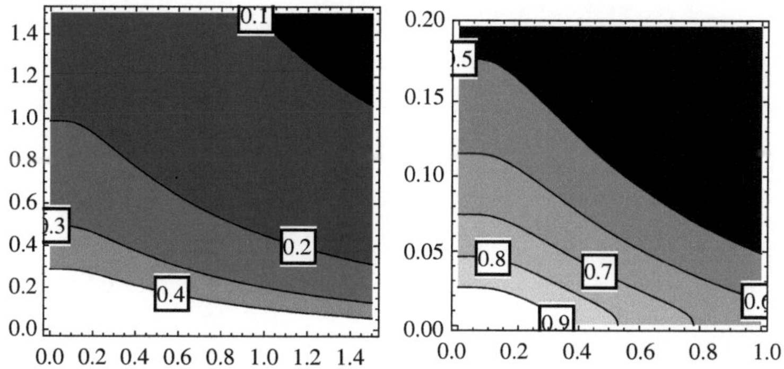

Fig. 17.2 Illustration of the feasible set $M(p)$ in (17.9) for different probability levels in a numerical example. Convexity appears for levels of approximately 0.8 and larger

convexity holds true for levels p larger than approximately 0.8. Not surprisingly, our theoretical result is more conservative. One reason for the gap is explained in Remark 1.

Acknowledgement The work of the first author was supported by the DFG Research Center MATHEON *Mathematics for key technologies* in Berlin.

References

Henrion, R., Strugarek, C.: Convexity of chance constraints with independent random variables. Comput. Optim. Appl., **41**, 263–276 (2008)

Klaassen, C.A.J., Wellner, J.A.: Efficient estimation in the bivariate normal copula model: Normal margins are least favourable. Bernoulli, **3**, 55–77 (1997)

Nelsen, R.B.: An Introduction to Copulas. Springer, New York, NY (2006)

Prékopa, A.: Stochastic Programming. Kluwer, Dordrecht (1995)

Prékopa, A.: Probabilistic programming. In: Ruszczynski, A. and Shapiro, A. (eds.): Stochastic Programming. Hamdbooks in Operations Research and Management Science, Vol. 10, pp. 267–351. Elsevier, Amsterdam (2003)

Chapter 18
Portfolio Choice Models Based on Second-Order Stochastic Dominance Measures: An Overview and a Computational Study

Csaba I. Fábián, Gautam Mitra, Diana Roman, Victor Zverovich, Tibor Vajnai, Edit Csizmás, and Olga Papp

Abstract In this chapter we present an overview of second-order stochastic dominance-based models with a focus on those using dominance measures. In terms of portfolio policy, the aim is to find a portfolio whose return distribution dominates the index distribution to the largest possible extent. We compare two approaches, the unscaled model of Roman et al. (Mathematical Programming Series B 108: 541–569, 2006) and the scaled model of Fabian et al. (Quantitative Finance 2010). We constructed optimal portfolios using representations of the future asset returns given by historical data on the one hand, and scenarios generated by geometric Brownian motion on the other hand. In the latter case, the parameters of the GBM were obtained from the historical data. Our test data consisted of stock returns from the FTSE 100 basket, together with the index returns. Part of the data were reserved for out-of-sample tests. We examined the return distributions belonging to the respective optimal portfolios of the unscaled and the scaled problems. The unscaled model focuses on the worst cases and hence enhances safety. We found that the performance of the unscaled model is improved by using scenario generators. On the other hand, the scaled model replicates the shape of the index distribution. Scenario generation had little effect on the scaled model. We also compared the shapes of the histograms belonging to corresponding pairs of in-sample and out-of-sample tests and observed a remarkable robustness in both models. We think these features make these dominance measures good alternatives for classic risk measures in certain applications, including certain multistage ones. We mention two candidate applications.

Keywords Second-order stochastic dominance · Portfolio optimization · Scenario generation

C.I. Fábián (✉)
Institute of Informatics, Kecskemét College, Kecskemét, Hungary; Department of OR, Eötvös Loránd University, Budapest, Hungary
e-mail: fabian.csaba@gamf.kefo.hu

M. Bertocchi et al. (eds.), *Stochastic Optimization Methods in Finance and Energy*, International Series in Operations Research & Management Science 163,
DOI 10.1007/978-1-4419-9586-5_18,

18.1 Introduction

The importance of second-order stochastic dominance (SSD) as a choice criterion in portfolio selection as well as the difficulty in applying it in practice has been widely recognised (Hadar and Russel 1969; Whitmore and Findlay 1978; Kroll and Levy 1980; Ogryczak and Ruszczyński 2001, 2002). SSD is a meaningful choice criterion, due to its relation to risk averse behaviour which is the general assumption about investment behaviour. The computational difficulty of the SSD-based models arises from the fact that finding the set of SSD-efficient portfolios is a model with a continuum of objectives.

Only recently, SSD-based portfolio models have been proposed: Dentcheva and Ruszczyński (2003, 2006) introduce *SSD constraints* and explore mathematical properties of the resulting optimisation problems. Roman et al. (2006) introduce a *dominance measure* and formulate portfolio choice problems that maximise this dominance measure. This approach requires discrete finite distributions with equally probable outcomes. Fábián et al. (2010) propose a scaled version of this dominance measure. The scaled measure has attractive properties: its negative is a *convex risk measure*. Moreover, the dominance-measure approach is extended to any discrete finite distribution. Meantime effective methods were developed for the solution of SSD-based problems (Rudolf and Ruszczyński 2008; Luedtke 2008; Fábián et al. 2009).

In this chapter we present an overview of SSD-based models with a focus on the above-mentioned dominance measures. We formulate dominance maximisation problems and describe solution methods. Moreover we present a computational study. We examine the return distributions belonging to the optimal portfolios of dominance maximisation problems. The aim is to explore characteristics of these return distributions and also to test whether the performance of these models is improved by using scenario generators. The studies known to us do not report such tests: Dentcheva and Ruszczyński (2003, 2006) and Roman et al. (2006) use historical data. Fábián et al. (2010) use scenario generation based on geometric Brownian motion, for testing the scale-up properties of the solution methods proposed. The latter paper contains an analysis of the (in-sample) return distributions of the solutions obtained, but no comparison is made with solutions obtained using historical data.

The present computational study demonstrates the robust nature of the portfolio choice models based on these dominance measures. We think these dominance measures are good alternatives for classic risk measures in many applications, including certain multistage ones.

This chapter is organised as follows. Section 18.2 presents on overview of SSD-based models. First, we discuss the concept of second-order stochastic dominance (SSD) and its application to portfolio choice. Then we describe the above-mentioned SSD-based portfolio choice models, with a focus on those models that use dominance measures. The dominance-maximisation model of Roman et al. (2006) will be called *unscaled model* in this chapter, and the model of Fábián et al. (2010) will be called *scaled model*. We describe the connection of the scaled model with convex risk minimisation. Ways of formulating SSD-based portfolio choice models are discussed in Section 18.3. We formulate the unscaled problem and the scaled problem as dominance maximisation problems, and describe a cutting-plane method for the solution of these problems.

Sections 18.4, 18.5, and 18.6 contain a computational study. In this study, we constructed optimal portfolios representing future asset returns by historical data on the one hand and by scenarios generated using geometric Brownian motion (GBM) on the other hand. In the latter case, the parameters of the GBM were obtained from the historical data using a robust estimation method. We also performed out-of-sample tests using historical data set aside for this purpose. Section 18.4 presents the set-up of the computational study, including a description of the scenario sets used in constructing and testing portfolios. We define the portfolio choice problems solved and introduce notation for the histograms used to characterise the return distributions of the different portfolios when tested against the different scenario sets. The implementation issues are detailed in Section 18.5. Section 18.6 presents the test results. Conclusions are drawn in Section 18.7.

18.2 Portfolio Choice Models Based on the SSD Criteria

This section is an overview of SSD-based models with a focus on those models that use dominance measures. In Section 18.2.1 we discuss the concept of second-order stochastic dominance (SSD) and its application to portfolio choice. In Section 18.2.2 we describe SSD-based portfolio choice models. The relationship between the scaled model and convex risk minimisation is reviewed in Section 18.2.3.

18.2.1 Second-Order Stochastic Dominance

Let R and R' denote random returns. For a precise description, let $(\Omega, \mathcal{M}, P)$ be an appropriate probability space. (The set Ω is equipped with the probability measure P, the field of measurable sets being $\mathcal{M}$.) We consider integrable random variables, formally, let $R, R' \in \mathcal{L}^1 = \mathcal{L}^1(\Omega, \mathcal{M}, P)$.

Second-order stochastic dominance (SSD) is defined by the following equivalent conditions:

(a) $\mathrm{E}(U(R)) \geq \mathrm{E}(U(R'))$ holds for any nondecreasing and concave utility function U for which these expected values exist and are finite.
(b) $\mathrm{E}([t - R]_+) \leq \mathrm{E}([t - R']_+)$ holds for each $t \in \mathbb{R}$.
(c) $\mathrm{Tail}_\alpha(R) \geq \mathrm{Tail}_\alpha(R')$ holds for each $0 < \alpha \leq 1$, where $\mathrm{Tail}_\alpha(R)$ denotes the unconditional expectation of the least $\alpha * 100\%$ of the outcomes of R.

Concavity of the utility function in (a) characterises risk-averse behaviour. For the equivalence of (a) and (b), see for example Whitmore and Findlay (1978) (theorem 2.2 on page 65). The equivalence of (b) and (c) is shown in Ogryczak and Ruszczyński (2002): the authors consider $\mathrm{Tail}_\alpha(R)$ as a function of α, and $E\left([t-R]_+\right)$ as a function of t; and observe that these functions are convex conjugates. In general, SSD relations can be described with a continuum of constraints.

If (a) or (b) or (c) above hold, then we say that R dominates R' with respect to SSD and use the notation $R \succeq_{\mathrm{SSD}} R'$. The corresponding strict dominance relation is defined in the usual way: $R \succ_{\mathrm{SSD}} R'$ means that $R \succeq_{\mathrm{SSD}} R'$ and $R' \not\succeq_{\mathrm{SSD}} R$.

In this chapter we deal with portfolio returns. Let n denote the number of the assets into which we may invest at the beginning of a fixed time period. A portfolio $\boldsymbol{x} = (x_1, \ldots x_n) \in \mathbb{R}^n$ represents the proportions of the portfolio value invested in the different assets.

Let the n-dimensional random vector $\boldsymbol{R}$ denote the returns of the different assets at the end of the investment period. We assume that the components of $\boldsymbol{R}$ belong to $\mathcal{L}^1$. It is usual to consider the distribution of $\boldsymbol{R}$ as discrete, described by the realisations under various scenarios. The random return of portfolio $\boldsymbol{x}$ is $R_{\boldsymbol{x}} := x_1 R_1 + \ldots + x_n R_n$.

Let $X \subset \mathbb{R}^n$ denote the set of the feasible portfolios. In this chapter we simply consider $X = \{\boldsymbol{x} \in \mathbb{R}^n \mid \boldsymbol{x} \geq \boldsymbol{0},\ x_1 + \cdots + x_n = 1\}$. A portfolio $\boldsymbol{x}^\star$ is said to be SSD efficient if there is no feasible portfolio $\boldsymbol{x} \in X$ such that $R_{\boldsymbol{x}} \succ_{\mathrm{SSD}} R_{\boldsymbol{x}^\star}$.

18.2.2 Portfolio Choice Models

The SSD-based models reviewed here assume that a reference random return $\widehat{R}$, with a known (discrete) distribution, is available; $\widehat{R}$ could be for example, the return of a stock index or of a benchmark portfolio.

Dentcheva and Ruszczyński (2006) propose an SSD-constrained portfolio optimisation model:

$$\begin{aligned} &\max\ f(\boldsymbol{x}) \\ &\text{such that}\quad \boldsymbol{x} \in X, \\ &R_{\boldsymbol{x}} \succeq_{\mathrm{SSD}} \widehat{R}, \end{aligned} \tag{18.1}$$

where f is a concave function. In particular, they consider $f(\boldsymbol{x}) = \mathrm{E}(R_{\boldsymbol{x}})$. They formulate the problem using criterion (b) (Section 18.2.1) and prove that, in case of finite discrete distributions, the SSD relation can be characterised by a finite system of inequalities from those in (b). For finite discrete distributions the authors also develop a duality theory in which dual objects are nondecreasing concave utility functions.

Roman et al. (2006) use criterion (c) (Section 18.2.1). They assume finite discrete distributions with equally probable outcomes and prove that, in this case, the SSD relation can be characterised by a finite system of inequalities, namely, $R_{\boldsymbol{x}} \succeq_{\mathrm{SSD}} \widehat{R}$ is equivalent to

$$\mathrm{Tail}_{\frac{i}{S}}\left(R_{\boldsymbol{x}}\right) \;\geq\; \mathrm{Tail}_{\frac{i}{S}}\left(\widehat{R}\right) \quad (i = 1, \ldots, S), \tag{18.2}$$

where S is the number of the (equally probable) scenarios. A portfolio $\boldsymbol{x}$ is chosen whose return distribution $R_{\boldsymbol{x}}$ comes close to, or emulates, the reference return $\widehat{R}$ in a uniform sense. Uniformity is meant in terms of differences among tails; the 'worst' tail difference $\vartheta = \min_{i=1...S}\left(\mathrm{Tail}_{i/S}\left(R_{\boldsymbol{x}}\right) - \mathrm{Tail}_{i/S}\left(\widehat{R}\right)\right)$ is maximised:

$$\begin{aligned} &\max \ \vartheta \\ &\text{such that} \quad \vartheta \in \mathbb{R}, \quad \boldsymbol{x} \in X, \\ &\mathrm{Tail}_{\frac{i}{S}}\left(R_{\boldsymbol{x}}\right) \;\geq\; \mathrm{Tail}_{\frac{i}{S}}\left(\widehat{R}\right) + \vartheta \quad (i = 1, \ldots, S). \end{aligned} \tag{18.3}$$

This can be considered a multi-objective model whose Pareto optimal solutions are SSD-efficient portfolios. (The origin of this multi-objective formulation can be traced back to Ogryczak (2000, 2002)). If the reference distribution is not SSD efficient (which is often the case) then the model improves on it until SSD efficiency is reached. The authors implemented the model outlined above and made extensive testing on problems with 76 real-world assets using 132 possible realisations of their joint return rates (historical data). Powerful modelling capabilities were demonstrated by in-sample and out-of-sample analysis of the return distributions of the optimal portfolios.

Fábián et al. (2010) proposed an enhanced SDD model in the form

$$\begin{aligned} &\max \ \vartheta \\ &\text{such that} \quad \vartheta \in \mathbb{R}, \quad \boldsymbol{x} \in X, \\ &R_{\boldsymbol{x}} \;\succeq_{\mathrm{SSD}}\; \widehat{R} + \vartheta. \end{aligned} \tag{18.4}$$

The relation $R_{\boldsymbol{x}} \succeq_{\mathrm{SSD}} \widehat{R} + \vartheta$ means that we prefer portfolio return $R_{\boldsymbol{x}}$ to the combined return of the stock index $\widehat{R}$ and a 'certain' return ϑ from a riskless asset (usually cash).

Assuming finite discrete distributions with S equally probable outcomes, the constraint $R_{\boldsymbol{x}} \succeq_{\mathrm{SSD}} \widehat{R} + \vartheta$ can be formulated as $\mathrm{Tail}_{i/S}\left(R_{\boldsymbol{x}}\right) \geq \mathrm{Tail}_{i/S}\left(\widehat{R} + \vartheta\right)$ $(i = 1, \ldots, S)$. Obviously we have $\mathrm{Tail}_{i/S}\left(\widehat{R} + \vartheta\right) = \mathrm{Tail}_{i/S}\left(\widehat{R}\right) + \frac{i}{S}\,\vartheta (i = 1, \ldots, S)$ with $\vartheta \in \mathbb{R}$. Hence (18.4) can be equivalently formulated as

$$\begin{aligned} &\max \ \vartheta \\ &\text{such that} \quad \vartheta \in \mathbb{R}, \quad \boldsymbol{x} \in X, \\ &\mathrm{Tail}_{\frac{i}{S}}\left(R_{\boldsymbol{x}}\right) \;\geq\; \mathrm{Tail}_{\frac{i}{S}}\left(\widehat{R}\right) + \frac{i}{S}\,\vartheta \quad (i = 1, \ldots, S). \end{aligned} \tag{18.5}$$

The difference between the above model and the original multi-objective model of Roman et al. (18.3) is that the tails are scaled in (18.5). In this chapter, we refer to (18.3) as unscaled model, and to (18.4) or (18.5) as scaled model.

The quantity ϑ in (18.4) measures the preferability of the portfolio return $R_{\boldsymbol{x}}$ relative to the reference return $\widehat{R}$. An opposite measure is

$$\widehat{\rho}(R) := \min\left\{\varrho \in \mathbb{R} \,\middle|\, R + \varrho \succeq_{\text{SSD}} \widehat{R}\right\} \quad \text{for any return } R. \tag{18.6}$$

In words, $\widehat{\rho}(R)$ measures the amount of the 'certain' return whose addition makes R preferable to $\widehat{R}$. The following re-formulation of $\widehat{\rho}$ may offer some insight into the nature of this measure: Let us characterise the SSD relation in (18.6) using criterion (a) (Section 18.2.1). Then we obtain the formula

$$\widehat{\rho}(R) = \sup_{U \in \mathcal{U}} \min\left\{\varrho \in \mathbb{R} \,\middle|\, \mathrm{E}(U(R+\varrho)) \geq \mathrm{E}\big(U(\widehat{R})\big)\right\}, \tag{18.7}$$

where $\mathcal{U}$ denotes the set of nondecreasing and concave utility functions. (Assumed discrete finite distributions, expected values always exist and are finite.) In this respect, $\widehat{\rho}$ can be considered a worst-case risk measure, and hence the quantity ϑ in the scaled model (18.4) can be also considered as a worst-case measure. (Connection with convex risk measures is discussed in Section 18.2.3, and this results a further, dual-type characterisation.)

Using $\widehat{\rho}$, model (18.4) can be equivalently formulated as

$$\min_{\boldsymbol{x} \in X} \widehat{\rho}(R_{\boldsymbol{x}}), \tag{18.8}$$

and an extension of the SSD-constrained model (18.1) of Dentcheva and Ruszczyński can be formulated as

$$\begin{aligned} &\max \ f(\boldsymbol{x}) \\ &\text{such that} \quad \boldsymbol{x} \in X, \\ &\widehat{\rho}(R_{\boldsymbol{x}}) \leq \gamma, \end{aligned} \tag{18.9}$$

where $\gamma \in \mathbb{R}$ is a parameter. (In an application, the decision maker sets the value of the parameter γ.)

A computational study comparing the unscaled model (18.3) and the scaled model (18.5) is reported in Fábián et al. (2010). Scenarios were generated using geometric Brownian motion. Parameters were derived from historical monthly returns of stocks that belonged to the FTSE 100 and from historical monthly FTSE 100 index returns. Unscaled and scaled problems were solved with scenario sets of cardinalities up to 30,000. In an unscaled problem, the bottleneck was typically the extreme left tail, i.e., the constraint belonging to $i = 1$. Formally, the constraints were typically satisfied as

$$\mathrm{Tail}_{1/S}\left(R_{\boldsymbol{x}^\star}\right) = \mathrm{Tail}_{1/S}\left(\widehat{R}\right) + \vartheta^\star \quad \text{and} \quad \mathrm{Tail}_{i/S}\left(R_{\boldsymbol{x}^\star}\right) > \mathrm{Tail}_{i/S}\left(\widehat{R}\right) + \vartheta^\star \quad (2 = 1, \ldots, S), \tag{18.10}$$

where $\boldsymbol{x}^\star$ and $\vartheta^\star$ denote an optimal portfolio and the optimal objective value, respectively, of problem (18.3). On the other hand, many of the constraints of the scaled problem (18.5) were typically active in an optimal solution.

Fábián et al. (2010) also compare the return distributions belonging to the respective optimal portfolios of the unscaled model (18.3) on the one hand and of the scaled model (18.5) on the other hand. In each case, both the unscaled distribution and the scaled distribution dominated the corresponding index distribution. There was no dominance relation between the unscaled distribution and the corresponding scaled distribution. The scaled distribution had larger mean value but worst-scenario outcomes were slightly better in the case of the unscaled distribution.

18.2.3 Connection with Risk Measures

In view of the importance of risk measures, we include a brief characterisation of $\widehat{\rho}$ in this context.

Let $\mathcal{R}$ denote the space of legitimate returns. In this section we consider $\mathcal{R} = \mathcal{L}^2(\Omega, \mathcal{M}, P)$. A *risk measure* is a mapping $\rho : \mathcal{R} \rightarrow [-\infty, +\infty]$. The concept of *coherent risk measures* was developed in Artzner et al. (1999) and Delbaen (2002). These are mappings that satisfy the following four criteria:

Subadditivity: $\rho(R + R') \leq \rho(R) + \rho(R')$ holds for $R, R' \in \mathcal{R}$.
Positive homogeneity: $\rho(\lambda R) = \lambda\rho(R)$ holds for $R \in \mathcal{R}$ and $\lambda \geq 0$.
Monotonicity: $\rho(R) \leq \rho(R')$ holds for $R, R' \in \mathcal{R},\ R \geq R'$.
Translation equivariance: $\rho(R + \varrho) = \rho(R) - \varrho$ holds for $R \in \mathcal{R},\ \varrho \in \mathbb{R}$.

A well-known example for a coherent risk measure is *conditional value at risk* (CVaR), characterised by Rockafellar and Uryasev (2000, 2002) and by Pflug (2000). In words, $\mathrm{CVaR}_\alpha(R)$ is the conditional expectation of the upper α-tail of $-R$. (In our setting, R represents gain, hence $-R$ represents loss.) We have the relation

$$\mathrm{CVaR}_\alpha(R) = -\frac{1}{\alpha}\,\mathrm{Tail}_\alpha(R) \quad (0 < \alpha \leq 1). \tag{18.11}$$

(Rockafellar and Uryasev define CVaR_α using $(1 - \alpha)$-tails, making the definition compatible with that of VaR. Since we do not use VaR in this chapter, we define CVaR_α using α-tails, also a customary way.)

The concept of *convex risk measures* is a natural generalisation of coherency. The concept was introduced in Heath (2000), Carr et al. (2001) and Föllmer and Schied (2002). A mapping ρ is said to be a convex risk measure if it satisfies monotonicity, translation equivariance, and

Convexity: $\rho\left(\lambda R + (1-\lambda)R'\right) \le \lambda\rho(R) + (1-\lambda)\rho(R')$ holds for $R, R' \in \mathcal{R}$ and $0 \le \lambda \le 1$.

Rockafellar et al. (2002, 2006) develop another generalisation of the coherency concept, which also includes the *scalability criterion*: $\rho(R_\varrho) = -\varrho$ for each $\varrho \in \mathbb{R}$ (here $R_\varrho \in \mathcal{R}$ denotes the return of constant ϱ). An overview can be found in Rockafellar (2007). Ruszczyński and Shapiro (2006) develop optimality and duality theory for problems with convex risk functions.

The above-cited works also develop dual representations of risk measures. A coherent risk measure can be represented as

$$\rho(R) = \sup_{Q \in \mathcal{Q}} \mathrm{E}_Q(-R), \tag{18.12}$$

where $\mathcal{Q}$ is a set of probability measures on Ω. A risk measure that is convex or coherent in the extended sense of Rockafellar et al. (2002, 2006) can be represented as

$$\rho(R) = \sup_{Q \in \mathcal{Q}} \left\{ \mathrm{E}_Q(-R) \;-\; \nu(Q) \right\}, \tag{18.13}$$

where $\mathcal{Q}$ is a set of probability measures, and ν is a mapping from the set of the probability measures to $(-\infty, +\infty]$. (Properties of $\mathcal{Q}$ and ν depend on the type of risk measure and also on the space $\mathcal{R}$, and the topology used.) On the basis of these dual representations, an optimisation problem that involves a risk measure can be interpreted as a robust optimisation problem.

A dual representation of the coherent risk measure CVaR_α is constructed in Rockafellar et al. (2002, 2006). The space $\mathcal{R}$ of returns is $\mathcal{L}^2(\Omega, \mathcal{M}, P)$, and these authors focus on those probability measures that can be described by density functions with respect to P. Moreover the density functions need to fall into $\mathcal{L}^2$. Let Q be a legitimate probability measure with density function d_Q. Under these conditions $E_Q(R) = E(R\, d_Q)$ holds. Rockafellar et al. show that the dual representation of CVaR_α in the form of (18.12) is

$$\mathrm{CVaR}_\alpha(R) = \sup_{d_Q \le \alpha^{-1}} E_Q(-R) \quad (0 < \alpha \le 1). \tag{18.14}$$

The above formula has an intuitive explanation: the CVaR measure focuses on the worst $\alpha * 100\%$ of the cases, i.e., it 'amplifies' the bad event that such a case occur. Such an amplification is described by a density d_Q. Of course the magnitude of the amplification cannot be larger than $1/\alpha$.

The dual representation of $\widehat{\rho}$ in the form of (18.13) is constructed in Fábián et al. (2010):

$$\widehat{\rho}(R) = \sup_{Q \in \mathcal{Q}} \left\{ E_Q(-R) \;-\; \mathrm{CVaR}_{s(Q)}(\widehat{R}) \right\}, \tag{18.15}$$

with $\mathcal{Q} := \{Q \mid \sup d_Q < +\infty\}$ and $s(Q) := (\sup d_Q)^{-1}$ $(Q \in \mathcal{Q})$. Minimising $\widehat{\rho}(R)$ can be interpreted as a worst-case minimisation, possible cases being elements of $\mathcal{Q}$.

The above dual representation can be specialised to the case of discrete finite distributions with equiprobable outcomes. Let $S \in \mathbb{N}$ denote the cardinality of the sample space. The probability measure P assigns the weight $1/S$ to each of the S elements of the sample space. A measure $Q \in \mathcal{Q}$ is defined by the weights (probabilities) $q_1, \ldots, q_S$. The density d_Q of Q with respect to P takes the values $\frac{q_1}{1/S}, \ldots, \frac{q_S}{1/S}$. We have $s(Q) = 1/\left(S \max_{1 \le j \le S} q_j\right)$, and hence (18.15) can be written as

$$\widehat{\rho}(R) = \max_{\substack{q_1, \ldots, q_S \ge 0 \\ q_1 + \cdots + q_S = 1}} \left\{ \sum_{j=1}^{S} -q_j \, r^{(j)} - \mathrm{CVaR}_{1/(S \max q_j)}(\widehat{R}) \right\}, \tag{18.16}$$

where $r^{(1)}, \ \ldots, \ r^{(S)}$ denote the realisations of the random return R. As observed in Fábián et al. (2010), the above maximum is always attained with vectors $(q_1, \ldots, q_S)$ that can be obtained from subsets of the scenario set in the following manner: Given $\mathcal{J} \subset \{1, \ldots, S\}$, let

$$q_j = \begin{cases} 1/|\mathcal{J}| & \text{if } j \in \mathcal{J}, \\ 0 & \text{otherwise.} \end{cases} \tag{18.17}$$

Other vectors $(q_1, \ldots, q_S)$ in (18.16) are redundant from a cutting-plane point of view. It means that if we consider the minimisation of $\widehat{\rho}$ in (18.16) as a robust optimisation model, then all critical probability measures will have the form (18.17).

18.3 Formulation of the SSD-Based Portfolio Choice Problems

In this section we discuss different SSD-based problem formulations. We focus on the unscaled model (18.3) and on the scaled model (18.5) and formulate them as dominance maximisation problems (18.24) and (18.25), respectively. We also describe a cutting-plane method for the solution of these problems. Cutting plane methods are considered fairly efficient for quite general problem classes. However, according to our knowledge, efficiency estimates only exist for the continuously differentiable, strictly convex case. An overview of cutting plane methods and a geometrically convergent version is presented in Dempster and Merkovsky (1995). Below we cite computational studies demonstrating that the cutting plane approach is very effective for the present types of problems.

We assume that the joint distribution of $\boldsymbol{R}$ and $\widehat{R}$ is discrete finite, having S equally probable outcomes. Let $\boldsymbol{r}^{(1)}, \ \ldots, \ \boldsymbol{r}^{(S)}$ denote the realisations of the random $\boldsymbol{R}$ vector of asset returns. Similarly, let $\widehat{r}^{(1)}, \ \ldots, \ \widehat{r}^{(S)}$ denote realisations of the reference return $\widehat{R}$. For the reference tails, we will use the brief notation $\widehat{\tau}_i := \mathrm{Tail}_{i/S}\left(\widehat{R}\right)$ $(i = 1, \ldots S)$.

18.3.1 Formulation Using Tails

Adapting the CVaR optimisation formula of Rockafellar and Uryasev (2000, 2002), tails can be computed in the following form:

$$\mathrm{Tail}_{i/S}(R_{\boldsymbol{x}}) = \max_{t_i \in \mathbb{R}} \left\{ \frac{i}{S} t_i - \frac{1}{S} \sum_{j=1}^{S} \left[t_i - \boldsymbol{r}^{(j)T} \boldsymbol{x} \right]_+ \right\}.$$

This formula was used in Roman et al. (2006) to represent tails in the unscaled model (18.3). The resulting convex problem was transformed to a linear programming problem, by introducing new variables for the positive parts $\left[t_i - \boldsymbol{r}^{(j)T} \boldsymbol{x} \right]_+$. These linear programming problems were found to be computationally demanding, though.

An alternative formulation was used in Fábián et al. (2009). Adapting the approach of Künzi-Bay (2006), the following cutting-plane representation was proposed:

$$\begin{aligned} &\mathrm{Tail}_{i/S}(R_{\boldsymbol{x}}) = \min \ \tfrac{1}{S} \sum_{j \in \mathcal{J}_i} \boldsymbol{r}^{(j)T} \boldsymbol{x} \\ &\text{such that} \quad \mathcal{J}_i \subset \{1, \ldots, S\}, \quad |\mathcal{J}_i| = i. \end{aligned} \tag{18.18}$$

In its original form, the Künzi-Bay–Mayer representation includes a cut for each subset of $\{1, \ldots, S\}$. However, in the equiprobable case, only cuts of cardinality i are required for the computation of $\frac{i}{S}$-tails.

The above formulation enables a cutting-plane approach for the solution of tail minimisation or tail-constrained problems. Künzi-Bay and Mayer present a computational study demonstrating the effectiveness of this approach. We mention that the Künzi-Bay–Mayer representation is the CVaR analogue of the Klein Haneveld–Van der Vlerk (2006) representation for integrated chance constraints. The two approaches employ the same idea, originally developed by Klein Haneveld and Van der Vlerk. (Künzi-Bay and Mayer obtained their representation independently.)

Using (18.18), the unscaled model (18.3) is formulated as

$$\begin{aligned} &\max && \vartheta \\ &\text{such that} && \vartheta \in \mathbb{R}, \quad \boldsymbol{x} \in X, \\ & && \vartheta + \widehat{\tau}_i \leq \tfrac{1}{S} \sum_{j \in \mathcal{J}_i} \boldsymbol{r}^{(j)T} \boldsymbol{x} \quad \text{for each} \quad \mathcal{J}_i \subset \{1, \ldots, S\}, \quad |\mathcal{J}_i| = i, \\ & && \qquad \text{where} \quad i = 1, \ldots, S. \end{aligned} \tag{18.19}$$

In a similar manner, the scaled model (18.5) is formulated as

$$
\begin{aligned}
&\max \quad && \vartheta \\
&\text{such that} \quad && \vartheta \in \mathbb{R}, \quad \boldsymbol{x} \in X, \\
& && \tfrac{i}{S}\vartheta + \widehat{\tau}_i \;\le\; \tfrac{1}{S} \sum_{j \in \mathcal{J}_i} \boldsymbol{r}^{(j)T}\boldsymbol{x} \quad \text{for each} \quad \mathcal{J}_i \subset \{1, \ldots, S\}, \; |\mathcal{J}_i| = i, \\
& && \qquad \text{where} \quad i = 1, \ldots, S.
\end{aligned} \tag{18.20}
$$

As reported in Fábián et al. (2009, 2010), the cutting-plane approach proved effective for the solution of both the unscaled and the scaled problems.

18.3.2 Formulation Using Integrated Chance Constraints

In their SSD-constrained model (18.1), Dentcheva and Ruszczyński (2006) characterise stochastic dominance with criterion (b) (Section 18.2.1). They prove that if $\widehat{R}$ has a discrete finite distribution with realisations $\widehat{r}^{(1)}, \ \ldots, \ \widehat{r}^{(S)}$, then $R_{\boldsymbol{x}} \succeq_{\text{SSD}} \widehat{R}$ is equivalent to a finite system of inequalities

$$
E\left(\left[\widehat{r}^{(i)} - R_{\boldsymbol{x}}\right]_+\right) \le E\left(\left[\widehat{r}^{(i)} - \widehat{R}\right]_+\right) \qquad (i = 1, \ldots, S). \tag{18.21}
$$

The right-hand sides of the above constraints do not depend on $\boldsymbol{x}$. In case of scenario i, the left-hand side represents expected shortfall of portfolio return $R_{\boldsymbol{x}}$ relative to the constant $\widehat{r}^{(i)}$. Such a constraint is called integrated chance constraint after Klein Haneveld (1986).

Dentcheva and Ruszczyński transform (18.21) into a set of linear constraints by introducing new variables to represent positive parts. The resulting linear programming problem has a specific structure. For such specific problems, Dentcheva and Ruszczyński develop a duality theory in which the dual objects are utility functions. Based on this duality theory, they construct a dual problem that consists of the minimisation of a weighted sum of polyhedral convex functions. Dentcheva and Ruszczyński adapt the regularized decomposition method of Ruszczyński (1986) to these special dual problems.

Klein Haneveld and Van der Vlerk (2006) present a cutting-plane representation for integrated chance constraints in case of discrete finite distributions. Based on this representation, they develop a cutting-plane method for integrated chance-constrained optimisation problems. In this method, cuts are generated in the space of the variables. Klein Haneveld and Van der Vlerk present a computational study demonstrating the effectiveness of their cutting-plane approach. This cutting-plane approach is adapted to system (18.21) in Fábián et al. (2009). This latter paper presents the method in a somewhat different perspective: instead of generating cuts in the variable space, a constraint function is constructed, and cuts are generated to the graph of this function. This view enables regularisation.

Independently, Rudolf and Ruszczyński (2008) also develop cutting-plane methods for the solution of SSD-constrained problems. They propose an extension of the Klein Haneveld–Van der Vlerk representation to integrable random variables. They formulate a primal cutting-plane method which, in case of finite distributions, is equivalent to the application of the Klein Haneveld–Van der Vlerk method. Rudolf and Ruszczyński also develop a duality theory for SSD-constrained problems and propose a dual column-generation method for the solution of such problems. Their computational study demonstrates that the primal cutting-plane method is computationally efficient. New formulations of stochastic dominance constraints are proposed by Luedtke (2008). His computational study demonstrates that his schemes are effective, but do not surpass the efficiency of the Klein Haneveld–Van der Vlerk formulation.

The constraint $R_{\boldsymbol{x}} \succeq_{\text{SSD}} \widehat{R} + \vartheta$ of the scaled model (18.4) can be formulated in the manner of (18.21). Under the present equiprobability assumption, it takes the form

$$\sum_{j=1}^{S} \frac{1}{S}\left[\widehat{r}^{(i)} + \vartheta - \boldsymbol{r}^{(j)T}\boldsymbol{x}\right]_+ \le \sum_{j=1}^{S} \frac{1}{S}\left[\widehat{r}^{(i)} - \widehat{r}^{(j)}\right]_+ \qquad (i = 1, \ldots, S). \tag{18.22}$$

The Klein Haneveld–Van der Vlerk cutting-plane representation of the ith constraint is

$$\sum_{j\in\mathcal{J}_i} \frac{1}{S}\left\{\widehat{r}^{(i)} + \vartheta - \boldsymbol{r}^{(j)T}\boldsymbol{x}\right\} \le \sum_{j=1}^{S} \frac{1}{S}\left[\widehat{r}^{(i)} - \widehat{r}^{(j)}\right]_+ \quad \text{for each } \mathcal{J}_i \subset \{1, \ldots, S\}.$$

Using the above representation, the scaled model (18.4) can be formulated as

$$\begin{aligned}
&\max \quad \vartheta\\
&\text{such that} \quad \vartheta \in \mathbb{R}, \quad \boldsymbol{x} \in X,\\
&\qquad \sum_{j\in\mathcal{J}_i} \tfrac{1}{S}\left\{\widehat{r}^{(i)} + \vartheta - \boldsymbol{r}^{(j)T}\boldsymbol{x}\right\} \le \sum_{j=1}^{S} \tfrac{1}{S}\left[\widehat{r}^{(i)} - \widehat{r}^{(j)}\right]_+ \quad \text{for each } \mathcal{J}_i \subset \{1, \ldots, S\},\\
&\qquad\qquad \text{where } i = 1, \ldots, S.
\end{aligned} \tag{18.23}$$

The scaled problem formulations (18.23) and (18.20) are equivalent. As observed in Fábián et al. (2010), the constraints belonging to $|\mathcal{J}_i| \neq i$ are redundant in (18.23). (If we drop the equiprobability assumption but keep the discrete finite assumption, then both the tail formulation and the ICC formulation remain applicable, and the latter will be more convenient.)

18.3.3 Formulation as Dominance Maximisation Problems and Solution by the Cutting Plane Method

We will use formulation (18.19) for the unscaled problem and formulation (18.20) for the scaled problem. We formulate these problems as dominance maximisation problems. The respective dominance measures are polyhedral concave functions. We adapt the cutting plane method to these problems.

The unscaled problem (18.19) can be stated as maximisation of the dominance measure $\phi(\boldsymbol{x})$:

$$\max_{\boldsymbol{x}\in X} \phi(\boldsymbol{x}) \quad \text{where} \quad \phi(\boldsymbol{x}) := \min\left\{ \frac{1}{S} \sum_{j\in\mathcal{J}_i} \boldsymbol{r}^{(j)T}\boldsymbol{x} \; - \widehat{\tau}_i \right\}$$
$$\text{such that} \quad \mathcal{J}_i \subset \{1,\ldots,S\}, \quad |\mathcal{J}_i| = i, \qquad (18.24)$$
$$\text{where} \quad i = 1,\ldots,S.$$

The scaled problem (18.20) can be stated as maximisation of the dominance measure $\varphi(\boldsymbol{x})$:

$$\max_{\boldsymbol{x}\in X} \varphi(\boldsymbol{x}) \quad \text{where} \quad \varphi(\boldsymbol{x}) := \min\left\{ \frac{1}{i} \sum_{j\in\mathcal{J}_i} \boldsymbol{r}^{(j)T}\boldsymbol{x} \; - \frac{S}{i}\widehat{\tau}_i \right\}$$
$$\text{such that} \quad \mathcal{J}_i \subset \{1,\ldots,S\}, \quad |\mathcal{J}_i| = i, \qquad (18.25)$$
$$\text{where} \quad i = 1,\ldots,S.$$

We have $\varphi(\boldsymbol{x}) = -\widehat{\rho}(R_{\boldsymbol{x}})$ $(\boldsymbol{x} \in \mathbb{R}^n)$ with the equiprobable formulation (18.16) of the risk measure $\widehat{\rho}$.

Obviously $\phi(\boldsymbol{x})$ and $\varphi(\boldsymbol{x})$ are polyhedral concave functions. The feasible domain X is a convex polyhedron. Applied to such a convex programming problem, the cutting plane method generates a sequence of iterates from X. At each iterate, supporting linear functions (cuts) are constructed to the objective function. A cutting-plane model of the objective function is then maintained as the upper cover of known cuts. The next iterate is obtained by minimising the current model function over X.

Supporting linear functions to the objective functions $\phi(\boldsymbol{x})$ and $\varphi(\boldsymbol{x})$ are easily constructed at any given point $\boldsymbol{x}^\star \in X$:

- Let $r_{\boldsymbol{x}^\star}^{(j_1^\star)} \leq \cdots \leq r_{\boldsymbol{x}^\star}^{(j_S^\star)}$ denote the ordered outcomes of $\boldsymbol{R}^T\boldsymbol{x}^\star$. Using this ordering of the scenarios, let us construct the sets $\mathcal{J}_i^\star := \left\{ j_1^\star, \ldots, j_i^\star \right\}$ $(i = 1,\ldots,S)$.
- Let us select

$$\iota^\circ \in \arg\min_{1\le i\le S} \left\{ \frac{1}{S} \sum_{j\in\mathcal{J}_i^\star} \boldsymbol{r}^{(j)\,T}\boldsymbol{x}^\star \; - \widehat{\tau}_i \right\}. \tag{18.26}$$

A supporting linear function to $\phi(\boldsymbol{x})$ at $\boldsymbol{x}^\star$ is then $\ell(\boldsymbol{x}) := (1/S)\sum_{j\in\mathcal{J}_{\iota^\circ}^\star} \boldsymbol{r}^{(j)\,T} \boldsymbol{x} - \widehat{\tau}_{\iota^\circ}$.

- Let us select

$$\iota^\star \in \arg\min_{1\le i\le S} \left\{ \frac{1}{i} \sum_{j\in\mathcal{J}_i^\star} \boldsymbol{r}^{(j)\,T}\boldsymbol{x}^\star \; - \frac{S}{i}\widehat{\tau}_i \right\}. \tag{18.27}$$

A supporting linear function to $\varphi(\boldsymbol{x})$ at $\boldsymbol{x}^\star$ is then

$$l(\boldsymbol{x}) := \frac{1}{\iota^\star} \sum_{j\in\mathcal{J}_{\iota^\star}^\star} \boldsymbol{r}^{(j)\,T}\boldsymbol{x} \; - \frac{S}{\iota^\star}\widehat{\tau}_{\iota^\star}.$$

This procedure requires little computational effort beyond sorting the outcomes.

18.4 Set-Up of the Computational Study

In this section we describe the composition of the scenario sets used in constructing and testing portfolios. We define the portfolio choice problems solved and the respective optimal portfolios. We introduce notation for the histograms used to characterise the return distributions of the different portfolios when tested against the different scenario sets. Finally we describe the composition of consolidated histograms characterising the aggregated results of repeated experiments.

18.4.1 Input Data

The data set consists of weekly returns of $n = 68$ stocks from the FTSE 100 basket, together with the weekly returns of the FTSE 100 index for the same period. We consider $T = 835$ weeks from the period January 1993 to January 2009.

Given the tth week ($0 \le t \le T$), we have $n + 1$ values: the respective prices of each component stock and also the stock index. Let s_0^t represent stock index and s_k^t ($1 \le k \le n$) represent the price of the kth stock. (These are end-of-week data in case of $t \ge 1$; starting data will be considered as of week 0.)

The returns of week $t\,(1 \le t \le T)$ are then computed as $r_k^t := \left(s_k^t - s_k^{t-1}\right)/s_k^{t-1}$ ($0 \le k \le n$). These will be arranged into an $(n + 1)$-dimensional vector $\boldsymbol{r}^t = \left(r_0^t, r_1^t, \dots, r_n^t\right)$. Our investigation is based on the data set $\mathcal{R} := \left\{\boldsymbol{r}^t \mid t = 1, \dots, T\right\}$.

18.4.2 An Experiment

18.4.2.1 Scenario Sets

We assume that stock prices evolve in such a manner that return distribution is the same for each week. We partition the data set $\mathcal{R}$ into subsets $\mathcal{H}$ and $\mathcal{T}$. (Formally, $\mathcal{H} \cup \mathcal{T} = \mathcal{R}$, $\mathcal{H} \cap \mathcal{T} = \emptyset$.) Partitioning is done in a random manner. The subset $\mathcal{H}$ is used for selecting a portfolio. (A portfolio is selected either directly using $\mathcal{H}$ or using a larger sample $\mathcal{G}$ constructed from $\mathcal{H}$.) The subset $\mathcal{T}$ is used for out-of-sample tests on portfolio returns. We use the following scenario sets:

Historical ($\mathcal{H}$): Elements of $\mathcal{H}$ are considered as scenarios of equal probability.

GBM ($\mathcal{G}$): Returns are assumed to follow geometric Brownian motion. (This is a standard assumption in finance, see, e.g., Ross (2002).) Formally, the return vector $\boldsymbol{r} \in \mathbb{R}^{1+n}$ follows GBM if the transformed vector $(\log(r_0 + 1), \log(r_1 + 1), \ldots, \log(r_n + 1))$ has normal distribution. Parameters are derived from the historical set $\mathcal{H}$. A sample is then drawn according to this lognormal distribution. Scenarios are sample elements occurring with equal probabilities.

Test set ($\mathcal{T}$): Elements of $\mathcal{T}$ are considered as scenarios of equal probability.

In our computational study, cardinalities of these sets are $|\mathcal{H}| = 668$ and $|\mathcal{T}| = 167$ (we have $|\mathcal{H}| : |\mathcal{T}| = 4 : 1$). Cardinality of the generated set is $|\mathcal{G}| = 10{,}000$.

18.4.2.2 Portfolio Choice Problems and Optimal Portfolios

Investment period in our computational study is 1 week, and asset returns are represented by one of the above scenario sets. In case a scenario $\boldsymbol{r} \in \mathbb{R}^{1+n}$ realises, return of a portfolio $\boldsymbol{x} \in \mathbb{R}^n$ will be $< \boldsymbol{x}, \boldsymbol{r} > := \sum_{i=1}^{n} r_i x_i$, and index return will be r_0.

We solve the unscaled problem (18.24) and the scaled problem (18.25), using the scenario sets $\mathcal{H}$ and $\mathcal{G}$. This results 4 problems. Let $\boldsymbol{x}_{\mathcal{H}}, \boldsymbol{x}_{\mathcal{G}}, \boldsymbol{x}'_{\mathcal{H}}, \boldsymbol{x}'_{\mathcal{G}}$ denote the respective optimal portfolios, as follows:

	Historical scenario set	GBM scenario set
Unscaled problem	$\boldsymbol{x}_{\mathcal{H}}$	$\boldsymbol{x}_{\mathcal{G}}$
Scaled problem	$\boldsymbol{x}'_{\mathcal{H}}$	$\boldsymbol{x}'_{\mathcal{G}}$

18.4.2.3 Diagrams

We will test the above optimal portfolios against different scenario sets.

We introduced notation $< \boldsymbol{x}, \boldsymbol{r} >$ for the return of portfolio $\boldsymbol{x}$ in case scenario $\boldsymbol{r}$ realises. Let us extend this notation to scenario sets, as follows: $< \boldsymbol{x}, \mathcal{S} >$ will consist of the values $< \boldsymbol{x}, \boldsymbol{r} >$ $(\boldsymbol{r} \in \mathcal{S})$, where the scenario set $\mathcal{S}$ may stand for

$\mathcal{H}, \mathcal{G}$, or $\mathcal{T}$. The object $< \boldsymbol{x}, \mathcal{S} >$ is an unsorted list; it may contain a value with multiplicity. We construct histograms based on these objects.

Let us also introduce notation for index values corresponding to a scenario set: The unsorted list $< \mathcal{S} >$ will consist of the values r_0 ($\boldsymbol{r} \in \mathcal{S}$), where the scenario set $\mathcal{S}$ may stand for $\mathcal{H}, \mathcal{G}$, or $\mathcal{T}$.

18.4.3 Repeated Experiments

The construction described in Section 18.4.2 was repeated 12 times with independently selected partitions $\mathcal{H}^{(\ell)} \cup \mathcal{T}^{(\ell)}$ ($\ell = 1, \ldots, 12$) of the data set $\mathcal{R}$. Let $\mathcal{G}^{(\ell)}$ denote the GBM scenario set corresponding to the ℓth experiment. We introduce notation for consolidated data.

Let $[\mathcal{H}, \mathcal{T}]$ denote the concatenation of the unsorted lists $< \boldsymbol{x}_{\mathcal{H}^{(\ell)}}, \mathcal{T}^{(\ell)} >$ ($\ell = 1, \ldots, 12$), where $\boldsymbol{x}_{\mathcal{H}^{(\ell)}}$ is the optimal portfolio of the unscaled model on the scenario set $\mathcal{H}^{(\ell)}$.

Similarly, let $[\mathcal{H}', \mathcal{T}]$ denote the concatenation of the unsorted lists $< \boldsymbol{x}'_{\mathcal{H}^{(\ell)}}, \mathcal{T}^{(\ell)} >$ ($\ell = 1, \ldots, 12$), where $\boldsymbol{x}'_{\mathcal{H}^{(\ell)}}$ is the optimal portfolio of the scaled problem on the scenario set $\mathcal{H}^{(\ell)}$. The objects $[\mathcal{G}, \mathcal{T}]$, $[\mathcal{G}', \mathcal{T}]$, etc., will be defined in the same manner.

Finally, let $[\mathcal{S}]$ denote the concatenation of the unsorted lists $< \mathcal{S}^{(\ell)} >$ ($\ell = , \ldots, 12$), where $\mathcal{S}$ may stand for $\mathcal{H}, \mathcal{G}$, or $\mathcal{T}$.

18.5 Implementation Issues

In this section we present details of the construction of the scenario sets. We describe the way of setting aside data for out-of-sample tests and method used to estimate GBM parameters.

We also present solver implementation details.

18.5.1 The Construction of Scenario Sets

The construction of scenario sets was implemented in MATLAB. Partitioning of $\mathcal{R}$ to $\mathcal{H} \cup \mathcal{T}$ is done by using a random number generator. A practical description of random number generators can be found, e.g., in Deák (1990).

The GBM scenario sets were generated using Gaussian copulas with lognormal marginal distributions. Correlations between marginals were estimated using Spearman's rank correlations. Rank correlation fits the copula approach; moreover, it is considered a more robust dependence measure than standard correlation, see e.g., Cherubini et al. (2006, Chapter 7.5.4). We sketch the copula approach and the way we use it in this study.

Sklar (1959) proves that given an m-dimensional distribution function F with marginal distribution functions $F_1, \ldots, F_m$, there exits a function $C : \mathbb{R}^m \to \mathbb{R}$ such that

$$F(z_1, \ldots, z_m) = C\,(F_1(z_1),\ \ldots,\ F_m(z_m)) \quad (z_1, \ldots, z_m \in \mathbb{R})$$

holds. (If, moreover, $F_1, \ldots, F_m$ are continuous then C is unique.) C is called *copula function*, and it can be used to construct multivariate distributions with given margins. A Gaussian copula function has the form

$$C_\rho\,(F_1(z_1),\ \ldots,\ F_m(z_m)) = \Phi_{m,\rho}\left(\Phi^{-1}F_1(z_1),\ \ldots,\ \Phi^{-1}F_m(z_m)\right),$$

where Φ is the (one-dimensional) standard normal distribution function, and $\Phi_{m,\rho}$ is the m-dimensional normal distribution function having expectation vector $\mathbf{0}$ and correlation matrix ρ. The usual approach is to use experimental marginals and to calibrate the correlation matrix ρ to the experimental data. As the transformation $\Phi^{-1}F_k$ is non-linear, rank correlation is used for calibration. (Rank correlation is preserved under monotonic transformations.) Spearman's rank correlation ρ^{S} is widely used in this context. There is a one-to-one mapping between the Spearman rank correlation coefficient and the linear correlation coefficient for bi-variate normal random variables. This mapping can be characterised by a formula. The inverse formula is then used to compute the linear correlation matrix ρ of $\Phi_{m,\rho}$ from the Spearman's rank correlation matrix ρ^{S}. Theory and details of the copula approach can be found in Cherubini et al. (2006) and McNeil et al. (2005). Gaussian copula is discussed in Joe (1997). A brief description of Gaussian copulas together with an interesting application can be found in AitSahlia et al. (2009).

In the present computational study we constructed Gaussian copulas with log-normal marginals. Parameters of the marginal distributions were estimated from the historical data. To be more precise, let us consider a single experiment. Parameters of the distribution function $F_k (k = 0, \ldots, n)$ were computed from the shifted components $r_k + 1$ of the vectors $\boldsymbol{r} \in \mathcal{H}$. We then computed Spearman's rank correlation coefficients between any two components k, k'. When applying the above-mentioned transformation on the Spearman's rank correlation matrix ρ^{S} in order to obtain linear correlation matrix ρ, we always obtained a positive definite ρ. We then generated a sample of cardinality 10,000 according to the copula distribution. The scenario set $\mathcal{G}$ was derived from this sample by taking the exponents of the components and subtracting 1. This approach indeed results in a GBM sample, because rank correlation is invariant to a monotone transformation like the log function.

18.5.2 Solution of the Portfolio Choice Problems

For the solution of unscaled problem (18.24) and the scaled problem (18.25), we implemented the cutting-plane method sketched in Section 18.3.3.

Our implementation use the AMPL modelling system Fourer et al. (2002) integrated with C functions. Under AMPL we use the FortMP solver Ellison et al.

(2008). FortMP was developed at Brunel University and NAG Ltd., the project being co-ordinated by E.F.D. Ellison.

In our cutting-plane system, cut generation is implemented in C, and cutting-plane model problem data are forwarded to AMPL in each iteration. Hence the bulk of the arithmetic computations is done in C, since the number of the scenarios is typically large as compared to the number of the assets. (Moreover, our experience is that acceptable accuracy can be achieved by a relatively small number of cuts in the master problem. Hence the sizes of the model problems do not directly depend on the number of scenarios.)

18.6 Test Results

We compare portfolio choice models in Section 18.6.1 and examine the effect of scenario generation in Section 18.6.2. The analysis is illustrated by histograms. These histograms are based on consolidated results of repeated experiments because we think they give a better characterisation than the individual experiments would. (Consolidation here means concatenating the 12 unsorted lists that describe the individual return distributions obtained for the 12 specific experiments we performed. A formal description and notation can be found in Section 18.4.3). The shapes of the consolidated histograms are very similar to the shapes of the histograms belonging to individual experiments. The same range and scale is used on the horizontal axes of all the histograms included. The range is $[-0.2, +0.2]$ and step size is 0.01.

18.6.1 Comparison of Portfolio Choice Models

In this study, an optimal portfolio is constructed either on the basis of a historical scenario set or on the basis of a GBM scenario set. Each optimal portfolio is tested against the scenario set on whose basis it was constructed (in-sample test) and against the corresponding test set (out-of-sample test). This results in four possible tests. In each test, we compared the relevant index distributions, and the return distributions belonging to the respective optimal portfolios of the unscaled problem (18.24) and of the scaled problem (18.25). We will use the term *unscaled distribution* for the return distribution belonging to the optimal portfolio of an unscaled problem, and the term *scaled distribution* for the return distribution belonging to the optimal portfolio of a scaled problem. The following observations hold for each comparison of the index, unscaled, and scaled distribution, in each individual experiment we performed:

- The index distribution has a longish left tail. The unscaled distribution is curtailed on the left. The tail of the scaled distribution is similar to that of the index.
- Both the unscaled and the scaled distributions have larger expectations and smaller standard deviations than the index distribution has. The scaled

distribution has larger expectation then the unscaled one, though its standard deviation is also larger.

In the unscaled problems, the bottleneck is always the extreme left tail, i.e., the constraint belonging to $i = 1$ in (18.3). (Moreover, other tail constraints are typically inactive; formally, (18.10) typically holds with the optimal portfolio $\boldsymbol{x}^\star$ and the optimal objective value $\vartheta^\star$.) Although we solved the unscaled problems in the form (18.24), this feature is more conveniently expressed in terms of tails.

In the following sections we present details of the four possible tests. Each test consists of 12 experiments as described in Section 18.4.3. Each test is illustrated by two histograms based on consolidated results of the experiments. One histogram compares the consolidated index distribution and the consolidated unscaled distribution, the other histogram compares the consolidated index distribution and the consolidated scaled distribution.

18.6.1.1 Testing Portfolios Constructed from Historical Scenario Sets, In-Sample Tests

In this test, optimal portfolios were constructed on the basis of historical scenario sets, and each portfolio was tested against its respective historical scenario set.

In each experiment, the index distribution was dominated by both the unscaled distribution and the scaled distribution.

Figure 18.1 depicts the consolidated index distribution, the consolidated unscaled distribution, and the consolidated scaled distribution. (To be precise, the histograms are based on $[\mathcal{H}]$, $[\mathcal{H}, \mathcal{H}]$, and $[\mathcal{H}', \mathcal{H}]$, using the notation of Section 18.4.3.) Expectations and standard deviations of the consolidated distributions are the following:

	Expect.	St. dev.
Index	0.0006	0.0239
Unscaled	0.0023	0.0210
Scaled	0.0045	0.0231

18.6.1.2 Testing Portfolios Constructed from Historical Scenario Sets, Out-of-Sample Tests

In this test, optimal portfolios were constructed on the basis of historical scenario sets, and each portfolio was tested against its respective test set.

The unscaled distribution often dominated the index distribution. (Dominance occurred in 7 experiments out of the 12 performed.) On the other hand, the scaled distribution typically did not dominate the index distribution. (Dominance occurred in 4 experiments out of the 12 performed. Even in the remaining 8 experiments, all but a few tail constraints were satisfied. The unsatisfied constraints belonged to the extreme left tails.)

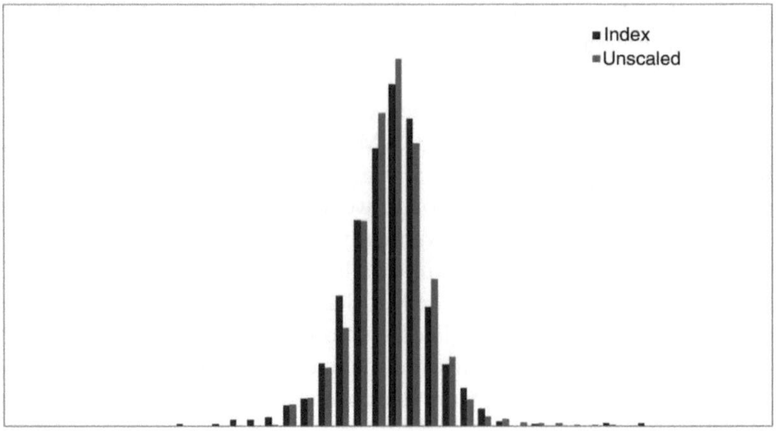

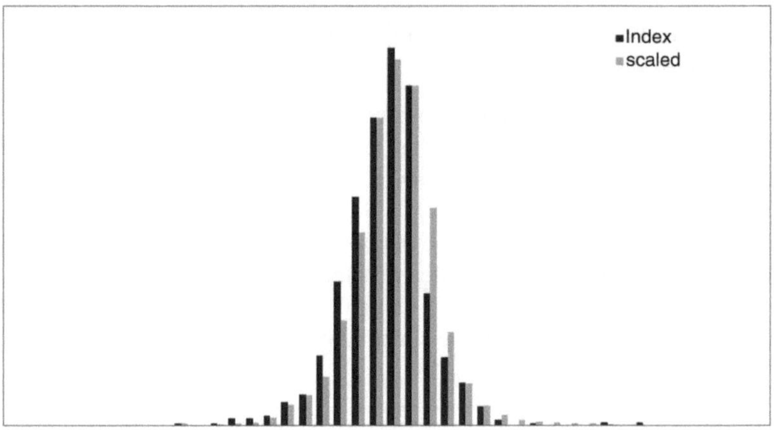

Fig. 18.1 Return distributions of portfolios constructed from historical scenario sets, in-sample tests. *Upper histogram*: consolidated index distribution and consolidated unscaled distribution. *Lower histogram*: consolidated index distribution and consolidated scaled distribution

Figure 18.2 depicts the consolidated index distribution, the consolidated unscaled distribution, and the consolidated scaled distribution. (To be precise, the histograms are based on $[\mathcal{T}]$, $[\mathcal{H}, \mathcal{T}]$, and $[\mathcal{H}', \mathcal{T}]$, using the notation of Section 18.4.3.) Expectations and standard deviations of the consolidated distributions are the following:

	Expect.	St. dev.
Index	0.0014	0.0235
Unscaled	0.0023	0.0219
Scaled	0.0032	0.0234

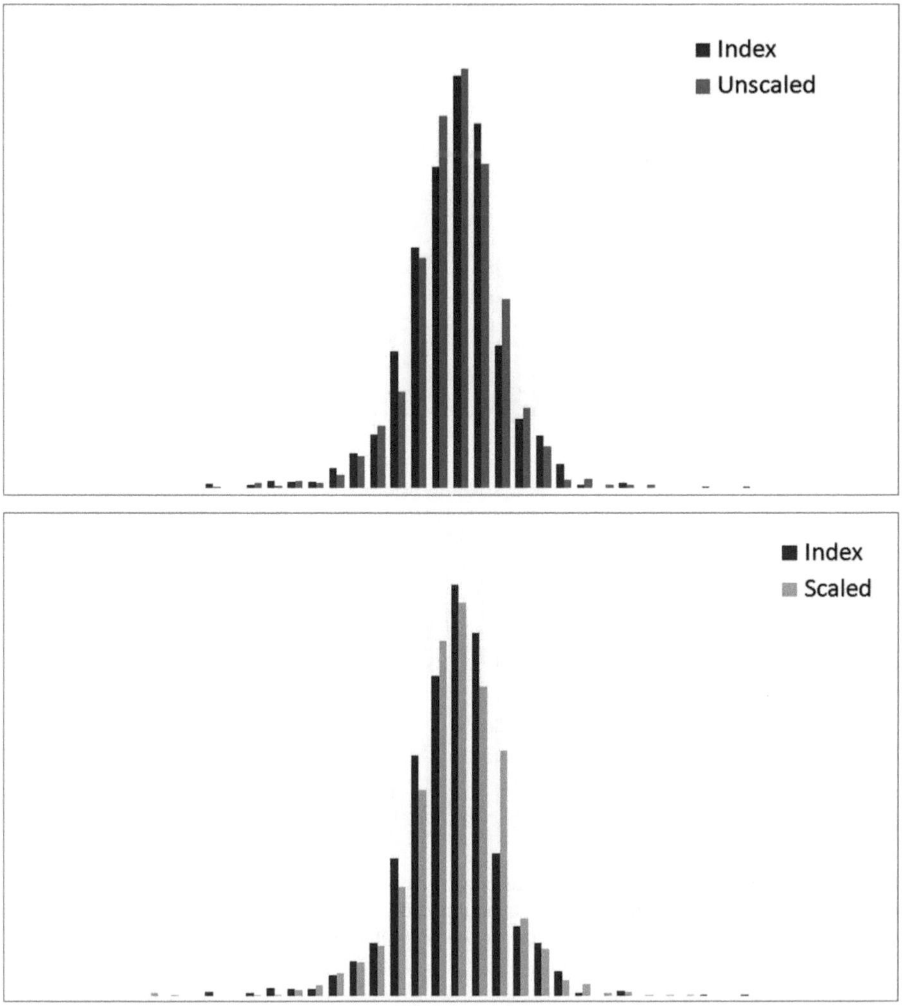

Fig. 18.2 Return distributions of portfolios constructed from historical scenario sets, out-of-sample tests. *Upper histogram*: consolidated index distribution and consolidated unscaled distribution. *Lower histogram*: consolidated index distribution and consolidated scaled distribution

18.6.1.3 Testing Portfolios Constructed from GBM Scenario Sets, In-Sample Tests

In this test, optimal portfolios were constructed on the basis of GBM scenario sets, and each portfolio was tested against its respective GBM scenario set.

In each experiment, the index distribution was dominated by both the unscaled distribution and the scaled distribution.

Figure 18.3 depicts the consolidated index distribution, the consolidated unscaled distribution, and the consolidated scaled distribution. (To be precise, the histograms

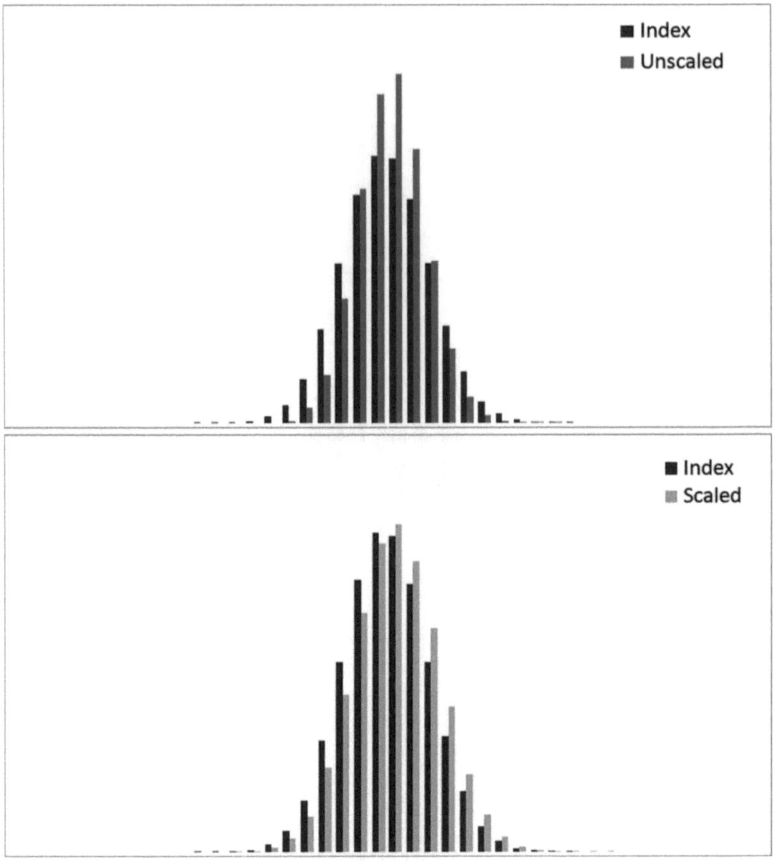

Fig. 18.3 Return distributions of portfolios constructed from GBM scenario sets, in-sample tests. *Upper histogram*: consolidated index distribution and consolidated unscaled distribution. *Lower histogram*: consolidated index distribution and consolidated scaled distribution

are based on $[\mathcal{G}]$, $[\mathcal{G}, \mathcal{G}]$, and $[\mathcal{G}', \mathcal{G}]$, using the notation of Section 18.4.3.) Expectations and standard deviations of the consolidated distributions are the following:

	Expect.	St. dev.
Index	0.0006	0.0239
Unscaled	0.0021	0.0184
Scaled	0.0046	0.0237

18.6.1.4 Testing Portfolios Constructed from GBM Scenario Sets, Out-of-Sample Tests

In this test, optimal portfolios were constructed on the basis of GBM scenario sets, and each portfolio was tested against its respective test set.

The unscaled distribution typically dominated the index distribution. (Dominance occurred in 9 experiments out of the 12 performed.) On the other hand, the scaled distribution typically did not dominate the index distribution. (Dominance occurred in 3 experiments out of the 12 performed. Even in the remaining 9 experiments, all but a few tail constraints were satisfied. The unsatisfied constraints belonged to the extreme left tails.)

Figure 18.4 depicts the consolidated index distribution, the consolidated unscaled distribution, and the consolidated scaled distribution. (To be precise, the histograms are based on $[\mathcal{T}]$, $[\mathcal{G}, \mathcal{T}]$, and $[\mathcal{G}', \mathcal{T}]$, using the notation of Section 18.4.3.) Expectations and standard deviations of the consolidated distributions are the following:

	Expect.	St. dev.
Index	0.0014	0.0235
Unscaled	0.0022	0.0188
Scaled	0.0028	0.0235

18.6.2 The Effect of Scenario Generation

In this section we compare optimal portfolios, constructed on the basis of historical scenario sets on the one hand, and on the basis of GBM scenario sets on the other hand. For short, an optimal portfolio constructed on the basis of a historical scenario set will be called a *historical portfolio* in this section and that constructed on the basis of a GBM scenario set will be called a *GBM portfolio*.

We compare the historical and the GBM portfolios as vectors. We also test the portfolios against their respective test sets and compare return distributions.

In the following two sections we examine the effect of portfolio generation on unscaled problems and on scaled problems, respectively. It turns out that the effect is quite different in these two cases.

18.6.2.1 Unscaled Problems

For each experiment $\ell (\ell = 1, \dots, 12)$, we compared the respective historical and GBM portfolios $\boldsymbol{x}_{\mathcal{H}^{(\ell)}}$ and $\boldsymbol{x}_{\mathcal{G}^{(\ell)}}$. We found these portfolios quite different: considering them as vectors, the angle $\angle \left(\boldsymbol{x}_{\mathcal{H}^{(\ell)}}, \, \boldsymbol{x}_{\mathcal{G}^{(\ell)}} \right)$ was about 70° at the average, the largest being about 80°.

We also tested the historical and GBM portfolios against their test sets. In each experiment we found that the return distribution of the GBM portfolio had a standard deviation remarkably smaller than the return distribution of the historical portfolio. Particularly, the left tails are shorter and less heavy in case of the GBM portfolio. The effect of scenario generation was a 10–20% decrease in standard deviations. We illustrate this effect on consolidated histograms of the respective return distributions of the historical and the GBM portfolios. Figure 18.5 depicts the consolidated return distribution of the historical portfolio and the consolidated return distribution of the

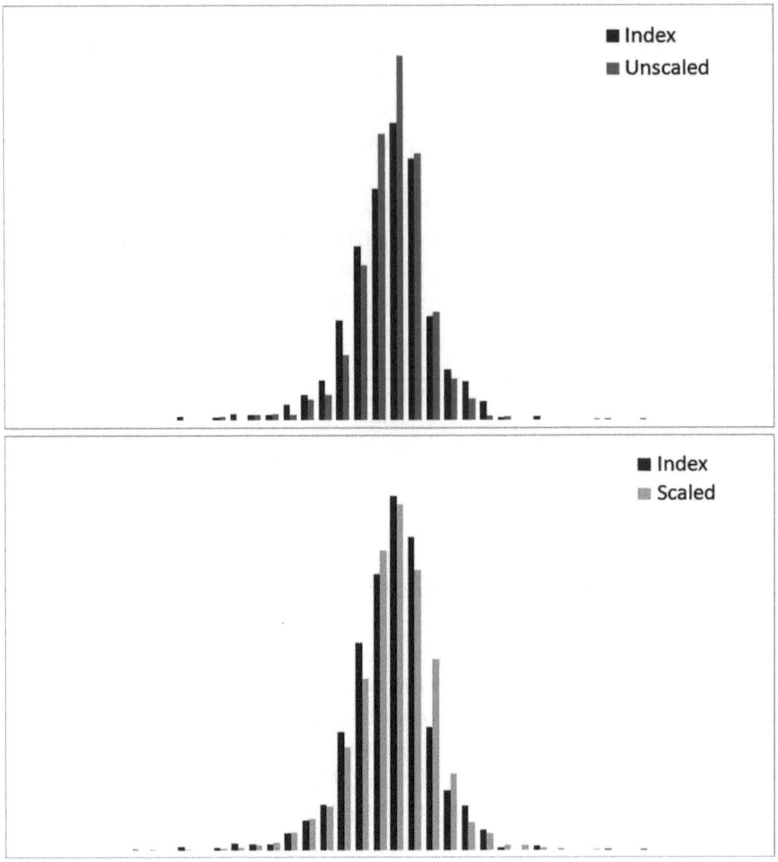

Fig. 18.4 Return distributions of portfolios constructed from GBM scenario sets, out-of-sample tests. *Upper histogram*: consolidated index distribution and consolidated unscaled distribution. *Lower histogram*: consolidated index distribution and consolidated scaled distribution

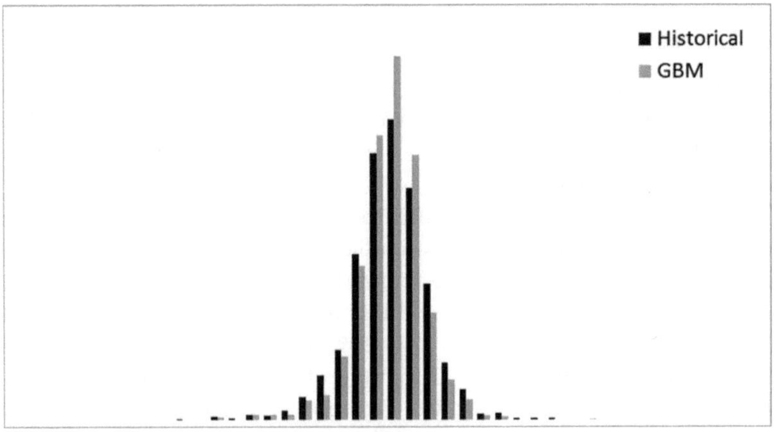

Fig. 18.5 The effect of scenario generation on the unscaled model, demonstrated in out-of-sample tests. Consolidated return distribution of the historical portfolio and that of the GBM portfolio

GBM portfolio. (To be precise, the histograms are based on $[\mathcal{H}, \mathcal{T}]$ and $[\mathcal{G}, \mathcal{T}]$, using the notation of Section 18.4.3.)

18.6.2.2 Scaled Problems

For each experiment $\ell(\ell = 1, \ldots, 12)$, we compared the respective historical and GBM portfolios $\boldsymbol{x}'_{\mathcal{H}(\ell)}$ and $\boldsymbol{x}'_{\mathcal{G}(\ell)}$. We found these portfolios quite similar: considering them as vectors, the angle $\angle\left(\boldsymbol{x}'_{\mathcal{H}(\ell)}, \boldsymbol{x}'_{\mathcal{G}(\ell)}\right)$ was about 30° at the average, the largest being about 45°.

We also tested the historical and GBM portfolios against their test sets. In each experiment we found a remarkable similarity between the respective return distributions of the GBM portfolio and of the historical portfolio. We illustrate this similarity on consolidated histograms of the respective return distributions of the historical and the GBM portfolios. Figure 18.6 depicts the consolidated return distribution of the historical portfolio and the consolidated return distribution of the GBM portfolio. (To be precise, the histograms are based on $[\mathcal{H}', \mathcal{T}]$ and $[\mathcal{G}', \mathcal{T}]$, using the notation of Section 18.4.3.)

18.6.3 Convergence

In an additional test we examined the convergence of the optimal solution vectors with respect to expanding scenario sets. This examination is more theoretical in nature than the previous ones, hence we avoid using financial terminology here.

Using a 69-dimensional geometric Brownian motion with fixed parameters, we independently generated 10 samples of 60,000 elements each. From each

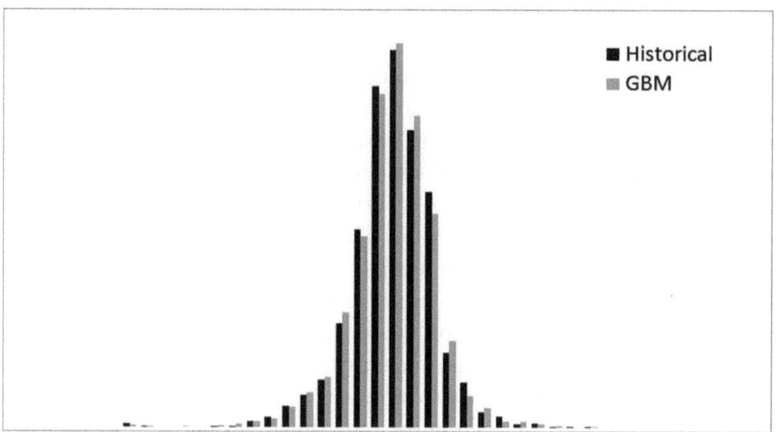

Fig. 18.6 The effect of scenario generation on the scaled model, demonstrated in out-of-sample tests. Consolidated return distribution of the historical portfolio and that of the GBM portfolio

sample, we constructed an expanding sequence of scenario sets by considering 10,000, 20,000, ..., 60,000 elements of that sample, as scenarios of equal probability. This way we obtained 10 independent sequences, each consisting of 6 scenario sets. We solved the resulting 60 unscaled and 60 scaled problems.

Given scenario set cardinality $\kappa \in \{10{,}000, 20{,}000, \ldots, 60{,}000\}$, let $\boldsymbol{x}_{\kappa 1}, \ldots, \boldsymbol{x}_{\kappa 10}$ denote the respective optimal solution vectors of the corresponding unscaled problems. For each cardinality κ, we found the smallest ball containing the vectors $\boldsymbol{x}_{\kappa\, 1}, \ldots, \boldsymbol{y}_{\kappa\, 10} \in \mathbb{R}^{68}$.

We did the same with the scaled problems. For each cardinality, we found the respective smallest ball containing the optimal solutions of the corresponding 10 scaled problems.

These are the radii belonging to the different scenario set cardinalities:

	10,000	20,000	30,000	40,000	50,000	60,000
Unscaled	0.216	0.213	0.201	0.199	0.210	0.189
Scaled	0.151	0.101	0.085	0.081	0.070	0.066

The unscaled data show a slowly decreasing trend. The scaled data exhibit an apparent monotone convergence.

18.7 Conclusion and Discussion

In this chapter we present an overview of second-order stochastic dominance-based models with a focus on those using a dominance measure. In terms of portfolio policy, the aim is to find a portfolio whose return distribution dominates the index distribution to the largest possible extent. We examine the unscaled model of Roman et al. (2006) and the scaled model of Fábián et al. (2010). We formulate dominance maximisation problems and describe solution methods.

In the computational study we examine the return distributions belonging to the respective optimal portfolios of the unscaled and the scaled problem. According to our experience, the unscaled model focuses on the worst cases and the corresponding return distributions have curtailed left tails. On the other hand, the shapes of the scaled return distributions are similar to the corresponding index distributions. The scaled distribution has larger expectation than the corresponding unscaled one, though its standard deviation is also larger.

We find that the performance of the unscaled model is improved by using scenario generators. By using large scenario sets, the standard deviations of the unscaled distributions are decreased. Moreover, large scenario sets help preserving dominance in out-of-sample tests. On the other hand, scenario generation seems to have little effect on the return distributions belonging to the optimal portfolios of the scaled model. Further, this model exhibits a definite convergence of the optimal solution vectors with respect to expanding scenario sets.

Comparing the shapes of the histograms and the moments belonging to corresponding pairs of in-sample and out-of-sample tests, we find that the examined models show a remarkable robustness. Moreover, the unscaled model typically preserves dominance in out-of-sample tests, provided a large scenario set was used in the portfolio choice problem. In the scaled model, all but a few tail constraints are typically preserved. (The unsatisfied constraints belong to extreme left tails.)

The unscaled measure focuses on extreme tails and hence enhances safety. On the other hand, the scaled model replicates the shape of the index distribution (i.e., the scaled model produces an optimal distribution that dominates the index distribution in a uniform sense). We think these features make these dominance measures good alternatives for classic risk measures or utility functions in certain applications, including certain multistage ones. We mention two candidate applications:

The dynamic international portfolio management model of Topaloglou et al. (2008) applies the classic CVaR measure in a multistage risk minimisation model, namely, the shortfall risk is minimised at the end of the time horizon. Alternatively, one can identify a *reference strategy* of dynamic portfolio management and find a strategy that dominates it to the largest possible extent.

The CALM dynamic asset–liability management model of Consigli and Dempster (1998) maximises expected utility of the terminal wealth. This model assumes a fund manager possessing a utility function which expresses his/her attitude to possible distributions of the terminal wealth.
Alternatively, one can identify a reference strategy of dynamic asset–liability management and find a strategy that dominates it to the largest possible extent. In view of characterisation (18.7), the scaled dominance measure takes into account all nondecreasing and concave utility functions. Hence maximising the scaled dominance measure can be considered as robust optimisation that takes into account possible attitudes of fund managers.

References

AITSAHLIA, F., C-J. WANG, V.E. CABRERA, S. URYASEV, and C.W. FRAISSE (2009). Optimal crop planting schedules and financial hedging strategies under ENSO-based climate forecasts. *Annals of Operations Research,* published online, DOI:10.1007/s10479-009-0551-2

ARTZNER, PH., F. DELBAEN, J.-M. EBER, and D. HEATH (1999). Coherent measures of risk. *Mathematical Finance* **9**, 203–227.

CARR, P., H. GEMAN, and D. MADAN (2001). Pricing and hedging in incomplete markets. *Journal of Financial Economics* **62**, 131–167.

CHERUBINI, U., E. LUCIANO, and W. VECCHIATO (2006). *Copula Methods in Finance.* Wiley, New York, NY.

CONSIGLI, G. and M.A.H. DEMPSTER (1998). The CALM stochastic programming model for dynamic asset-liability management. In *Worldwide Asset and Liability Modeling,* edited by W.T. Ziemba and J.M. Mulvey, pp 464–500. Cambridge University Press, Cambridge.

DEÁK, I. (1990). *Random Number Generators and Simulation.* Akadémiai Kiadó, Budapest.

DELBAEN, F. (2002). Coherent risk measures on general probability spaces. *Essays in Honour of Dieter Sondermann.* Springer, Berlin, Germany.

DEMPSTER, M.A.H. and R.R. MERKOVSKY (1995). A practical geometrically convergent cutting plane algorithm. *SIAM Journal on Numerical Analysis* **32**, 631–644.

DENTCHEVA, D. and A. RUSZCZYŃSKI (2003). Optimization with stochastic dominance constraints. *SIAM Journal on Optimization* **14**, 548–566.

DENTCHEVA, D. and A. RUSZCZYŃSKI (2006). Portfolio optimization with stochastic dominance constraints. *Journal of Banking & Finance* **30**, 433–451.

ELLISON, E.F.D., M. HAJIAN, H. JONES, R. LEVKOVITZ, I. MAROS, G. MITRA, and D. SAYERS (2008). *FortMP Manual.* Brunel University, London and Numerical Algorithms Group, Oxford. http://www.optirisk-systems.com/manuals/FortmpManual.pdf

FÁBIÁN, C.I., G. MITRA, and D. ROMAN (2009). Processing Second-Order Stochastic Dominance models using cutting-plane representations. *Mathematical Programming, Ser A.* DOI:10.1007/s10107-009-0326-1.

FÁBIÁN, C.I., G. MITRA, D. ROMAN, and V. ZVEROVICH (2010). An enhanced model for portfolio choice with SSD criteria: a constructive approach. *Quantitative Finance.* DOI: 10.1080/14697680903493607.

FÖLLMER, H. and A. SCHIED (2002). Convex measures of risk and trading constraints. *Finance and Stochastics* **6**, 429–447.

FOURER, R., D. M. GAY, and B. KERNIGHAN. (2002). *AMPL: A Modeling Language for Mathematical Programming*. Brooks/Cole Publishing Company/Cengage Learning.

HADAR, J. and W. RUSSEL (1969). Rules for ordering uncertain prospects. *The American Economic Review* **59**, 25–34.

HEATH, D. (2000). Back to the future. *Plenary Lecture at the First World Congress of the Bachelier Society,* Paris, June 2000.

JOE, H. (1997). *Multivariate Models and Dependence Concepts.* Chapman & Hall, London.

KLEIN HANEVELD, W.K. (1986). Duality in stochastic linear and dynamic programming. *Lecture Notes in Economics and Math. Systems* **274**. Springer, New York, NY.

KLEIN HANEVELD, W.K. and M.H. VAN DER VLERK (2006). Integrated chance constraints: reduced forms and an algorithm. *Computational Management Science* **3**, 245–269.

KROLL, Y. and H. LEVY (1980). Stochastic dominance: A review and some new evidence. *Research in Finance* **2**, 163–227.

KÜNZI-BAY, A. and J. MAYER (2006). Computational aspects of minimizing conditional value-at-risk. *Computational Management Science* **3**, 3–27.

LUEDTKE, J. (2008). New formulations for optimization under stochastic dominance constraints. *SIAM Journal on Optimization* **19**, 1433–1450.

MCNEIL, A.J., R. FREY, and P. EMBRECHTS (2005). *Quantitative Risk Management.* Princeton University Press, Princeton, NJ.

OGRYCZAK, W. (2000). Multiple criteria linear programming model for portfolio selection. *Annals of Operations Research* **97**, 143–162.

OGRYCZAK, W. (2002). Multiple criteria optimization and decisions under risk. *Control and Cybernetics* **31**, 975–1003.

OGRYCZAK, W. and A. RUSZCZYŃSKI (2001). On consistency of stochastic dominance and mean- semideviations models. *Mathematical Programming* **89**, 217–232.

OGRYCZAK, W. and A. RUSZCZYŃSKI (2002). Dual stochastic dominance and related mean-risk models. *SIAM Journal on Optimization* **13**, 60–78.

PFLUG, G. (2000). Some remarks on the value-at-risk and the conditional value-at-risk. In *Probabilistic Constrained Optimization:Methodology and Applications*, edited by S. Uryasev, pp. 272–281. Kluwer, Norwell, MA.

ROCKAFELLAR, R.T. (2007). Coherent approaches to risk in optimization under uncertainty. *Tutorials in Operations Research* INFORMS 2007, 38–61.

ROCKAFELLAR, R.T. and S. URYASEV (2000). Optimization of conditional value-at-risk. *Journal of Risk* **2**, 21–41.

ROCKAFELLAR, R.T. and S. URYASEV (2002). Conditional value-at-risk for general loss distributions. *Journal of Banking & Finance* **26**, 1443–1471.

ROCKAFELLAR, R. T., S. URYASEV, and M. ZABARANKIN (2002). Deviation measures in risk analysis and optimization. *Research Report* 2002-7, Department of Industrial and Systems Engineering, University of Florida.

ROCKAFELLAR, R. T., S. URYASEV, and M. ZABARANKIN (2006). Generalised deviations in risk analysis. *Finance and Stochastics* **10**, 51–74.

ROMAN, D., K. DARBY-DOWMAN, and G. MITRA (2006). Portfolio construction based on stochastic dominance and target return distributions. *Mathematical Programming* Series B **108**, 541–569.

ROSS, S.M. (2002). *An Elementary Introduction to Mathematical Finance.* Cambridge University Press, Cambridge.

RUDOLF, G. and A. RUSZCZYŃSKI (2008). Optimization problems with second order stochastic dominance constraints: duality, compact formulations, and cut generation methods. *SIAM Journal on Optimization* **19**, 1326–1343.

RUSZCZYŃSKI, A. (1986). A Regularized Decomposition Method for Minimizing the Sum of Polyhedral Functions. *Mathematical Programming* **35**, 309–333.

RUSZCZYŃSKI, A. and A. SHAPIRO (2006). Optimization of convex risk functions. *Mathematics of Operations Research* **31**, 433–452.

TOPALOGLOU, N., H. VLADIMIROU, and S. ZENIOS (2008). A dynamic stochastic programming model for international portfolio management. *European Journal of Operational Research* **185**, 1501–1524.

SKLAR A. (1959). Fonctions de répartition à n dimensions et leurs marges. *Publications de l'Institut de Statistique de l'Universit de Paris* 1959/8, 229–231.

WHITMORE, G.A. and M.C. FINDLAY (1978). *Stochastic Dominance: An Approach to Decision-Making Under Risk.* D.C.Heath, Lexington, MA.

Index

M. Bertocchi et al. (eds.), *Stochastic Optimization Methods in Finance and Energy*, International Series in Operations Research & Management Science 163, DOI 10.1007/978-1-4419-9586-5, © Springer Science+Business Media, LLC 2011

MIX
Papier aus verantwortungsvollen Quellen
Paper from responsible sources
FSC® C105338

If you have any concerns about our products,
you can contact us on
ProductSafety@springernature.com

In case Publisher is established outside the EU,
the EU authorized representative is:
Springer Nature Customer Service Center GmbH
Europaplatz 3, 69115 Heidelberg, Germany

Printed by Libri Plureos GmbH
in Hamburg, Germany